WORKING WITH HEAT EXCHANGERS: Questions and Answers

J. P. Gupta

Chemical Engineering Department
Indian Institute of Technology
Kanpur (U.P.), India

HEMISPHERE PUBLISHING CORPORATION
A member of the Taylor & Francis Group
New York   Washington   Philadelphia   London

To my parents, Sh. Niader Mal and Smt. Kripa Devi
sisters, Pushpa, Krishna, and Suvira
wife, Prabha
children, Sonjai, Swati, and Jaideep

WORKING WITH HEAT EXCHANGERS: Questions and Answers

1 2 3 4 5 6 7 8 9 0 EBEB 8 9 8 7 6 5 4 3 2 1 0 9

This book was previously published in a hardcover edition entitled *Fundamentals of Heat Exchanger and Pressure Vessel Technology* (Hemisphere Publishing Corporation, 1986).

This book was set in Press Roman by Hemisphere Publishing Corporation. The editor was Carolyn V. Ormes. Cover design by Debra Eubanks Riffe.
Edwards Brothers was printer and binder.

Library of Congress Cataloging in Publication Data

Gupta, J. P. (Jai Prakash)
[Fundamentals of heat exchanger and pressure vessel technology]
Working with heat exchangers: questions and answers / J. P. Gupta.
p. cm.
Originally published as: Fundamentals of heat exchanger and pressure vessel technology. 1986.
Includes bibliographical references.
1. Heat exchangers—Miscellanea. I. Title. II. Title: Fundamentals of heat exchanger and pressure vessel technology.
TJ263.G87 1989
621.402′5—dc20 89-27224
ISBN 1-56032-054-0 CIP

CONTENTS

PREFACE

Heat exchangers are the workhorse of industry. Variously known as condensers, coolers, evaporators, heaters, reboilers, vaporizers, and so forth, they are used extensively in chemical, fertilizer, petrochemical, petroleum, power, refrigeration, and other industries. With the current awareness and concern about energy and resource conservation as well as efficient and safe operation of plants, it is imperative that the heat exchangers be designed and operated optimally. Optimal design has become still more important as the size of heat exchangers continues to increase to take advantage of the economy of scale.

With the above in mind, I have taught a one-semester course on heat exchanger design at the Indian Institute of Technology, Kanpur, since 1976; I also taught it during the winter semester of 1983 at the University of Michigan, Ann Arbor. An abbreviated version has been used in short courses for practicing engineers in India and Argentina. During these courses, students have asked specific questions about thermal and mechanical designs, vibration analysis, and operation and maintenance of heat exchangers. Some questions were received by mail. The questions were compiled and many more were added (along with answers). The questions were then circulated among students in class and among leaders in academia and industry in India, the United Kingdom, and the United States. The responses received were very positive regarding both the usefulness of a book with a question-and-answer approach and the material covered in the questions.

The main advantage of a question-and-answer approach, which this book follows, is that it focuses the attention of the reader on a specific point, which facilitates learning the answer quickly since the reader's mind is not cluttered with much unrelated material, as in a conventional text.

Although the literature on heat exchangers is extensive and growing, a newcomer to the field confronts difficulties at a number of points and does not know where to turn since no one has enough time to answer all questions. Sometimes he or she is hesitant to ask "simple" questions. Until now, there was no single book that directly told a person about such items as tie-rods and various types of baffles; what to do if the calculated pressure drop is higher than the allowable value; comparison of various types of multi-pass exchangers, multiple segmental baffles, and tube layouts; effects of laminar and turbulent flows; minimum information required for thermal and mechanical design; flow stratification; when a stress anlaysis report is required and which factors are to be taken into account during a stress analysis; mechanical design and failure analysis; nondestructive testing methods; maintenance and repair; types of corrosion and fouling and steps to counter them; causes of vibration, vibrational damage and steps to prevent it; and so on.

This book is intended to help newcomers, those with a few years of experience in design and operation, and senior-level students in design courses find answers to over a thousand such questions. It is not meant for experts in the field, although they might use it to review information. The book is self-explanatory, self-contained, and easy to read. It brings out subtle details that do much to aid one's understanding of the topics.

As work on the book progressed, the scope of the topics covered was expanded. And so to keep this a reasonably sized volume, the actual thermal and mechanical design calculations have been left out. This decision was aided by two related factors. First, a number of books exist that deal with thermal design, while mechanical design is governed by codes and standards that are easy to follow. Second, in most industries the designers have access to in-house computer programs or those developed by co-operative research organizations, whose membership is continuously increasing. Hence, this book is not to be taken as a design book but one that would help a designer come up with better designs, in terms of both cost and performance, and an operator in understanding and controling the performance of a unit. In addition, this book is not meant to be an up-to-date and exhaustive literature survey of all the varied topics covered, since that would be impossible. It does cover much of the latest information available and gives over 540 references.

One drawback to using the question-and-answer approach is that the reader may unintentionally compare an answer with his or her own previous knowledge of the topic and become very critical of the answer given. It is hoped that readers who disagree with any of the answers in this book will contact the author so that improvements to the book can be made.

In an undertaking of this kind, some duplication of subject matter occurs. This has generally been handled by cross-referencing questions, figures, tables, and equations. In a cross-reference, the first number indicates the chapter number and the second indicates the question, figure, table, or equation number, as the case may be. There are over 680 such cross-references in this book.

Due to space restrictions, a number of questions and answers had to be deleted and the number of figures restricted. It is likely that some important questions have been missed altogether. Readers are encouraged to contact the author regarding any suggested questions they might have.

An author's experiences and training inevitably bias his or her treatment of topics chosen for inclusion in a book and this one is no exception. While direct-fired boilers, cooling towers, and fluidized bed combustors are outside the scope of this volume, the decision not to treat regenerators was a difficult one to make, it being a rather vast topic in itself. Some readers may find that some of their favorite topics are not dealt with in as much detail as others. Their indulgence is sought in communicating such topics to the author.

ACKNOWLEDGMENTS

This book has benefited greatly by the comments of my mentor, Professor Stuart W. Churchill, University of Pennsylvania, Philadelphia, and Dr. Krishna P. Singh, Joseph Oat Corp., Camden, New Jersey. They provided critical suggestions and encouragement and spent countless hours reading the manuscript.

Each chapter was reviewed by several internationally renowned experts in academia and industry. Their comments and criticism greatly improved the book. These individuals, as well as a few companies, are noted below:

B. D. Agarwal
M. Atkins
A. E. Bergles
B. C. Bhattacharyya
S. S. Biswas
D. Butterworth
T. C. Carnavos
S. S. Chen
S. W. Churchill
J. G. Collier
R. C. Conrad
A. Cooper
Jim Cunningham
Curwen and Newberry
Davy Powergas (India) Ltd.
P. Dayaratnam
M. J. Dholakia
K. Elayaperumal
N. Epstein
Pierre Le Goff
D. J. Gorman
O. P. Goyal
Hugh Goyder
Graphite Equipment Ltd.
B. N. Greene
G. F. Hewitt
A. Hussain
Kishore Jhalaria
A. E. Jones
M. V. Joshi
S. Kakaç
H. Kanoh
J. H. Kissler
P. C. Knott
Prashant Kumar
Linde A. G.
A. L. London
K. K. Mahajan
K. Marcinauskas
Jan Marriott
F. Mayinger
A. Moliterno
Y. Mori
A. C. Mueller
John S. Mueller
G. Muthukumarasamy
C. V. Ramakrishna
Sam A. Scaglione
S. N. Sengupta
B. H. Shah
R. K. Shah
S. Sideman
K. P. Singh
S. Sivasankaran
A. Slanciauskas
A. Stephen
Sulzer Brother Ltd.
V. Sundararajan
J. Taborek
V. J. Tennery
M. L. Vaidya
R. L. Webb
A. Žukauskas

The comments of two anonymous reviewers also proved quite helpful. Much encouragement and help also were provided, directly or indirectly, by Om Goyal, Dr. Virendra Bansal, Dr. D. N. Saraf, Mr. G. Stanko, Mr. N. K. Metia, Mr. S. Reginald, Yogendra Singh, Rajendra Sharma, my sisters Dr. Krishna Garg and Dr. Suvira Gupta, my wife Prabha, and children Sonjai, Swati, and Jaideep.

Mr. R. N. Srivastava, Mr. U. S. Misra, and Mr. B. S. Pandey typed and retyped several versions of the manuscript. The final version was typed by Ms. Cynthia Miller at the University of Michigan, Ann Arbor. The drawings were done by Mr. D. S. Panesar and Mr. D. K. Misra. Reproduction of drawings and associated work was done by Mr. D. Chakraborty and at the Graphic Arts, I.I.T., Kanpur, under the supervision of Mr. R. Mahesh and Mr. K. C. Kapoor.

Financial support was provided by the Quality Improvement Programme, Government of India, Ministry of Education and Social Welfare, at I.I.T., Kanpur. Use of necessary facilities at the University of Michigan was provided by the Chemical Engineering Department.

Several organizations gave permission to reproduce their copyrighted materials. These are acknowledged in the text. The following organizations permitted unlimited use of their publications without the need to seek permission for individual items: American Society of Mechanical Engineers, Hitachi-Zosen, Morgan Grampian Ltd., NALCO Chemical Co., Oak Ridge National Laboratory, and Patterson-Kelley Co. C. S. Enterprises, Publications Division, Delhi, granted complete freedom to reproduce over a hundred questions that appear (without answers) in my earlier book* and agreed to this summary acknowledgement.

Over the many years it took to write this book, and the unending changes that were made in the numerous drafts, the files for each chapter became thick and dog-eared and my memory started dimming. It is certain, therefore, that a few names are missing in the list of reviewers as well as a few acknowledgments for material used in this book. I beg the forgiveness of such people and organizations and wish to assure that it was not deliberate.

Both William Begell, president, and Mary Phillips Born, former editor-in-chief, of Hemisphere Publishing, were very helpful and generous in the publication of this book.

To all the individuals and organizations that helped me, I owe sincere thanks, and I feel a sense of deep gratitude toward them.

J. P. Gupta

*Gupta, J. P., *Heat Exchanger Design–A Practical Look,* 2d ed. C. S. Enterprises: Delhi, India. 1979.

NOTE ABOUT UNITS

Both SI and British units are used in this book. To a large extent, the general pattern follows the recommendations given in the *ASME Orientation and Guide for Use of SI (Metric) Units* (8th ed., ASME: New York, 1978). Generally, when an original source referred to in this book uses British units, the same units are used here, and the SI equivalents are given in parentheses, and vice versa. In several cases the conversions are not exact, and an engineering judgment about rounding off has been carefully made, in keeping with the recommendations of the above publication–for example, 34 ft = 10.4 m, instead of 10.3632 m. Furthermore, it is not implied that the material given in equivalent dimensions is standard and commercially available. For example, if a 2-in thickness is written as 51 mm, a flange of 51 mm thickness may not be a standard item.

CHAPTER

ONE

SHELL-AND-TUBE HEAT EXCHANGERS—I: GENERAL

INTRODUCTION

• Shell-and-tube heat exchangers can handle a variety of services in a vast range of allowable design pressures and temperatures, from full vacuum to 6000 psig (41.5 MPa) and from cryogenic temperature to 2000°F (1100°C). Hence, these are the most versatile heat transfer equipment used in the process, power, and refrigeration industries.

In order to understand the design and operation of the shell-and-tube heat exchangers, it is important to know the nomenclature and terminology used to describe them and the various parts that go into their making. Only then can we understand without any anomaly, the design, communications, and reports on these units put out by researchers, designers, fabricators, and users.

Therefore, in this chapter we will introduce the nomenclature and parts of a shell-and-tube exchanger. This is the first chapter of a series discussing these units. Chapter 2 discusses the process (thermal) design without phase change, followed by Chapter 3, which covers the situations involving phase change. In the second half of this book we will discuss the mechanical design and other items related to some of the important parts of the exchangers discussed in this chapter.

NOMENCLATURE AND PARTS

1.1 How are the size and type of a heat exchanger designated?

TEMA Standards (Tubular Exchanger Manufacturers Assoc., 1978) and HEI Standards (Heat Exchange Institute, 1980) have specified the nomenclature for the size and type of a shell-and-tube heat exchanger. Those according to the TEMA Standards are given below, while the HEI Standards' nomenclature is discussed in the next question.

Size is designated by the nominal shell diameter and the tube length. In the British system of units, the nominal shell diameter is the inside diameter of the shell in inches rounded off to the nearest integer. However, for the kettle-type reboiler, the nominal port diameter precedes the nominal shell diameter. The tube length for a straight tube unit is the distance between the outermost tubesheet faces. For a U-tube unit, it is the distance between the outermost face of the tubesheet and the bend tangent, i.e., only the straight leg. In the SI and the metric system, all the dimensions are given in millimeters (mm). The type of the exchanger is designated by a set of letters associated with the front end, the shell, and the rear end as given in Fig. 1.1 and illustrated by the following examples.

1. Split-ring floating head (S) exchanger with removable channel and cover (A), single-pass shell (E), $23\frac{1}{4}$ in (590 mm) inside diameter with tubes 16 ft (4880 mm) long. Size 23-192 (590-4880), Type AES.
2. U-tube exchanger (U) with bonnet-type stationary head (B), split flow shell (G), 19 in (480 mm) inside diameter with tubes 7 ft (2130 mm) straight length. Size 19-84 (480-2130), Type BGU.
3. Fixed tubesheet exchanger with removable channel and cover (A), bonnet-type rear head (M), two-pass shell (F), $33\frac{1}{8}$ in (840 mm) inside diameter with tubes 8 ft (2440 mm) long. Size 33-96 (840-2440), Type AFM.

Special designs may be specified as best suits the manufacturer.

● In the discussion above, terms such as single-pass and two-pass shell have been used. These, respectively, mean that the shell-side fluid travels only once through the shell (single-pass) or twice, i.e., it enters at one end, travels to the other, and then returns to the end where it entered the shell (two-pass). Similarly, one can have multiple tube passes also. The number of tube passes is equal to or greater than the number of shell passes. In general, the multi-shell and tube passes are usually designated by two numerals separated by a hyphen, with the first numeral indicating the number of shell passes and the second numeral indicating the number of tube passes. Thus, a one-shell pass and two-tube pass AEL exchanger will be written as 1-2 AEL. It should be emphasized that this is not according to the TEMA Standards. They require the number of shell and tube passes to be spelled out as in the above examples. In a heat exchanger specification sheet (Q. 1.116), there is a separate space indicating the number of shell and tube passes.

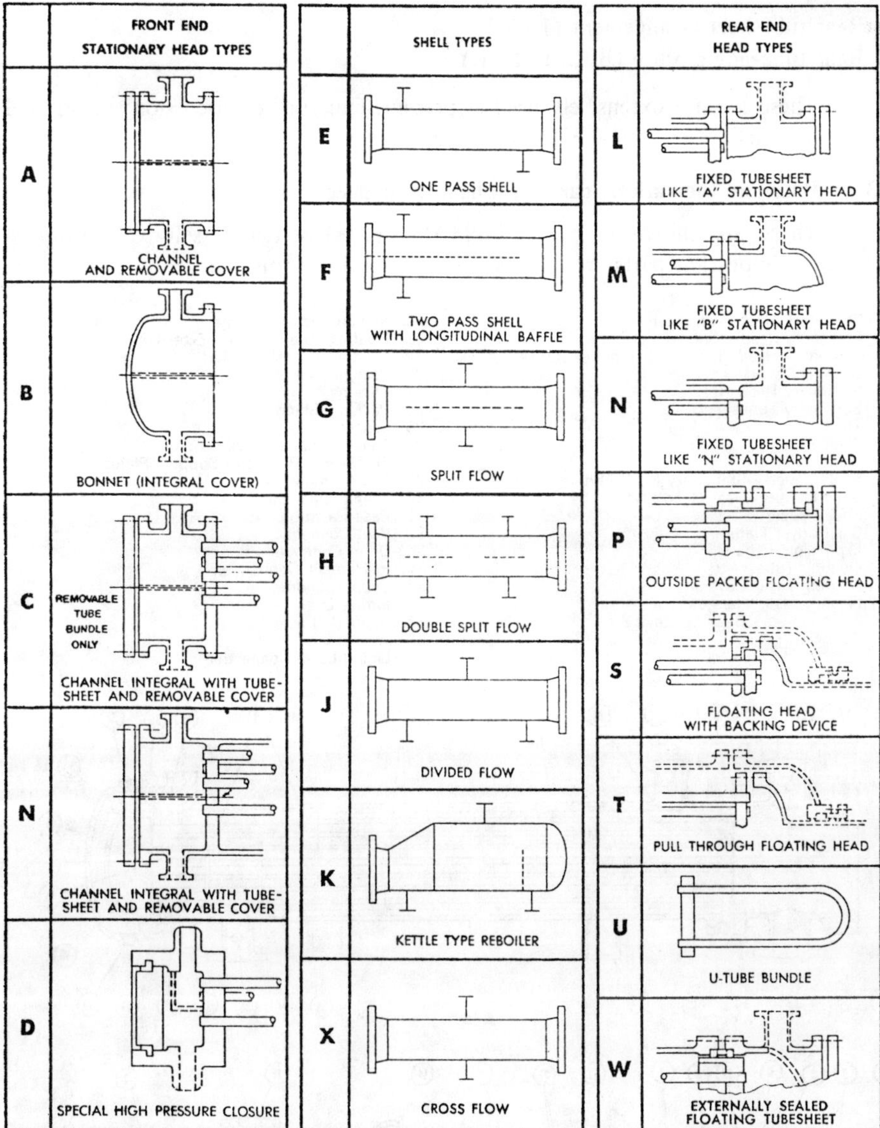

Figure 1.1 Heat exchanger nomenclature. *(© 1978 by Tubular Exchanger Manufacturers Association. All rights reserved.)*

1.2 How is the nomenclature specified by the HEI Standards (1980)?

These require five symbols to denote a heat exchanger as detailed below:

1. Front tube-side closure (symbols used: B, C, R)
2. Front tubesheet arrangement (1, 2, 3, 4)
3. Shell-side arrangement (S_1, S_2, S_d, S_k, S_s)

4. Rear tubesheet arrangement (1 to 8)
5. Rear tube-side closure (B, C, F, R, U)

These are not extensively used at present. Our further discussion is restricted to the nomenclature of the TEMA Standards.

1.3 What are the various parts of a heat exchanger?

These are shown in Fig. 1.2 (TEMA, 1978), which is self-explanatory. A number of important parts are discussed subsequently in this chapter.

1. Stationary Head—Channel
2. Stationary Head—Bonnet
3. Stationary Head Flange—Channel or Bonnet
4. Channel Cover
5. Stationary Head Nozzle
6. Stationary Tubesheet
7. Tubes
8. Shell
9. Shell Cover
10. Shell Flange—Stationary Head End
11. Shell Flange—Rear Head End
12. Shell Nozzle
13. Shell Cover Flange
14. Expansion Joint
15. Floating Tubesheet
16. Floating Head Cover
17. Floating Head Flange
18. Floating Head Backing Device
19. Split Shear Ring
20. Slip-on Backing Flange
21. Floating Head Cover—External
22. Floating Tubesheet Skirt
23. Packing Box
24. Packing
25. Packing Gland
26. Lantern Ring
27. Tierods and Spacers
28. Transverse Baffles or Support Plates
29. Impingement Plate
30. Longitudinal Baffle
31. Pass Partition
32. Vent Connection
33. Drain Connection
34. Instrument Connection
35. Support Saddle
36. Lifting Lug
37. Support Bracket
38. Weir
39. Liquid Level Connection

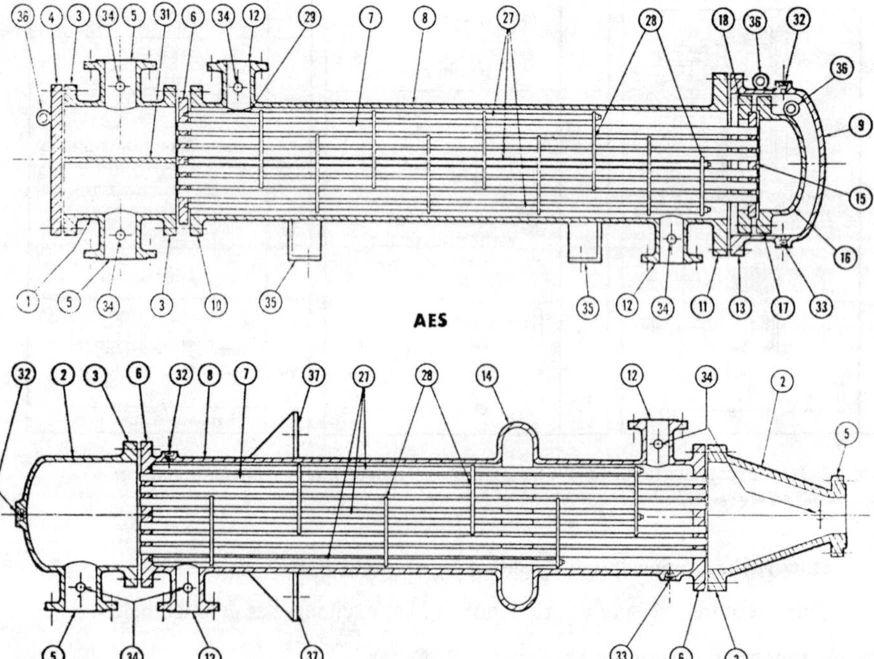

Figure 1.2 Heat exchanger parts. *(© 1978 by Tubular Exchanger Manufacturers Association. All rights reserved.)*

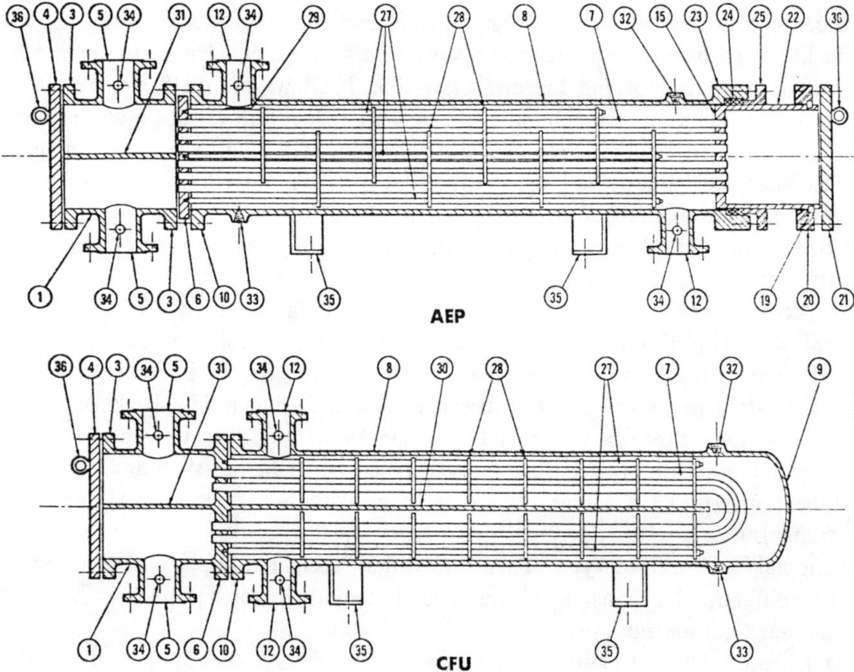

Figure 1.2 Heat exchanger parts. *(© 1978 by Tubular Exchanger Manufacturers Association. All rights reserved.)* *(Continued)*

PRINCIPAL TYPES OF SHELL-AND-TUBE CONSTRUCTIONS

1.4 What are the principal types of shell-and-tube constructions?

These basically fall into three groups:

1. Fixed-tubesheet type (Fig. 1.1C and N)
2. U-tube type (Fig. 1.1U)
3. Floating-head type (Fig. 1.1P, S, T, and W)

1.5 What are the advantages of the principal types of shell-and-tube constructions?

The advantages are the following.

1. Fixed-tubesheet type: These find the maximum use in industry for moderate services where the thermal stresses are low. These are easy to fabricate and cost less. There can be any number of tube passes, although shell passes are usually limited to two. Tubes can completely fill the shell. Although the tube side is accessible for cleaning and repair, the shell side is not. Hence, clean and nonfouling fluids should be used on the shell side.

2. U-tube type: These are used in high-pressure and high-temperature cases as well as in kettle reboilers, evaporators, tank suction heaters, etc. Each tube is free to expand to take care of the thermal expansion. It allows the shell inlet nozzle to be located beyond the tube bundle U-bends to reduce the vibration and erosion damage. The tube bundle is removable for cleaning, although it is not possible to replace the inner tubes since single tubes cannot be removed from the inside of the bundle without first removing the outer tubes.
3. Floating-head type: There are several kinds of floating-head constructions. Their relative advantages are discussed below.
   a. Packed-lantern-ring type: These are generally used at below 375°F (190°C) and 300 psi (2070 kPa) for use with steam, water, air, lube oil, etc. These are the least costly removable bundle-type exchangers. Separate packing rings at the floating tubesheet contain the shell- and tube-side fluids. Leakage at the packing falls through weep holes to the floor and does not result in the intermixing of the two fluids. The clearance between the shell and the outer tube limit (OTL) is larger than that in a fixed tubesheet or a U-tube-type exchanger.
   b. Internal-floating-head type: These are widely used in the petroleum industry. The differential expansion of the tubes is accounted for by the movement of the rear floating tubesheet.
   c. Outside-packed-floating-head type: These find the maximum use among the removable bundle-type constructions in the chemical industry. The shell-side pressure and temperature can be 600 psi (4137 kPa) and 600°F (316°C), respectively, whereas there is no restriction on the tube side. The shell-side fluid is contained by packing rings in a stuffing box. The shell-to-OTL clearance is the maximum in this type of exchangers.
   d. Pull-through-floating-head type: These are similar to the internal-floating-head type except that the floating head cover is directly bolted on to the floating tubesheet. The downtime for repair and maintenance is minimized. The shell-to-OTL clearance is very large, and seal strips are used to reduce the bypassing of the shell-side fluid.

In some of the floating-head-type exchangers, the problem of tilting of the floating tubesheet with the resultant leakage needs to be overcome. For this reason, various ways are adopted by different manufacturers, some of which are the precise machining and accurate fit-up, spacer-ring, or a second clamp ring 90° from the first to attain a leak tight joint.

Table 1.1 summarizes the other features of the principal types of shell-and-tube constructions (Perry and Chilton, 1973).

1.6 What are the advantages of the different types of tube headers?

Tube headers refer to the front end where the tube-side fluid enters the exchanger. These are primarily of three kinds (Fig. 1.1):

1. Channel (Fig. 1.1A, C, and N): It can either be integral with the shell or bolted to it. It has a cover that permits access to the tube without disturbing the piping connections.

Table 1.1 Features of principal shell-and-tube-type exchangers

Type of design	Fixed tube-sheet	U-tube	Packed lantern ring floating head	Internal floating head (split backing ring)	Outside-packed floating head	Pull-through floating head
TEMA rear-head type	L or M or N	U	W	S	P	T
Relative cost increases from A (least expensive) through E (most expensive)[a]	B	A	C	E	D	E
Provision for differential expansion	Expansion joint in shell	Individual tubes free to expand	Floating head	Floating head	Floating head	Floating head
Removable bundle	No	Yes	Yes	Yes	Yes	Yes
Replacement bundle possible	No	Yes	Yes	Yes	Yes	Yes
Individual tubes replaceable	Yes	Only those in outside row[b]	Yes	Yes	Yes	Yes
Tube cleaning by chemicals inside and outside	Yes	Yes	Yes	Yes	Yes	Yes
Interior tube cleaning mechanically	Yes	Special tools required	Yes	Yes	Yes	Yes
Exterior tube cleaning mechanically:						
Triangular pitch	No	No[c]	No[c]	No[c]	No[c]	No[c]
Square pitch	No	Yes	Yes	Yes	Yes	Yes
Hydraulic jet cleaning:						
Tube interior	Yes	Special tools required	Yes	Yes	Yes	Yes
Tube exterior	No	Yes	Yes	Yes	Yes	Yes
Double tubesheet feasible	Yes	Yes	No	No	Yes	Yes
Number of tube passes	No practical limitations	Any even number	Limited to one or two passes	No practical limitations[d]	No practical limitations	No practical limitations[d]
Internal gaskets eliminated	Yes	Yes	Yes	No	Yes	No

Source: Perry and Chilton (1973), p. 11.5.

[a]Relative costs A and B are not significantly different and interchange for long lengths of tubing.

[b]U-tube bundles have been built with tube supports which permit the U-bends to be spread apart and tubes inside of the bundle replaced.

[c]Normal triangular pitch does not permit mechanical cleaning. With a wide triangular pitch, which is equal to 2 (tube diameter plus cleaning lane)/$\sqrt{3}$, mechanical cleaning is possible on removable bundles. This wide spacing is infrequently used.

[d]For odd number of tube-side passes, floating head requires packed joint or expansion joint.

2. Bonnet (Fig. 1.1B): It is bolted to the shell but has no separate cover. Hence it has to be removed to have access to the tubes. This disturbs the piping connections also. It is, however, generally cheaper than a removable channel with cover.
3. High-pressure header (Fig. 1.1D): It is specially made for very high pressures. The channel barrel and the tubesheet are generally forged.

1.7 What factors decide the type of tube header to be chosen?

The major considerations are the following:

1. Accessibility to the tubes for cleaning, inspection, and replacement
2. Piping convenience so that the connections do not have to be opened for the inspection of the tubes
3. Design conditions of pressure, temperature, and corrosion
4. Cost

1.8 What are the uses for different types of shells?

These are shown in Fig. 1.1 and are known by the initials E, F, G, H, J, K, X, and L. A brief description of each type is provided below. It is suggested that flow-paths for the shell-side fluid be traced for each case as done in Fig. 2.1 for some cases. Relative comparison of the performance of some types is given in Q. 2.101 (see Chapter 2).

1. E-Type: It has one shell pass with the shell-side fluid entry and exit nozzles attached at the two opposite ends of the shell. It is easy to fabricate and is the most-used type. Other shells are used only when an E-type proves less effective.
2. F-Type: A longitudinal baffle (Q. 1.49) divides the shell into two passes. Both the entry and exit nozzles are attached at the same end. Fluid enters the shell at one end, travels to the other end where it enters the second pass, and travels to the exit nozzle located in-line with the entry nozzle. The amount of heat transferred is more than in an E-type but at the cost of an increased pressure drop.
3. G-Type: It differs from the F-type in that the nozzles are in the middle, and the shell-side fluid splits into two with the two halves flowing in opposite directions. Hence, it is also known as a split-flow unit. It is used when the available pressure drop is limited, as in condensers. It is a 2-shell pass unit.
4. H-Type: It is similar to the G-type but with two inlet nozzles and two outlet nozzles and two horizontal baffles resulting in a double split-flow unit. It is used when the available pressure drop is very limited. It is a 2-shell pass unit.
5. J-Type: It has one nozzle in the middle used for entry (or exit) and two nozzles at opposite ends used for exit (or entry) of the fluid in a pressure-drop limited case. It is a 1-shell pass unit.
6. K-Type: It is known as a kettle reboiler with the tube bundle being far smaller than the shell diameter and the shell-side fluid vaporizing. A weir helps to maintain the liquid level over the tube bundle, while the empty space above the

liquid helps in the disengagement of the entrained liquid droplets in the outgoing vapor. It is used underneath the distillation towers to provide vapor reflux and energy for distillation (see Q. 3.86, Chapter 3, for further discussion).

7. X-Type: It has the nozzles in the middle as in the G-type. However, the shell-side fluid is divided into multiple fractions or substreams, and each substream flows over the tube bundle only once (see Fig. 2.1, Chapter 2, for comparison of E-, G-, and X-types). It is used when the pressure drop is limited.
8. L-Type: In this (not shown in Fig. 1.1), the shell-side flow is parallel to the tubes instead of being in cross-flow as in the other units. No cross-baffles are used, only support plates with a cut approaching 50 percent are used to support the tube bundle. This shell is used when the permissible pressure drop is very low (see also Q. 2.98 and 2.154, Chapter 2).

1.9 Which is more economical: a large diameter, small length shell or a small diameter, long shell?

A small diameter, long shell is more economical because of the following:

1. In some cases it may be fabricated from standard pipes.
2. The thicknesses of the shell, flanges, covers, etc., decrease with a decrease in the diameter of the exchanger for a given set of operating conditions.
3. The tubesheet is a very costly item because of the work needed to drill and machine it accurately. With a smaller diameter shell, the tubesheet diameter and thickness decrease. Furthermore, the number of tube-end connections (rolling, welding, etc.) also decreases. This reduces the cost and the probability of tube rupture or leakage at the tube-to-tubesheet joints.
4. The cost per unit length of the longer tubes may be less than that for the shorter tubes.

Of course, the selection of a small diameter, long shell depends upon its satisfying the heat transfer requirements within the allowable pressure drop, complying with the plant layout, installation, and servicing capabilities and being able to push-in and take-out the tube bundle in the available space.

1.10 What are the current limits on the shell diameter, length, and surface area?

The limits keep changing with larger and larger units being constructed. TEMA Standards (1978) have firm guidelines up to 60 in (1500 mm) shell diameter, while the "Recommended Good Practice" section in the TEMA Standards suggests guidelines up to 100 in (2500 mm) shell diameter. People are already talking about 120 in (3000 mm) shell diameter (Taborek, 1979), and soon the size may be limited only by the capability to transport the units.

1.11 What are the limitations on the number of tube passes in different types of exchangers?

Table 1.2 gives the limitations on the number of tube passes in different types of exchangers.

Table 1.2 Limitations on number of tube passes in different types of exchangers

Type of exchanger	Limitations on tube passes
U-tube	Any practical even number; pass partition plates are built in the front end only
Fixed tubesheet	Any practical number, odd or even; pass partition plates are built in both the front and the rear heads
Floating-head-pull-through type	Any practical even number; for single pass, a packed joint has to be installed on the floating head; it is generally not used as a single pass exchanger
Floating-head-outside-packed-lantern-ring type	Only single or 2-pass since no pass partition plate can be installed in the rear head
Floating-head-split-backing-ring type	Any practical even number; for single pass, a packed joint has to be installed on the floating head
Floating-head-outside-packed-stuffing-box type	Any practical number, odd or even

Source: Patterson-Kelley Manual 700A.

TYPES OF TUBE, TUBE THICKNESS, PITCH, LAYOUT ANGLE, AND TUBE COUNT

1.12 What are the different kinds of tubes used in the heat exchangers?

They are the following:

1. Plain tubes
2. Finned tubes
3. Duplex or bimetallic tubes
4. Enhanced surface tubes

Representative sketches of these are shown in Figs. 1.3–1.8 and their uses are discussed below.

1.13 When are the plain tubes preferred?

Plain tubes are the cheapest per unit length and are the easiest to handle. Hence they are the first choice unless the special requirements of a particular situation prove them to be unsuitable or inefficient.

1.14 In what forms are the plain tubes used?

These are used as follows (Fig. 1.3):

1. Straight tubes
2. U-tubes with a U-bend
3. Coiled tubes

1.15 Tubes can be welded or seamless type. How are they made?

The welded tubes are rolled into cylindrical shape from strip material and welded automatically under precisely controlled conditions. The seamless tubes are either extruded or hot pierced and drawn over a mandrill. While tubes of most materials are available in both types, those of copper and copper alloys are available only as seamless. The seamless tubes have better structural integrity and are costlier than the welded ones.

1.16 What are the finned tubes and when are they used?

As the name implies, finned tubes have fins attached to the tubular surface. Fins can be longitudinal, radial, or helical and may be on the outside or inside or on both sides of the tubes (some are shown in Figs. 1.4 and 1.5). These are generally used when at least one of the two fluids is a gas. Their manufacture and applications are discussed in detail in Chapter 4.

1.17 What are the duplex or bimetallic tubes? How are they made?

Duplex or bimetallic tubes are in reality two tubes of different metals, one closely fitting over the other with no gap in between the two tubes. These are made by drawing the outer tube onto the inner one or by shrink fitting or by other mechanical means.

1.18 Where are the bimetallic tubes used?

These are used where the corrosive nature of the tube-side and shell-side fluids is such that no one metal or alloy is compatible with the fluids. Thus, while the tube-side fluid touches one metal, the shell-side fluid is in contact with the other metal. The metals selected should be such that they will corrode at similar rates. Over 100 material combinations are available. An example is the use of steel outside and admirality or cupro-nickel inside the ammonia condensers using water in the tubes as a coolant.

1.19 How is the thickness of the bimetallic tubes chosen?

Since perfect contact is made between the outer and the inner tubes, the total thickness is chosen on the basis of the strength of the softer of the two materials.

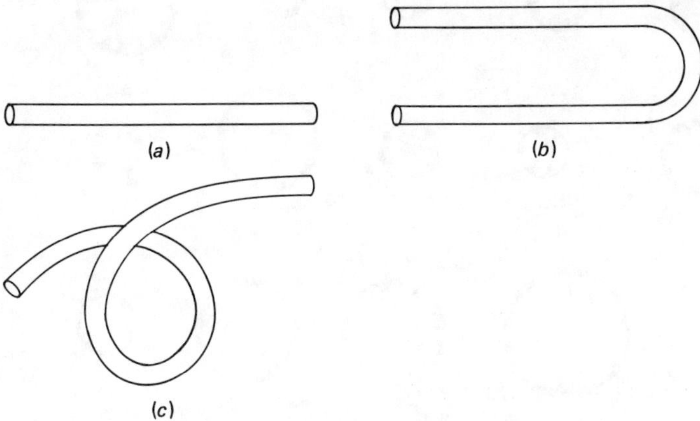

Figure 1.3 Types of plain tubes. (*a*) Straight tube. (*b*) U-tube. (*c*) Coiled tube.

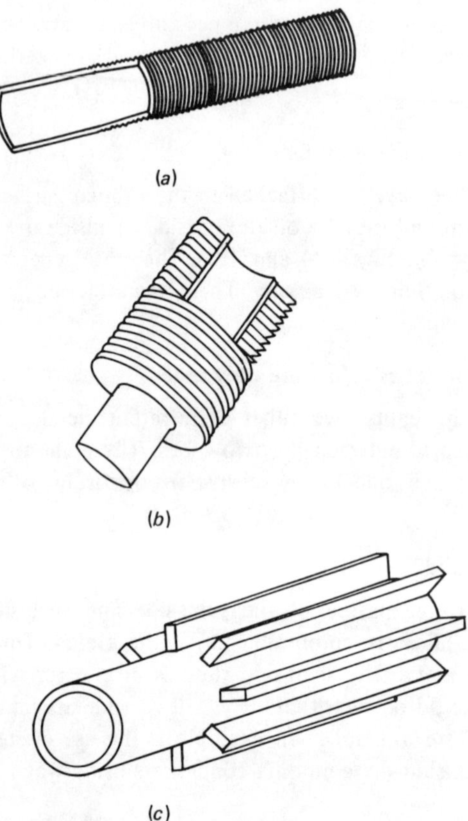

Figure 1.4 Outside fin tubes. (*a*) Radial low-finned tube. (*b*) Radial high-finned tube. (*c*) Longitudinal finned tube.

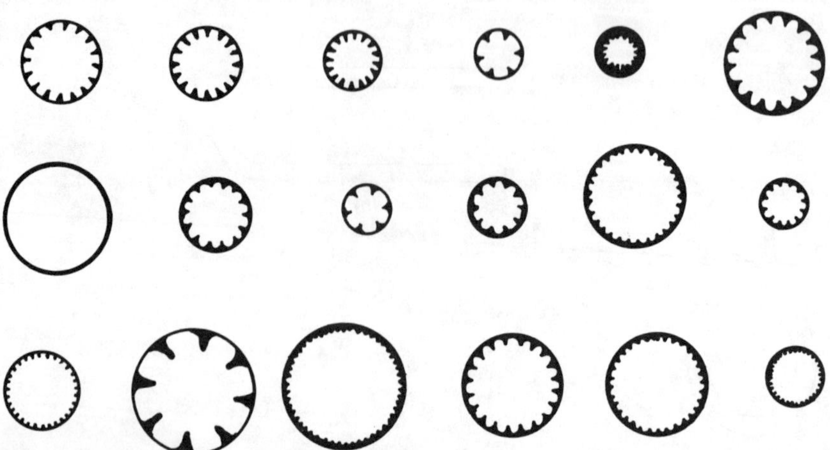

Figure 1.5 Inside fin tubes. *(N. H. Afgan and E. U. Schlünder.* Heat Exchangers: Design and Theory. *McGraw-Hill. 1974. By permission.)*

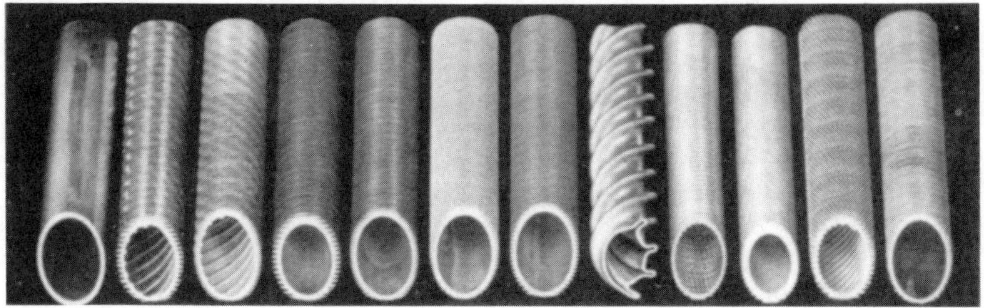

Figure 1.6 Enhanced surface tubes. *(Courtesy of Noranda Metal Industries, Inc., Newtown, Conn.)*

1.20 Are both the inner and the outer tubes of the same thickness?

The thickness is the same for both the tubes for the thin gauges. For thicker gauge tubes, the more expensive material may be one-fifth to one-third of the total thickness to reduce the cost.

1.21 What are the enhanced surface tubes? Where are they used?

These tubes enhance the heat transfer coefficient. It is achieved by two techniques:

1. The surface is contoured (or grooved) in a variety of ways forming valleys and ridges (Fig. 1.6). These are generally used in condensers where the condensate drains into valleys (or troughs) leaving only a very thin film on the ridges (or peaks) resulting in an increased heat transfer coefficient. Some such tubes are also used in double-phase-change situations, where vapor phase condenses on one side and its latent heat is used to evaporate a liquid on the other side (Fig. 1.7).
2. The surface is prepared with a special coating to provide a large number of nucleation sites for use in boiling operations. The bubbles form at a low superheat.

Enhanced surfaces are further discussed in Chapter 4.

1.22 For maximum heat transfer in a given sized unit, the tubes should have the largest possible surface area per unit volume. Why are the cylindrical tubes usually preferred even though some noncylindircal tubes, such as triangular tubes, have larger area per unit volume? Consider only the plain tubes.

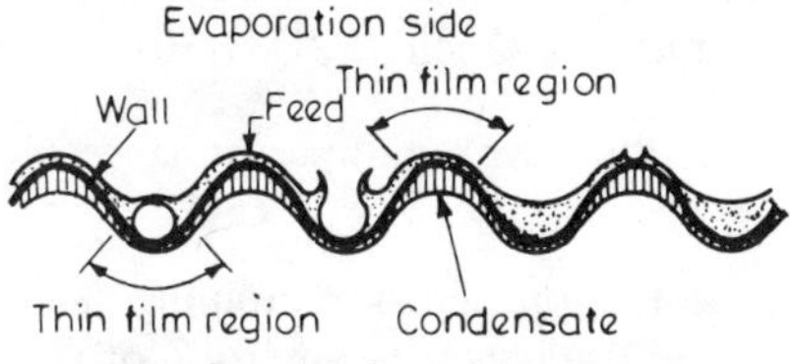

Figure 1.7 Double phase change on an enhanced surface tube.

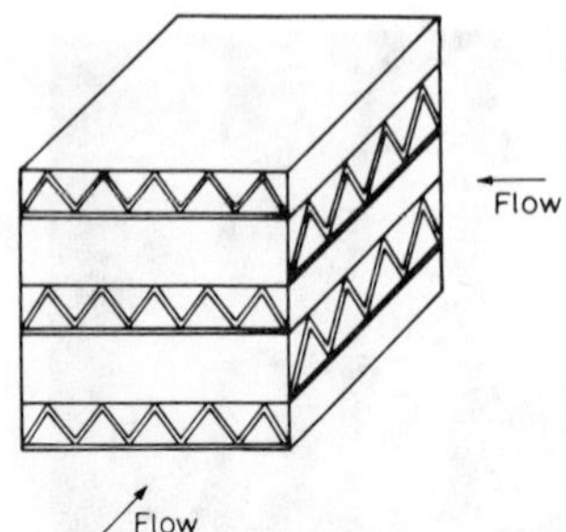

Figure 1.8 Plate fin exchanger.

Cylindrical tubes are the best for withstanding the fluid pressure. Thus, for a given pressure, the cylindrical tubes have the minimum wall thickness. They are also easy to manufacture, handle, and roll into the tubesheet. However, at low pressures, specially with gases, noncylindrical tubes or flow passages are used as in the plate-fin exchangers (Fig. 1.8 and Chapter 4). The surface area per unit volume for the cylindrical tubes (diameter $= D$) is $4/D$, while that for the equilateral triangular flow passages (side $= D$) is $4\sqrt{3}/D$ or almost 73 percent larger than the cylindrical tubes.

1.23 What caution should be exercised in selecting the tube material, tube thickness, and length?

Tubes should be able to withstand the following:

1. Pressure on the inside and the outside
2. Temperature on both the sides
3. Thermal stresses due to the differential expansion of the shell and the tube bundle
4. Corrosive nature of both the shell-side and the tube-side fluids

The thermal conductivity (or thermal resistance) of the tube plays an important role in equipment involving phase change (condensers, reboilers, etc.). Hence a tube material with large thermal conductivity and compatible with the above four points should be selected. The thickness should be specified. This is to take care of the thinning at the U-bend. The length and the number of tubes should be so selected as to give the required heat transfer area (based on the tube outer diameter) clear off the tubesheet, since the tube length in the tubesheet does not contribute to the heat transfer. The tube length embedded in the tubesheet can be significant in high-pressure exchangers. Table 1.3 (TEMA, 1978) gives the characteristic dimensions of the tubes most used in heat transfer equipment. For U-tube bundles, the surface area is calculated based on the mean tube length (Q. 1.30).

1.24 What effects does the tube material have on the design and performance of an exchanger?

The tube material affects the extent of corrosion, erosion, fouling, and microbial deposits (Chapter 13); vibration and natural frequency of vibration of the

Table 1.3 Characteristics of tubing

Tube O.D. (in)	B.W.G. gauge	Thickness (in)	Internal area (in²)	Sq. ft. external surface per foot length	Sq. ft internal surface per foot length	Weight per foot length steel (lb)[a]	Tube I.D. (in)	Moment of inertia (in)⁴	Section modulus (in)³	Radius of gyration (in)	Constant C[b]	O.D./I.D.	Transverse metal area (in²)
$\frac{1}{4}$	22	0.028	0.0295	0.0655	0.0508	0.066	0.194	0.00012	0.00098	0.0792	46	1.289	0.0195
$\frac{1}{4}$	24	0.022	0.0333	0.0655	0.0539	0.054	0.206	0.00011	0.00083	0.0810	52	1.214	0.0159
$\frac{1}{4}$	26	0.018	0.0360	0.0655	0.0560	0.045	0.214	0.00009	0.00071	0.0824	56	1.168	0.0131
$\frac{1}{4}$	27	0.016	0.0373	0.0655	0.0570	0.040	0.218	0.00008	0.00064	0.0829	58	1.146	0.0117
$\frac{3}{8}$	18	0.049	0.0603	0.0982	0.0725	0.171	0.277	0.00068	0.0036	0.1164	94	1.354	0.0502
$\frac{3}{8}$	20	0.035	0.0731	0.0982	0.0798	0.127	0.305	0.00055	0.0029	0.1213	114	1.233	0.0374
$\frac{3}{8}$	22	0.028	0.0799	0.0982	0.0835	0.104	0.319	0.00046	0.0025	0.1227	125	1.176	0.0305
$\frac{3}{8}$	24	0.022	0.0860	0.0982	0.0867	0.083	0.331	0.00038	0.0020	0.1248	134	1.133	0.0244
$\frac{1}{2}$	16	0.065	0.1075	0.1309	0.0969	0.302	0.370	0.0022	0.0086	0.1556	168	1.351	0.0888
$\frac{1}{2}$	18	0.049	0.1269	0.1309	0.1052	0.236	0.402	0.0018	0.0072	0.1606	198	1.244	0.0694
$\frac{1}{2}$	20	0.035	0.1452	0.1309	0.1126	0.174	0.430	0.0014	0.0056	0.1649	227	1.163	0.0511
$\frac{1}{2}$	22	0.028	0.1548	0.1309	0.1162	0.141	0.444	0.0012	0.0046	0.1671	241	1.126	0.0415
$\frac{5}{8}$	12	0.109	0.1301	0.1636	0.1066	0.602	0.407	0.0061	0.0197	0.1864	203	1.536	0.177
$\frac{5}{8}$	13	0.095	0.1486	0.1636	0.1139	0.537	0.435	0.0057	0.0183	0.1903	232	1.437	0.158
$\frac{5}{8}$	14	0.083	0.1655	0.1636	0.1202	0.479	0.459	0.0053	0.0170	0.1938	258	1.362	0.141
$\frac{5}{8}$	15	0.072	0.1817	0.1636	0.1259	0.425	0.481	0.0049	0.0156	0.1971	283	1.299	0.125
$\frac{5}{8}$	16	0.065	0.1924	0.1636	0.1296	0.388	0.495	0.0045	0.0145	0.1993	300	1.263	0.114
$\frac{5}{8}$	17	0.058	0.2035	0.1636	0.1333	0.350	0.509	0.0042	0.0134	0.2016	317	1.228	0.103
$\frac{5}{8}$	18	0.049	0.2181	0.1636	0.1380	0.303	0.527	0.0037	0.0118	0.2043	340	1.186	0.089
$\frac{5}{8}$	19	0.042	0.2298	0.1636	0.1416	0.262	0.541	0.0033	0.0105	0.2068	358	1.155	0.077
$\frac{5}{8}$	20	0.035	0.2419	0.1636	0.1453	0.221	0.555	0.0028	0.0091	0.2089	377	1.126	0.065

Table 1.3 Characteristics of tubing (*Continued*)

Tube O.D. (in)	B.W.G. gauge	Thickness (in)	Internal area (in^2)	Sq. ft. external surface per foot length	Sq. ft. internal surface per foot length	Weight per foot length steel (lb)[a]	Tube I.D. (in)	Moment of inertia $(in)^4$	Section modulus $(in)^3$	Radius of gyration (in)	Constant C[b]	O.D./I.D.	Transverse metal area (in^2)
$\frac{3}{4}$	10	0.134	0.1825	0.1963	0.1262	0.884	0.482	0.0129	0.0344	0.2229	285	1.556	0.260
$\frac{3}{4}$	11	0.120	0.2043	0.1963	0.1335	0.809	0.510	0.0122	0.0326	0.2267	319	1.471	0.238
$\frac{3}{4}$	12	0.109	0.2223	0.1963	0.1393	0.748	0.532	0.0116	0.0309	0.2299	347	1.410	0.220
$\frac{3}{4}$	13	0.095	0.2463	0.1963	0.1466	0.666	0.560	0.0107	0.0285	0.2340	384	1.339	0.196
$\frac{3}{4}$	14	0.083	0.2679	0.1963	0.1529	0.592	0.584	0.0098	0.0262	0.2376	418	1.284	0.174
$\frac{3}{4}$	15	0.072	0.2884	0.1963	0.1587	0.520	0.606	0.0089	0.0238	0.2410	450	1.238	0.153
$\frac{3}{4}$	16	0.065	0.3019	0.1963	0.1623	0.476	0.620	0.0083	0.0221	0.2433	471	1.210	0.140
$\frac{3}{4}$	17	0.058	0.3157	0.1963	0.1660	0.428	0.634	0.0076	0.0203	0.2455	492	1.183	0.126
$\frac{3}{4}$	18	0.049	0.3339	0.1963	0.1707	0.367	0.652	0.0067	0.0178	0.2484	521	1.150	0.108
$\frac{3}{4}$	20	0.035	0.3632	0.1963	0.1780	0.269	0.680	0.0050	0.0134	0.2532	567	1.103	0.079
1	8	0.165	0.3526	0.2618	0.1754	1.462	0.670	0.0392	0.0784	0.3009	550	1.493	0.430
1	10	0.134	0.4208	0.2618	0.1916	1.237	0.732	0.0350	0.0700	0.3098	656	1.366	0.364
1	11	0.120	0.4536	0.2618	0.1990	1.129	0.760	0.0327	0.0654	0.3140	708	1.316	0.332
1	12	0.109	0.4803	0.2618	0.2047	1.037	0.782	0.0307	0.0615	0.3174	749	1.279	0.305
1	13	0.095	0.5153	0.2618	0.2121	0.918	0.810	0.0280	0.0559	0.3217	804	1.235	0.270
1	14	0.083	0.5463	0.2618	0.2183	0.813	0.834	0.0253	0.0507	0.3255	852	1.199	0.239
1	15	0.072	0.5755	0.2618	0.2241	0.714	0.856	0.0227	0.0455	0.3291	898	1.167	0.210
1	16	0.065	0.5945	0.2618	0.2278	0.649	0.870	0.0210	0.0419	0.3314	927	1.149	0.191
1	18	0.049	0.6390	0.2618	0.2361	0.496	0.902	0.0166	0.0332	0.3366	997	1.109	0.146
1	20	0.035	0.6793	0.2618	0.2435	0.360	0.930	0.0124	0.0247	0.3414	1060	1.075	0.106

1¼	7	0.180	0.6221	0.3272	0.2330	2.057	0.890	0.0890	0.1425	0.3836	970	1.404	0.605
1¼	8	0.165	0.6648	0.3272	0.2409	1.921	0.920	0.0847	0.1355	0.3880	1037	1.359	0.565
1¼	10	0.134	0.7574	0.3272	0.2571	1.598	0.982	0.0741	0.1186	0.3974	1182	1.273	0.470
1¼	11	0.120	0.8012	0.3272	0.2644	1.448	1.010	0.0688	0.1100	0.4018	1250	1.238	0.426
1¼	12	0.109	0.8365	0.3272	0.2702	1.329	1.032	0.0642	0.1027	0.4052	1305	1.211	0.391
1¼	13	0.095	0.8825	0.3272	0.2775	1.173	1.060	0.0579	0.0926	0.4097	1377	1.179	0.345
1¼	14	0.083	0.9229	0.3272	0.2838	1.033	1.084	0.0521	0.0833	0.4136	1440	1.153	0.304
1¼	16	0.065	0.9852	0.3272	0.2932	0.823	1.120	0.0426	0.0682	0.4196	1537	1.116	0.242
1¼	18	0.049	1.042	0.3272	0.3016	0.629	1.152	0.0334	0.0534	0.4250	1626	1.085	0.185
1¼	20	0.035	1.094	0.3272	0.3089	0.456	1.180	0.0247	0.0395	0.4297	1707	1.059	0.134
1½	10	0.134	1.192	0.3927	0.3225	1.955	1.232	0.1354	0.1806	0.4853	1860	1.218	0.575
1½	12	0.109	1.291	0.3927	0.3356	1.618	1.282	0.1159	0.1546	0.4933	2014	1.170	0.476
1½	14	0.083	1.398	0.3927	0.3492	1.258	1.334	0.0931	0.1241	0.5018	2181	1.124	0.370
1½	16	0.065	1.474	0.3927	0.3587	0.996	1.370	0.0756	0.1008	0.5079	2299	1.095	0.293
2	11	0.120	2.433	0.5236	0.4608	2.410	1.760	0.3144	0.3144	0.6660	3795	1.136	0.709
2	12	0.109	2.494	0.5236	0.4665	2.201	1.782	0.2904	0.2904	0.6697	3890	1.122	0.647
2	13	0.095	2.573	0.5236	0.4739	1.934	1.810	0.2586	0.2586	0.6744	4014	1.105	0.569
2	14	0.083	2.642	0.5236	0.4801	1.699	1.834	0.2300	0.2300	0.6784	4121	1.090	0.500

Source:

[a]Weights are based on low carbon steel with a density of 0.2833 lb/in³. For other metals multiply by the following factors:

Aluminum	0.35	Aluminum Brass	1.06
Titanium	0.58	Nickel–chrome–iron	1.07
A.I.S.I. 400 series stainless steels	0.99	Admiralty	1.09
A.I.S.I. 300 series stainless steels	1.02	Nickel and nickel–copper	1.13
Aluminum Bronze	0.97	Copper and cupro–nickels	1.14

[b]Liquid velocity = (lb per tube per hr/C × Sp. Gr. of liquid) in ft/s; (Sp. Gr. of water at 60° F = 1.0).

tube (Chapter 15); thermal stresses induced due to the differential thermal expansion of the tube bundle and the shell (Chapter 9) and the thermal resistance offered by the tube (Chapter 2); ease of joining the tubes to the tubesheet (Chapter 10); and the cost. Furthermore, in designs utilizing finned, duplex, or enhanced tubes, the total design is influenced by the fact that these are not available in all the materials.

1.25 What are the advantages of standardizing the dimensions of tubes used in various heat exchangers in a plant?

It reduces the number of different sizes of the tubes to be stocked for replacement. The number of tools to be carried for the maintenance and repair is also reduced. The "standard" diameters used most are $\frac{3}{4}$--1 in (19.2-25.4 mm).

1.26 How is the tube thickness designated?

Tube thickness is given in terms of the gauge numbers in the British Units. Table 1.3 shows some of the common gauge numbers and the corresponding tube thicknesses used in the heat exchangers. In the SI units, the thickness is given in millimeters. The inner diameter of tube is obtained by subtracting twice the thickness from the outer or the nominal diameter of the tube.

1.27 What is the difference between the nominal diameter of a pipe and that of a tube?

The nominal diameter of a pipe does not relate directly to either its outer diameter or to its inner diameter up to 12 in (305 mm) nominal diameter. Above that size, the nominal diameter of the pipe is the same as the outer diameter.

The nominal diameter of a tube is its outer diameter for all tube sizes. It should be mentioned that the pipes are not used as tubings in the shell-and-tube heat exchangers. They are used as shells in small-diameter shells [generally up to 24 in (610 mm)] and in double pipe heat exchangers (Q. 2.181, Chapter 2).

1.28 What is meant by minimum wall tubing?

As per the ASME Code (ASME, 1980), the minimum wall tubing has a zero permissible under-tolerance and a maximum of 18-22 percent over-tolerance.

1.29 What is meant by average wall tubing?

In this, the under- and the over-tolerance are usually identical and approximately half of the over-tolerance for a minimum wall tube, or ±10 percent, approximately.

1.30 What is the mean or effective tube length of a U-tube bundle?

It is the arithmetic mean of the tube lengths of the outermost and the innermost tubes taking the U-bends into account. The outermost tube is the longest, and the innermost tube is the shortest in the whole U-tube bundle.

1.31 Why are the tubes omitted at the center of a U-bundle?

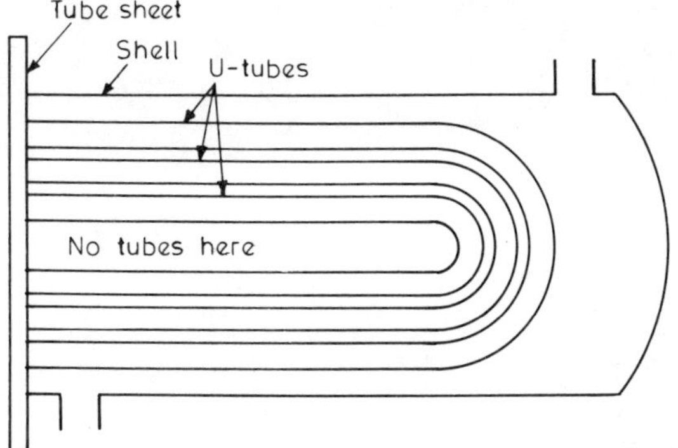

Figure 1.9 U-tube layout.

When a tube is bent into a U-shape, the outer surface expands and gets thinned down in the process. In order not to produce excessive thinning (more than 20–25 percent of the tube thickness), which will damage the tube, the minimum bend radius is kept approximately 1.5–2 times the tube diameter. Thus, there can be no tubes in the space within the minimum bend radius tube (Fig. 1.9).

1.32 What is tube pitch?

The center-to-center distance between any two adjacent tubes is called the tube pitch. The ratio of the tube pitch to the tube outer diameter is called the pitch ratio. The value of the pitch ratio normally varies from 1.1 to 1.5, with the most common value being around 1.25 (Fig. 1.10). To avoid any ambiguity in the specification of the tube pitch, generally, a sketch should be provided along with the tube layout (Q. 1.34, Fig. 1.11).

1.33 What factors decide the pitch in a particular case?

The heat transfer coefficient and the limit on the pressure drop on the shell side, the number of tubes in a given diameter shell, the need to avoid vibration damage (if the tubes are very close together), and the need to access the outer

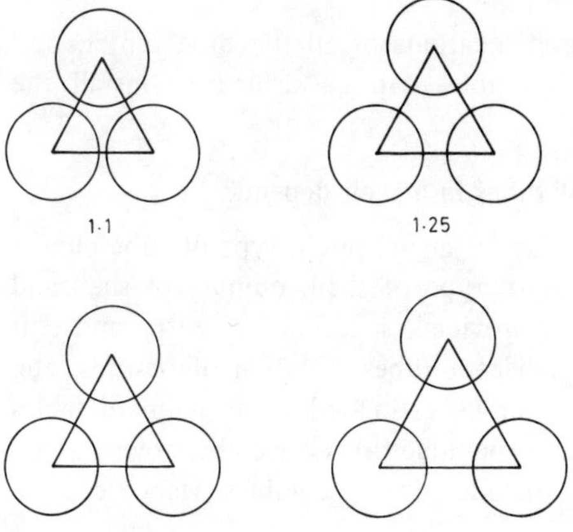

Figure 1.10 Tube pitch ratios.

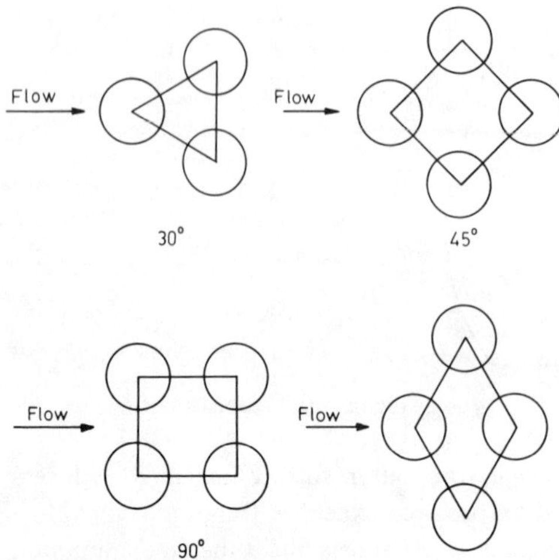

Figure 1.11 Tube layout angles.

surface of the tubes for cleaning are among the reasons that determine the value of the pitch (see also Q. 2.103, Chapter 2). Increased tube pitch usually means a lower heat transfer coefficient and a costlier heat exchanger.

1.34 What is tube layout angle?

The tube layout angle is defined with respect to the flow direction and not with respect to any fixed coordinate system. For example, in any equilateral triangular tube layout, the flow along the apex gives a 30° layout angle, while the flow along the base gives a 60° layout angle (Fig. 1.11). The other two layout angles are 45° and 90°, also shown in Fig. 1.11. A 30° layout angle is generally referred to as the triangular layout, 60° as the rotated triangular layout, 90° as the square layout, and 45° as the rotated square or the diamond layout (see also Q. 2.104 and 2.105, Chapter 2).

1.35 What is meant by tubesheet layout?

It refers to the marking of the exact locations of all the tube centers and tie-rods (Q. 1.82) on the tubesheet so that holes can be drilled to install the requisite number of tubes and tie-rods.

1.36 On what factors does the number of tubes in a shell depend?

It depends upon the tube diameter, layout angle, pitch, type of tube bundle (fixed, U-tube, floating head), outer tube limit, type of shell, number of shell and tube passes, whether impingement baffle is provided, locations of entry and exit nozzles and pass partition plates and the adjacent tubes, location of tie-rods, etc. The number of tubes in a particular case can be estimated from standard tables (Perry and Chilton, 1973; Kern, 1950). Most manufacturers have their own tables. Usually, allowance for the tie-rods is not included in these tables. Hence caution needs to be exercised in using them.

• Some of the factors mentioned above which have not already been discussed, will be discussed later in this chapter.

1.37 In a multi-pass exchanger, do all the passes have the same number of tubes?

The number of tubes per pass is usually kept the same within the constraints imposed by the tube layout and locations of the pass partition plates, impingement baffle, and the tie-rods. This results in the same velocity and the same heat transfer area in each pass for a no-phase-change situation. These are important assumptions in the derivation of the log mean temperature difference (Q. 2.23, Chapter 2) and in the relations for the heat transfer coefficient and the pressure drop.

If the flow volume of the tube-side fluid changes due to phase change, the number of tubes in different passes may be different in order to provide a reasonably high velocity in each pass despite the change in the flow volume. For example, in a horizontal multi-pass in-tube condenser, the condensate produced in one pass will flow full through the lowermost tubes of the subsequent passes. Allowance should, therefore, be made for such tubes since these are not available for the condensing duty.

1.38 Are the tubes laid out symmetrically around the center of the tubesheet in a multi-pass exchanger?

Since the tube layout depends upon a number of factors as noted in Q. 1.36, the tubes are not symmetrically laid out in a multi-pass exchanger. Sometimes it is not possible to do so even in a single-pass unit. However, this does not result in any appreciable penalty in heat transfer.

1.39 How does the number of passes affect the number of tubes in a given sized shell?

Since the pass partition plates have to be provided to increase the number of shell and/or tube passes, the tubes have to be deleted from the part of the tubesheet where the pass partition plate groove is made. Furthermore, larger space has to be provided between the pass partition plate and the adjacent tube row compared to the distance between two consecutive tube rows. For example, the number of straight tubes, [$\frac{3}{4}$ in (19.2 mm) O.D., $\frac{15}{16}$ in (23.8125 mm) pitch in a 25 in (635 mm) shell] in a 1-1 unit is 532; in a 1-2 unit is 506; in a 2-4 unit is 468, and in a 1-8 unit is 434 (Kern, 1950).

1.40 For a given heat load, the least expensive exchanger is obtained when the smallest diameter tubes having the longest possible length are used on a close, triangular pitch. Why are these not always used?

Such units are not always used because the above advantage in the initial cost of the exchanger may be offset by the following factors.

1. The pressure drop both on the shell side and the tube side is high in such a unit. This requires larger investment in the pumping capacity and higher operating costs.
2. The maintenance and cleaning of the small diameter tubes is rather difficult and

so is their replacement in case of a tube failure. For close, triangular spacing, the cleaning on the outside of the tubes also becomes difficult (see Chapter 14).

3. For small shell diameter and close pitch tube layout, the vibration problem becomes more acute since the cross-flow velocity is very high (see Chapter 15).
4. Long tubes require large space not only for the installation of the exchanger but also for the removal and replacement of a ruptured tube. Often, such large, unhindered spaces are not easily available in complex chemical plants.

BAFFLES, BAFFLE CUT, SPACING, AND ORIENTATION

1.41 Why are baffles used?

Baffles are used to direct the shell-side and tube-side flows so that the fluid velocity is increased sufficiently (within the limits imposed by the allowable pressure drop) to obtain high heat transfer coefficients and to reduce fouling deposits. In horizontal units, the baffles also provide support to the tubes against sagging and vibration damage.

1.42 What are the principle kinds of baffles?

The baffles for the shell-side fluid are the following:

1. Orifice (Fig. 1.12)
2. Disk-and-doughnut (Fig. 1.13)
3. Segmental (Figs. 1.14 to 1.17)
4. Rod type (Fig. 1.18)
5. Nest type (Fig. 1.19)
6. Longitudinal (Fig. 1.20)
7. Impingement (Fig. 1.21)

The baffles for the tube-side fluid are known as the pass partition plates (Fig. 1.23).

- Now we shall discuss the uses of these baffles.

1.43 What are the uses of orifice baffles?

These are shown in Fig. 1.12. These are made of full circle plates with holes $\frac{1}{16}$–$\frac{1}{8}$ in (1.6–3.2 mm) larger in diameter than the tube O.D. The fluid flows parallel to the tubes and passes through the annular orifice between the tube and the baffle hole. There is no significant cross-flow over the tube bundle. These are not used often because of several reasons, important among which are the following:

1. Since the fluid has to pass through the annular orifice, the pressure drop is high.
2. Since the flow is parallel to the bundle, the boundary layer on the tubes is rather thick, resulting in low heat transfer coefficient.

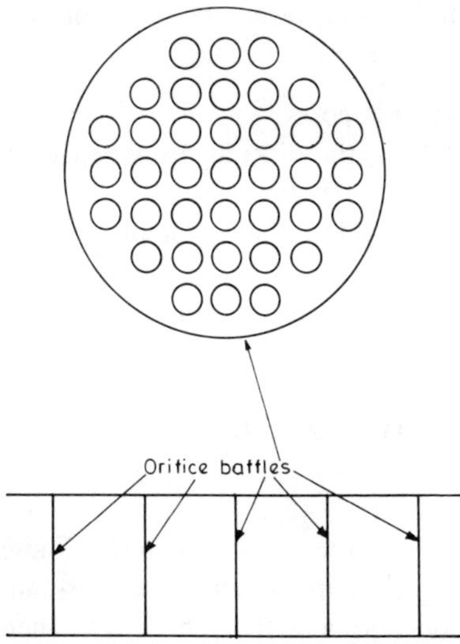

Figure 1.12 Orifice baffles.

3. It is not easy to clean the scale from the outside of the tubes.
4. Since there are no tube supports, vibration damage is highly probable.

1.44 What are the uses of disk-and-doughnut baffles?

These are shown in Fig. 1.13. The disk and the doughnut are cut from the same circular plate and are arranged alternately along the length of the tube bundle. The clearance between the baffle holes and the tube O.D. is not as large as in the orifice baffles. It is approximately $\frac{1}{32}$ in (0.8 mm) or less, as is the case with the segmental baffles discussed next. The flow is partly parallel to the tubes and partly cross-flow, closer to the flow pattern in the double segmental baffles discussed later.

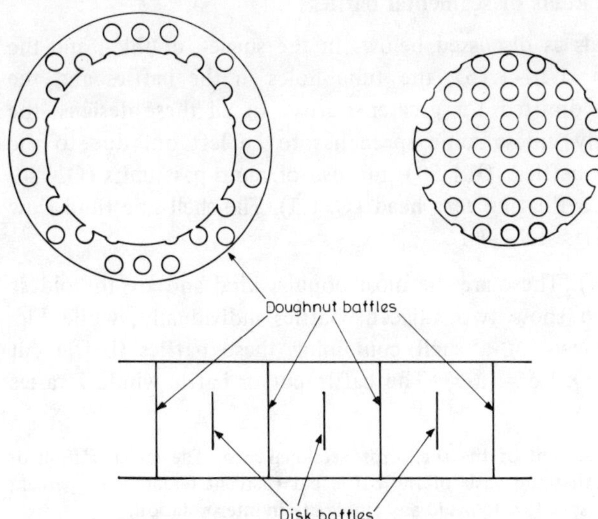

Figure 1.13 Disk-and-doughnut baffles.

However, the disk-and-doughnut baffles are not used often because of the following:

1. The tube support in the central zone of the bundle is poor.
2. One needs a separate set of internal tie-rods to keep the disks at proper spacing, and this reduces the number of tubes in a given diameter shell.
3. There is a greater possibility of scale depositing behind the lower part of the doughnut that blankets part of the heat transfer area.
4. Difficulties are encountered in venting any released gases and in draining the unit.
5. It is difficult to use the unit for alternate services.
6. The round edges of the disks require milling and that adds to the cost.

1.45 What are the uses of segmental baffles?

These are shown in Figures 1.14–1.17 and are the most used ones. These give higher heat transfer coefficient compared to the orifice and the disk-and-doughnut baffles. The clearance between the baffle hole and the tube is governed by the TEMA Standards. These baffles are made out of round plates, one or two segments of which have been cut out. Hence these are called the segmental baffles. The size of the segment removed, known as the baffle cut, is usually specified as percent of the shell diameter.* This cut results in a window between the shell and the baffle. For given shell and tube diameters, tube layout, and the shell-side flow rate, the size of this cut determines the flow velocity through the baffle window, whereas the spacing between two consecutive baffles determines the cross-flow velocity over the tube bundle. Thus, both the baffle cut and spacing influence the fluid velocity and hence the pressure drop and heat transfer coefficient vary significantly (Chapter 2). Segmentally baffled exchangers can be used for varied services.

1.46 What are the different kinds of segmental baffles?

These are of four kinds as discussed below. In the single-, double-, and the triple-segmental designs (Figs. 1.14–1.16), the tube holes in the baffles and the tubes in the shell have been omitted for greater clarity. In all these designs, the tubes fill the shell completely unless some space has to be left out due to the locations of the impingement baffles (Q. 1.50), the use of multi-pass units (Q. 1.49 and 1.51), and/or the use of a floating rear head (Q. 1.5). The shell-side fluid flow is shown by dashed lines in Figs. 1.14–1.16.

1. Single Segmental (Fig. 1.14): These are the most popular kind and are the oldest in use. Figure 1.14*a* and *b* show two adjacent baffles individually, while Fig. 1.14*c* and *d* show two views of a shell containing these baffles (baffle cut orientation is discussed in Q. 1.55–1.61). The baffle cut or baffle window varies

*It is sometimes defined as percent of the shell cross-sectional area. The latest edition of TEMA Standards (1978) requires that the basis of the baffle cut (percent diameter or percent shell cross-sectional area) should be specified to avoid any ambiguity in interpretation.

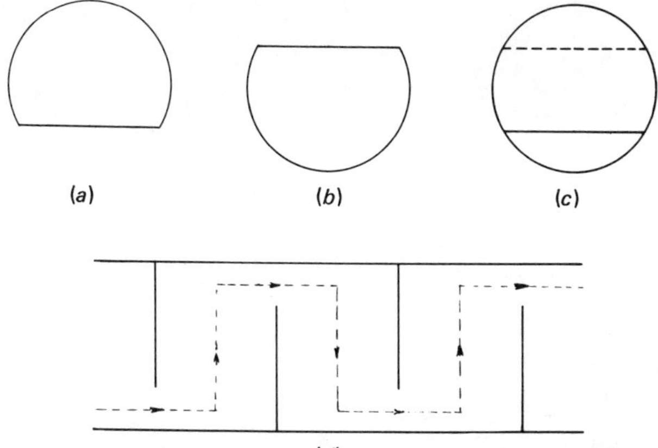

Figure 1.14 Single segmental baffles.

from 15 to 45 percent of the shell diameter, although 20–25 percent is the optimum to achieve the highest heat transfer coefficient for a given pressure drop. The pressure drop, however, is rather high in the single-segmental design.

The performance of different baffles can be compared by using an index to be called C_{ph}. It is a measure of the amount of heat transfer achieved per unit pressure drop in a given system. Thus, in a single-segmental design, C_{ph} is the highest for a 20–25 percent baffle cut.

2. Double Segmental (Fig. 1.15): In situations where moderate restrictions on pressure drop are applicable, double-segmental baffles are used. Figure 1.15*a* and *b* show one pair of adjacent baffles, while Fig. 1.15*c* and *d* show two views of a shell containing these baffles. Since there are more baffle windows compared to the single-segmental case, the fluid flows parallel to the tubes for a greater length of the shell in the double-segmental case than in the single segmental with a resultant decrease in the heat transfer coefficient. The pressure drop is also comparatively low, but so is C_{ph}. For ease of calculations, the fluid is treated as divided into two equal substreams, one passing through the top half of the exchanger above the

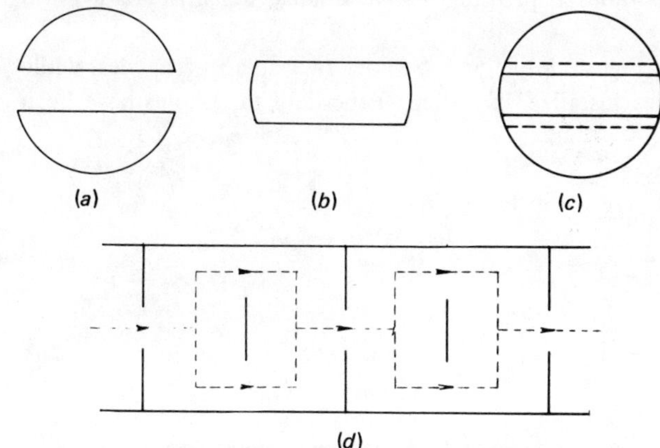

Figure 1.15 Double segmental baffles.

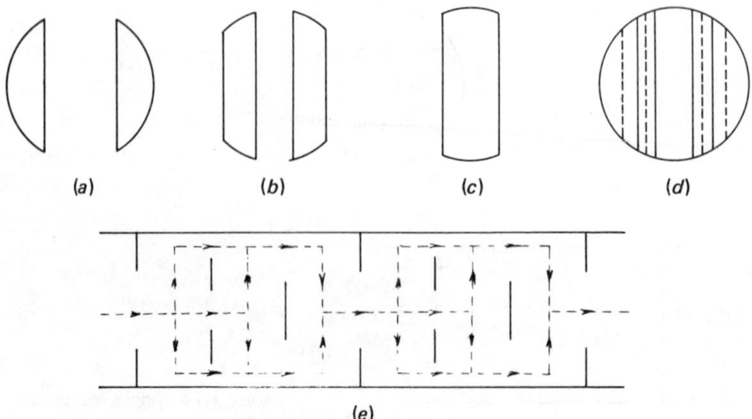

Figure 1.16 Triple segmental baffles.

center plane, the other passing through the bottom half. Each of the two halves of the shell can be approximated as separate units with single-segmental baffles. Parallel flow through the baffle windows should be accounted for. There is a slight overlap, generally corresponding to two tube rows, in adjacent baffles as a guard against vibration damage (Chapter 15).

3. Triple Segmental (Fig. 1.16): These are used in situations where the pressure drop is severely restricted, as might happen in large power plants. Figure 1.16*a–c* shows one set of adjacent baffles, while Fig. 1.16*d* and *e* show two views of a shell containing these baffles. The fluid flows parallel to the tubes for a greater length of the exchanger since the baffle cut is very large. The heat transfer coefficient, pressure drop, as well as C_{ph} are lower than in the single- and the double-segmental baffle cases. The design, fabrication, and installation of these baffles is more complicated and hence costly. Damaging vibrations can easily be caused because of the long gap between the baffle segments. Parallel flow in the baffle window is very important and should be accounted for in the design. Usually, an overlap of two tubes is provided between adjacent baffles as a guard against vibration damage.
4. No-tube-in-window design: Figure 1.17*a* and *b* show two adjacent baffles, while Fig. 1.17*c* shows the baffles installed in a shell. Tubes and tube holes have been

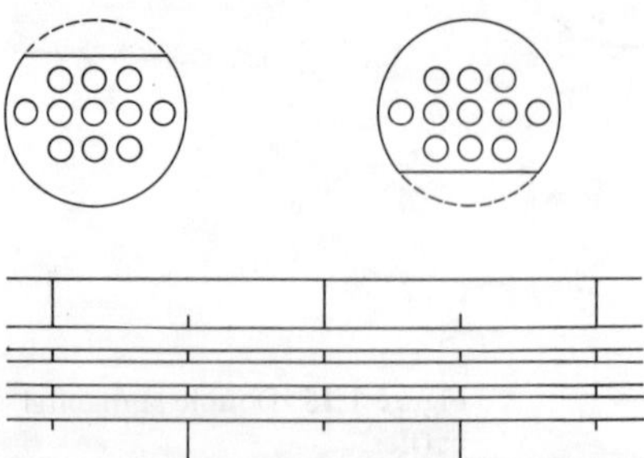

Figure 1.17 No-tube-in-window design.

shown to bring out the fact that, as the name implies, there are no tubes in the baffle window, otherwise they look similar to the single-segmental baffles. Thus, the fluid is always in cross-flow over the tube bundle, giving a higher value of C_{ph}. These baffles are used in case of severe pressure drop restrictions. Even though the C_{ph} is high, the heat transfer area is reduced because there are no tubes in the baffle windows. Hence, larger-sized shells are needed. Since all the tubes are supported by each baffle, the baffle spacing is made larger to lower the pressure drop. However, to reduce the possibility of vibration damage (Chapter 15), a support plate is generally introduced between two adjacent baffles. The support plate looks like the central piece of the double-segmental baffle (Fig. 1.15*b*) and does not result in any appreciable pressure drop (Q. 1.73).

1.47 What are the uses of the rod baffles?

These are among the latest developments in heat exchanger design (Fig. 1.18) (Small and Young, 1979). Each tube is supported at four or more circumferential points. The flow is parallel to the tube bundle, and hence it saves against vibration damage caused principally by the cross-flow over the tube bundle in the segmental baffle designs (Chapter 15). It does not have the disadvantages of other parallel flow designs (orifice baffles, disk-and-doughnut baffles), such as excessive pressure drop, since the flow area is large; preferential scale deposit, since no stagnant zones exist; or thick boundary layer, since the boundary layer is frequently disturbed by the rod baffles, thus giving a good heat transfer coefficient. One problem is the fluid bypass between the shell and the tube bundle (stream C in Q. 2.3, Chapter 2).

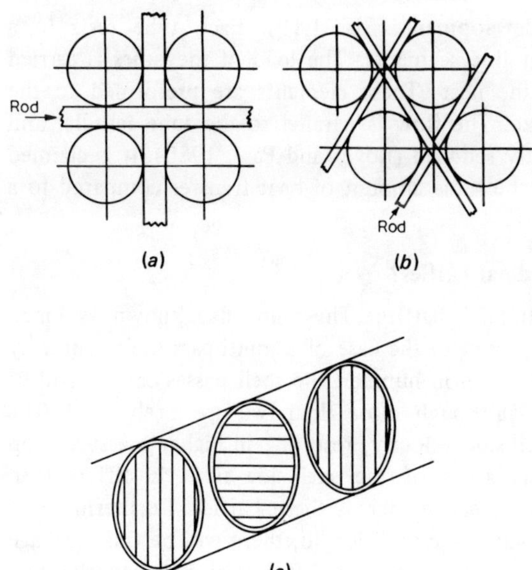

Figure 1.18 Rod baffles. (*a*) Square layout. (*b*) Triangular layout. (*c*) Shell and rod baffle arrangement.

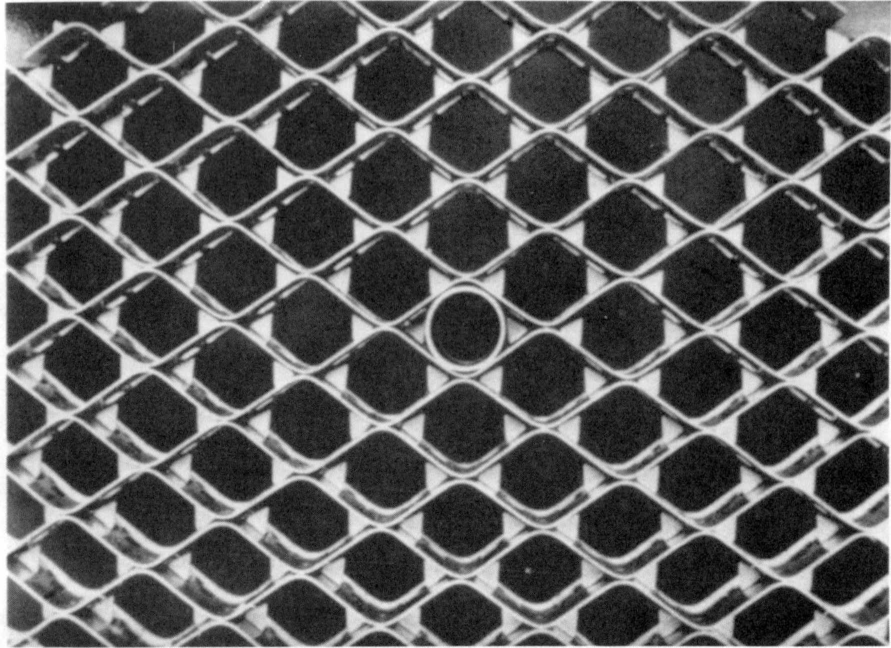

Figure 1.19 NEST–type baffles. *(Courtesy of Ecolaire Heat Transfer Company, Easton, Pa.)*

1.48 What are the uses of the NESTS*-type baffles?

These are also a recent development (Fig. 1.19). Each tube rests in a V-shaped cradle and is supported at line segments. The load of the tubes is carried by the smooth elements forming the nest. These elements are preformed to the desired tube pitch and ligament size. The flow is parallel to the tube bundle, and hence the vibration problem is greatly reduced (Boyer and Pase, 1980). It is claimed that the pressure drop is lower for the same amount of heat transfer compared to a segmentally baffled exchanger.

1.49 What are the uses of longitudinal baffles?

Figure 1.1F–H show longitudinal baffles. These are also known as longitudinal pass partition plates and are used in the case of a multi-pass shell. Contrary to the tube passes that are normally even in number, the shell passes can be odd or even, usually up to six in number, although more than two are rarely used. The longitudinal baffles increase the shell-side velocity, resulting in higher pressure drop and heat transfer. Since the temperature of the shell-side fluid is different in different shell passes, there is thermal leakage across these baffles. Furthermore, if the baffle is not effectively sealed against the shell length, there will be fluid leakage along the entire baffle length due to the pressure differential between the two

*NESTS is a patented trademark of Ecolaire Heat Transfer Company, Easton, Pa.

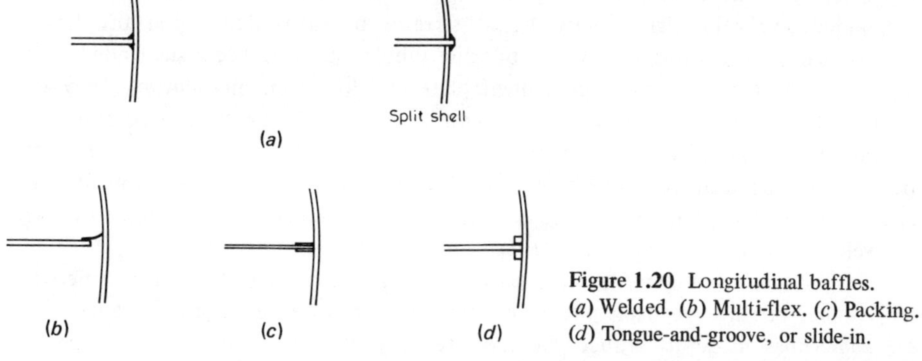

Figure 1.20 Longitudinal baffles. (*a*) Welded. (*b*) Multi-flex. (*c*) Packing. (*d*) Tongue-and-groove, or slide-in.

passes. This adversely affects the heat transfer coefficient and makes its accurate prediction rather difficult. The sealing can be done in numerous ways shown in Fig. 1.20:

1. Welding to the shell: In large shells a welder can go in to weld the baffle. In small shells, the baffle is first welded on the one-half of the shell. Then the other half of the shell is welded on the baffle.
2. Sealing strips or multi-flex arrangement
3. Packing arrangement
4. Slide-in or tongue-and-groove arrangement

The baffle is shorter than the shell length so that the shell-side fluid can enter the next pass. The cut-out area must be at least equal to the shell-side flow area for one pass so that there is no excessive pressure drop in it. The longitudinal baffle material is generally the same as the transverse baffles, since it has to be compatible with the shell-side fluid. TEMA Standards (1978) specify the minimum thickness of the longitudinal baffles for different materials.

1.50 What are the uses of impingement baffles?

These are also known as impingement plates and are shown in Fig. 1.21. These can be flat or curved, taking the contour of the shell. These are located at the inlet to the heat exchanger between the tube bundle and the nozzle opening in

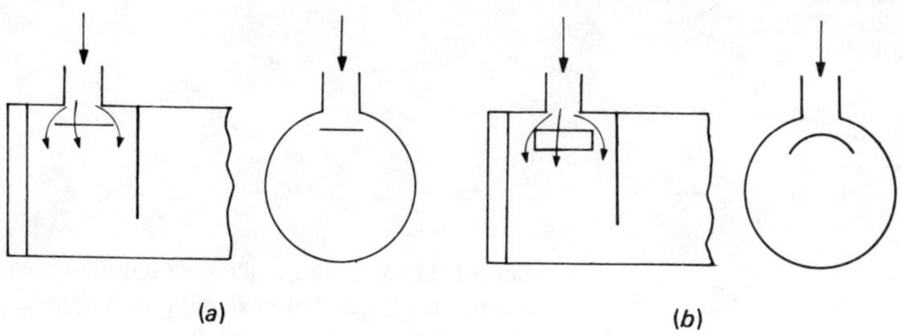

Figure 1.21 Impingement baffles. (*a*) Flat-plate type. (*b*) Curved-plate type.

the shell. The aim is to prevent damage to the tubes due to the carry-over of the suspended solids in high-velocity liquid streams or suspended liquid droplets in high-velocity gas streams that will erode the tubular surface. These suspended solids and liquid droplets hit the impingement plate and lose their momentum. They also help in the better distribution of the incoming fluid over the tube bundle in order to prevent vibration damage to the tubes and to make better use of the heat transfer surface near the tubesheet. The flow area between the impingement plate and the shell is kept between 1.25 and 1.5 times the nozzle cross section to reduce the velocity at the entrance to the tube bundle.

Since an impingement plate takes up space inside the shell where otherwise more tubes could have been put, given more heat transfer area, other arrangements are sometimes made to reduce the flow velocity at the entrance. Three such are shown in Fig. 1.22. In Fig. 1.22*a*, the nozzle is expanded at the entrance. This decreases the fluid velocity. In Fig. 1.22*b*, a jacket is placed over the shell at the point where the nozzle comes in, whereas one or two larger openings are made further along the circumference of the shell for the fluid inlet. These arrangements can be made even on the existing exchangers if found to be facing vibration damage by the high-entrance velocity, without modifying the tube arrangement inside. In Fig. 1.22*c*, the shell construction is modified to have fluid entering all around the tube bundle.

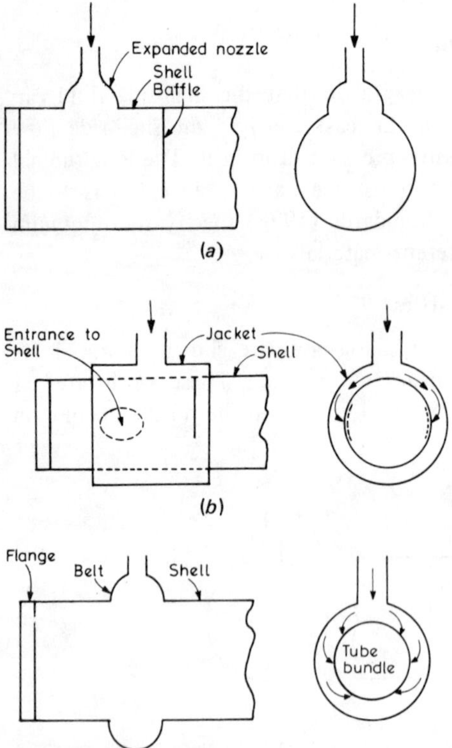

Figure 1.22 Ways to reduce shell inlet velocity. (*a*) Expanded or flared nozzle type. (*b*) Jacket type. (*c*) Modified shell construction.

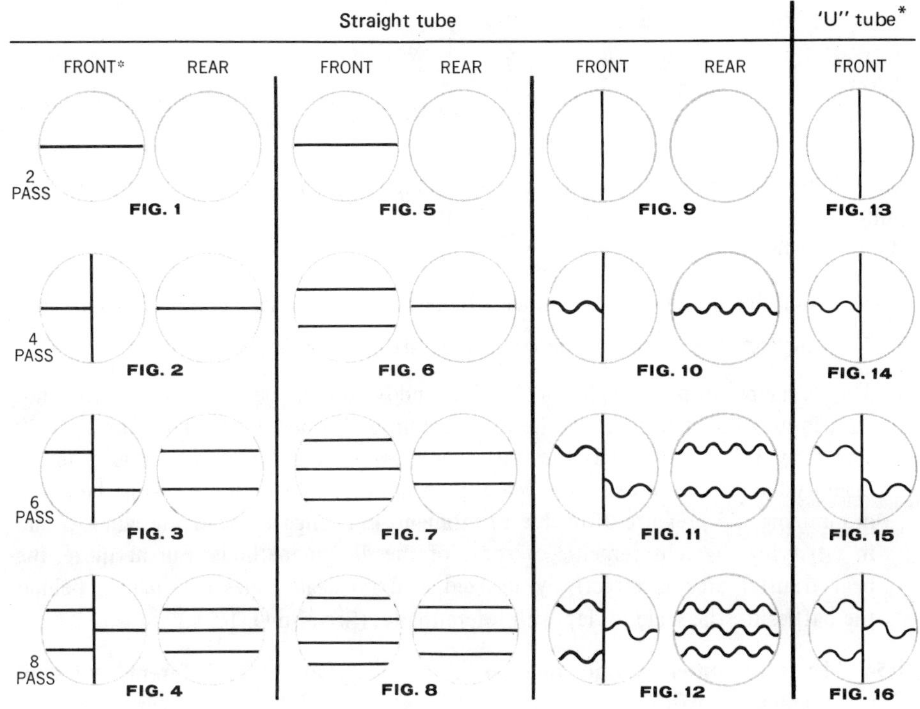

Figure 1.23 Tube-side pass partition plates. *(Courtesy of Patterson-Kelley Co., Division of HARSCO Corp., East Stroudsburg, Pa.)*

Sometimes the tubes are strengthened in the entry region by putting a short segmental baffle on the tubes that are directly in line of the entering fluid. Alternately, heavier gauge tubing can be used in the first few rows.

1.51 What are the uses of the tube-side pass partition plates?

These are the baffles used to direct the tube-side fluid into different tube passes and are located in the front and rear ends of the exchanger (Fig. 1.23). These pass partition plates can be horizontal and/or vertical depending upon the case. These are generally flat plates, although curved, disk-type may be used when the pressure differential across the pass partition plate is very high, which would result in a rather thick flat plate with capability to take cyclic thermal gradients as in a nuclear steam generator (Schroeder, 1979). These may be laid differently for the straight tube and U-tube bundles. These are generally welded in place in the channel and set in grooves in the tubesheet by means of integral gaskets.

1.52 What is a continuous baffle into the window?

As shown in Fig. 1.24, it is a continuous plate from which a baffle window has been cut out. The continuous part (rim) acts as a seal strip to force the fluid into

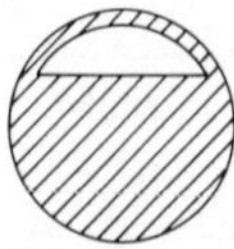

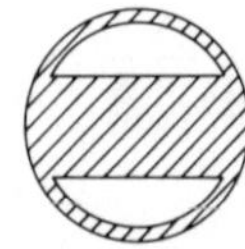

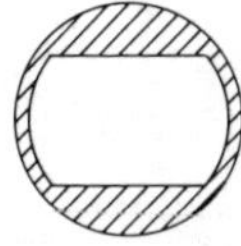

Figure 1.24 Continuous baffle into the window.

the bundle. It can be done with single- or multi-segmental baffles, preferably in a pull-through-floating-head-type tube bundle where the seal strips are a must (Q. 1.87–1.89).

1.53 What factors affect the choice of the type of the segmental baffles to use?

The factors that should be considered are the following:

1. The pressure drop through the baffle window should be low since it does not contribute to heat transfer to the same extent as it does in the cross-flow.
2. The flow velocity through the baffle window should be of the same order of magnitude as the cross-flow velocity over the tube bundle. This helps in minimizing the pressure drop due to momentum change through the window and in satisfying the requirements of some of the design methods. Furthermore, the heat transfer area is effectively utilized, and no dead zones are formed behind the baffles for the scale to deposit preferentially. (See also Q. 1.54.)

1.54 In what order should one go about evaluating the different types of segmental baffles?

First, the single-segmental baffle should be tried. Then the double-segmental or the no-tube-in-window baffle and finally the triple-segmental baffle should be tried. The overall aim is to obtain the required heat transfer within the allowable pressure drop in the most economical, yet safe, way. Possibility of vibration damage should also be kept in mind.

1.55 How is the baffle cut sized in a multi-segmental baffle design?

It is based on the equality of the free window area for the different cut portions. Figure 1.25 explains it for the double-segmental baffle. The baffle cut is assigned on the basis of the central or solid baffle. Then the half-moon or split baffle pieces are cut to get the same free window area, A. The mensuration formulae are available in any standard handbook (Perry and Chilton, 1973). For the exact calculations, trial-and-error may be involved since the area occupied by the tubes has to be accounted for.

1.56 How are the baffle cuts in the adjacent baffles oriented with respect to one another in a single-segmental design?

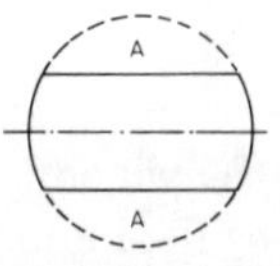

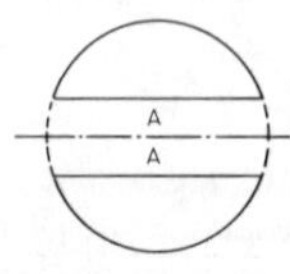

Figure 1.25 Multi-segmental baffle cut sizing.

The cuts of the adjacent baffles are oriented at 180° from one another (Fig. 1.26). The advantages are the following:

1. It allows fluid to flow over the bundle in cross-flow and make contact with the tubes several times in an exchanger.
2. There is no damming up of the fluid that would result if all the baffles were oriented in the same fashion. Most of the surface area would not be available for heat transfer then.
3. It provides support to the tubes in the window at alternate baffles, or else they would sag and vibrate to failure quickly.
4. Finally, the design correlations used are derived based on the experiments conducted using such orientations of the baffles.

1.57 How is the horizontal or vertical orientation of the baffle cuts determined?

For Single-Pass Shell Designs

1. Horizontal units: (a) For no-phase-change cases, the cut may be horizontal or vertical. No significant difference in heat transfer or pressure drop is encountered. (b) For condensing duty, the cut should be vertical for the free flow of the condensate without it damming up behind the baffle.
2. Vertical units: The baffle cut orientation has no meaning since all the baffles are horizontal.

For Multi-Pass Shell Designs

1. Horizontal units: (a) For no-phase-change cases, the shell-side cross-flow velocity, pressure drop, and heat transfer are all increased by the vertical orientation of the cut compared to the horizontal one because the cross-flow area is less in the former compared to the latter. (b) For condensing duty, the cut should be

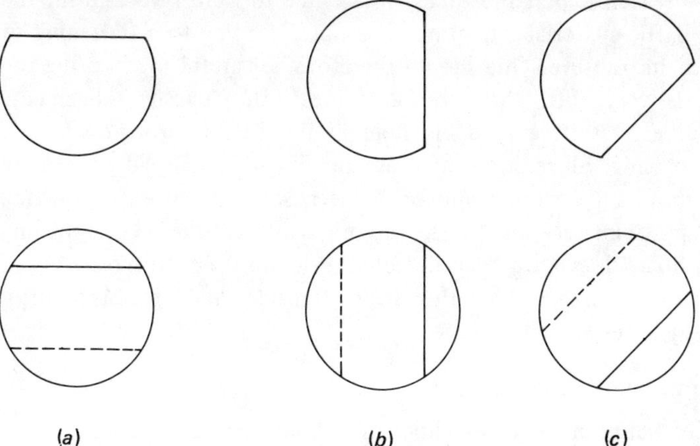

Figure 1.26 Baffle cut orientation. (*a*) Horizontal cut. (*b*) Vertical cut. (*c*) Rotated or 45° cut.

vertical for the same reasons as for the single shell pass horizontal unit discussed above.

2. Vertical units: The baffle cut orientation perpendicular to the longitudinal baffle results in higher cross-flow velocity, heat transfer coefficient, and pressure drop compared to the baffle cut orientation parallel to the longitudinal baffle. It does not matter whether there is any phase change or not in the vertical unit.

1.58 How should a single-segmental baffle cut be oriented with respect to the nozzles for the most efficient heat transfer in the entry and exit zones?

The baffle cut should be perpendicular to the nozzle axis and be located farther from the nozzle so that the fluid flows over the whole tube bundle in the entry and exit zones.

1.59 What is meant by the percent baffle cut in a two-pass shell using single-segmental baffles?

It is the percentage of the shell diameter. Thus, the meaning is the same as in a one-pass shell.

1.60 Does the baffle cut influence the cross-flow velocity?

No, only the baffle pitch or baffle spacing influences the cross-flow velocity for a given exchanger with specified tube size, pitch, layout angle, baffle cut orientation, and number of shell passes.

1.61 Why are the notches made at the bottom of the baffles in horizontal cut orientation?

These are made so that the exchanger can be drained completely when required for repair, storage, or change of service, and no liquid is held up behind the baffles.

1.62 Is the baffle material harder or softer than the tube material and why?

It is softer. The reason is that the tubes vibrate due to fluid flow and hit the baffle hole. A harder baffle material will produce a notch leading to a flattening of the tube and finally to its rupture. One should therefore be careful in choosing the baffle material. If it is very soft, the tubes can break the tubesheet ligaments between the nearby tubes, resulting in a big hole in the baffle through which a number of tubes are passing. Since some of these tubes will not be supported by the baffle any more, they could vibrate and break because their natural frequency would be reduced to approximately one-fourth the original value due to one missing support (Chapter 15). Such cases have been detected. Prior to tube rupture, a lot of rattling noise and some changes in the heat transfer and pressure drop generally occur that may forewarn an operator.

1.63 What is the central baffle spacing?

It is the distance between the center lines of two adjacent baffles.

1.64 What is the inlet baffle spacing?

It is the distance between the tubesheet and the first baffle where the fluid enters the shell. However, in J- and G-shells, it refers to the distance between the two baffles straddling the central inlet nozzle.

1.65 What is the outlet baffle spacing?

It is the distance between the tubesheet and the baffle next to it where the fluid leaves the exchanger in a fixed tubesheet or a floating tubesheet exchanger. For a U-tube unit, it is the distance between the last baffle and the end closure. For F- and G-shells, it is the same as the inlet baffle spacing.

1.66 Why are the inlet and outlet baffle spacings usually different than the central baffle spacing?

The main reason is that extra space has to be provided for the inlet and outlet shell-side nozzles. The increased baffle spacing does not result in excessive vibration of the tubes in the end zones since these tubes are rigidly anchored to the tubesheet by rolling, grooving, and/or welding, thus increasing their natural frequency (Chapter 15).

1.67 What factors determine the minimum inlet and outlet baffle spacings?

These are dependent, among others, upon the type of shell, front and rear heads, shell-side nozzle sizes, type of nozzle [self-reinforcing or reinforcing pad used (Chapter 9)], distance from the end flanges and the type of the end flanges used, code requirements for welding close to another weld, etc.

1.68 How does the maximum baffle spacing vary with the type of the segmental baffle?

Single-segmental baffles: It depends upon the tube material, tube temperature limit, and the tube O.D. TEMA Standards are followed.

Double-segmental baffles: Same as for the single-segmental baffles.

Triple-segmental baffles: Two-thirds of the value for the single-segmental baffles since each baffle supports about one-third of the tube bundle.

No-tube-in-window design: The maximum possible spacing to give the minimum number of cross-passes. Support plates are inserted in between the baffles to support the bundle and guard against the damaging vibrations.

The above are for the plain and bimetallic straight tubings. For U-bend and finned tubes, the TEMA Standards should be followed.

1.69 How is the actual baffle spacing or baffle pitch decided?

It is decided based upon the required heat duty, allowable pressure drop, and the possibility of vibration damage. It has also to conform to the stipulations of the TEMA Standards.

1.70 Baffles aid in heat transfer by forcing the fluid to flow over the tube

bundle. In what heat transfer equipment are they not needed? How are tubes supported in such equipment?

Baffles are not needed in reboilers where all the boiling occurs above the tube bundle level. In condensers, the design is generally based on vapor shear being absent. Hence the cross-baffles are not needed in equipment whose performance does not depend upon the cross-flow of the fluid over the tube bundle. Tubes are supported in such equipment by tube support plates, which may be significantly smaller than the shell I.D. as in kettle reboilers. These are located at the maximum possible distance from one another, compatible with the prevention of tube vibrations. In certain cases, a baffled heat exchanger may be used as a condenser as long as its baffle orientation does not dam up the liquid condensate. This requires vertical orientation of the baffle cut in a horizontal condenser, whereas the orientation does not matter in a vertical unit (see Chapter 3 for discussion of reboilers, condensers, etc.).

1.71 In a 2-4 exchanger, which baffle cut is preferred, horizontal or vertical?

A 2-4 exchanger is used when a large temperature cross is required ($t_2 > T_2$). Usually such fluids also have long temperature ranges and sufficient pressure drop is available. Hence, vertical cuts are preferred that result in a higher shell-side velocity and heat transfer coefficient and hence a more compact unit. The fouling deposit is also less.

1.72 What are the kinds of baffles used in U-bends?

There are several kinds of baffles that can be used in U-bends to save them from vibration damage. Probably, the best is the one made in sections with holes drilled to fit around the tubes (Fig. 1.27). These sections are then welded at the top and the bottom. Other kinds include thin metal strips used in triangular layouts and rods used in square layouts. These provide less grip over the tubes than the one shown in Fig. 1.27.

1.73 Why are support plates used?

Support plates are used for supporting the tubes against vibrations and/or for flow division. They are not considered to affect the heat transfer and pressure drop in the exchanger. With single-segmental baffles, they may occasionally be used in the end zones. With multi-segmental baffles, the support plates are not used. With no-tubes-in-window design, the support plates are used and have windows cut on both the sides. For X-shells, the support plates are full circles so that the fluid does not enter regions other than where it is put by the distributor plates or multiple nozzles. For G-shells, a full circle support plate is placed at the middle of the entry nozzle to divide the flow into two streams that flow in opposite directions and then combine at the outlet nozzle. The support plate is a must in a vertically mounted G-shell. It is also used in J-shells to divide the flow into two streams. In L-shells, the flow is laminar and the support plates are similar to 50 percent cut segmental baffles.

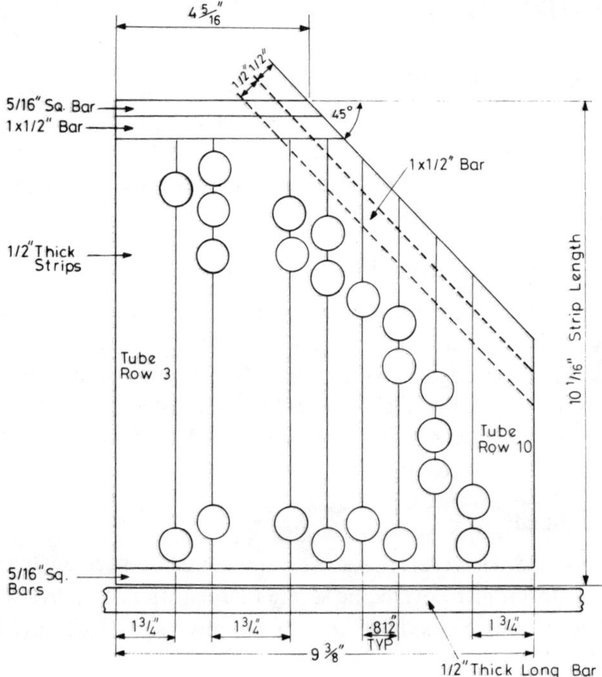

Figure 1.27 U-bend baffles. *(Courtesy of Joseph Oat Corp., Camden, N.J.)*

USES OF MULTI-PASS UNITS

1.74 When are multi-tube passes used?

These are used to increase the tube-side velocity so that lower fouling and higher heat transfer coefficient are obtained within the allowable pressure drop.

1.75 When are multi-shell passes used?

These are used to increase the shell-side velocity so that lower fouling and higher heat transfer coefficient are obtained. These also result in an increased heat transfer and a larger temperature cross ($t_2 > T_2$) in the unit.

1.76 In a plant usually a standard 1-2 exchanger is used in most places. Why?

Using a standard 1-2 exchanger does sometimes result in a little inefficiency since it cannot be optimum for all the different uses in a plant. However, having a standard unit (tube diameter, number of tubes, baffle cut and spacing, shell type, and number of passes) helps in reducing the inventory of the replacement tubes, parts, and tools. Also, a unit can be kept spare for use in case another unit has to be taken out of service. Two such units connected in series both for the shell- and tube-side fluids will result in a 2-4 unit. Similarly, higher-order arrangements (3-6, 4-8) may be obtained if desired.

TUBESHEET, DOUBLE TUBESHEET

1.77 What is the importance of the tubesheet?

The tubesheet is the most important part of a heat exchanger. Its primary purpose is to separate the shell-side and the tube-side fluids. Another important purpose is to provide fixed support to the tubes and tie-rods and to supply incidental support to the pass partition plates.

1.78 How are the tubesheets used in different kinds of exchangers?

For a fixed tubesheet exchanger, there are two similar tubesheets, one each at the front and the rear ends. For a floating-head exchanger, the front end tubesheet is bigger than the floating-head one by 2-5 in (50–125 mm) in diameter since it has to cover part or whole of the shell flange for effective sealing. For a U-tube exchanger, only the front end tubesheet is required. It also has to cover part or whole of the shell flange for effective sealing.

1.79 Are tubesheets always round?

In most cases they are round. However, in some surface condensers rectangular shapes are also used (Fig. 3.2). In order to overcome severe thermal stresses in certain systems, curvilinear (nonplanar) tubesheets have also been proposed (Fig. 9.15) (Joseph Oat, 1980).

1.80 What are double tubesheets?

These refer to the use of two tubesheets at the same end. Thus, in a fixed tubesheet exchanger, a total of four tubesheets will be used, two at each end (Fig. 1.28). In a U-tube design, only two tubesheets will be required, both at the front end. Sometimes an intergral double tubesheet is made out of a single thick plate. It has fewer problems in its use but is very costly compared to two separate tubesheets.

1.81 Where is double tubesheet construction used?

This is used wherever it is dangerous to let the shell-side and tube-side fluids mix due to leakage. Some such examples are given in Table 1.4.

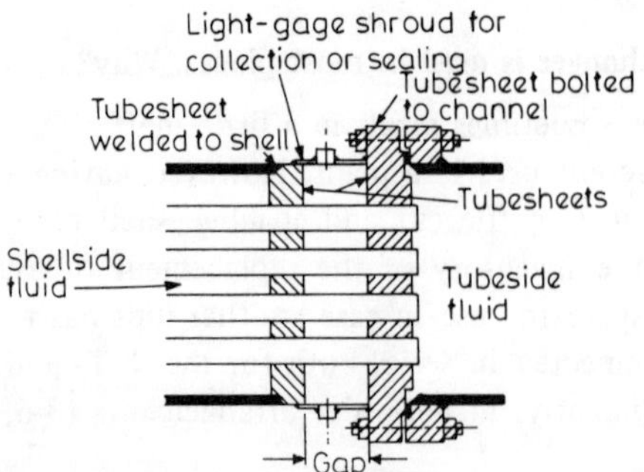

Figure 1.28 Double tubesheet. *(Excerpted by special permission from* Chemical Engineering, *May 14 © 1973, by McGraw-Hill, Inc., New York, N.Y. 10020.)*

Table 1.4 Process conditions that might require double tubesheet construction

Conditions	Process factors	Examples
Corrosion	Fluid streams are not corrosive separately but the mixture of streams is highly so	HCl gas cooler, oleum coolers, Cl_2 vaporizers
Health	One fluid stream is lethal. If it mixes with the second stream, the lethal material may get carried to other areas of the plant not designed to handle lethal material	Phosgene reboilers, phosgene condensers, phosgene coolers, cyanogen coolers
Safety	Streams ignite or explode when mixed	Alkali metal coolers
Equipment fouling	Resinous substances or polymers are generated when the streams mix	Methacrylic acid coolers, latex heaters
Catalyst	Catalyst form may be altered or its chemical composition changed if the catalyst contacts the second stream	Fluidized bed regenerating coolers
Yield reduction	A chemical reaction may be stopped or reversed if the fluid streams mix	Esterification overhead condensers
Impure product	Product may become contaminated with the fluid from the second stream	Purification column condenser, distillation column economizers

Source: Yokell (1973).

TIE-RODS AND SPACERS

1.82 What are the functions of tie-rods and spacers?

These are used to keep the baffles and the tube support plates at their appropriate locations as determined during the process design of the heat exchanger. Without them, the baffles and the support plates can move along the tubes, thus adversely affecting the performance of the exchanger.

1.83 How many tie-rods are used in an exchanger?

Their number and diameter vary with the shell diameter. TEMA Standards (1978), paragraphs R,C,B-4.71 and RGP-RCB-4.71, provide guidance in this matter, which can be varied to suit special needs or individual's preferences as long as a stipulated minimum is adhered to.

1.84 How are the tie-rods and spacers put in place?

These are put in the outer-tube-limit (OTL, a contour around the outermost tubes in the tube layout) by replacing a few tubes for each set of a tie-rod and spacer. One end of the tie-rod is threaded into a tubesheet, while the other end terminates at the last baffle where it is appropriately fixed by lock nuts or is welded. Concentric spacers are placed on the tie-rod in each baffled section to keep the baffles at their predetermined positions. Alternately, the tie-rod could be welded to each baffle.

1.85 When during the assembly of an exchanger are the tie-rods and spacers put in place?

These are put in place in the very beginning to make a cage along with the baffles and the tubesheets. Thereafter, the tubes are threaded into the baffle holes.

1.86 How do the tie-rods and spacers affect the number of tubes in an exchanger?

The number of tubes in an exchanger, also known as the tube count, is reduced because of the space occupied by the tie-rods and spacers. Usually each set of a tie-rod and spacer requires the removal of four to six tubes around it. Thus, the total number of tubes removed is dependent upon the number of tie-rods and spacers, which, in turn, is dependent upon the shell diameter (Q. 1.83). This should be kept in mind while looking up the tables for the tube count in the tubesheet layout (Q. 1.36).

SEAL STRIPS

1.87 What are the seal strips and seal rods? Describe their functions.

These are strips of metals used to reduce the amount of fluid by-passing the tube bundle in a floating-head heat exchanger where the clearance between the shell and the OTL is large. Seal rods are used in tube pass partition lanes to reduce the bypass stream (Q. 2.3).

1.88 How are the seal strips installed?

These are installed in pairs on the baffles usually by welding (Fig. 1.29). These seal off the space between the shell and the OTL, thus forcing the bypassing liquid to flow over the tube bundle. Generally, one pair of seal strips is used for every five to seven tube rows in cross-flow. Sometimes dummy tubes are used in place of the seal strips.

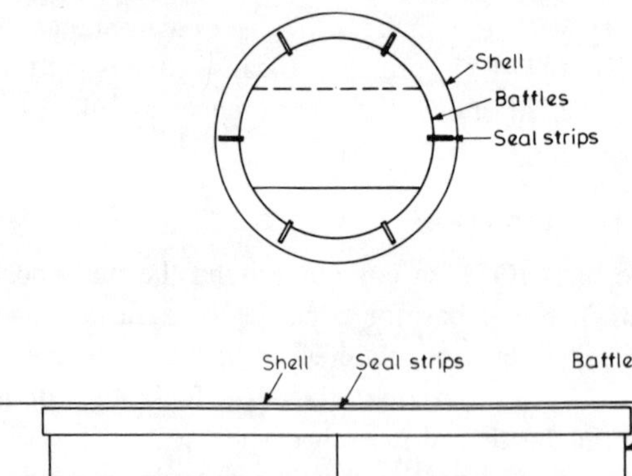

Figure 1.29 Seal strips.

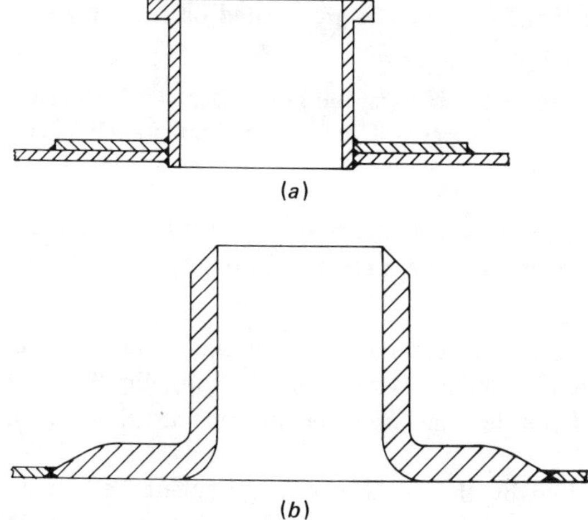

Figure 1.30 Nozzles. (*a*) Reinforcing pad. (*b*) Self-reinforcing nozzle.

1.89 When are the seal strips not used?

These are not used in the fixed tubesheet or U-tube bundle units where the clearance between the shell and the OTL is not large. Furthermore, these are also not used when a phase change is occurring (boiling, condensation) even if the clearance is large because the seal strips might affect the phase separation adversely. The performance of reboilers, condensers, etc., do not depend on the cross-flow stream.

NOZZLES

1.90 What are the functions of nozzles?

Nozzles provide the ports of entry and exit for the shell-side and the tube-side fluids. They also provide the exit points for draining and venting a unit for the pressure relief as well as for installing any monitoring and/or measuring instruments.

1.91 What are the different types of nozzles?

The nozzles are of two types (Fig. 1.30):

1. Nozzles with reinforcing pads
2. Self-reinforcing nozzles

See Chapter 9 for more details.

1.92 Why are some tubes removed from underneath the nozzles?

In order to reduce the pressure drop at the entrance to the shell, to facilitate the entry of the fluid into the bundle, and to reduce the erosion and vibration damage, some tubes are removed from underneath the nozzles even if no impingement plate is used. This allows for a larger entry area for the shell-side fluid.

1.93 What difference does it make if the two nozzles are located on the same side or on the opposite sides of the shell?

Tubes have to be removed from both sides in the latter case. The effect on the heat exchanger area is only one to two percent in large diameter exchangers, but could be significant in small exchangers.

1.94 Since the pressure drop in a nozzle is a waste, why is a very large diameter nozzle not usually chosen to reduce the pressure drop? What limits the nozzle diameter?

The nozzle diameter has broad limits both on the lower and the upper side.

On the lower side, it is limited by the excessive pressure drop, vibration and erosion of the tubes under the shell nozzles, and possible fluid maldistribution on the tube side.

On the upper side, it is limited by the fabrication requirements, effect on the shell strength, discontinuity stresses introduced into the system, connecting pipe to nozzle head loss, nozzle loads, and cost.

FLANGES

1.95 What is a flange?

It is a rib or rim used for joining two sections by bolting. Both the sections must have a flange attached to their joining end. The joint is made leakproof by using gaskets.

1.96 What are the different kinds of flange?

There are three kinds of flanges (Fig. 1.31):

Intergral-type flanges: The hub and the flange are a continuous structure either as originally manufactured or made so by a full penetration weld. Welding neck flange is an integral flange. This type are the costliest.

Loose-type flanges: These either do not have any attachment to the shell or the attachment is not considered integral with the shell.

Optional-type flanges: These are attached to the vessel wall or nozzle such that the assembly acts as an integral unit. These differ from the integral-type flanges in not having any hub or in having a very small hub. The vessel wall acts as a hub. The name is "optional" since these can also be designed as a loose-type flange with certain restrictions.

There is yet another kind called the blind flange, used to close a flanged opening. When used over the channels in an exchanger, it is known as the channel cover.

EXPANSION JOINTS

1.97 What are expansion joints?

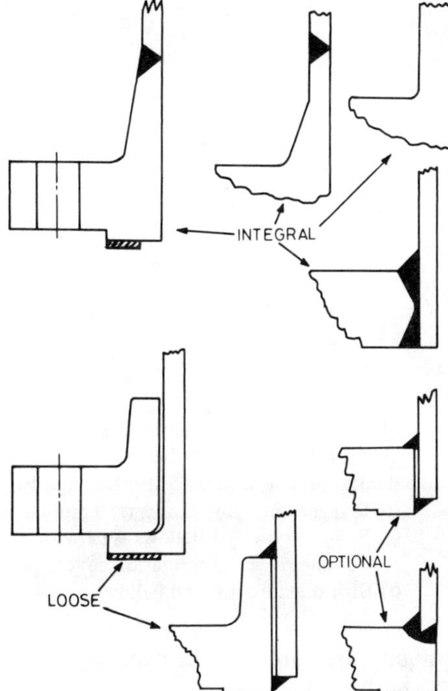

Figure 1.31 Flanges. *(Reprinted by permission from* Modern Flange Design, *Bulletin 502, Ed. 7, G&W Taylor-Bonney Division, Southfield, Mich.)*

These are thin, flexible joints put in the shell to eliminate the excessive stresses caused by the differential thermal expansion between the shell and the tube bundle due to cyclic temperature conditions in a fixed tubesheet exchanger.

1.98 What are the principal kinds of expansion joint?

These are shown in Fig. 1.32 (Perry and Chilton, 1973).

1.99 Do expansion joints make a weak joint?

These do not make a weak joint if properly designed, keeping the pressure, temperature, corrosion, and thermal cyclic conditions in mind. Their design factors are discussed in Chapter 9.

1.100 Where else are the expansion joints used?

These are used in piping systems to provide relief from stresses due to piping movement induced by pressure and/or temperature changes (Bedson, 1980). These are outside the purview of this book.

PACKING GLANDS AND GASKETS

1.101 What are packing glands? Where are they used?

Packing glands are generally circular pieces used to prevent leakage between the shell-side and the tube-side fluids and still allow some movement of the

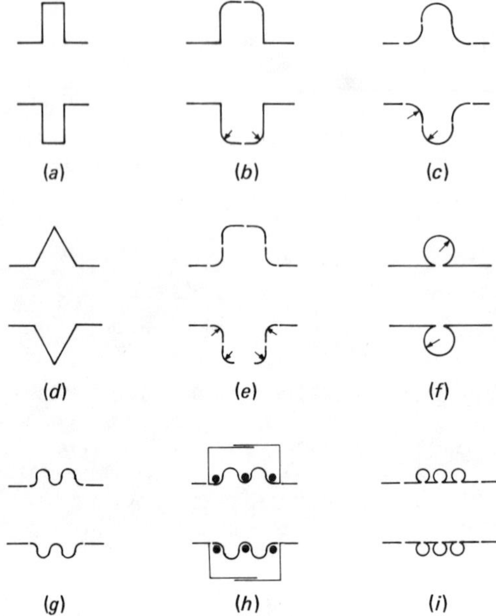

Figure 1.32 Exapnsion joints. (*a*) Flat plates. (*b*) Flanged only heads. (*c*) Flared shell or pipe segments. (*d*) Formed heads. (*e*) Flanged and flued heads. (*f*) Toroidal. (*g*) Bellows type. (*h*) Bellows type with reinforcing rings and insulation cover. (*i*) High-pressure toroidal bellows.

tubesheet to account for differential expansion. These are used in many types of floating rear-head constructions (type P, S, T, and W in Fig. 1.1).

1.102 What are the characteristics of a good packing gland?

These are the following:

1. Its material should be compatible with both the shell-side and the tube-side fluids. It should be able to withstand the operating temperature and pressure and the corrosive nature of the fluids. Fluids should not diffuse through it.
2. The machining of the surfaces on which it is installed should be of a high standard.
3. Its width and thickness should be properly sized and no undue pressure should be applied in installing it.
4. TEMA Standards (1978), paragraphs R,C,B-5.31, put limits on the use of square braided asbestos packing in lantern rings. These limits should be obeyed.

1.103 Where are the gaskets used in an exchanger?

These are used in all the bolted joints between two metal surfaces to make the joints leak tight. Such joints are the shell-to-channel, channel-to-channel cover, nozzle-to-piping, etc. (see also Q. 9.37).

VENT AND DRAIN CONNECTIONS

1.104 Why are the vent and drain connections provided?

The vent connection is provided to allow all the air in the exchanger to leave the unit as it gradually fills up with the fluid at the start of an exchanger, or else air binding will occur. It also helps to equalize the pressure when the liquid is drained at the shutdown.

The drain connection is provided to drain the liquid at the shutdown for maintenance or storage of the unit or for a change of the process fluids. Any liquid left standing in the exchanger causes it to foul and corrode early.

1.105 Where are the vent and drain connections located? What are their sizes?

The vent connections are located at the top of the channel, shell, and the bonnet, whereas the drain connections are located at the bottom in each section. Their sizes usually range from 1 to 3 in (25–75 mm) diameter depending on the size of the unit and the time available to vent and drain it.

1.106 Do the vent and drain connections open into the atmosphere?

While the vent connections generally open into the atmosphere, the drain connections are usually piped to a storage unit or safe disposal unit depending on whether the liquid is costly and is to be stored or whether the municipal regulations restrict its discharge into the open. If the liquid is water, it may be discharged into an open drain.

PRESSURE RELIEF DEVICES

1.107 Why are pressure relief devices necessary on any pressure vessel?

Pressure vessels are liable to fail if the pressure rises above the maximum allowable working pressure due to sudden changes in the operating conditions or blockage of the flow passages downstream. Such a failure may cause both economic and personnel losses. Besides, replacements take a long time. Hence, the pressure relieving devices are used so that there is no human life lost and the economic losses are minimized, often restricted to the material present in the vessel at the time of the over-pressure release. Furthermore, minimum interruption is caused to the process at large. ASME Code Section VIII, Div. 1, paragraphs UG-125 to 134, govern the pressure relief devices. The responsibility to provide over-pressure protection lies with the fabricator.

1.108 What are the different kinds of pressure relief devices?

These can be either safety relief valves or rupture disks (Figs. 1.33 and 1.34). The pressure at which the valves open or disks rupture is set a little higher than the operating pressure. Valves are costlier than the rupture disks, but they relieve the excess pressure above the set pressure without seriously interrupting the process. Valves, being spring loaded, close by themselves after the excess pressure is released. On the other hand, when a disk ruptures, the pressure inside is equalized with the outside, and the process is interrupted and has to be restarted with

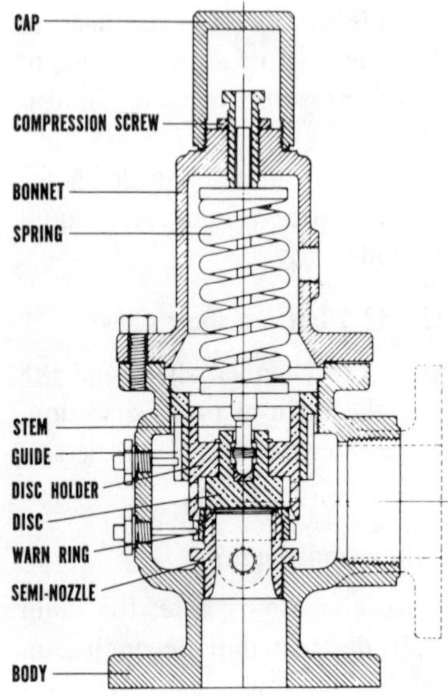

Figure 1.33 Safety valve. *(Courtesy of J. E. Lonergan Co., Philadelphia, Pa.)*

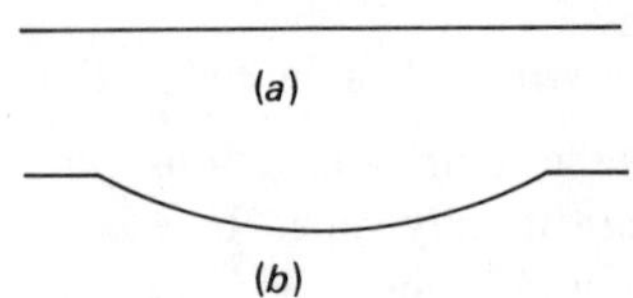

Figure 1.34 Rupture disks. (*a*) Plain flat. (*b*) Pre-bulged.

consequential loss of material, time, and money. The rupture disk also has to be replaced. Sometimes a combination of rupture disk and relief valve is used in case of corrosive fluids (Crozier, 1980). The advantage is that the rupture disk is made of corrosion-resistant material, while the valve is made of carbon steel, which is much cheaper.

1.109 Are safety relief and vent valves necessary on the shell side only?

The need of the safety relief valves is established depending upon the design conditions and the types of fluids handled. Safety relief valves are usually installed on the vapor/gas side. The vent valves are necessary on both the shell and the tube side to vent the inerts and other undesirable materials.

1.110 What happens to the fluid released by the safety valve or rupture disk?

If it is air, water, or any other harmless fluid, it is released into the atmosphere. If it is a hazardous or polluting fluid, the valve/rupture disk outlet is piped to a proper storage so that the fluid is not released into the atmosphere.

SUPPORTS AND LIFTING LUGS

1.111 How are the heat exchangers and pressure vessels supported?

The units can be placed either horizontally or vertically.

In a horizontal unit, saddle supports are used. Two of these are welded to

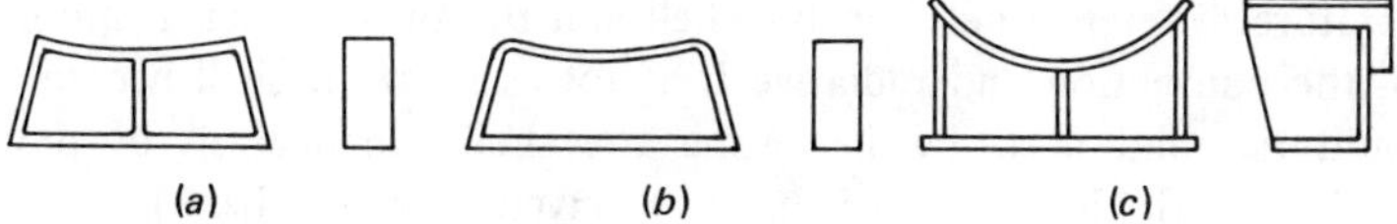

Figure 1.35 Supports for horizontal exchangers. (*a*) Heavy duty-loose cast iron. (*b*) Medium duty-loose pressed steel. (*c*) Heavy duty-loose or welded to shell-fabricated steel. *(Courtesy of Patterson-Kelley Co., Division of HARSCO Corp., East Stroudsburg, Pa.)*

the shell. The saddles can be part circle (Figs. 1.35 and 9.23) or full circle using stiffening rings (Fig. 9.24) depending upon the loading and stress analysis. One of the supports is rigidly held on the ground, while the other is placed on rollers to allow for the thermal expansion of the shell.

In a vertical unit, cylindrical skirt, brackets, or columns are used for supporting it (Fig. 1.36). A cylindrical skirt is used for small vessels. One has to account for the bending stresses produced by the skirt by thickening the vessel head near the joint. Brackets are used as supports where flanges exist and the vessel wall has already been thickened to account for the bending stresses due to the flanged joints. Columns are used for large, thin-walled vessels so that the load is transmitted in shear and local stress concentration is avoided.

- Supports are further discussed in Chapter 9.

1.112 What is a lifting lug?

It is an ear or loop-like projection that is used as a handle to move heavy removable parts of an exchanger while the shell is stationary. Such parts include the channel, channel cover, bonnet, and removable tubesheet along with tube bundle, etc. In the last case, the lug may be screwed into the tubesheet when necessary, but a provision for it should be made at the design stage itself. It should be strong enough to bear the weight of the part and the impact at the time of lifting it. One end of a steel chain or rope is connected to the lug when the part is to be moved. The other end is connected to a crane or pulley system.

INSULATION

1.113 When should the outside of the shell be insulated?

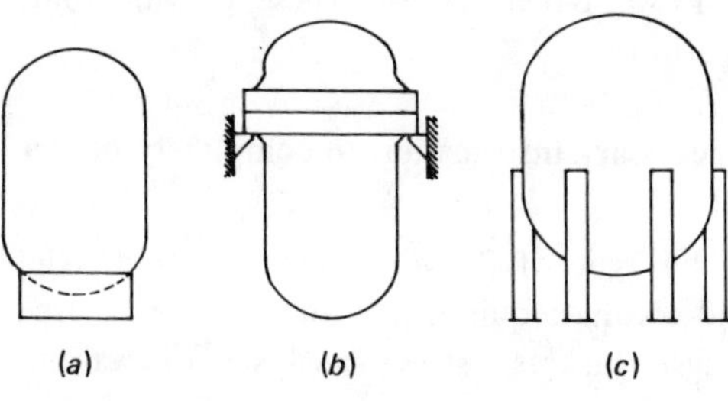

Figure 1.36 Supports for vertical exchangers. (*a*) Cylindrical skirt. (*b*) Bracket attached to flange. (*c*) Columns.

If the temperature difference between the shell and the surroundings is more than 100°F (38°C), the convection and radiative heat losses from the shell become significant and should be minimized by insulating the shell. An analysis of the uninsulated shell considering the losses from it has been given by Kern (1976).

1.114 What are the materials used as insulation?

Some of these are the following:

1. Alumina
2. Diatomaceous earth
3. Mica
4. Polyurethane and other foams
5. Silica

HEAT EXCHANGER SPECIFICATION SHEET

1.115 What is a heat exchanger specification sheet?

It is a data sheet that contains the information provided by the customer to the vendor for the process and mechanical designs of an exchanger. After the process design is done, the engineer fills in some further information. Finally, when the mechanical design is done, the rest of the information is filled in. Thus, the specification sheet, generally abbreviated as the spec-sheet, is a medium of communication between different parties involved in the procurement, design, and fabrication of a heat exchanger. It is also used to compare the performance of the installed unit with the design conditions.

1.116 What information does it contain?

This is best illustrated by a figure of the spec-sheet. Although each company has its own version, the most popular one is that of the TEMA Standards (Fig. 1.37). It is similar to the one in the API Standard 660 (American Petroleum Institute, 1976). It contains all the pertinent information about the shell-side and tube-side fluid flow rates and properties, heat duty, heat transfer coefficient, fouling resistance, details about the shell and tube sizes, materials, baffles, nozzles, etc. Some variations include information for alternate designs and different systems of units (British, SI, metric). Occasionally, articles appear about how to specify a heat exchanger (Rubin, 1968; Devore et al., 1980; Baker, 1966). These provide some detailed guidelines.

1.117 Does the spec-sheet have *all* the necessary information to completely design, fabricate, and install the exchanger?

No. The specifications regarding the type of flanges, studs, vent and relief valves, drain lines, welding, inspection and testing requirements of the material used and the fabricated unit, flange facings, gaskets, stress analysis of various

1				Job No.		
2	Customer			Reference No.		
3	Address			Proposal No.		
4	Plant Location			Date	Rev.	
5	Service of Unit			Item No.		
6	Size	Type	(Hor/Vert)	Connected In	Parallel	Series
7	Surf/Unit (Gross/Eff.)	Sq Ft: Shells/Unit		Surf/Shell (Gross/Eff.)		Sq Ft
8	PERFORMANCE OF ONE UNIT					
9	Fluid Allocation		Shell Side		Tube Side	
10	Fluid Name					
11	Fluid Quantity, Total	Lb/Hr				
12	Vapor (In/Out)					
13	Liquid					
14	Steam					
15	Water					
16	Noncondensable					
17	Temperature (In/Out)	°F				
18	Specific Gravity					
19	Viscosity, Liquid	Cp				
20	Molecular Weight, Vapor					
21	Molecular Weight, Noncondensable					
22	Specific Heat	Btu/Lb °F				
23	Thermal Conductivity	Btu Ft/Hr Sq Ft °F				
24	Latent Heat	Btu/Lb @ °F				
25	Inlet Pressure	Psig				
26	Velocity	Ft/S				
27	Pressure Drop, Allow. Calc.	Psi	/		/	
28	Fouling Resistance (Min.)					
29	Heat Exchanged		Btu/Hr: MTD (Corrected)			°F
30	Transfer Rate, Service		Clean			Btu/Hr Sq Ft °F
31	CONSTRUCTION OF ONE SHELL				Sketch (Bundle/Nozzle Orientation)	
32			Shell Side	Tube Side		
33	Design/Test Pressure	Psig	/	/		
34	Design Temperature	°F				
35	No. Passes per Shell					
36	Corrosion Allowance	In.				
37	Connections	In				
38	Size &	Out				
39	Rating	Intermediate				
40	Tube No.	OD	In.;Thk (Min/Avg)	In.; Length	Ft; Pitch	In. ◁ 30 △ 60 □ 90 ◇ 45
41	Tube Type			Material		
42	Shell	ID	OD In.	Shell Cover		(Integ.) (Remov.)
43	Channel or Bonnet			Channel Cover		
44	Tubesheet-Stationary			Tubesheet-Floating		
45	Floating Head Cover			Impingement Protection		
46	Baffles-Cross	Type		% Cut (Diam/Area)	Spacing: c/c	Inlet In.
47	Baffles-Long			Seal Type		
48	Supports-Tube		U-Bend		Type	
49	Bypass Seal Arrangement			Tube-Tubesheet Joint		
50	Expansion Joint			Type		
51	ρv^2-Inlet Nozzle		Bundle Entrance		Bundle Exit	
52	Gaskets-Shell Side			Tube Side		
53	-Floating Head					
54	Code Requirements				TEMA Class	
55	Weight/Shell		Filled with Water		Bundle	Lb
56	Remarks					
57						
58						
59						
60						
61						

Figure 1.37 Heat exchanger specification sheet. *(© 1978 by Tubular Exchanger Manufacturers Association. All rights reserved.)*

components, loading, thermal cycling, emergency and upset conditions, painting, shipment, storage, installation, foundations, base plate, start-up, shutdown, etc., are not given in the spec-sheet.

1.118 What are the various conditions regarding the exchanger operation that should be known to a thermal designer for critical applications?

These are the following:

1. Start-up conditions and procedure
2. Normal operating conditions
3. Upset and emergency conditions
4. Shutdown conditions and procedure
5. Required overdesign in heat transfer area
6. Possibility of switching the shell-side and tube-side fluids for better design.
7. Possibility of increasing the allowable pressure drop so that less fouling and more compact unit can be designed
8. Besides the above, the spec-sheet (Fig. 1.37) should also be provided with all the available information, such as the composition of the streams, their thermal and physical properties, and any phase change occurring.

1.119 Why should the minimum operating conditions be also considered along with the maximum or the worst ones?

Whereas the consideration of the worst operating conditions (maximum heat load, maximum fouling, and maximum cooling water temperature) ensures sufficient heat transfer area under all conditions, it is important to consider the minimum operating conditions (minimum heat load, clean tubes, and the lowest cooling water temperature) so that there will not be any adverse effects on the control of the whole plant and no process stream would freeze.

1.120 Is the spec-sheet (Fig. 1.37) for continuous or batch service?

It is for units used in continuous service.

1.121 What information is needed for batch and semi-batch services?

For these, the following information is needed:

1. Batch size at the start
2. Any addition to or removal from the batch during the operation
3. Flow rates of the inlet/outlet streams
4. Properties of the fluids and variations in them during the operation
5. Temperatures at the start and finish of the batch
6. Heat load as it varies with time during the operation
7. Extent of agitation or mixing that affects the heat transfer coefficient
8. Time allowed per batch for operation, filling, emptying, and cleaning

BID EVALUATION

1.122 What factors should be kept in mind while evaluating the bids submitted by competitive bidders?

For ease in evaluation, all pertinent data from each bid should be put on a large data sheet. During evaluation, the following factors should be kept in mind:

1. The design (and any alternate design submitted by the bidders) should meet the heat transfer and pressure drop requirements. In order to avoid wasteful overdesign, consider specifying to the bidders both an upper and a lower limit on the pressure drop.
2. If the designs offered by different bidders vary a lot, the spec-sheet provided to them should be checked to see if any anomalies exist.
3. Adequate vents, drainage, and safety valves should be provided.
4. Unit should have no hot spots and dead zones.
5. Vibration analysis should have been done by the bidders and design modifications made to account for any damaging vibrations predicted by such an analysis.
6. For the shell-side fluids that foul, the tube layout should permit easy cleaning. In general, the inspection, cleaning, and repair of the unit should be easily possible.
7. The fabrication shop should be regularly inspected by a reputable inspection agency. The vendor should have the reputation of providing quality products and of providing good service should a complaint arise.
8. The materials of construction should preferably be available within the country or their import should not pose any difficulty.
9. The delivery schedule should match with the plant erection schedule for a new plant or the unit replacement schedule for an existing plant.
10. The cost should be low. Cost escalation factors should also be considered.
11. The payment, penalty, and guarantee clauses in the contract should be evenly balanced and not be unduly favorable to the bidder.

One or more of the above may tilt the balance in favor of a bidder whose cost is not necessarily the lowest. Occasionally, articles appear on bid evaluation. Some of these provide good information (Baker, 1966).

DEVELOPMENTS AND FUTURE TRENDS IN HEAT EXCHANGERS

1.123 What factors have affected the developments in heat exchanger design during the post-World War II period?

A very significant development was the book by Kern (1950), which gave conservative, safe, and simple design methods for numerous kinds of heat exchangers with and without phase change. Another was the analysis by Tinker (1947, 1958) of various streams within the shell into which the shell-side fluid divides (Chapter 2). This analysis has been further improved, and detailed computer programs have been made out by the cooperative research institutions, such as HTRI and HTFS, which have the backing of their member companies and are involved in doing industry-related work. Increasing numbers of users are relying on computer programs for design and analysis. The most significant applied work was done at the University of Delaware from 1947 to 1962, which produced approximately 25 theses and numerous authoritative papers and reports and culminated in the Bell-Delaware method (Bell, 1981), which is a very detailed and useful one. Some computer programs are based on this or a combination of this and the Tinker's method. A recent listing of 390 available computer programs for all kinds of chemical engineering operations included 39 on heat exchangers (Peterson et al., 1978a-d). Many companies have their own in-house programs, too.

1.124 What are the major unresolved problems in the heat exchanger design?

The following is a list of such problems:

1. Accurate data on the thermophysical properties: These are needed both for the pure fluids and fluid mixtures in single- and two-phase systems under extremes of temperatures and pressures. It would be best if more reliable predictive correlations could be obtained.
2. Fouling of units in various services: How to predict it, the effects of design parameters on fouling, and ways of reducing the fouling.
3. Flow-induced vibrations: How to predict them, and ways to reduce them.
4. Two-phase flow systems: How to predict the flow regime, how flow regimes change due to phase change and pressure drop, and the effects of the design parameters.
5. Boiling of mixtures: Calculations of the heat transfer coefficients in pool and forced-convection boiling of mixtures need to be better evaluated.
6. Behavior of turbulence: Since generally turbulent flow of fluids is encountered in heat exchangers, condensers, reboilers, etc., a better understanding of it would aid in the better design and evaluation of the units.

Butterworth has discussed items 2–6 in greater details and has given ideas about avoiding some of these problems and also about future research (Butterworth, 1981).

1.125 What are the future trends in heat exchanger design?

In future, more and more sophisticated computer programs will be developed. These will

1. Do stepwise calculations instead of assuming an overall heat transfer coefficient
2. Supply physical and thermodynamic properties from built-in subroutines
3. Do most of the mechanical design as per the applicable Codes and Standards. A workshop fabrication drawing will also be plotted by it. These facilities are already available in some programs to a limited extent.
4. Restrict larger diameter shells only by the ability to transport, with substantially higher surface area being designed. This will place a lot of strain on the fabrication, testing and inspection facilities, transport, erection, etc. Besides, the Codes and Standards may have to be modified accordingly, and special attention will have to be paid to the baffle spacing, welding details, vibration analysis, and seismic analysis.
5. Designers will accept the computed results without critical analysis and this might lead to uneconomic designs, and sometimes lead to disastrous

consequences due to errors in data input and/or limitations of the computer programs used (Taborek, 1979).

6. Researchers are also working in the area of improved heat transfer to pressure drop ratio by changing the tube layout angles to other than the ones currently in use (Q. 1.34) and using unequal pitch values.

Hence, the need for experienced designers will only increase, and not decrease, with increasing sophistication in the design of heat exchangers and pressure vessels.

• In this chapter, we have introduced the nomenclature and parts of the shell-and-tube heat exchangers. In the following chapter, we shall discuss their process design without involving any phase change.

REFERENCES

American Petroleum Institute. Standards 660, *Heat Exchangers for General Refinery Services,* 3d ed. API: Washington, D.C. 1976.

ASME. Code Section II, *Material Specifications.* ASME: New York. 1980.

Baker, W. J. How to Rate Heat Exchanger Bids. *Hydrocarbon Process.* 45(6):141–143, 1966.

Bedson, J. H. Elastomeric Expansion Joints. *Chem. Eng.* 87(26):34–40. 1980.

Bell, K. J. Final Report of the Cooperative Research Program on Shell and Tube Heat Exchangers. Bulletin No. 5, University of Delaware Engineering Experiment Station. 1963. An improved version is given by Bell in his paper Delaware Method for Shell Side Design. In *Heat Exchangers: Thermal-Hydraulic Fundamentals and Design.* S. Kakac, A. E. Bergles, and F. Mayinger (eds.). Hemisphere: Washington, D.C. 1981.

Boyer, R. C., and G. K. Pase. The Energy-Saving NESTS Concept. *Heat Transfer Eng.* 2(2):19–27. 1980.

Butterworth, D. Unresolved Problems in Heat Exchanger Design. In *Heat Exchangers–Thermal-hydraulic Fundamentals and Design.* S. Kakac, A. E. Bergles, and F. Mayinger (eds.). Hemisphere, Washington, D.C. 1981. An earlier version in *Inst. Chem. Eng. (Lond.) Symp. Ser. No. 60.* 1980.

Crozier, R. A., Jr. Pressure Relief to Prevent Heat-Exchanger Failure. *Chem. Eng.* 87(25):79–93. 1980.

Devore, A., G. J. Vago, and G. J. Picozzi. Specifying and Selecting (Heat Exchangers). *Chem. Eng.* 87(10):133–148. 1980.

Heat Exchange Institute. *Standards for Steam Surface Condensers,* 7th ed. HEI: Cleveland, Ohio. 1978.

Heat Exchange Institute. *Standards for Power Plant Heat Exchangers.* HEI: Cleveland, Ohio. 1980.

Joseph Oat Corp. U.S. Patent No. 6207944, June 17, 1980.

Kern, D. Q. *Process Heat Transfer.* McGraw-Hill: New York. 1950.

Kern, J. On the Feasibility of Heat-Exchange Processes in Uninsulated Equipment. *Warme und Stoffubertragung.* 9:267–272. 1976.

Patterson-Kelley Co. Manual 700A, *Heat Exchangers.* Patterson-Kelley: E. Stroudsburg, Pa.

Perry, R. H., and C. H. Chilton (eds.). *Chemical Engineers' Handbook,* 5th ed. McGraw-Hill: New York. 1973.

Peterson, J. N., C. C. Chen, and L. B. Evans. Computer Programs for Chemical Engineers. Parts 1–4, *Chem. Eng.* (1978). 85(13):145–154, 85(15):69–82, 85(17):79–86, 85(19):107–115, respectively.

Rubin, F. L. How to Specify Heat Exchangers. *Chem. Eng.* 75(8):130–136. 1968.

Schroeder, J. W. Explosively Free-Formed Pass Partition Plate for a Nuclear Steam Generator. In *Explosive Welding, Forming, Plugging, and Compaction-PVP44.* I. Berman and J. W. Schroeder (eds.). ASME: New York. 1979.

Small, W. M., and R. K. Young. The RODbaffle Heat Exchanger. *Heat Transfer Eng.* 1(2):21–27. 1979.

Taborek, J. Evaluation of Heat Exchanger Design Techniques. *Heat Transfer Eng.* 1(1):15–29, 1979.

Tinker, T. Shell Side Heat Transfer Characteristics of Segmentally Baffled Shell-and-Tube Heat Exchangers, paper presented at the ASME Annual Meeting, New York City, 1947.

Tinker, T. Shell Side Characteristics of Shell-and-Tube Heat Exchangers: A Simplified Rating System for Commercial Heat Exchangers. *Trans. ASME.* 80:36–52. 1958.

Tubular Exchanger Manufacturers Assoc. *TEMA Standards,* 6th ed. TEMA: Tarrytown, New York. 1978. Addenda, 1982. Known as TEMA Standards.

Union Carbide Corp. Manual F-3635. Union Carbide: Tonawanda, New York.

Yokell, S. Double-Tubesheet Heat Exchanger Design Stops Shell-Tube Leakage. *Chem. Eng.* 80(10):133–136, 1973.

CHAPTER
TWO

SHELL-AND-TUBE HEAT EXCHANGERS—II: PROCESS DESIGN WITHOUT PHASE CHANGE

INTRODUCTION

• In this chapter, we will discuss the process design of shell-and-tube heat exchangers not involving any phase change.* A large fraction of the exchangers used in various industries do not involve any phase change. With increasing importance of conserving energy and the development of nonconventional energy sources (ocean, geothermal, solar, etc.), such exchangers are becoming more and more important. This chapter is intended to bring about a better understanding of the working of such units.

FLOW THROUGH SHELL AND TUBES

2.1 How does fluid flow through the shell side of a shell-and-tube exchanger?
Depending upon the type of shell (Fig. 1.1), the shell-side fluid enters the shell through one or more nozzles and, after passing through the shell one or more times (depending upon the number of shell passes), exits from it through one or

*Reference to phase change does arise in a few places in this chapter with regard to LMTD. This has been done to complete the discussion of LMTD in this chapter itself for all the different situations.

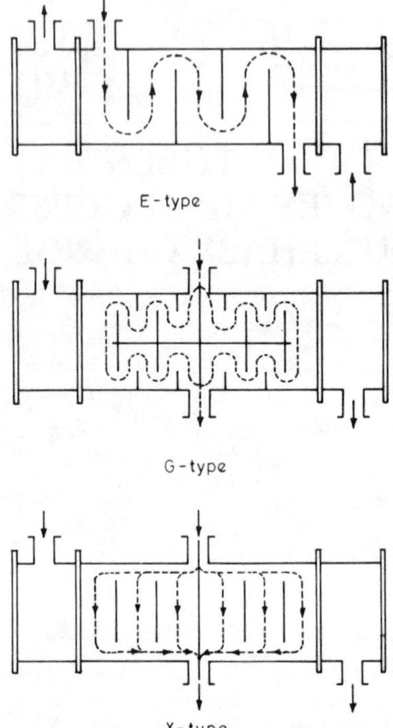

Figure 2.1 Shell-side fluid flow profiles.

more nozzles. In the shell, the fluid flows over the tube bundle several times depending upon the number of baffles, except in the X-type shell or when the rod baffles are used (Fig. 1.18). The flow over the tube bundle is cross-flow between consecutive baffles and parallel flow in the baffle window (Fig. 2.1). In the X-type shell, fluid is distributed in each baffled zone by a manifold and flows over the tube bundle only once. In the units using the rod baffles, the fluid flows in parallel over the tube bundle. The boundary layer formed is disturbed periodically by the rod baffles (see Q. 1.41 to Q. 1.73 about various types of baffles and their uses).

This is a gross description of the overall shell-side flow, not taking into account the effects of the manufacturing clearances.

2.2 What are the manufacturing clearances?

For ease in threading the tubes through the baffles, a slight clearance is allowed between the tube holes in the baffles and the tube O.D. Similarly, for ease in pushing a tube bundle into the shell, a slight clearance is allowed between the shell I.D. and the baffle. These are called the manufacturing clearances.

The clearance between the baffle holes and the tube O.D. depends upon the tube O.D. and the unsupported tube length (baffle spacing). It is governed by the TEMA Standards paragraphs R,C,B-4.2 (Tubular Exchanger Manufacturers Assoc., 1978; referred to as TEMA Standards, 1978 throughout the chapter). It is generally $\frac{1}{32}$ in (0.8 mm) in diameter, when the unsupported tube length is 36 in (910 mm) or less, and $\frac{1}{64}$ in (0.4 mm) when the unsupported tube length is longer.

The clearance between the shell I.D. and the baffles depends upon the shell diameter. It is governed by the TEMA Standards, paragraphs R,C,B-4.3, RGP-RCB-4.3 and varies from a minimum of 0.100 in (2.5 mm) for a shell of 6 in (150 mm) diameter to a maximum of 0.438 in (11 mm) for a shell of 100 in (2500 mm) diameter.

Besides the above, the outermost tubes in the tube layout (called outer tube limit or OTL) also require some tubesheet ligament, or else they will break loose from the baffle and will vibrate to failure. The minimum ligament is governed by the TEMA Standards paragraphs R,C,B-7.42. Its value depends upon the tube O.D., wall thickness, pitch, and the tubesheet thickness. It varies from a minimum of 0.090 in (2.3 mm) for a $\frac{3}{4}$ in (19 mm) tube to a maximum nominal value of 0.478 in (12 mm) for a 2 in (50 mm) tube. Thus, there is some clearance between the shell I.D. and the OTL. The values given in the TEMA Standards are diametral clearances and hold good for fixed tubesheet designs and the U-tube bundles. For floating-head constructions, the diametral clearance is of the order of $1\frac{1}{2}$-3 in (38-76 mm) because of the design of the tubesheet and packings to prevent leakage of the fluids. This reduces the number of tubes (tube count) in a given sized shell.

2.3 How do the manufacturing clearances affect the shell-side flow?

Tinker (1951, 1958) was the first one to show the effects of the manufacturing clearances on the shell-side flow. He divided the flow into four streams, namely, A, B, C, and E. (Figs. 2.2-2.5). We will describe these briefly.

1. A-stream (Fig. 2.2). Since the fluid loses pressure as it flows through the exchanger, there is a net pressure difference between any two consecutive baffled zones. This forces a part of the fluid to leak through the annular space between the tube O.D. and the tube holes in the baffle. This is known as a leakage stream.
2. B-stream (Fig. 2.3). This is the stream in cross-flow over the tube bundle between any two baffles. It comes closest to the theoretical analysis of flow over

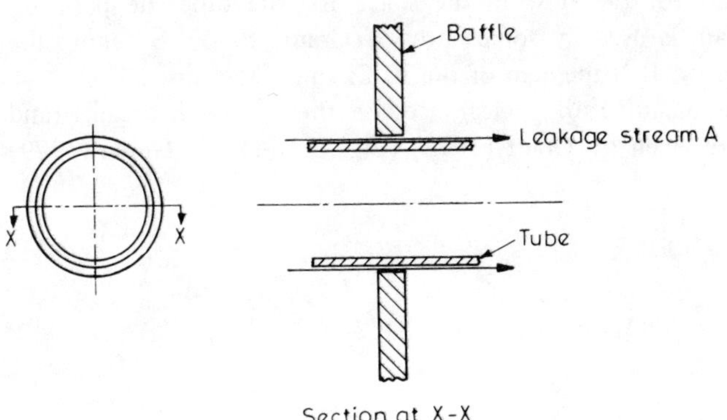

Figure 2.2 Tube-to-baffle hole leakage stream, A-stream.

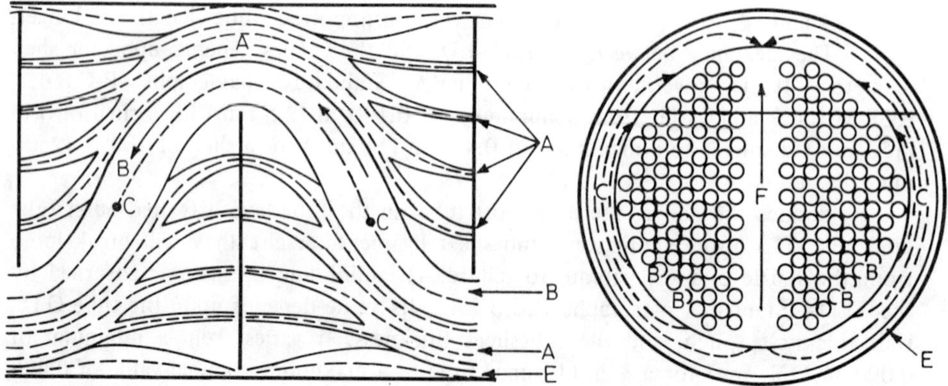

Figure 2.3 Shell-side flow streams in a typical unit. B-stream is cross-flow. *(From Tinker, 1958.)*

an ideal tube bank. The main effort in the design should be to maximize the amount of the B-stream within the allowable pressure drop. This is known as the cross-flow stream.

3. C-stream (Fig. 2.4). The resistance to fluid flow is the maximum in cross-flow over the tube bundle and the least in the space between the shell I.D. and the OTL. Since any fluid prefers to take the path of least resistance, a part of the total stream flows through this space. This is known as a bypass stream, since it bypasses the tube bundle. This stream is very small in the fixed tubesheet designs and in the U-tube bundles, but is significant in the floating-head designs because the space between the shell I.D. and the OTL is large (Q. 2.2). To reduce the amount of the C-stream in the latter case, seal strips or dummy tubes are used (Q. 1.87–1.89).
4. E-stream (Fig. 2.5). Just like the A-stream, there is leakage through the annular space between the shell I.D. and the baffle due to the difference in the pressure of the fluid on either side of a baffle. This is known as a leakage stream.

Besides the above, another stream F, has been introduced (Palen and Taborek, 1969) to account for the flow in the space for the tube-side partition plate (Fig. 2.6). A W-stream is used by some authors (Grant, 1973) to denote the flow through a baffle window. It is the sum of the B, C, and F streams.

The effects of the manufacturing clearances on the A- and E-streams (and hence on the B-stream) are given by Grant (1973) (Fig. 2.7) and by Gupta (1979)

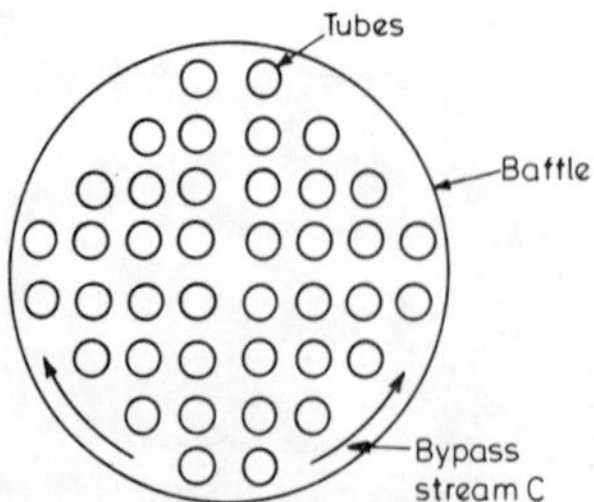

Figure 2.4 Shell-to-tube bundle bypass stream, C-stream.

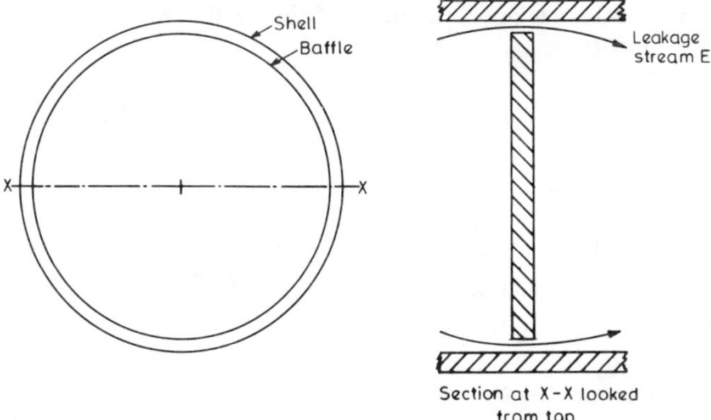

Figure 2.5 Shell-to-baffle leakage stream, E-stream.

for specific cases. These show that the manufacturing clearances have a significant influence on the amounts of various streams.

It is important to identify these streams by their alphabets since these will be used frequently in our further discussions.

2.4 How effective are the various streams as far as the shell-side heat transfer is concerned?

The cross-flow B-stream is the most effective. A-Stream comes close to it since it is also in effective contact with the tubes. C- and F-streams contact the tubes only partly, while the E-stream does not contact them at all. Thus, the relative effectiveness toward heat transfer is

$$B > A > (C, F) > E$$

Hence, in the design and fabrication of an exchanger, effort should be made to increase the B-stream and reduce the C-, F-, and E-streams.

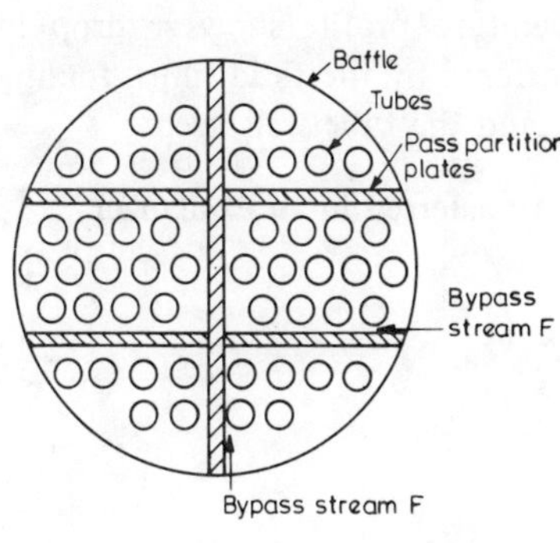

Figure 2.6 Pass partition bypass stream, F-stream.

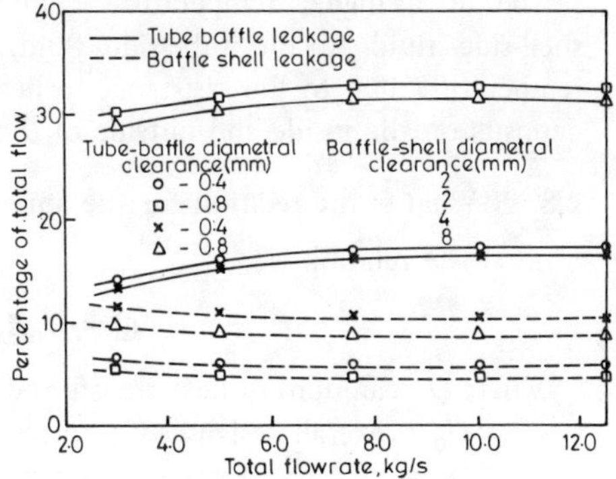

Figure 2.7 Influence of manufacturing clearances on leakage streams, A- and E-streams. *(I. D. R. Grant. In NEL Report 590 (1973) with permission of Director, NEL. British Crown Copyright.)*

2.5 When are the C- and E-streams larger fractions of the total flow?

This happens when

1. The manufacturing clearances are large or floating-head design is used without the provision of an adequate number of seal strips (Q. 1.87–1.89).
2. The shell-side flow is laminar.

2.6 How does fluid flow through the tube side of a shell-and-tube exchanger?

The fluid enters the tubes through a channel or bonnet (Fig. 1.1), travels the length of the exchanger one or more times depending upon the number of tube-side passes, then exits from the unit. In a multi-tube-pass unit using straight tubes, the fluid leaves the tubes corresponding to one pass, mixes in the end zone, then enters the tubes for the next pass. In a U-tube bundle, it travels the length of the exchanger twice before exiting and mixing in the end zone.

A common view held by a majority of heat exchanger designers and users is that the fluid flow rate through each tube in a single pass is the same, at least for a single phase, no-phase-change case (pure liquid or pure gas). However, recent preliminary studies show it to be far from being so (M. Walsh and K. P. Singh, personal communication, 1981). Water was used as a test fluid and dual-beam laser velocity meter was used to measure the point velocities. It was found that the velocity (and hence the amount) of fluid entering a particular tube depended on the location of the tube, the type of the front head (channel, bonnet), and whether the water entered the front head radially or axially. Further studies are needed.

OVERALL HEAT TRANSFER COEFFICIENT

2.7 How does the heat transfer occur from one fluid to the other?

This is illustrated in Fig. 2.8. The shell-side fluid has arbitrarily been taken to be at a higher temperature than the tube-side fluid. Heat passes from the shell-side fluid to the tube-side fluid. The temperature profile shows a drop in temperature due to the resistance to heat transfer offered by the fluid films, fouling deposits on the inside and outside of the tube wall, and the tube wall itself.

2.8 What is the relation for the amount of heat transferred in an exchanger?

The relation is

$$Q = U_d A \, \Delta T_{LMTD} \, F_T \tag{2.1}$$

where Q = amount of heat transferred per unit time
U_d = overall design heat transfer coefficient
A = heat transfer area based upon tube O.D.
ΔT_{LMTD} = log mean temperature difference
F_T = correction factor to ΔT_{LMTD} for multi-pass arrangements
It is 1 for 1-1 E-type or 2-2 F-type exchangers.

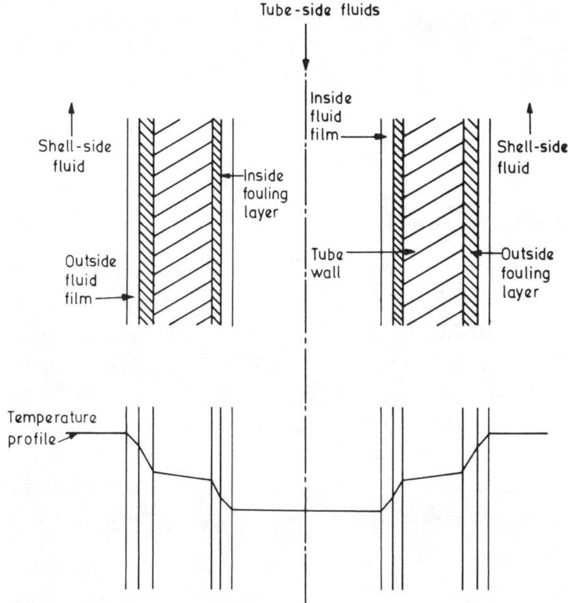

Figure 2.8 Heat transfer process in a heat exchanger.

2.9 Give the relation for U_d.

U_d is given by

$$\frac{1}{U_d} = \frac{1}{h_o} + R_{fo} + \frac{\Delta x}{k_w} + R_{fi}\frac{A_o}{A_i} + \frac{1}{h_i}\frac{A_o}{A_i} \tag{2.2}$$

where h_o = outside heat transfer coefficient ($1/h_o$ is the resistance per unit area to heat transfer offered by the outside fluid film)
R_{fo} = fouling resistance on the outside of the tube
Δx = tube wall thickness
k_w = tube wall thermal conductivity ($\Delta x/k_w$ is the resistance to heat transfer offered by the tube wall)
R_{fi} = fouling resistance on the inside of the tube
A_o = outside tube surface area per unit tube length
A_i = inside tube surface area per unit tube length (A_o/A_i transfers the inside resistance to the common area based on the outside of the tube)
h_i = inside heat transfer coefficient

For finned tubes, the tube surface area includes the area of fins multiplied by the fin efficiency (Chapter 4). For plain tubes, $A_o/A_i = d_o/d_i$ and Eq. (2.2) becomes

$$\frac{1}{U_d} = \frac{1}{h_o} + R_{fo} + \frac{\Delta x}{k_w} + R_{fi}\frac{d_o}{d_i} + \frac{1}{h_o}\frac{d_o}{d_i} \tag{2.3}$$

where d_o is the tube O.D. and d_i is the tube I.D.

For a clean tube and for nonfouling fluids, the fouling resistances do not enter into the picture and, in Eq. (2.1), U_d is replaced by U_c, the overall clean heat transfer coefficient. It is given by

$$\frac{1}{U_c} = \frac{1}{h_o} + \frac{\Delta x}{k_w} + \frac{1}{h_i}\frac{d_o}{d_i} \tag{2.4}$$

2.10 Is U_d uniform all through the exchanger?

Although Eq. (2.1) assumes U_d to be uniform all through the exchanger, in actual practice it is not. It depends upon h_o and h_i [Eq. (2.3)]. h_i (and also h_o as will be discussed later) is dependent upon the fluid properties, viz. heat capacity (C_p), thermal conductivity (k), viscosity (μ), and density (ρ), which are themselves temperature dependent. Hence, as the fluid temperature changes along the heat exchanger length, the properties also change, which affects h_i and h_o and hence U_d. As discussed in Q. 1.125 one of the future developments is likely to be the design of the heat exchangers based upon local values of U_d instead of an overall U_d. The total heat duty of the unit will then be the sum of heat transferred over small segments.

2.11 Is U_d invariant with time?

Although Eq. (2.1) assumes U_d to be invariant with time, in actual practice it is not. This is because initially a tube is clean when a unit is started (whether new unit or after one of the periodic cleaning operations), and the fouling resistances, R_{fi} and R_{fo}, are zero and Eq. (2.4) holds. Then, with use, the unit starts to get fouled up, and Eq. (2.3) holds with R_{fi} and R_{fo} gradually increasing. Finally, R_{fi} and R_{fo} reach their asymptotic values and Eq. (2.3) becomes independent of time (for fouling, see Chapter 13). As discussed in Q. 1.125, one of the future developments is expected to be a better prediction of the fouling resistances.

To sum up, U_d is neither uniform with heat exchange path length nor invariant with time. It actually varies both with length and time. However, currently Eq. (2.3) is the only way to calculate the heat duty in most cases, although the situation is expected to change with future developments.

2.12 What does the term "controlling film resistance" imply?

Equation (2.4) uses the film resistances, $1/h_o$ and $d_o/h_i d_i$. If one is significantly higher compared to the other (conversely, if one of h_o and h_i is significantly lower compared to the other), its value dominates and it is called the controlling film resistance. Thus, for water being heated in tubes by steam condensing on the outside, the film resistance for water ($d_o/h_i d_i$) governs the overall heat transfer coefficient and is called the controlling resistance.

2.13 How are the various terms for the calculation of U_d in Eq. (2.3) obtained?

Equation (2.3) requires the values of d_i, d_o, h_i, h_o, k_w, R_{fi}, R_{fo}, and Δx for calculating U_d. In this set of terms

1. d_i, d_o, and Δx are obtained from the characteristics of the tubing (Table 1.3). Also $\Delta x = (d_o - d_i)/2$.
2. h_i and h_o are obtained from the various correlations available in the literature.
3. k_w is obtained from the standard tables for the properties of the metals and alloys at an average temperature.
4. R_{fi} and R_{fo} are obtained from standard references (TEMA Standards, 1978; Kern, 1950) or as specified by the buyer.

2.14 How is the area, A, in Eq. (2.1) obtained?

It is usually obtained by rewriting Eq. (2.1) as

$$A = \frac{Q}{U_d \, \Delta T_{LMTD} \, F_T} \tag{2.5}$$

This assumes that for a required heat duty, Q, one should know U_d, ΔT_{LMTD}, and F_T independently. A major part of this chapter is devoted to the discussions regarding these three terms and the effects of various design parameters on them. We have already discussed the evaluation of U_d, although we shall come back to it as the occasion demands. Now we shall discuss ΔT_{LMTD} and F_T but, before that, we need to know about the co-current (or parallel) flow and the countercurrent flow.

2.15 What are the possible ways in which the two fluids can flow in an exchanger?

Some of the various possibilities are summarized in Fig. 2.9 along with their representative temperature profiles. T_1 and T_2 denote the inlet and outlet temperatures for the shell-side fluid, while t_1 and t_2 denote the inlet and outlet temperatures for the tube-side fluid. Further, it is assumed that $T_1 > t_1$. Thus

1. Figure 2.9*a* shows a 1-1 E-type shell where both the fluids are flowing in the same direction. This is called co-current or parallel flow. Tube-side fluid has not been traced in the figure to preserve clarity.
2. Figure 2.9*b* shows a 1-1 E-type shell where the fluids are flowing in the opposite directions. This is called countercurrent flow. Both such flows are possible in a 2-2 F-type shell also.

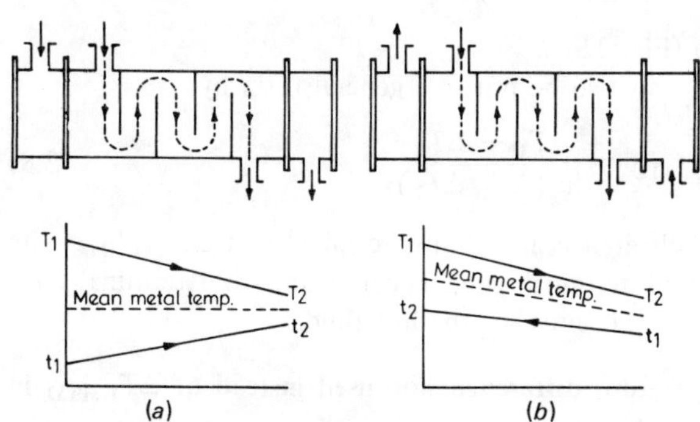

Figure 2.9 Co-current and countercurrent flows. (*a*) 1-1E-type co-current flow. (*b*) 1-1E-type countercurrent flow.

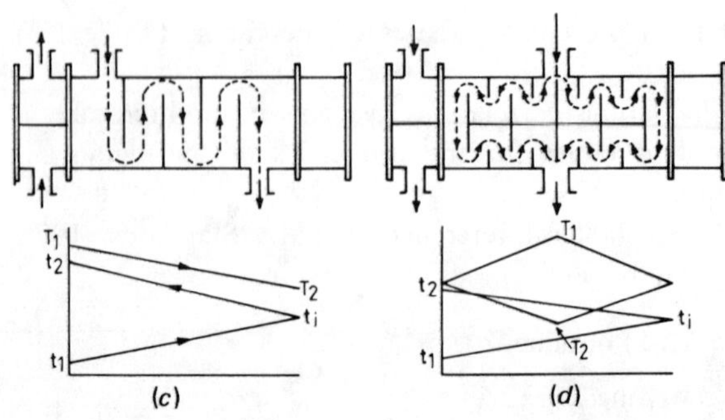

Figure 2.9 Co-current and countercurrent flow. (*Continued*) (*c*) 1-2E-type. (*d*) G-type.

3. Figure 2.9*c* shows a 1-2 E-type shell where the tube-side fluid flows once parallel to and once counter to the shell-side fluid.
4. Figure 2.9*d* shows a G-type shell where the tube-side fluid flows parallel to part of the shell-side fluid and counter to the remainder of the shell-side fluid.

It is highly recommended that the fluid flow and the temperature profiles be drawn for other types of shells (Fig. 1.1) and multi-shell and multi-tube passes as an exercise for a better understanding of the subject matter.

LOG MEAN TEMPERATURE DIFFERENCE ΔT_{LMTD} AND CORRECTION FACTOR F_{T}

2.16 The temperature profiles in Fig. 2.9 show that the driving potential, ΔT, between the hot and cold fluids varies along the length of the exchanger. However, Eq. (2.1) uses ΔT_{LMTD} for the whole unit. What is ΔT_{LMTD}?

ΔT_{LMTD} is the log mean temperature difference and is given by

$$\Delta T_{\mathrm{LMTD}} = \frac{(T_1 - t_1) - (T_2 - t_2)}{\ln\,[(T_1 - t_1)/(T_2 - t_2)]} \tag{2.6}$$

for the co-current flow case (Fig. 2.9*a*) and by

$$\Delta T_{\mathrm{LMTD}} = \frac{(T_1 - t_2) - (T_2 - t_1)}{\ln\,[(T_1 - t_2)/(T_2 - t_1)]} \tag{2.7}$$

for the countercurrent flow case (Fig. 2.9*b*).

Equations (2.6) and (2.7) can also be put in a general form by

$$\Delta T_{\mathrm{LMTD}} = \frac{\Delta T_1 - \Delta T_2}{\ln\,(\Delta T_1/\Delta T_2)} \tag{2.8}$$

where ΔT_1 is the temperature difference at the hot terminal end and ΔT_2 is the temperature difference at the cold terminal end. The hot and cold terminal ends are, respectively, the entry and the exit ends for the hot fluid.

2.17 Why is an average temperature difference not used instead of ΔT_{LMTD} in Eq. (2.1)?

While the average temperature difference

$$\Delta T_{av} = \frac{\Delta T_1 + \Delta T_2}{2} \tag{2.9}$$

is based only upon the two terminal temperature differences, ΔT_{LMTD} takes into account the flow pattern in the exchanger. The derivation of the relations for ΔT_{LMTD} [Eqs. (2.6)–(2.8)] is beyond the scope of this book but is given in detail by Kern (1950). It clearly shows the influence of the flow pattern.

For our purposes, a simple example will suffice: Let $T_1 = 100$, $T_2 = 60$, $t_1 = 20$, and $t_2 = 40$ in a 1-1 E-type exchanger (Fig. 2.9*a* and *b*). While ΔT_{av} comes out to be 50 regardless of the flow being co-current or countercurrent [Eq. (2.9)], the ΔT_{LMTD} is 43.281 for the co-current flow [Eq. (2.6)] and 49.326 for the counter-current flow [Eq. (2.7)]. This shows the sensitivity of ΔT_{LMTD} toward the flow pattern.

2.18 If two fluids undergo the same temperature change in countercurrent flow and in co-current flow, is the ΔT_{LMTD} also the same in the two cases? What can you say about the surface area, A, needed in the two cases, assuming Q and U_d to be the same?

Equations (2.6) and (2.7) as well as the example in Q. 2.17 show that ΔT_{LMTD} is not the same in the two cases. As a matter of fact (ΔT_{LMTD}) countercurrent $>$ (ΔT_{LMTD}) co-current. Furthermore, from Eq. (2.1), for a 1-1 E-type exchanger with $F_T = 1$,

$$Q = U_d A \ \Delta T_{LMTD}$$

When Q and U_d are the same in both the cases

$$(A \ \Delta T_{LMTD})_{\text{co-current}} = (A \ \Delta T_{LMTD})_{\text{countercurrent}}$$

or, the area ratio for the two cases is inverse of the ratio of ΔT_{LMTD} driving potentials. Thus, the area required for the countercurrent flow is less than that for the co-current flow. Hence the countercurrent flow should be used in most situations.

2.19 What are the advantages of co-current flow?

The rate of change in the temperature of a fluid is higher in co-current flow than in the countercurrent. Hence, the former should be used for highly viscous fluids where pumping is otherwise difficult. The high value of $(T_1 - t_1)$ in the beginning reduces the viscosity and increases the Reynolds number and the average heat transfer coefficient.

If acid vapors are present in a gaseous stream, the tube wall temperature should not reach the dew point, or else acid vapors will condense and corrode the exchanger (Scaccia and Theoclitus, 1980). The tube wall temperature changes less in the co-current flow than in the countercurrent flow (Fig. 2.9*a* and *b*). Hence, the chances of reaching the dew point are less in the co-current flow than in the countercurrent flow. Therefore, the co-current flow should be seriously considered in such a case (see also Q. 2.24). Furthermore, the differential thermal stresses are lower than in the countercurrent flow, since the mean wall temperature does not change significantly in the co-current flow (Fig. 2.9*a* and *b*).

2.20 Can ΔT_{LMTD} be the same both in the co-current and the countercurrent flows for any special case?

It will occur when one of the two fluids is condensing or vaporizing isothermally (Fig. 2.10*a* and *b*). As an example, let $T_{\mathrm{cond}} = 100°\mathrm{C}$ (212°F, saturated steam at 1 atm pressure) and $t_1 = 30°\mathrm{C}$ (86°F), $t_2 = 50°\mathrm{C}$ (122°F), then Eqs. (2.6) and (2.7) give

$$(\Delta T_{\mathrm{LMTD}})_{\text{co-current}} = 59.44°\mathrm{C}\ (139°\mathrm{F})$$

$$(\Delta T_{\mathrm{LMTD}})_{\text{countercurrent}} = 59.44°\mathrm{C}\ (139°\mathrm{F})$$

2.21 Can ΔT_{LMTD} ever be the same as ΔT_{av}?

$(\Delta T_{\mathrm{LMTD}})_{\text{countercurrent}}$ [Eq. (2.7)] is the same as ΔT_{av} [Eq. (2.9)] in the case when the temperature profiles are parallel (Figs. 2.11 and 2.12), i.e., $\Delta T_1 = \Delta T_2$, then

$$(\Delta T_{\mathrm{LMTD}})_{\text{countercurrent}} = \frac{\Delta T_1 - \Delta T_2}{\ln (\Delta T_1/\Delta T_2)} \tag{2.8}$$

It is indeterminate when $\Delta T_1 = \Delta T_2$. Hence L'Hospital's rule (Pipes, 1958) is applied. Equation (2.8) can be written as

$$(\Delta T_{\mathrm{LMTD}})_{\text{countercurrent}} = \frac{\Delta T_2\,[(\Delta T_1/\Delta T_2) - 1]}{\ln (\Delta T_1/\Delta T_2)} \tag{2.10}$$

Let

$$\frac{\Delta T_1}{\Delta T_2} = A \tag{2.11}$$

Then, the right-hand side of Eq. (2.10) becomes

$$\frac{\Delta T_2\ (A - 1)}{\ln A} \tag{2.12}$$

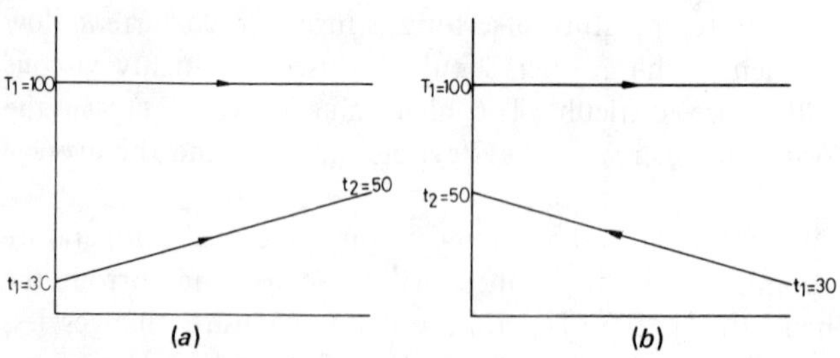

Figure 2.10 Temperature profiles when one fluid condenses isothermally. (*a*) Co-current flow. (*b*) Countercurrent flow.

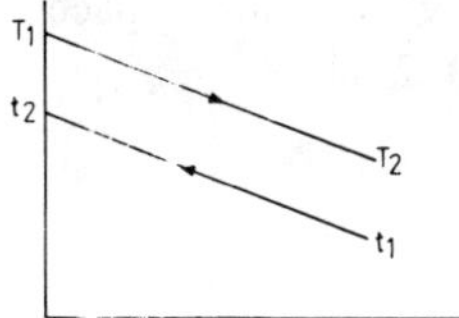

Figure 2.11 Parallel temperature profiles (equal volumetric heat capacities).

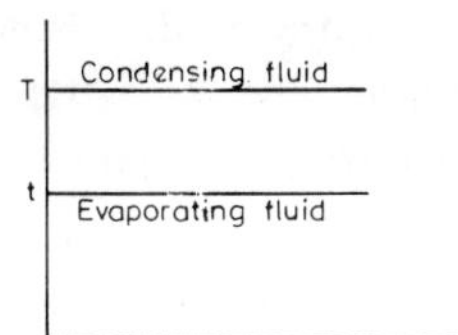

Figure 2.12 Parallel temperature profiles (condensing and evaporating fluids).

As $A \to 1$, Eq. (2.12) is written as

$$\Delta T_2 \lim_{A \to 1} \frac{(A-1)}{\ln A} \tag{2.13}$$

Applying the L'Hospital's rule, and differentiating with respect to A gives

$$\Delta T_2 \lim_{A \to 1} A$$

Taking the limit, we get

$$(\Delta T_{\mathrm{LMTD}})_{\mathrm{countercurrent}} = \Delta T_2 \tag{2.14}$$

which is the same as ΔT_{av} from Eq. (2.9) in this case since

$$\Delta T_2 = \Delta T_1$$

Similarly, one can prove that $(\Delta T_{\mathrm{LMTD}})_{\text{co-current}}$ [Eq. (2.6)] is the same as ΔT_{av} [Eq. (2.9)] in the case when one fluid is isothermally condensing and the other is isothermally evaporating (Fig. 2.12). As a matter of fact, in such a case

$$(\Delta T_{\mathrm{LMTD}})_{\text{co-current}} = (\Delta T_{\mathrm{LMTD}})_{\mathrm{countercurrent}}$$
$$= \Delta T$$

2.22 List the assumptions in the derivation of the relations for ΔT_{LMTD}.

The assumptions are the following (Kern, 1950; Gardner and Taborek, 1977):

1. The overall heat transfer coefficient, U_c (or U_d), is constant. This assumption will be removed later in Q. 2.26.
2. There are no phase changes in a part of the exchanger. Equations (2.6) and (2.7) are applicable when there is only sensible heat transfer or isothermal phase change over the entire heat exchange path. This assumption will be removed later in Q. 2.29.
3. All of the shell-side fluid participates equally in heat exchange. This ignores the effects of the bypass and leakage streams A, C, E, and F. Their effects on ΔT_{LMTD} are discussed later in Q. 2.32.
4. There is a large number of baffles that makes the cross-flow over the tube bundle between each baffle space approach the countercurrent flow. This implies

that the change in temperature of the shell-side fluid in any one baffle space is small. The effect of a small number of baffles is discussed in Q. 2.34.

5. The rate of flow of each fluid is constant.
6. The specific heat of each fluid is constant.
7. The heat losses to the surroundings are negligible.

2.23 What are the assumptions in the calculation of ΔT_{LMTD} for a multi-pass exchanger?

Besides those noted in Q. 2.22, the following assumptions are also made:

1. The surface area in each pass is equal.
2. The temperature of the shell-side fluid at any cross section in each pass of the shell is uniform.
3. There is no leakage of heat or fluid past the longitudinal baffle in a multi-shell pass exchanger. The effects of these leakages are discussed in Q. 2.33.
4. In J- and G-shells, flow rate of the shell-side fluid in each half is equal.

ΔT_{LMTD} for a multi-pass exchanger is sometimes called ΔT_{TMTD} or the true mean temperature difference.

2.24 An important assumption in the derivation of ΔT_{LMTD} is that U_c is constant all through. Since U_c depends upon the fluid properties that change with temperature, it is argued that this assumption holds better in a co-current flow than in a countercurrent flow exchanger. Discuss why.

Figures 2.9*a* and *b* show the temperature profiles for 1-1, E-type co-current and countercurrent flow exchangers. The temperatures of both the fluids are decreasing along the exchanger length in the countercurrent case resulting in a change in the property values in the same direction for both the fluids. Thus μ, c_p, ρ, and k for both the fluids increase or decrease together, depending upon their temperature coefficients. This influences U_c significantly. However, in a co-current flow, while the temperature of one fluid decreases, that of the other fluid increases along the length of the exchanger. So, the properties of the two fluids vary in the opposite directions, thus neutralizing to a large extent the effect on U_c. Hence, the assumption of U_c being uniform is more valid in the co-current case than in the countercurrent case.

2.25 Under what situations does U_c vary significantly along the exchanger path?

This happens specially when (1) the fluid is very viscous and its viscosity is a strong function of temperature as with some petroleum fractions; for example, the viscosity of SAE 10 lubricating oil (21° API) varies from 100 cP (0.1 Pa s) at 50°F (10°C) to 1 cP (0.001 Pa s) at 400°F (204°C) (Foust et al., 1980), and (2) the temperature change of one or both fluids is rather large, say 100°F (37.8°C), or more.

2.26 How is the variation in U_c accounted for in such cases?

For a linear variation in U_c, and with the rest of the assumptions in Q. 2.22 still valid, Colburn (1933) derived the relation

$$\frac{Q}{A} = \frac{U_{c1}\,\Delta T_2 - U_{c2}\,\Delta T_1}{\ln\left(U_{c1}\,\Delta T_2 / U_{c2}\,\Delta T_1\right)} \tag{2.15}$$

where U_{c1}, U_{c2} = overall heat transfer coefficients at the two ends
Q = total heat transfer
A = total surface area
ΔT_1, ΔT_2 = temperature difference between the two fluids at the two ends

This is for pure co-current or countercurrent flow, although it has also been used for the baffled exchangers ever since it was proposed. However, for the latter situations, which result in cross-flow over the bundle, an involved analysis has been done by Roetzel (1974). The case of variable U_c is also discussed by Gardner and Taborek (1977) along with the relevant literature survey.

2.27 Referring to Eq. (2.15), does one need to calculate both U_{c1} and U_{c2}?

Not necessarily so. For a linear variation in U_c, Colburn (1933) put Eq. (2.1) in a modified form:

$$\frac{Q}{A} = U_{cx}\,\Delta T_{\mathrm{LMTD}} \tag{2.16}$$

where U_{cx} is the overall clean coefficient valid for all the surface when it is transferring heat at ΔT_{LMTD}. h_i and h_o, used in the calculation of U_{cx} [Eq. (2.4)] are evaluated at a temperature, T_c, where the equality between Eq. (2.15) and Eq. (2.16) holds. T_c is called the caloric temperature. The calculation of T_c depends both on the kind of fluid and the process conditions (Kern, 1950). For the case of constant U_c, the properties are calculated at the arithmetic average temperature for each of the two fluids.

2.28 How would a nonlinear variation in U_c be accounted for?

Emerson (1973) and Butterworth (1973) have solved the above case for the multiple tube-pass exchangers in a stepwise manner. While U_c can vary along the length, it is assumed to be the same at any cross-section for all the tube passes. Furthermore, in their approaches, the specific heats of the fluids can be temperature dependent too, i.e., the enthalpy curves need not be linear.

2.29 How is ΔT_{LMTD} calculated in the case when a phase change is occurring over part of the exchanger surface?

A typical set of temperature profiles for such a case is shown in Fig. 2.13. It is for a desuperheater–condenser–subcooler for a pure vapor. To calculate the ΔT_{LMTD}, the figure is divided into zones such that the temperature profile in each zone is linear or can be closely represented by a straight line. ΔT_{LMTD} is calculated for each zone and a weighted ΔT_{LMTD} is obtained by the relation

$$\frac{Q}{\Delta T_{\mathrm{LMTD}}} = \frac{q_1}{\Delta T_{\mathrm{LMTD1}}} + \frac{q_2}{\Delta T_{\mathrm{LMTD2}}} + \cdots \tag{2.17}$$

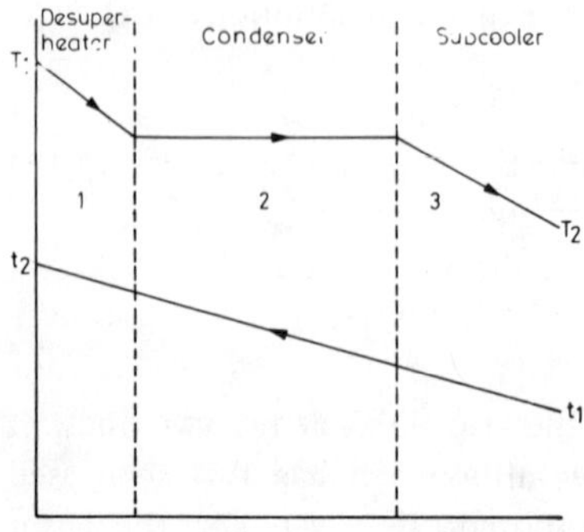

Figure 2.13 Temperature profiles when phase change occurs over part of a heat exchanger.

where Q = total enthalpy change in the process
q_1, q_2 = enthalpy change in zones 1, 2, . . .
$\Delta T_{\text{LMTD1}}, \Delta T_{\text{LMTD2}}$ = log mean temperature difference in zones 1, 2, . . .
$\overline{\Delta T_{\text{LMTD}}}$ = weighted log mean temperature difference for the entire unit

2.30 How is the overall clean heat transfer coefficient calculated in such a case?

It is based upon the heat transfer coefficient and area for each zone and is given by

$$U_c = \frac{U_{c1}A_1 + U_{c2}A_2 + \cdots}{A_1 + A_2 + \cdots} \tag{2.18}$$

2.31 Do Eqs. (2.17) and (2.18) hold for all phase-change situations?

Equations (2.17) and (2.18) refer to a pure component undergoing a phase change over a part of the exchanger surface. The situation is different with multi-component condensation occurring over a wide temperature range. The idea of $\overline{\Delta T_{\text{LMTD}}}$ does not hold in that case. Local temperature driving potential needs to be determined by stepwise calculations.

2.32 What are the effects of the bypass and leakage streams on ΔT_{LMTD}?

As discussed in Q. 2.4, the effectiveness toward heat transfer is the highest for the cross-flow B-stream and the least for the leakage E-stream. Thus, the temperature profiles for the individual streams in the shell are not identical (Gardner and Taborek, 1977) (Fig. 2.14). All the streams mix in the last zone

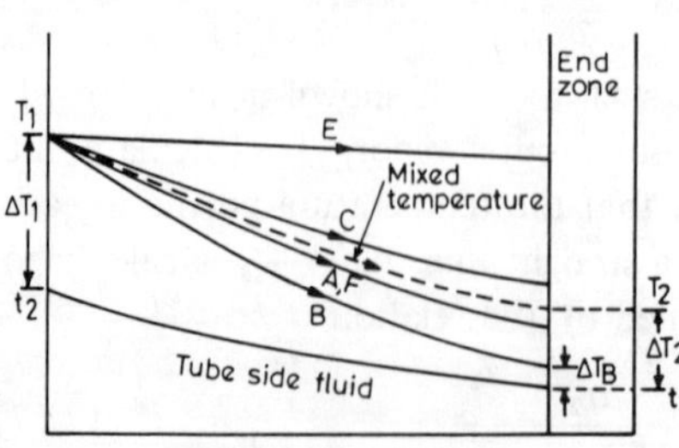

Figure 2.14 Effects of bypass and leakage streams on temperature profiles.

before exiting. ΔT_{LMTD} calculated using the mixed exit temperature is more than the actual value calculated using the exit temperature of the B-stream. For the same reason, the overall transfer coefficient calculated is less than the actual value. A correction factor has been proposed to account for the reduction in the temperature driving potential (Gardner and Taborek, 1977; Gotovski et al., 1979).

Attempts to increase the amount of heat transferred by decreasing the shell-side (mixed) exit temperature will not succeed much since ΔT_{B} (Fig. 2.14) will approach zero or negligibly small value. This is an important factor that should be kept in mind while designing a heat exchanger, particulating if the clearances are large. However, if TEMA specified clearances and seal strips are used, the temperature profile distortion is negligible.

2.33 What are the effects of the leakages past the longitudinal baffle on ΔT_{LMTD} in a multi-shell pass exchanger?

There is only heat leakage across the longitudinal baffle when it is welded to the shell. It occurs due to the temperature difference between the shell-side streams on either side of the baffle. In many cases, it does not make any significant difference, and nothing can be done about it.

There is fluid leakage past the longitudinal baffle when it is not welded to the shell and the seal or packing provided is inadequate. It occurs due to the pressure difference between the shell-side streams on either side of the baffle. It results in serious malfunctioning of the exchanger and should be avoided by the proper design and installation of the longitudinal baffle. ΔT_{LMTD} is significantly lowered. It is difficult to calculate its effects exactly, *a priori*, because they depend upon the fluid properties, temperature, and pressure profiles in the different shell passes and the leakage area between the shell and the baffle. The effects can be calculated in an existing unit by comparing the actual heat transferred with the design value.

2.34 What is the effect of fewer number of baffles on ΔT_{LMTD}?

It reduces the value of ΔT_{LMTD}. A detailed analysis has been given by Gardner and Taborek (1977). They point out that the effect of the number of baffles is felt in a 1-1 counter-flow exchanger for baffles <9 (or tube bundle crossings <10) and in a 1-2 exchanger, for baffles <3 (or tube bundle crossings <4). These factors should be kept in mind.

2.35 What does the ΔT_{LMTD} correction factor F_{T} depend upon?

It depends upon the type of shell, number of shell and tube passes, and the inlet and outlet temperatures of the two fluids.

2.36 How does F_{T} vary with the type of shell?

The various types of shells are shown in Fig. 1.1, and the temperature profiles for some of them are given in Fig. 2.9. As is evident, the flow pattern and the temperature profiles and hence the correction factor F_{T} vary with the type of shell and the number of shell and tube passes.

2.37 How does F_T vary with the number of shell passes?

For the same terminal temperatures of the two fluids, the value of F_T increases with the increasing number of shell passes since the fluids are able to transfer heat more effectively in a multi-shell pass unit. The plots of F_T for different numbers of shell passes are available in most standard references (TEMA Standards, 1978; Kern, 1950).

2.38 The value of F_T increases with the number of shell passes. Why then are the multi-shell passes not used very frequently?

The fabrication of a multi-shell pass exchanger poses a lot of difficulties. The installation of the longitudinal baffles becomes difficult. The leakage of the fluid and heat across the longitudinal baffles cannot be calculated exactly and causes uncertainties in the design (Q. 2.33). It is not easily possible to design and service removable bundle units for more than two-shell passes. Hence, if the shell-side fluid scales the outer surface of the tubes, which entails the removal of the bundle for periodic cleaning, one cannot generally use more than two-shell passes. Further, the pressure drop also increases tremendously. Multi-shell passes in such cases may be obtained by using a series of 1-2 or 2-4 units.

2.39 Is the F_T factor for each shell pass the same?

Fischer (1938) has proved that for a multi-shell pass, the F_T factor is the same for each shell pass and, further, that it is also the same for the total shell, provided the area per shell pass is uniform.

2.40 Why is an odd number of tube passes not generally used?

These are liable to create mechanical problems in the fixed tubesheet exchangers (Kern, 1950). However, such units are sometimes fabricated to take advantage of their higher F_T value (Singh, personal communications, 1981).

2.41 How does F_T vary with the number of tube passes?

The difference in the F_T value for a 1-2 and 1-8 unit is only about 2 percent (Kern, 1950). Hence, in actual practice, this difference is neglected, and the same F_T plot is used for a given number of shell passes regardless of the number of tube passes. This holds for an even number of tube passes (Q. 2.43).

2.42 Why does F_T vary with the number of shell passes and not with the number of tube passes?

In the discussion here, it will be assumed that an even number of tube passes are used.

In a 1-2 or 1-8 unit, the shell-side fluid temperature at any cross section is uniform over all the tube passes, and, as discussed in Q. 2.41, the net difference in F_T is negligible. However, in a 2-8 unit, the shell-side fluid temperature has two very different values at any shell cross section, with one value holding for the first four tube passes and the other value for the remaining four tube passes. Thus, a larger temperature cross, $(t_2 - T_2)$, is possible. This affects the F_T factor significantly, and its value is appreciably different than that for the 1-8 unit.

2.43 For a given shell, how does F_T for an even number of tube passes differ from that for an odd number of tube passes?

The analysis for the odd number of tube passes has been done by Fischer (1938). He shows that the value of F_T for an odd number of tube passes per shell pass is the greatest when the countercurrent flow tube passes exceed the co-current flow tube passes. Further, in such a case, the F_T value is higher than for any even number of tube passes. Thus, for a 1-3 unit with two tube passes in countercurrent flow with respect to the shell-side flow, the F_T is greater than that for a 1-2 unit. If, however, two tube passes in a 1-3 unit are in co-current flow with respect to the shell-side flow, the F_T value is less than that for a 1-2 unit.

2.44 Can an exchanger be designed to operate at any value of F_T?

The plots for F_T have a steep slope at values around 0.75. Any built-in error in predicting F_T is likely to be disastrous if the design is in the steep slope region. The slope of the F_T curve should also be checked by interchanging the shell-side and tube-side fluids and steeper of the two slopes considered in deciding whether or not the F_T value should be used. Furthermore, any variation in the inlet temperatures or the flow rates of the streams during the operation, which will affect the exit temperatures, is likely to bring F_T down very significantly, thus affecting the heat transfer adversely. Therefore, exchangers are generally not designed to operate at $F_T < 0.75$. In certain cases, the minimum F_T used is 0.8 or even higher, since it depends as to whether the point lies above the steep slope region or not for the particular case under consideration.

2.45 If F_T for an exchanger lies in the steep slope region, what should be done to increase the F_T?

Since F_T increases with an increasing number of shell passes (Q. 2.37), the next higher number of shell passes should be tried. Usually that results in an appreciable change in the value of F_T and brings it significantly above 0.75.

2.46 Since true counterflow gives the maximum driving potential (ΔT_{LMTD}), why are the multi-tube pass arrangements sometimes preferred even though these result in a penalty on the value of ΔT_{LMTD}?

Limitations on space, higher tube-side allowable pressure drop, a minimum limit on the tube-side velocity to reduce the fouling deposit and to achieve high heat transfer coefficient, limit on the available tube lengths, etc., result in the choice of a multi-tube pass arrangement instead of a true counterflow.

2.47 Compare a 1-1 counterflow exchanger with a 1-2 exchanger.

In a 1-1 counterflow exchanger, F_T is unity, and the driving potential is the full ΔT_{LMTD} value. The largest temperature cross $(t_2 > T_2)$ is possible. However, the tube-side fluid velocity is low, and hence h_i is low and fouling deposit is high. These reduce U_c, which results in a high heat transfer area requirement.

In a 1-2 exchanger, F_T is less than unity, and the driving potential is proportionately reduced. However, the tube-side fluid velocity is high and so is h_i. This increases U_c, which results in a lower heat transfer area requirement. In order to utilize the available pressure drop fully, it may be economical to use a 1-2 unit instead of a 1-1 unit, since the decrease in the driving potential may be more than compensated by the increase in U_c. This type of unit is useful when the tube-side film resistance is controlling or the fouling conditions require a minimum tube-side velocity to be maintained or when the U-tubes are used.

2.48 Compare a 1-2 exchanger with a 2-4 exchanger.

The characteristics of a 1-2 exchanger have been given in Q. 2.47.

A 2-4 exchanger has a higher F_T value than a 1-2 exchanger, hence the driving potential is higher than that for a 1-2 unit. Furthermore, the velocities of both the shell-side and the tube-side fluids are higher than in a 1-2 or 1-1 unit. Hence, both h_i and h_o are higher, and the fouling deposits are lower, resulting in a higher U_c. Thus, increases in both F_T and U_c result in more heat transfer in a 2-4 unit or less area requirement for a given heat load. Of course, the pressure drop is also higher than in a 1-2 or 1-1 unit. A 2-4 unit is used when plenty of pressure drop is available and a large temperature cross is required.

As the number of shell passes increases, more and more heat transfer will occur. However, both the fabrication costs and the pressure drop rise tremendously. Hence, units with more than two shell passes are not used often (Q. 2.38).

2.49 In a 1-2 exchanger, the surface area in each of the two tube passes is assumed to be equal. Does it mean that the actual amount of heat transferred in each of the two passes is also equal?

Figure 2.15 shows the temperature profile for a 1-2 exchanger with typical numbers. Area in each of the two passes is equal. Assume that U_c is also equal. The value of ΔT_{LMTD} for the first pass is

$$\Delta T_{LMTD} = \frac{(200 - 85) - (140 - 120)}{\ln [(200 - 85)/(140 - 120)]} = \frac{115 - 20}{\ln (115/20)} = 54.31$$

while ΔT_{LMTD} for the second pass is

$$\Delta T_{LMTD} = \frac{(200 - 160) - (140 - 120)}{\ln [(200 - 160)/(140 - 120)]} = \frac{40 - 20}{\ln (40/20)} = 28.85$$

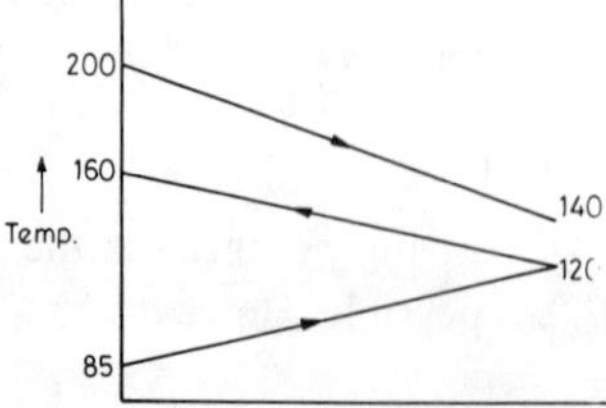

Figure 2.15 Temperature profiles for a 1-2 heat exchanger.

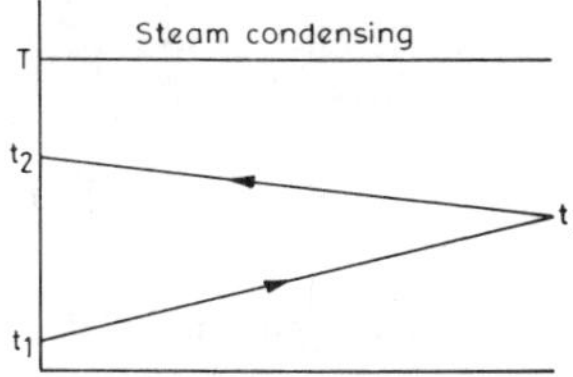

Figure 2.16 Temperature profiles for a 1-2 condenser.

Hence, the ratio of the amount of heat transferred in the first pass to that in the second pass is 54.31/28.85 = 1.88, even though the fluids are in co-current flow in the first pass and in countercurrent flow in the second pass. Thus, although the area in the two passes is equal, the heat transferred is not. It can be argued that U_c in the first pass will be less than that in the second pass because the average temperature at which the properties of the tube-side fluid are evaluated are different. However, the difference in U_c will not be enough to offset the effect of the difference in ΔT_{LMTD} in the two passes.

2.50 Comment on whether a 2-4 exchanger can be treated as two 1-2 units in series, or parallel.

A 2-4 exchanger can be treated as two 1-2 exchangers in series. A sketch can easily be drawn to distinguish between the series and parallel arrangements of the 1-2 units.

2.51 In the above, what can you say about the relative size of the 1-2 units compared to the 2-4 units?

The tube length will be the same in each case, and the number of tubes in one 1-2 unit will be half of that in the 2-4 unit. The 1-2 unit shell diameter will be approximately 70 percent of the 2-4 unit.

2.52 What is the F_T factor for the isothermal condensation of steam in the shell of a 1-2 exchanger?

Such a situation is often encountered in surface condensers (Chapter 3). The temperature profiles are as shown in Fig. 2.16. There is no subcooling of the condensate. Hence, F_T is unity.

2.53 Does $F_T = 1$ mean that the exchanger is very effective?

It only means that the full ΔT_{LMTD} driving potential is available. The effectiveness (toward heat transfer) depends on several factors, such as the fluid properties, fouling tendencies, fluid velocities, and flow arrangement. A simple definition is given in Q. 2.94. As discussed in Q. 2.47, in certain cases, a 1-2 unit is likely to be more effective than a 1-1 unit even though the F_T value for the 1-2 unit is less than unity.

2.54 Define "temperature range," "temperature approach," and "temperature cross."

Let T_1 and T_2 denote the inlet and outlet temperatures of the hot fluid, and t_1 and t_2 the inlet and outlet temperatures of the cold fluid. Then the requisite definitions are as follows (Kern, 1950):

1. Temperature range is the actual rise or fall in the temperature which, for the hot fluid, is $T_1 - T_2$ and, for the cold fluid, is $t_2 - t_1$.
2. Temperature approach, in co-current flow, is the difference between the outlet temperatures of the hot and cold fluids, T_2 and t_2. In countercurrent flow, it is the smaller of $T_1 - t_2$ and $T_2 - t_1$. In multi-pass flow it is $T_2 - t_2$.
3. In countercurrent flow or in multi-pass flow arrangements, if the cold fluid outlet temperature is greater than the hot fluid outlet temperature, i.e., $t_2 > T_2$, then $t_2 - T_2$ is called the temperature cross. In such cases, the concept of temperature approach has no meaning.

AVERAGE AND PIPE WALL TEMPERATURES

2.55 What is average temperature and where is it used?

It is the arithmetic average of the inlet and the outlet temperatures of the fluids. It is calculated separately for the shell-side and the tube-side fluids.

$$T_{av} = \frac{T_1 + T_2}{2} \quad \text{for the shell-side fluid} \tag{2.19a}$$

$$t_{av} = \frac{t_1 + t_2}{2} \quad \text{for the tube-side fluid} \tag{2.19b}$$

It is used to calculate the fluid properties in a case where U_c can be taken to be constant (Q. 2.27).

2.56 What is pipe wall temperature?

In the calculation of the heat transfer coefficient, the term μ/μ_w appears. μ_w is the viscosity at the pipe wall temperature that is obtained by equating the heat flux across each individual resistance to that across the total resistance comprising of the inner film, tube wall, and the outer film. In most cases, the tube wall resistance is negligible. Then,

$$\frac{T_c - T_p}{1/h_o} = \frac{T_p - t_c}{1/h_{io}} = \frac{T_c - t_c}{(1/h_o) + (1/h_{io})} \tag{2.20}$$

where T_p = pipe wall temperature
T_c, t_c = caloric temperatures for the shell-side and tube-side fluids, respectively
h_o, h_{io} = heat transfer coefficient for the shell-side fluid and for the tube-side fluid, respectively, referred to the outside surface. $h_{io} = h_i(d_i/d_o)$

This gives

$$T_p = T_c - \frac{h_{io}}{h_{io} + h_o}(T_c - t_c) \tag{2.21}$$

and
$$T_p = t_c + \frac{h_o}{h_{io} + h_o}(T_c - t_c) \tag{2.22}$$

In a case where U_c is assumed constant, the caloric temperatures in Eqs. (2.21) and (2.22) are replaced by the average temperatures [Eqs. (2.19*a*) and (2.19*b*)].

2.57 What is the effect of fouling on the pipe wall temperature?

When there is fouling on the outside and inside surfaces, Eq. (2.20) becomes

$$\frac{T_c - T_p}{(1/h_o) + R_{fo}} = \frac{T_p - t_c}{(1/h_{io}) + R_{fio}} = \frac{T_c - t_c}{(1/h_o) + R_{fo} + (1/h_{io}) + R_{fio}} \tag{2.23}$$

where R_{fo} is the fouling resistance on the outside surface and R_{fio} is the fouling resistance on the inside surface referred to the outside surface.

$$R_{fio} = R_{fi}\frac{d_o}{d_i}$$

From Eq. (2.23), relations for T_p along the lines of Eqs. (2.21) and (2.22) can be obtained.

2.58 What is the effect of wall resistance on the pipe wall temperature?

In such a case, the pipe wall temperature on the outside surface, T_{po}, is different than that on the inside surface, t_{pi}. Equation (2.20) becomes

$$\frac{T_c - T_{po}}{(1/h_o)} = \frac{T_{po} - T_{pi}}{(\Delta x/k_w)} = \frac{t_{pi} - t_c}{(1/h_{io})} = \frac{T_c - t_c}{(1/h_o) + (\Delta x/k_w) + (1/h_{io})} \tag{2.24}$$

where Δx is the wall thickness and k_w is the wall thermal conductivity. From Eq. (2.24), relations for T_{po} and t_{pi} along the lines of Eqs. (2.21) and (2.22) can be obtained. The wall resistance may be important in some pure vapor condensation cases where h_o is high and the film resistance, $1/h_o$, is comparable to the wall resistance, $\Delta x/k_w$.

ALLOCATION OF FLUIDS TO SHELL SIDE AND TUBE SIDE

2.59 How is the allocation of fluids to shell side and tube side decided?

Among the various factors involved are the following:

1. Corrosive nature: A more corrosive fluid is placed on the tube side, and alloy tubings are used. If placed on the shell side, both the tubes and the shell will have to be of alloy materials to combat the corrosion.
2. Fouling tendency: In a fixed tubesheet design, the fouling fluid should be placed on the tube side. In the removable bundle designs, other factors, such as the ease of removal of the scale, tube pitch, and the layout angle, also influence the choice.

3. Fluid pressure: A fluid at a very high pressure is put on the tube side, or else the shell will be very thick.
4. Fluid temperature: Very high fluid temperature greatly reduces the allowable stresses. So it should not be placed in the shell since its thickness would increase. Further, convection and radiation losses and safety of personnel may require an additional expense of insulating the shell. An analysis of uninsulated heat exchanger has been done by Kern (1976).
5. Viscosity: Highly viscous fluids and those flowing in the laminar zone should be placed on the shell side, since the continual change in cross-sectional flow area in the shell induces turbulence at low Reynolds number.
6. Allowable pressure drop: If it is low, the calculations may be done by switching the fluids from the shell to the tube side and determining which way the pressure drop requirement is met.
7. Flow rate: Fluid with the lower flow rate may be placed on the shell side since the turbulence is produced at a lower Reynolds number there resulting in a higher heat transfer coefficient.

If there are no restrictions imposed by the fluid characteristics, operating conditions, or the buyer, then one may switch fluids to see which arrangement gives higher U_c without exceeding the allowable pressure drop.

2.60 Gases at moderate pressures, to be cooled or heated, are placed in the shell, while at high pressures they are placed in the tubes. Why?

At moderate pressures, the shell-side thickness is small. The gases are placed in the shell while the water used for cooling or steam used for heating is placed in the tubes so as to localize their corrosive behavior to the tubes only.

At high pressures, the shell-side thickness becomes large, and the equipment becomes very costly. Hence, the gases are placed in the tubes since the tube thickness is not excessive.

2.61 Does a switch of fluid from the shell to the tube side affect the correction factor to the log mean temperature difference, F_T?

It is assumed that the heat duty and hence the terminal temperatures are the same in the two cases. Then F_T is not affected due to the switch mentioned above. The factors on which F_T depends are given in Q. 2.35. These are independent of the location of the fluids.

RATING AND DESIGN

2.62 What is the difference between the rating and the design of a heat exchanger?

Generally speaking, rating is the evaluation of the performance of a fully specified heat exchanger to do a job, while design is the determination of the specifications for manufacturing an exchanger to fulfill the specified process conditions (Taborek, 1979a, 1979b). The rating of an exchanger does not imply the best possible utilization of the given exchanger. It simply evaluates whether or to what degree a given job can be performed within the constraints of the flow rates, heat duty, fouling resistances, and the allowable pressure drop. The exchanger, if acceptable, almost always is larger than actually needed, giving a large over design. On the other hand, in design, one can reach an optimum so that the over design is minimal. Design essentially results in rating a number of exchangers with varying geometrical details until the right unit is arrived at.

2.63 List the steps in the rating of a given exchanger.

The steps are the following:

1. Determine the heat duty, Q.
2. Calculate the ΔT_{LMTD} [Eq. (2.8)] and find the value of F_T from an appropriate plot.
3. If temperatures of the streams change significantly, making the assumption of U_c = constant invalid, calculate the caloric temperature (Kern, 1950).
4. Depending upon the method used for the shell-side heat transfer coefficient, calculate h_o.
5. Depending upon the method used for the tube-side heat transfer coefficient, calculate h_i and h_{io}. Usually Kern's (1950) method is used for this purpose.
6. Calculate U_c from Eq. (2.4).
7. Calculate U_d from Eq. (2.3).
8. Again calculate U_d from Eq. (2.1). The area is known for the given exchanger.
9. If U_d in step 8 is equal to or greater than U_d in step 7, the desired heat duty, Q, can be obtained from the given unit.
10. Now calculate the pressure drop to check if it is within the allowable pressure drop limits or not. If it is within the limits, the exchanger will work. If not, then the exchanger can not be used. The only way out is either to modify the unit or to increase the allowable pressure drop to meet the requirement.

2.64 In general, what is the aim in the design of a heat exchanger?

The aim is to transfer heat from one fluid to the other within the allowable pressure drop of the two fluid streams and with no hazards to plant or life. This should be done at a minimum cost, considering the reliability of the unit for its entire expected life.

2.65 What is the minimum information necessary for the process design (thermal design) of an exchanger?

The minimum information necessary is the following:

1. Flow rates of the process fluids
2. Inlet (or outlet) temperatures of the two fluids
3. Heat load
4. Phase change, if any
5. Composition of mixture, if any fluid has more than one component

6. Pressure drop allowed
7. Inlet pressures of the fluids, specially if a gas is involved
8. Any minimum velocity requirements
9. Fouling conditions
10. Size restrictions, if any, due to the plant layout or due to customers' preferences to minimize inventory of the spares, tools, etc.
11. Anticipated changes in the flow rates, heat load, cooling water temperature, etc.

Of course, there can be some variations in the above depending upon the situation at hand.

2.66 What factors should be considered while evaluating the final computed design?

The following factors should be considered while evaluating the final design for its economic workability (Taborek, 1979b):

1. Thermal resistances: The total thermal resistance is made up of the shell-side and tube-side films, the shell-side and tube-side fouling layers, and the metal wall. If any of the resistances dominates, the reasons should be analyzed and steps taken to reduce it if possible.
2. Pressure drop utilization and distribution: The allowable pressure drop should be utilized to the fullest extent. If a substantial difference exists between the allowed and the calculated pressure drop, a change in the design factors, viz. shell type, number of passes, tube length, pitch and layout angle, baffle type and pitch, etc., should be tried (Q. 2.152).

   Pressure drop distribution should be such that it mostly occurs where the heat transfer is high, viz. in cross-flow over the tube bundle. In areas where the heat transfer is nil or minimal, the pressure drop should be the minimum possible. Such areas are the nozzles, baffle window in no-tube-in-window design, etc.

   Another factor to keep in mind is that often the allowable pressure drop is arbitrarily fixed. Hence, if in a design the total allowable pressure drop is being utilized and the designer feels that with a little more pressure drop a more economic design can be obtained, he should do such a design and discuss it with the client.
3. Overdesign factor: This is the difference between $Q_{required}$ and $Q_{calculated}$ based upon the design. It is usually kept at 10 percent (or less) to take care of the design uncertainties, plugging of the tubes, fouling, etc.
4. Flow velocities: The fluid velocity should be high to reduce fouling deposit but not so high as to cause erosion and/or vibration damage. The latter should be checked separately, especially in the baffle window and the inlet and outlet regions (Chapter 15). Erosion damage is more likely when suspended solids are present in the inlet liquid stream or suspended liquid droplets are present in the inlet gaseous stream (Chapter 13).
5. Shell-side flow distribution: This can be calculated by the Tinker's method or by some of the proprietary programs. The effort should be to keep the cross-flow stream as high as possible.

6. Baffle design analysis: This can affect the design significantly. Various types of baffles are discussed in Chapter 1. The attempt should be to keep (a) the baffle cut-baffle spacing ratio close to unity so as to use the whole of the heat transfer area and avoid dead zones, and (b) the cross-flow and window flow velocities of the same order.
7. ΔT_{LMTD} correction factor: It should be properly analyzed as discussed earlier (Q. 2.44).
8. Heat transfer coefficient: If the fluid viscosity varies significantly, U_c may vary greatly too and hence, instead of an average value of U_c, a stepwise calculation may be better (Q. 2.25 to 2.28).
9. Tube vibration analysis: As discussed in item 4 above, it is an important factor and should be carefully checked.
10. Overall design geometry: A rough sketch of the computed design should be made and important dimensions shown on it to determine any obvious errors in design and to check if it is compatible with the plant layout and the connecting piping.

2.67 What are some rules-of-thumb in the design of a shell-and-tube exchanger?

The following rules-of-thumb should be observed during the preliminary design (Frank, 1978):

1. Tube-side velocity for process fluid 3–6 ft/s (0.9–1.8 m/s).
2. Tube-side velocity for cooling water about 6–8 ft/s (1.8–2.5 m/s) in process heat exchangers and higher in surface condensers used in the power plants.
3. Shell-side liquid velocity 2–5 ft/s (0.6–1.5 m/s).
4. Vapor velocity 20–250 ft/s (6–76 m/s) depending on the pressure (high, atmospheric, or vacuum) and the molecular weight of the vapor (high or low).
5. More corrosive and/or high-pressure fluid should be put on the tube side.
6. Odd number of tube passes should be avoided.
7. The ΔT_{LMTD} correction factor F_T should not lie in the steep slope region.
8. For moderate temperatures and pressures, and for situations requiring alloy steel or nonferrous alloys, plate heat exchanger should be tried (Chapter 6).
9. For highly fouling fluids, spiral exchangers should be tried if the heat load is low and the operating conditions are mild (Chapter 6).

2.68 List the steps in the design of an exchanger.

The following steps are used though there can be some variations depending upon the available data and one's own experience:

1. By heat balance determine the terminal temperatures or heat load as the case may be.
2. Calculate ΔT_{LMTD}.
3. Assume an overall design heat transfer coefficient including the fouling resistances.
4. Select a tube size and length based upon available space, experience, etc., and determine the number of tubes needed.

5. Calculate the exchanger area required by Eq. (2.5).
6. Select a tube pitch and layout angle based upon the kinds of fluid, cleaning requirements, etc. (Q. 2.104). Determine the shell size for them by looking up the standard tables. Allowance should be provided for the tie rods and spacers.
7. Select the kind of baffles, their orientation, and spacing (Q. 1.41–1.73).
8. Calculate the shell-side heat transfer coefficient.
9. Calculate the tube-side heat transfer coefficient.
10. Calculate the overall design heat transfer coefficient by taking into account the wall resistance and the fouling resistances [Eq. (2.3)].
11. Compare the calculated value in step 10 with the assumed value in step 3. If the former is equal to or more than the latter, proceed to next step, or else repeat step 3 onward by assuming a different value for the overall design heat transfer coefficient.
12. Calculate the shell-side pressure drop.
13. Calculate the tube-side pressure drop.
14. If either of the calculated pressure drops does not satisfy the allowable pressure drop, and if the allowable limit cannot be increased, then the configuration of the exchanger has to be modified. One should check if the pressure drop in the nozzles or in the baffle windows is excessive, since these are almost a waste. The various modifications may include changing the shell type, baffle type, baffle spacing, baffle cut, number and length of tubes, pitch and layout of the tubes, and the number of passes. The whole procedure starting from step 3 has to be repeated.
15. An exchanger designed to satisfy the heat transfer and pressure drop requirements should also be checked to satisfy that the vibrations will not damage the tube bundle (Chapter 15). If they will, the design has to be modified by changing, among others, the baffle spacing or providing intermediate tube supports (Q. 15.60). The shell side should be rerated, since any change in the baffles will change the shell-side heat transfer coefficient.
16. Finally, stress analysis of critical parts should be done if so required by the Codes, local regulations, or by the customer (Chapter 9).

2.69 What are the important unknowns in the design of a no-phase-change heat exchanger?

These are the following:

1. The exact flow patterns and fluid velocities on the shell side
2. The variation of the local heat transfer coefficient
3. The extent and nature of fouling

2.70 What are the gross assumptions in most methods of heat exchanger design?

The following comments (Frank, 1978) apply to most design methods available in the literature:

1. Constant values for the shell-side and the tube-side heat transfer coefficients are assumed, although they vary due to the variation in properties and velocities in the exchanger.

2. Shell-side cross-flow area is taken at the center of the shell. The cross-flow area actually varies with the tube row and so does the cross-flow velocity (Q. 2.73).
3. The effects of the various bypass and leakage streams are neglected in many methods. These streams actually reduce the amount of heat transferred (Q. 2.3 and 2.4).
4. A constant and highly conservative value of the fouling factor is assumed. The fouling factor actually varies with the usage of the exchanger between two consecutive cleanings and depends upon the operating conditions.
5. Fluid properties are evaluated at the average temperature. Actually, the properties vary significantly, and the average temperature values may not always represent the best compromise. The problem is more difficult if mixtures are involved.

With continuing efforts of a large number of people, the above situation is expected to improve with time and some of the gross assumptions may not have to be made in future designs.

2.71 Discuss the use of safety factor in the design of an exchanger.

The safety factor allows for the uncertainties in the heat transfer equations that are normally up to ±20–30 percent in the best of the design methods. It is generally applied when the area required has been finally calculated. Generally, the area is increased by up to 10 percent.

The resulting increase in area should be accommodated by increasing the length of the tubes rather than their number, since the latter decreases both the shell-side and tube-side velocities and the heat transfer coefficients, thus reducing the effective safety factor as well (Q. 2.103). The increase in the tube length will proportionately increase the pressure drop, and this should be kept in mind.

2.72 What are the other reasons for providing some extra heat transfer surface in an exchanger?

Some extra heat transfer surface is provided to take care of the following:

1. The deviations in the operating conditions from the design conditions
2. The emergencies and the upset conditions when the flow rates, temperatures, and pressures might change for a short time
3. Excessive fouling, which may block off some of the tubes
4. Expansion of the plant or process changes with developing technology, which may require more throughput
5. Some tubes may have to be plugged because of leakage

The amount of extra surface to be provided depends on the particular situation and past experience with the fluids being handled.

2.73 Since the cross-flow area changes with the tube row, what area should be used for calculating the shell-side cross-flow velocity?

Some design methods, including Kern's (1950), use the maximum cross-flow area along the diameter. This results in a lower value of the cross-flow velocity and hence lower values of the heat transfer coefficient and the pressure drop. A better design may be obtained by taking some sort of average of the cross-flow areas at the center and at the baffle tip. Some of the proprietary design methods follow this technique.

2.74 How does one decide the flow area for the tubular fluid in the channel, floating-head cover or bonnet?

The above refers to the cross-sectional flow area of the channel, floating-head cover, or bonnet at their center line. It should not be less than the cross-sectional flow area of all the tubes in a single pass, otherwise excessive pressure drop will result when fluid enters or leaves the tubes, which in no way contributes to the heat transfer. The TEMA Standards' minimum requirements of the flow area are given in Table 2.1.

2.75 The relations for the heat transfer coefficient and the pressure drop hold for both liquids and noncondensing gases. However, the density of the gases is greatly dependent upon the operating pressure. How is this taken into account, since the pressure decreases along the exchanger length?

The Reynolds number is defined as Gd/μ, where G is the mass flux (ρv). As the gas pressure decreases in passage through an exchanger, ρ decreases and v increases, keeping G constant. Hence, there is no effect on the heat transfer coefficient. Furthermore, the pressure drop in gaseous flow is usually a small fraction of the total pressure. Hence, the inlet values of ρ and v may be used in calculating the pressure drop without any appreciable error.

2.76 How is the effect of varying temperature on liquid properties accounted for?

In a liquid, the primary change is produced in viscosity. One can account for it in two ways (Kays and London, 1964):

1. Using the film average temperature of Colburn (caloric temperature, Q. 2.27)
2. Using the ratio (μ/μ_w) with an appropriate exponent (Kern, 1950)

The latter is preferred because of its elegance and simplicity. It also points out the net difference produced by considering the effects of temperature variation.

Table 2.1 Minimum flow area for different TEMA constructions

	Floating head	Channel, bonnet
R-type	1.3 times	1.3 times
C-type	1 time	1.3 times
B-type	1 time	1.3 times

Source: TEMA Standards (1978), Sections R,C,B-5.11 and 8.12.

2.77 How is the effect of varying temperature on gas properties accounted for?

It is accounted for by a ratio of the wall to average absolute temperatures raised to an exponent, since the gaseous properties, C_p, μ, k, ρ, vary in a similar manner with the absolute temperature (Kays and London, 1964). The exponent is –0.5 for gas heating in turbulent flow through circular tubes and is zero for all other conditions encountered in a heat exchanger.

2.78 Where does the thermal conductivity play an important role in the design of heat exchangers?

Thermal conductivities of the two fluids comes into the picture in determining the individual heat transfer coefficients (Nusselt and Prandtl numbers) and hence in determining the overall heat transfer coefficient.

Thermal conductivity of the fouling deposits is very low, and hence even a thin layer of the deposit offers high resistance to the flow of heat.

Thermal conductivity of the heat transfer surface material is important in the following:

1. Condensation of pure steam or other vapors where the condensing coefficient is very high and the resistance offered by it is comparable to that of the wall material
2. Finned tube heat exchangers where the fin efficiency depends, among other factors, upon the thermal conductivity of the fin (Chapter 4)

2.79 For a given heat load, various types and sizes of heat exchangers are possible. What factors, besides the heat load, determine the type of the heat exchanger finally chosen?

The choice is usually between different kinds of shell-and-tube units; sometimes it is also between them and the plate heat exchangers and spiral heat exchangers.

Numerous factors play a role in decision making, some of which are the following:

1. Allowable pressure drop
2. Service life desired and corrosion conditions
3. Fouling conditions and cleanability of the unit
4. Operating pressure and temperature
5. Single-phase or two-phase flow and phase change (if any)
6. Influence of temperature on the bond between the tube and fin if the two are of different materials
7. Reliability and safety of the unit
8. Individual's experience in operating and maintaining such units in the past

USE OF COMPUTERS IN RATING AND DESIGN

2.80 Explain the trend toward computerization of the heat exchanger design methods.

With larger units being designed and competition increasing in the international market, the designs have to be cost effective and reliable. A lot of effort has gone into research and development, and many improvements have been made in the design methods that have become increasingly complicated for the manual calculations. Hence, the use of the computers in the design of the heat exchangers is increasing.

2.81 What are the drawbacks in manual calculations?

Some of the important drawbacks are (Gupta, 1979):

1. The increasingly complicated design methods are intractable without computer usage.
2. The inherent inertia in most people makes them take an easier path and bypass a complicated procedure, even if the latter is better.
3. To err is human, and some such errors might go undetected unless cross-checking is done by another experienced designer, thus increasing the cost of designing.
4. Repetitive manual calculations for the better sizing of a heat exchanger by varying some parameters get to be monotonous, highly time consuming, and costly.
5. If a large number of quotations have to be submitted, a company may not be able to spare the necessary manpower to design and quote for them in the limited time usually available for submitting such quotations.
6. Professional and/or family reasons may force a loyal design engineer to move away, leaving behind an untrained or half-trained engineer to do the future designs.

2.82 What are the advantages in using computer programs for the design?

All the disadvantages of the manual calculations discussed in Q. 2.81 are taken care of. Complicated methods can be used to produce designs within a short time, and calculations can be repeated easily by varying different parameters to obtain an optimum design. With future developments, local heat transfer coefficients may be used for the calculations (Q. 1.125).

2.83 What are the pitfalls in using the computer programs?

A computer program is only as good as the data or correlations on which it is based. Hence, the limits on the applicability of the programs must be known and respected in their use. The results obtained should be analyzed very carefully, and each item evaluated with one's experience. If the values are very different than expected, a critical analysis should be done or the calculations repeated. The computed results should never be accepted without such an analysis.

Proper use of a computer program also depends to a large extent on the user understanding the program and its limitations and being able to model the problem properly and provide adequate input data and interpret the intermediate and final results. Von Riesemann (1979) has proposed a "certification of the user" and has sought the involvement of the professional societies, universities, the management in charge of the user groups, users themselves, and the developers of the programs in reducing the risks in the use of the computer program packages.

2.84 What role does the documentation of a computer program play?

Proper documentation of a program is as important as having a correct program. The utility of the program is greatly increased with proper documentation. It aids in correct input and in interpreting the output. Also, other persons besides the engineer who programmed it can use it.

2.85 What items should be included in the documentation?

The important items to be documented are (Gupta, 1979):

1. Applicable ranges of important parameters, such as the Reynolds number, Prandtl number, etc.
2. The design logic or set of correlations used
3. Range of geometrical parameters, such as types of shell, number of shell-and-tube passes, types, cut and orientation of baffles, tube sizes, layout angles and pitch, etc.
4. Whether valid for single phase only or two phase flow and phase change (condensation) also can be used
5. All the abbreviations and nomenclature used along with their units
6. Whether applicable for British, metric, and/or SI units
7. Details of input and output should be given along with a few solved examples
8. If properties are evaluated by the program, the limits of applicability of such subroutines with regards to pressure, temperature, and composition should be discussed

SALIENT FEATURES OF SHELL–SIDE DESIGN METHODS

● We shall now discuss the salient features of the various methods for calculating the heat transfer and pressure drop for single-phase flow through the shell. The two-phase flow will be discussed in Chapter 3.

2.86 What are the various approaches used in heat exchanger design?

There are basically three approaches:

1. ΔT_{LMTD} approach: This is the most widely used one and a host of methods on this approach have been proposed (see Q. 2.87 and 2.90–2.93).
2. ϵ-NTU approach: Here, ϵ is the effectiveness, and NTU is the number of (heat) transfer units. This approach is used mostly in the design of compact heat exchangers (see Q. 2.94 and 2.95).
3. Irreversibility approach: This is a relatively new approach based on the number of entropy units, N_s, generated in a particular unit (Bejan, 1977, 1978). N_s is related to the amount of heat transferred and the pressure loss in each stream.

2.87 What are the most popular methods of calculating the shell-side heat transfer coefficient and pressure drop?

There exist a very large number of methods to calculate the shell-side heat transfer coefficient (h_o) and the shell-side pressure drop (Δp_s) for segmental baffles. Newer methods still appear in various journals. These generally give shortcut methods or improvements in the existing methods (Goyal, 1975, 1976; Spencer, 1978). However, among the most popular methods useful over a wide range of design conditions, are

1. Kern's method
2. Donohue's method
3. Bell–Delaware method
4. Tinker's method

Besides the above methods available in the open literature, there are some proprietary methods put out by various organizations for use by their member companies. A number of these are based on one of the above methods. Some are based upon a judicious combination of methods 3 and 4 above and supplemented by further research data. Among the most popular of the proprietary methods, judged by their large clientele, are

a. Heat Transfer Research Inc. (HTRI), Alhambra, California, method, also known as the stream analysis method
b. Heat Transfer and Fluid Flow Services (HTFS), Engineering Science Division, AERE, Harwell, United Kingdom method

Due to their proprietary nature, it is not possible to discuss these. However, a very large number of exchangers are still made by the open literature methods 1–4 mentioned above.

2.88 Why are fewer methods available for the tube-side calculations, whereas many methods are available for the shell-side calculations?

The flow through the tubes is well defined and amenable to analytical solutions, supplemented by experiments in some situations. However, the flow through the shell side over a bank of tubes is not well defined and hence is not amenable to analytical solutions. Therefore, the research for the shell-side calculations has primarily been experimental with the results being used to develop empirical correlations for a very wide range of parameters. This explains the multiple correlations for the shell-side calculations.

In a two-phase flow case, with or without a phase change occurring during the flow, the situation is not well defined even for flow through the tubes, and multiple empirical correlations are available for it.

2.89 Why is the flow through the shell side not well defined?

The factors responsible are too numerous, but the important ones are listed below:

1. The flow area between different tube rows is not constant.
2. The flow through the tube bundle is three-dimensional. Furthermore, it repeatedly suffers acceleration and deceleration because of the changing cross-flow area in the flow direction. Three-dimensional turbulent flow even in tubes is not rigorously understood, to say nothing about the flow over a tube bundle. However, significant efforts to understand the latter have been made by Butterworth both with regards to the flow (Butterworth, 1978) and the heat transfer (Butterworth, 1979), and by Zukauskas and Ulinskas (1979), who carried out a number of experiments. Spalding and co-workers (Spalding, 1974; Patankar and Spalding, 1974) have also used the three-dimensional flow analysis approach for the design of heat exchangers. Local values of the heat transfer coefficient and pressure drop are required for their method to be more successful. Massey et al. (1979) have solved the Navier–Stokes and energy equation using the finite difference method to predict fully developed turbulent flow and heat transfer in flow over a tube bundle. This does not cover all the work being done, but mentions just a few.
3. The overall flow pattern is different in different types of shells shown in Fig. 1.1. Some flow patterns are shown in Fig. 2.1.
4. The manufacturing clearances result in flow streams (Figs. 2.2–2.6) whose amounts and contribution to heat transfer and pressure drop cannot be evaluated exactly.
5. Fouling results in blocking part of the manufacturing clearances, thus affecting the flow rates of the individual streams (Q. 2.3).
6. The entrance and exit sections for the shell side behave differently than the central zones since the baffle spacings are different (Q. 1.63–1.66).
7. The multiple variety of segmental baffles (Figs. 1.14–1.17) and the range of baffle cuts result in cross-flow in the baffle overlap zone and parallel flow in the baffle window. Further, over certain tube rows near the baffle cut, the flow is neither parallel nor cross since it is changing the direction.
8. Depending upon the baffle type, cut, spacing, and orientation and on the fluid flow rates, there may be stagnant zones in the heat exchanger. The maldistribution of flow over a tube bundle cannot be analyzed easily. Its detrimental effects on the performance of heat exchangers have lately been attracting significant attention (Gardner and Taborek, 1977; Gotovski et al., 1979; Chiou, 1979). Also, local values of the heat transfer coefficient are being obtained by using an electrochemical mass transfer experimental technique (Gay et al., 1976; Macbeth, 1979). The effects of various geometrical parameters along the lines of the Delaware work (Bell, 1963) are being reevaluated using this technique (Gay et al., 1977). All these are expected to lead to better design methods.
9. When a phase change is involved, the relative amounts of liquid and vapor as well as their velocities change along the length of the exchanger. It is very difficult to characterize this flow.

All the above factors point toward the difficulty in defining the shell-side flow and analyzing it theoretically.

2.90 What are the salient features of the Kern's method?

It is probably a true statement that more heat exchangers have thus far been designed using the Kern's method than any other. His book (Kern, 1950) became a classic in the field after World War II. The salient features of his method are (Gupta, 1979):

1. It was proposed for the turbulent region ($2000 < \text{Re} < 1{,}000{,}000$) but has been used in the lower laminar regions also by numerous designers without any apparently serious inconsistency. The Reynolds number of 10 as the lower limit in his plot (Kern, 1950) is due to the use of an equivalent shell-side diameter.
2. It assumes that the shell is full of tubes, which, in practice, depends upon the type of the tube bundle, shell-to-OTL clearance, presence of impingement plate, etc.
3. Thus, it does not account for the bypass C-stream, which in certain designs can be significant. It does account for the normal manufacturing clearances for the A- and E-streams in its composite design equations.
4. Direct effect of the flow through the baffle window on heat transfer and pressure drop is not shown. It is included in the composite design equations.
5. It does not account for the fact that the inlet and the outlet baffle spacings are usually larger than the central baffle spacings (Q. 1.66). This gives less cross-flow velocity in the end zones and hence a lower heat transfer coefficient there.
6. The pressure drop correlation includes the entrance and exit nozzle losses for an average unit. Care should be exercised if the unit is built with special nozzle sizes and/or arrangements.
7. The pressure drop is for an industrial unit under average operating conditions, i.e., fouled conditions that result in the near plugging of the baffle hole-to-tube O.D. clearance and also decrease the cross-flow area. This important point should be kept in mind.
8. Although the initial plots given by Kern for the heat transfer (j_H) and the friction factor (f) were restricted to 25 percent baffle cut and a 30° triangular tube layout (Kern, 1950), Kern later gave plots for the baffle cuts from 15 to 45 percent and the tube layout angles of 30°, 45°, and 90° (Wolverine Truefin Engineering Data Book, 1968). These are not average curves but safe with a safety factor ranging from 0 to 20 percent.
9. Thus, the Kern's method gives a safe operating design with minimum information and simple calculation procedures meant for easy manual calculations.

2.91 What are the salient features of Donohue's method?

These are (Donohue, 1949, 1955, 1956):

1. It is based on data from real exchangers up to 23 in (580 mm) diameter, with a pitch to tube outer diameter ratio of 1.3 and 1.4 and the Reynolds number from low laminar region to the highly turbulent region (38 to 50,700). However, it has often been used for very different conditions without any apparently serious discrepancy.
2. It is almost as simple to use as the Kern's method (Q. 2.90).

3. It takes into account the shell-to-OTL bypass clearance (C-stream), but does not account for the use of the sealing devices.
4. It accounts for the contribution of the flow through the baffle windows to heat transfer by using a geometric mean of the cross-flow and window flow mass velocities.
5. It accounts for the contribution of the flow through the baffle windows to the pressure drop by separately calculating the pressure drop in window flow and cross-flow over the tube bundle.
6. As in the Kern's method, it also does not account for the fact that the inlet and the outlet baffle spacings are usually different than the central baffle spacing (Q. 1.66).
7. The values obtained are for a clean unit, unlike the case with the Kern's method, which gives values for an average fouled unit (Q. 2.90).

2.92 What are the salient features of Bell-Delaware method?

These are the following:

1. It is probably the only method in the literature that is based upon extensive research at the University of Delaware extending over a period of 15 years. It produced over 25 masters and doctoral theses and numerous authoritative papers and reports as the work progressed, and culminated with the final report (Bell, 1963) which is the main source of the discussion here.
2. The plots of j_H and f factor given in Bell (1963) are up to a Reynolds number of 10,000 based upon the tube O.D. and the maximum cross-flow velocity through the center of the tube bundle without any allowance for the bypass stream. The plots have been extended to Re = 1,000,000 (Mueller, 1973).
3. The method is called an evolutionary method since it uses the j_H and f factor plots for an ideal bank and corrects for the deviations from ideality introduced by the bypass and leakage streams, effect of sealing devices, number of tube rows in cross-flow in all the baffled zones, flow through the baffle window, etc.
4. It accounts for the fact that the number of tube rows in cross-flow is different in the entry and exit zones than in the central zone because there is only one baffle window in each of the entry and exit zones compared to two in the central zones.
5. It calculates the pressure drop in cross-flow in the baffled zone and in window flow separately.
6. In the calculation of the pressure drop in window flow, it accounts for the flow across the tubes in the window, flow parallel to those tubes, and flow turn-around losses. For this, a geometric average of the cross-flow and window flow velocities is used.
7. Even before the final report (Bell, 1963) was published, earlier results published by Bell (1960) had proved their worth in comparison with the Kern's method when both were used to predict the performance of several exchangers (Whitley, 1961). In each case, the Bell-Delaware method gave results very close to the values observed in operation whereas the Kern's method over-predicted, sometimes by a factor of 10. Thus, the Bell-Delaware method is by far the best

in the open literature. An updated version of this method has been given by Bell (1981).

2.93 What are the salient features of Tinker's method?

These are the following:

1. This method, when first proposed by Tinker (1951), was ahead of its times both in the requirements of the industry and the computational capabilities then existing. This was an iterative method because the flow resistance terms were dependent on the flow rates. Later, Tinker separated the two and, using values of some constants from industrial practices, proposed a noniterative method (Tinker, 1958) that sacrificed some accuracy for expediency in the calculations.
2. This method accounts for the bypass and leakage clearances, effects of sealing devices, baffle spacing and cut, tube size, pitch, and layout angle.
3. It accounts for the fact that the entry and exit zones have different lengths than the central baffled zones. This effect is accounted for only in the calculation of the heat transfer coefficient.
4. It accounts for the fouling only in the calculation of the pressure drop. Fouling blocks some manufacturing clearances that increases the cross-flow B-stream. This gives higher pressure drop. The results thus are conservative. If used in the calculation of the heat transfer coefficient also, the result would be an over-prediction and hence on the unsafe side.
5. It uses rating numbers for the heat transfer coefficient and the pressure drop. From these, cross-flow stream fraction is determined. Heat transfer coefficient and pressure drop are calculated based upon this stream. It includes the usual B-stream and a quarter of the A-stream, while it excludes the C- and E-streams as being totally ineffective in heat transfer. Prediction of the heat transfer coefficient is better than that of the pressure drop. Furthermore, the heat transfer coefficient is predicted better in the turbulent flow than in the laminar flow.
6. The window flow pressure drop is taken as proportional to the cross-flow pressure drop. This is a weakness of the method.
7. A further simplified version of the noniterative method of Tinker has been given by Devore (1961).

A comparison of the Kern, Bell–Delaware, and Tinker methods is given in Table 2.2.

2.94 Describe briefly the ϵ-NTU approach.

It is based upon the effectiveness (ϵ) of an exchanger to transfer heat as a function of its size called NTU (number of transfer units) and the ratio of the volumetric heat capacities (C_{min}/C_{max}) of the two fluids ($C = MC_p$). C_{min} refers to the fluid having the minimum value of MC_p, and C_{max} to that having the highest value of MC_p. The ratio C_{min}/C_{max} lies between 0 and 1. ϵ is the ratio of the larger temperature range and the difference between the inlet temperatures of the two fluids.

Table 2.2 Comparison of Kern, Bell–Delaware, and Tinker methods

	Kern	Bell–Delaware	Tinker
1. Ease of use	Simple	More involved	More involved
2. Resultant design	Very conservative	Not conservative	Not conservative
3. Cost of equipment	Very high	Relatively low	Relatively low
4. Does it account for			
a. Bypass and leakage streams	No	Yes	Yes
b. Inlet and outlet baffle spacing being different than the central one	No	Yes	Yes
c. Number of tube rows being different in inlet and outlet zones than in the center	No	Yes	No
d. Seal strips	No	Yes	Yes
e. Different tube layouts and baffle cuts	No (see Q. 2.90)	Yes	Yes
f. Effectiveness of tube rows in window by a separate calculation	No	Yes	Yes
g. Laminar flow	Original plots extended. Method remains unchanged	Δp_{lam} calculated differently	Method remains unchanged
h. Size of tube bundle	Assumes shell full of tubes	Accounts for it by number of tubes in cross-flow	Accounts for it by mean bundle width
i. Pressure drop in the nozzle	Yes	No	No
j. Δp due to gradual fouling of heat exchanger	Yes (takes an average working unit)	No	Yes
5. Basis of Reynolds number calculation	Equivalent diameter	Tube O.D.	Tube O.D.
6. Which mass velocity used	Cross-flow over the tube bundle	Geometric mean of cross-flow and window flow	Cross-flow mass velocity multiplied by a factor for tubes in baffle window
7. Multi-pass exchangers	All three were initially formulated for E-type shell. However, they can all be modified for use with any shell, and any number of passes		

Source: Gupta (1979).

$$\epsilon = \frac{(t_{out} - t_{in})_{max}}{T_{in} - t_{in}} = \frac{(T_{in} - T_{out})_{max}}{T_{in} - t_{in}} \tag{2.25a}$$

Some define it also as the ratio of the heat transferred to the thermodynamically maximum heat transfer possible in the unit

$$\epsilon = \frac{C_c(t_{in} - t_{out})}{C_{min}(T_{in} - t_{in})} = \frac{C_h(T_{in} - T_{out})}{C_{min}(T_{in} - t_{out})} \tag{2.25b}$$

where C_c is MC_p for the cold fluid and C_h is MC_p for the hot fluid.

The number of transfer units is related to the size of the exchanger and is equal to AU_d/C_{min}, where A is the exchanger area and U_d the overall design heat transfer coefficient.

A typical plot of ϵ-NTU is shown in Fig. 2.17.

2.95 What are the advantages of using the ϵ-NTU approach?

In situations where ΔT_{LMTD} cannot be calculated directly, the ΔT_{LMTD} approach requires trial-and-error. However, since no ΔT_{LMTD} and F_T are used in the ϵ-NTU approach, the latter gives the results without trial-and-error. It also shows directly the effects of varying mass flow rates on the effectiveness of the exchanger. It is a better method for transient analysis of heat exchanger performance. It finds the maximum use in designing compact heat exchangers.

2.96 What factors might cause an erroneous thermal design?

The thermal design can go wrong because of the following factors:

1. Some of the assumptions in the calculation of LMTD may not be valid for the particular case. For example, U_c may vary significantly between the inlet and the outlet, or phase change may not have been taken into account.
2. Use of wrong or misleading correlations.
3. Extrapolation of the correlations into inapplicable ranges.
4. Reverse temperature difference in part of the unit, i.e., the hot stream becoming cooler than the cold stream.
5. Neglect of fluid and thermal leakages across the longitudinal pass partition plate.
6. Use of wrong values for the physical properties of the fluids involved at the design temperature and pressure.
7. Fouling factor used being grossly wrong, i.e., grossly underestimated or overestimated.
8. Not accounted for the maldistribution of flow due to an inefficient combination of baffle cut, baffle spacing, shell-to-OTL clearance, tube-to-baffle hole clearance, low flow rate, etc.

EFFECTS OF VARIOUS GEOMETRICAL AND PROCESS PARAMETERS ON THE PERFORMANCE OF AN EXCHANGER

• We shall qualitatively discuss the effects of various geometrical and process parameters on the heat exchanger performance (on h_o, h_i, Δp_s, Δp_t). The parameters to be discussed are the following:

1. Type of shell
2. Number of shell-and-tube passes
3. Tube pitch, layout angles, and length
4. Baffle type, cut, and orientation
5. Laminar and turbulent flow

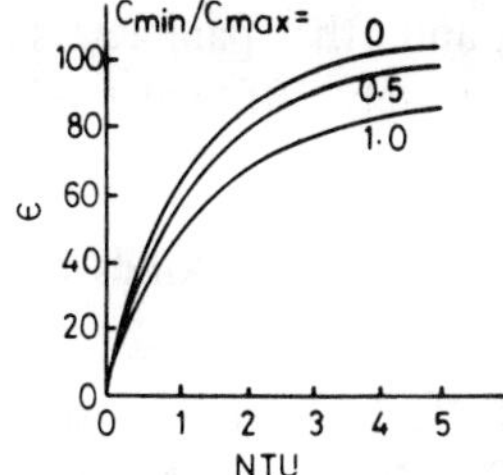

Figure 2.17 A typical ϵ-NTU plot.

6. Thermal and fluid leakages past the longitudinal baffle
7. Natural convection heat transfer
8. Fouling resistance

It will be assumed that, in general, when one parameter is varied, the rest of the geometrical and process parameters remain unchanged except as required for the sake of consistency. For our qualitative estimates, we will take the flow to be turbulent except in item 5 above and, further

$$h \propto v^{0.55} \quad \text{for shell side} \tag{2.26a}$$

$$h \propto v^{0.8} \quad \text{for tube side} \tag{2.26b}$$

$$\Delta p \propto Lv^2 \quad \text{for both the shell and tube sides} \tag{2.27}$$

where v is the flow velocity and L is the length of the flow path.

Effects of Type of Shell

2.97 How is the performance of an exchanger influenced by the type of shell?

The various types of shells are shown in Fig. 1.1. The comparison of h_i, h_o, Δp_t, and Δp_s is given in Table 2.3. It is for the minimum number of tube passes compatible with the type of the shell. The E-shell is taken as the basis. Numbers in other columns give a very rough estimate of the relative orders of magnitude. The actual numbers would be affected by the manufacturing clearances, type of rear

Table 2.3 Relative performance of different types of exchangers

Shell type	E	F	G	H	J
h_i	1	$2^{0.8}$	$2^{0.8}$	$2^{0.8}$	1
h_o	1	$2^{0.55}$	1	$(\frac{1}{2})^{0.55}$	$(\frac{1}{2})^{0.55}$
Δp_t	1	8	8	8	8
Δp_s	1	8	1	$\frac{1}{8}$	$\frac{1}{8}$
F_T	1	1	<1	<1	<1

Source: Gupta (1979).

head, return pressure loss in multi-pass straight tube bundles, and other parameters listed in the beginning of this section. The X-shell, having an entirely different kind of shell-side flow pattern, is not included.

2.98 What is the order of magnitude of the pressure drop in different kinds of shells?

This question further elaborates on some of the shells discussed in the previous question and includes X- and L-shells too. Taking E-shell as the basis for comparison:

1. The F-shell produces approximately 8 times the pressure drop as the E-shell, since the flow velocity and the flow path length are about double of those in an E-shell. Vertical baffle cut is assumed. Fluid leakage past the longitudinal baffle affects the heat transfer and pressure drop.
2. The G-shell produces approximately the same pressure drop as the E-shell. Although the flow area is halved, so is the volumetric flow rate, and hence the flow velocity remains about the same as in the E-shell. The net flow path length is also the same as in an E-shell. The F_T correction to LMTD is higher in G-shell compared to multi-tube pass E-shell.
3. The J-shell produces about an eighth of the pressure drop as in an E-shell because the flow velocity and the flow path length are both halved compared to the E-shell. F_T correction to LMTD results in a penalty on the heat transfer.
4. The X-shell cannot be compared directly with the E-shell. This is because the X-shell has flow distributed in each baffled zone, and the fluid does not pass from one baffled zone to the other but exits directly. There are no baffle windows present. A- and E-streams are not there either. The pressure drop is low. Stepwise analysis is required for an accurate prediction of the heat transfer.
5. The L-shell also cannot be compared directly with the E-shell. This is because there are no baffles and the shell-side fluid flows primarily parallel to the tubes all through. There are no A-, B-, and E-streams present. The pressure drop is rather low for this shell too. To prevent vibration damage, tube supports are used (Q. 1.73).

2.99 What precautions should be exercised in designing the J- and G-shells using U-tubes?

Usually it is assumed that the flow stream entering the J- or G-shell divides equally into two parts even if no full baffle is present in the middle of the entrance nozzle in a horizontal unit.

This is almost true in the straight tube bundles. However, in the U-tube bundles with flow over the U-bends, the pressure drop in this rear section will be less than that in the tubesheet end. This will result in an uneven split of the entering stream and hence in flow maldistribution. Therefore, in such cases, the U-bend region should be blanked off by a full baffle or a full baffle should be used in the middle of the entrance nozzle.

2.100 In a 1-1 counterflow exchanger mounted vertically, pressure drop for the

stream going upward is higher than it would be if the exchanger were mounted horizontally. Why?

It is because the fluid has to overcome the hydrostatic head equal to the effective height of the unit.

Effects of Number of Shell and Tube Passes

2.101 How is the performance of an exchanger influenced by the number of shell and tube passes?

The number of shell passes can be varied only for the E-shell and F-shell, while it is fixed for the G-, H-, and J-shells due to their construction features. The number of tube passes can be varied in all the shells though general practice limits them to lower values in G-, H-, and J-shells.

Increase in the number of shell and/or tube passes for a given type of shell results in an increase in the fluid velocity. The resultant increases in h_i, h_o, Δp_t, and Δp_s are governed by Eqs. (2.26) and (2.27).

2.102 What is meant by reheating in an exchanger?

In a multi-tube pass exchanger when a temperature cross occurs, the temperature of the cooler, tube-side fluid may become greater than that of the hot, shell-side fluid, resulting in a reverse heat transfer to the shell-side fluid. This is called reheating and is shown in Fig. 2.18 for a 1-2 exchanger.

Effects of Tube Pitch, Layout Angle, and Length

2.103 How is the performance of an exchanger affected by the tube pitch?

We shall assume that the required heat duty and the tube dimensions (outer diameter, thickness) are invariant.

When the tube pitch increases while the layout angle remains unchanged, the shell diameter increases and the shell-side velocity decreases with a resultant reduction in h_o and Δp_s [Eqs. (2.26) and (2.27)].

In order to compensate for the reduction in h_o, more surface area will need to be provided. This can be done in two ways: (1) By increasing the tube length: This will keep h_i the same while Δp_t will increase proportionately. It can be done if

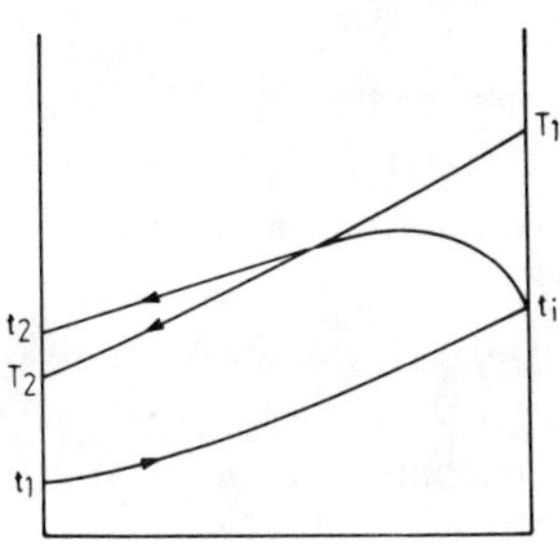

Figure 2.18 Temperature profiles when reheating occurs. *(After Kern, 1950.)*

the space exists to install and service a longer shell. (2) By increasing the number of tubes: This will further increase the shell diameter. Both the shell-side and the tube-side velocities will decrease and the values of all the terms, viz. h_i, h_o, Δp_t, and Δp_s will decrease.

2.104 What effects does the tube layout angle have on the performance of an exchanger?

The tube layout angle affects the shell-side flow area and velocity, number of tube rows in cross-flow, tube count in the exchanger of a given shell size, the heat transfer coefficient and pressure drop on the shell side if the same number of tubes are used in all the layouts, extent of fouling deposit and ease of its removal, etc. Thus, it influences the performance of the exchanger very significantly. In general, both h_o and Δp_s are the highest for a 30° layout and the least for a 90° layout angle.

2.105 Why is the triangular tube layout more effective in heat transfer compared to the square layout?

Figure 1.11 shows the triangular and the square layouts. As is evident, in the square layout, straight channels are formed, and fluid flows through them suffering some contraction and expansion between each tube row. The boundary layer on each tube is thick, and the heat transfer coefficient is low. On the other hand, in a triangular layout, the fluid not only undergoes contraction and expansion, but the fluid flowing between two tubes directly hits the tube below it, disturbing the boundary layer on the lower tube and giving a heat transfer coefficient that is approximately 25 percent higher than in the case of the square layout. Of course, the triangular layout also has a higher pressure drop. Recent studies show that significant mixing occurs in cross-flow of gases over a tube bundle with square layout due to transverse drift (Fallows et al., 1979; Jones et al., 1979). This will also adversely influence the heat transfer from the square layout.

2.106 What are the effects of doubling the straight tube length while keeping the heat transfer area and the volumetric flow rates the same?

The important effects are the following:

1. Reduction in the number of tubes to half the original number
2. Reduction in the shell diameter by approximately $\sqrt{2}$
3. Increase in the shell-side and tube-side velocities by a factor of 2
4. Increase in the shell-side and tube-side heat transfer coefficients by $2^{0.55}$ and $2^{0.8}$, respectively [Eq. (2.26)], and therefore more heat transfer will occur
5. Increase in the shell-side and tube-side pressure drops by approximately 8 times [Eq. (2.27)]
6. Decrease in the thickness of the shell, rear and front heads, flanges, covers, and tubesheet since the shell diameter has been reduced
7. Increase in the space needed to install and service the exchanger. Longer tools are needed to clean it.

Table 2.4 Relative performance of single-segmental, disk-and-doughnut, and orifice baffles

	Single-segmental[a]	Disk-and-doughnut[a]	Orifice[a]
Δp_s	3	1	~2
h_o	3	2	1
Fraction in parallel flow over the tubes	1	2	3
Vibration problem (Chapter 15)	3	2	1
Utility for alternate service	3	2	1
Overall size	1	2	3

Source: Gupta (1979).
[a]1 is the lowest relative value, while 3 is the highest.

Effects of Baffle Type, Cut, and Orientation

2.107 Compare a single-segmental baffle design with disk-and-doughnut and orifice baffle designs.

Relative comparison is shown in Table 2.4 for the shell side. A few extra terms are included since these are particularly influenced by the baffles. In the table, 1 means the lowest relative value, and 3 means the highest. The tube side is affected only when the number of tubes is increased to compensate for the reduction in h_o as discussed in Q. 2.103.

2.108 Compare a single-segmental baffle design with double-segmental, triple-segmental, and the no-tube-in-window baffle designs.

Relative comparison is given in Table 2.5. The lowest relative value is 1; 4 is the highest.

Table 2.5 Relative performance of single-segmental, double-segmental, triple-segmental, and no-tube-in-window baffles

	Single-segmental[a]	Double-segmental[a]	Triple-segmental[a]	No-tube-in-window[a]
Δp_s	3	2	1	~2[b]
h_o	3	2	1	4
Fraction in parallel flow over the tubes	2	3	4	1 (NIL)
Vibration problem (Chapter 15)	2	3	4	1
Utility for alternate service	4	2	1	3
Overall size	1	2	3	~2[b]

Source: Gupta (1979).
[a]1 is the lowest relative value, while 4 is the highest.
[b]It is difficult to judge accurately.

2.109 Compare a two-pass shell having double-segmental baffles with a one-pass shell having single-segmental baffles.

The pressure drop and heat transfer coefficient in a two-pass shell are higher than in a one-pass shell when the same baffle design is used in both, since the flow velocity in the former is twice that in the latter, assuming vertical orientation of the baffle cut in the two-pass shell. On the other hand, the pressure drop and heat transfer in a single-segmental baffle design are higher than in a double-segmental baffle design for equal number of shell passes for the same reason. It is evident then that the use of a two-pass shell with double-segmental baffle will be almost equal to a one-pass shell with single-segmental baffles. The latter is easier to fabricate and is more versatile in its applications.

2.110 In a two-pass shell, when are the double- and triple-segmental baffles used?

Such designs are used very rarely. It is because the two-pass shell is used to increase the heat transfer at an increased pressure drop, while the double- and triple-segmental baffles are used to reduce the pressure drop and hence the heat transfer is also low. Thus, the advantages of the two-pass shell are negated by the use of the double- and the triple-segmental baffles.

2.111 How does the baffle cut influence the performance of an exchanger?

As the baffle cut increases, the number of tube rows in cross-flow decreases and that in the parallel flow increases. The boundary layer is thicker in the parallel flow than in the cross-flow and hence h_o keeps decreasing with an increasing baffle cut. Also, there are less constrictions in the parallel flow compared to the cross-flow, and hence Δp_s decreases as the baffle cut increases.

2.112 How does the baffle cut orientation influence the performance of an exchanger?

In a single-shell pass involving no phase change, the baffle cut orientation makes no difference. If phase change is involved, vertical orientation is preferred for the unit to work effectively without the liquid damming behind the baffles (see also Q. 1.57).

2.113 In a 2-4 exchanger using single-segmental baffles, how is the shell-side flow affected by the horizontal and the vertical orientation of the baffle cut? Compare these with a 1-2 exchanger using identical baffle spacing and shell diameter.

Figure 2.19 shows the horizontal and vertical cut baffles in a 2-4 exchanger. In the horizontal cut case, the shell-side mass velocity at the center is the same as in a 1-2 exchanger. In the vertical cut case, the mass velocity is doubled since the cross-flow area is halved compared to the 1-2 exchanger. The pressure drop is approximately 8 times and the heat transfer coefficient is approximately 50 percent higher [Eqs. (2.26) and (2.27)] than for the horizontal cut case, or for the shell side of the 1-2 exchanger with either orientation.

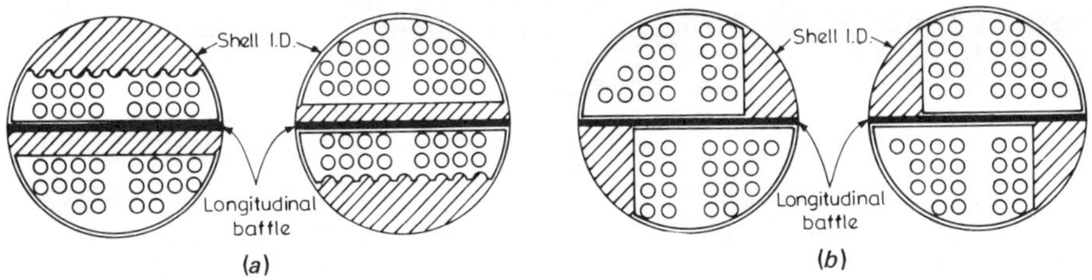

Figure 2.19 Segmental baffles in a 2-4 exchanger. (*a*) Horizontal cut. (*b*) Vertical cut. *(From Kern, 1950.)*

2.114 Name the situations in which the shell-side fluid flows parallel to the tubes (in contrast to the cross-flow).

Parallel flow occurs in the segmental baffle windows and in exchangers using orifice baffles, disk-and-doughnut baffles, rod baffles, and nest-type tube supports (see Q. 1.43–1.48). The leakage stream, A, is also parallel to the tubes (Q. 2.3). Parallel flow also occurs in L-type shells.

2.115 On what factors do the effective number of tube rows and the effective number of tubes in parallel flow in the different segmental baffle designs depend? Assume single-shell pass construction.

These depend upon the shell diameter, OTL, baffle cut (percent shell diameter), tube O.D., pitch, and layout angle or distance between two tube rows in the flow direction.

2.116 When is the longitudinal or parallel flow over the tubes in the segmental baffle design important?

It is always important in the triple-segmental baffles because the window area is up to 67 percent of the shell cross-sectional area. Hence, more fluid is likely to be in parallel flow than in cross-flow at any baffle location. In the double-segmental baffles, the window area can be up to 45 percent, and hence the contribution due to parallel flow should be taken into account, especially if the pressure drop in parallel flow is more than 10 percent of the total pressure drop. In the single-segmental baffles, the effects of parallel flow are usually negligible unless the window area is above 35 percent.

2.117 How does the heat transfer in cross-flow compare with that in parallel flow over the tube bundle?

It is more in the cross-flow since the boundary layer thickness is significantly less than that in the parallel flow.

2.118 How are the heat transfer and pressure drop in parallel flow over the tubes in the window calculated?

The relations as used in the tubular flow can be used with an appropriate value for the equivalent diameter.

Effects of Laminar and Turbulent Flow Regimes

2.119 How does the flow regime influence the performance of an exchanger?

The dependence of h_i, h_o, Δp_s, and Δp_t for the turbulent flow is given by Eqs. (2.26) and (2.27). For the fully developed laminar flow, we have

$$h = \text{constant in fully developed flow}^* \tag{2.28}$$

$$\Delta p \propto Lv^2 \tag{2.29}$$

where L is the tube length and v is the fluid velocity. We will assume that Eq. (2.28) holds both for h_i and h_o, and Eq. (2.29) both for Δp_s and Δp_t. As is evident from Eqs. (2.26)-(2.29), the heat transfer coefficients, h_i and h_o, and the pressure drop, Δp_s and Δp_t, are higher in the turbulent flow than in the laminar flow. Furthermore, the bypass and leakage streams (C, F, and E) are larger in the laminar flow. Hence, most designs are done in the turbulent flow regime. The transition flow region is ill-defined, and it, as well as the laminar flow, should be avoided in the shell-and-tube heat exchangers.

2.120 When is the laminar flow unavoidable?

When the stream sizes are very different, then the smaller of the two streams is likely to be in laminar flow. Further, very viscous liquids are sometimes in laminar flow since producing turbulent flow requires very high pressure drop. Such streams should generally be placed on the shell side, since the turbulence there starts at a lower Reynolds number due to the changing cross sections. However, one should be aware of possible maldistribution on the shell side (see Q. 2.171 also). On the other hand, if the fluid is corrosive, or if sufficient pressure drop is available, it may have to be placed on the tube side (see also Q. 2.19).

2.121 Under what conditions would the flow change from laminar to turbulent (or vice versa)? How would you design such a unit?

The flow regime depends upon the Reynolds number

$$N_{Re} = \frac{\rho v d}{\mu} = \frac{Gd}{\mu} \tag{2.30}$$

Hence, N_{Re} can change if any of the above terms changes. Usually d, the characteristic length, is constant for a given exchanger. G, the mass flux, can change if there is a phase change as in condensing or boiling. A vapor stream entering the condenser may be in turbulent flow, but as it condenses, the condensate may be in laminar flow. The reverse can happen in the case of boiling. Furthermore, μ, the viscosity, can change significantly due to phase change and/or a large change in the temperature and lead to a change in the flow regime, especially if the Reynolds

*Nusselt No. (hD/k) equals 3.66 for constant temperature at the tube wall as in pure vapor condensers, and equals 4.36 for constant heat flux at the tube wall as in equal volumetric heat capacity flow rate case.

number at the entrance is close to the limit between the laminar and turbulent flows.

Stepwise calculations have to be done in such a case by dividing the total length of the exchanger into a number of short segments.

2.122 Do most correlations for heat transfer on the shell side account for the laminar region also?

Usually not. However, the designers have generally extended them into the laminar region without any apparently wrong results, though large overdesigns have sometimes occurred. It could be due to the safety factor built into these correlations, which are all empirical in nature.

2.123 If the transition region cannot be avoided, how should the heat transfer coefficient and the pressure drop be calculated?

There are no hard and fast rules. One could

1. Extend the relations for the turbulent region into the transition one
2. Extend the relations for the laminar region into the transition one
3. Take a weighted average of the two values obtained in (1) and (2) above

2.124 Since roughness induces turbulence, which aids in heat transfer, why are only commercially smooth tubes used in exchangers?

Rough surface does aid in heat transfer by inducing turbulence and cutting down the thickness of the boundary layer. However, deposition of the scale will soon even out the rough surface, and the advantage to heat transfer would vanish. The rough surface with its peaks and valleys entraps a layer of liquid, resulting in a faster deposit of the scale. Furthermore, a rough surface is more prone to corrosion damage. Hence, the smooth tubes are preferred. On the other hand, for clean fluids, one could conceivably use rough surfaces for heat transfer augmentation (Zukauskas and Ulinskas, 1981).

2.125 Why is the heat transfer coefficient greater in turbulent flow than in the laminar flow?

In the laminar flow, the heat transfer in the radial direction is by conduction alone, whereas in the turbulent flow, it is both by conduction and turbulent motion of the fluid. Actually, at high Reynolds number, the heat transfer by turbulent motion is significantly more than that by pure conduction. This result can also be derived by an analogy between the momentum and heat transfer (Bird et al., 1960).

2.126 What are the various ways in which heat transfer may be increased?

Heat transfer may be increased by the following:

1. Increasing velocity
2. Promoting turbulence by using turbulence promotors (Evans and Churchill, 1962).

3. Using extended surfaces (Q. 1.16)
4. Using enhanced surfaces (Q. 1.21)

- Turbulence promotors, extended surfaces, and enhanced surfaces are discussed in Chapter 4.

2.127 What is flow stratification?

When a fluid stream is subjected to large temperature changes and the stream velocity is low (say, under 2 ft/s or 0.6 m/s) the whole of the fluid may not undergo the same temperature change. This produces a difference in density, and the colder liquid, being heavier, might flow along the bottom and not participate actively in the heat transfer. This affects the performance of the exchanger and should be avoided. Such a situation can occur in the laminar flow of viscous liquids through the shell. It is also encountered in large diameter steam lines during warm-up when flow velocity is kept low to avoid sudden large temperature changes and thermal stresses (Machacek and Zelenka, 1979).

2.128 Can the friction factor correlation used in hydraulics for flow through conduits be used for the tubular flow in heat exchangers?

No, since the former is for the isothermal conditions while there is heat transfer across the heat exchanger tubes. The friction factor under the non-isothermal conditions is different than that under the isothermal ones. All the design methods mentioned in Q. 2.87 have their own friction factor plots. As a fair approximation one can also use

$$f_{\text{nonisothermal}} = f_{\text{isothermal}} \left(\frac{\mu}{\mu_{\text{w}}} \right)^{0.14}$$

where f = friction factor
μ = fluid viscosity at bulk temperature
μ_{w} = fluid viscosity at tube wall temperature

2.129 For the same Reynolds number, is the friction factor higher for the shell side or the tube side?

The friction factor is more for the shell side by an order of magnitude because of the flow over the tube bundle, acceleration, deceleration, changing directions, etc.

2.130 Does the pressure drop on the two sides also follow the same order as the friction factor for equal Reynolds number?

Not necessarily so since the pressure drop involves the velocity, equivalent diameter, number of baffles, fluid density, etc. It could, therefore, be lower for the shell side than for the tube side for the same Reynolds number.

2.131 What effects does the fluid velocity have in a heat exchanger?

Fluid velocity has multiple effects in a heat exchanger:

1. It plays an important role in determining the heat transfer coefficient and pressure drop. Both increase with increasing velocity [Eqs. (2.26)-(2.29)]. The heat transfer coefficient along with ΔT_{LMTD} determines the exchanger area [Eq. (2.1)]. The calculated pressure drop, in comparison with the allowable value, influences the kind of shell and the number of shell and tube passes.
2. It affects the fouling deposit, since higher velocity gives less time for the scale to deposit and for any suspended material to settle down. Furthermore, it provides high shear that removes some scale deposit, thus reducing the overall fouling resistance.
3. High velocity of the shell-side fluid may result in tube erosion and damaging vibrations. Ways to counter these include impingement plate and flared nozzles. These have been discussed in Q. 1.50. Vibrations are discussed in Chapter 15.
4. High velocity of the tube-side fluid may result in tube end erosion. Smooth entry into the tubes reduces the erosion damage. Erosion is discussed in Chapter 13.

2.132 How should the fluid velocity be chosen?

The fluid velocity should be the highest possible within the allowable pressure drop to obtain a high heat transfer coefficient and a low fouling deposit. However, it should not result in the vibration or erosion damage to the tubes.

2.133 What factors influence the choice of velocity for a given mass flow of a gas on the shell side?

The factors are the molecular weight of the gas, pressure and the allowable pressure drop, vibrations of the tubes due to cross-flow, acoustic vibrations, presence of suspended droplets or particles that might cause erosion, and partial or complete condensation inside the shell, etc.

2.134 For equal mass velocities of a gas and a liquid, comment on the relative magnitudes of the Reynolds number and the heat transfer coefficient.

The Reynolds number for the gas flow is very high compared to that for the liquid flow for equal mass velocities, since the viscosity of the gases is a tenth or even less than that of the liquids. Hence, the Nusselt number is higher for the gaseous flow, neglecting the effects of the Prandtl number. However, the heat transfer coefficient is lower, since the thermal conductivity of the gases is one-twentieth or even less compared to that of the liquids.

2.135 How are the shell-side and tube-side velocities increased? How is pressure drop affected by an increase in velocity?

The shell-side velocity may be increased by the following:

1. Increasing the number of passes on the shell side
2. Decreasing the baffle spacing
3. Using single-segmental baffles instead of double- and triple-segmental ones

The tube-side velocity may be increased by increasing the number of tube-side passes. Usually even numbers of tube passes are used.

Increasing the velocity results in an increase in the pressure drop as given by Eqs. (2.27) and (2.29).

Effects of Thermal and Fluid Leakages Past the Longitudinal Baffle

2.136 How do the thermal and fluid leakages past the longitudinal baffle influence the performance of the exchangers?

The reasons for the thermal and fluid leakages have been discussed in Q. 1.49. They affect both the log mean temperature difference and the amount of fluid in cross-flow, resulting in a lowering of the values of both the h_o and Δp_s. The accurate prediction of the leakages and their effects is difficult, since the temperature and pressure keep changing along the entire length of the shell on both sides of the longitudinal baffle. Thus, the performance of the unit may become erratic. Of course, in a longitudinal baffle welded to the shell, there is no fluid leakage (except possibly some near the tubesheet) and the detrimental effects on h_o of only the thermal leakage are very small (see also Q. 2.33).

Effects of Natural Convection Heat Transfer

2.137 How does natural convection affect the heat transfer to fluid flowing through a tube?

When the fluid flows in a turbulent fashion, as is most often the case, the natural convection has negligible effect, since the fluid is very thoroughly mixed. However, in laminar flow, the heat transfer to an inner layer of fluid occurs by conduction from an outer layer that is closer to the tube surface. The superimposition of the natural convection over the laminar flow (when viscosity is not very high) increases the heat transfer coefficient by as much as 300 percent (Kern, 1950). Neglecting the natural convection effects results in a rather conservative value for the heat transfer coefficient.

2.138 What factors influence the heat transfer coefficient for natural convection in a pipe or tube?

The dependence of the natural convection coefficient is functionally given by

$$\frac{h_i D}{k} = C_1 \left(\frac{\beta g \rho^2 d^3 \ \Delta T}{\mu^2}\right)^a \left(\frac{c\mu}{k}\right)^b \tag{2.31}$$

where the term in the first set of parentheses on the right-hand side is the Grashof number and the term in the second set of parentheses is the Prandtl number; β is the coefficient of volumetric expansion of the fluid.

2.139 When is the free convection on the outside of a horizontal pipe encountered industrially?

The free convection on the outside of a horizontal pipe is encountered in the following:

1. Immersion heaters or tank heaters used to keep a stored liquid warm
2. Around steam pipes in process industries
3. Around pipes carrying hot flue gases from furnaces in metallurgical industries

2.140 How does the free convection outside a pipe differ from the natural convection inside a pipe?

The free convection currents outside a pipe are usually not restricted by the immediate surroundings and hence are free to rise high above the pipe. On the other hand, the natural convection currents inside a pipe are obviously contained within the pipe.

Furthermore, a pipe carrying a hot fluid under turbulent flow conditions will not have any appreciable natural convection currents within the pipe, but may have significant free convection currents outside the pipe if the surroundings are quiet.

2.141 Where do the thermal radiations play an important role in the heat transfer equipment?

Some such situations are the following:

1. Boiler furnace tube arrays
2. High temperature heat exchangers
3. Furnaces in the metallurgical and chemical industries

Effects of Fouling Resistance

2.142 Is the effect of the fouling resistance, R_f, the same regardless of the value of the overall clean heat transfer coefficient, U_c?

No, the effect varies with the value of U_c. Equations (2.3) and (2.4) can be written together as

$$\frac{1}{U_d} = \frac{1}{U_c} + R_f \tag{2.32}$$

where R_f includes the fouling resistance on the inside as well as on the outside of the tube. The percent change in U_c due to R_f is

$$\frac{U_c - U_d}{U_c} \times 100 \tag{2.33}$$

On substitution for U_d for Eq. (2.32), Eq. (2.33) becomes

$$\frac{100\,R_f U_c}{1 + R_f U_c} \tag{2.34}$$

For a constant R_f, the percent change in U_c increases with increasing U_c. For example, consider two cases with $U_c = 400$ and 60. For a constant $R_f = 0.003$, the change in the two cases is, respectively, 54.34 and 15.25 percent.

2.143 Comment on the statement: "Assumption of an excessive fouling can be a self-fulfilling prophecy." In other words, with fouling, you get what you ask for.

This is true when very conservative values are used for the fouling resistance. These result in an excessive overdesign, and the unit becomes very large. The resultant fluid velocity is greatly reduced. With low fluid velocity, there is heavy fouling deposit resulting in a high fouling resistance (Chapter 13). Thus, the statement of a self-fulfilling prophecy is true.

2.144 A clean heat exchanger has a higher heat transfer coefficient than the design value based on fouled tubes. It will, therefore, deliver a larger heat load, which might upset the downstream process conditions. How should such a situation be tackled?

The following can be done in such a case:

1. Bypass part of the stream until the design conditions are obtained in the exchanger, then send the total stream through the exchanger. If operations permit, the flow rate of one or both of the streams may be only gradually increased so that the process downstream is not upset. This would circumvent bypassing part of a stream.
2. Install another unit downstream to compensate for the changed conditions of the fluid and to bring them back into line. When the main unit reaches the design conditions, take the downstream unit out of operation. However, no one would like to incur this extra capital expenditure on an extra unit. Hence item (1) is preferable.

ALLOWABLE PRESSURE DROP

2.145 What is meant by allowable (or permissible) pressure drop?

It is the maximum value of the static fluid pressure drop that can be expended to drive the fluid through the exchanger to achieve the required heat transfer. It implies that provision exists in the pumping capacity of the plant to supply this pressure drop in the particular exchanger. The allowable pressure drops for the shell-side and tube-side fluids are specified separately and are independent of one another even though the same value is assigned to both in a large number of designs.

2.146 How does the allowable pressure drop affect the process design of the heat exchangers?

As discussed in Q. 2.132 and given by Eqs. (2.27) and (2.29), the allowable pressure drop determines the fluid velocity. Since the heat transfer coefficient is related to this velocity [Eq. (2.26)], the allowable pressure drop plays a very important role in the process design of the heat exchangers.

2.147 Does the allowable pressure drop affect the mechanical design of the heat exchangers?

No. The mechanical design is based on the absolute pressures of the entering streams and not on the pressure drop. Although the fluid pressure changes as the fluids pass through the exchanger, the change is usually small compared to the entering fluid pressure. To account for the changing pressure, a shell of decreasing thickness will have to be made which will be uneconomical and contrary to the Code requirements.

2.148 How should the allowable pressure drop be decided?

A high value of the allowable pressure drop results in a higher flow velocity and hence a more compact unit with lesser capital investment but higher operating costs. A low value of the allowable pressure drop has just the reverse consequences. Hence, an economic balance may be done between the capital and the operating costs. Usually, such planning can be done more easily in a new plant being set up than in an existing plant, since the pressure of the streams exiting from an exchanger may influence the design of the units downstream. Hence, a total integral approach for new plants should be adopted.

Sometimes it may be better to let the bidder decide the pressure drop within a certain range to permit the design of a cost effective unit. Also, to reduce the incidence of wasteful overdesign by very conservative methods, one may consider specifying both upper and lower pressure drop limits (Q. 1.122).

2.149 Some design methods predict a significantly higher pressure drop than the actual value encountered during operation. What are its disadvantages?

Both the capital investment in the exchanger and pumps and the operating costs of the pumping system increase unnecessarily. Usefulness of the optimization procedures where such design methods are used to evaluate the performance is rather limited.

2.150 What factors contribute to the shell-side pressure drop?

The shell-side pressure drop is made up of the pressure losses due to the cross-flow over the tube bundle, the bypass streams, C and F, turn around and parallel flow in the baffle window, and losses in the entrance and exit zones. The leakage streams, A and E, suffer lesser pressure drop compared to the other streams.

2.151 What factors contribute to the tube-side pressure drop?

The tube-side pressure drop is made up of the pressure losses due to flow through the tubes, entrance and exit losses, and mixing and turn around in the end region in a multi-tube pass unit.

2.152 What steps should be taken if the calculated pressure drop is less than the allowable pressure drop or more than the allowable pressure drop?

When the calculated pressure drop is less than the allowable value, one or more of the following should be done to make a fuller use of the allowable pressure drop and to get more heat transfer or smaller exchanger for the same duty:

1. Increase the tube length
2. Decrease the tube pitch and/or change the layout angle
3. Decrease the tube diameter thus accommodating more tubes in a given sized shell

When the calculated pressure drop is more than the allowable one, one or more of the following should be done:

1. See if the customer would increase the allowable pressure drop
2. Decrease the tube length
3. Increase the tube pitch and/or change the layout angle
4. Use double- or triple-segmental or no-tube-in-window baffle designs

2.153 If the calculated pressure drop for a fluid is higher than the permissible value, why is part of the fluid not bypassed and mixed with the rest downstream of the exchanger?

Bypassing will result in a reduced flow through the exchanger. As Eq. (2.26) shows, the heat transfer coefficient for that stream will reduce, and so will U_c [Eq. (2.4)]. Furthermore, the fractional stream through the exchanger will have to undergo a larger temperature change so that on mixing with the bypass fraction downstream of the exchanger, the required final temperature is achieved. This will result in a closer temperature approach at one end, reducing the ΔT_{LMTD}. Thus, both U_c and ΔT_{LMTD} will reduce if part of a stream is bypassed. The equipment will not deliver the required heat load. Hence, bypassing is not generally practiced for limiting the pressure drop.

2.154 If the permissible shell-side pressure drop is very small, how should the heat exchanger be constructed? How are the shell-side heat transfer coefficient and pressure drop calculated in such a case?

In such a case, baffles may have to be dispensed with (L-type shell, Q. 2.98). Instead, tube support plates with almost 50 percent cut may be used to support the tube bundle (Fig. 2.20). One or two tube rows in the center are supported by all the support plates. These half-circles provide protection against sagging and vibrations but do not contribute significantly to pressure drop. The fluid is taken to be in parallel flow instead of cross-flow over the tube bundle. The heat transfer coefficient and pressure drop on the shell side are calculated by using an equivalent

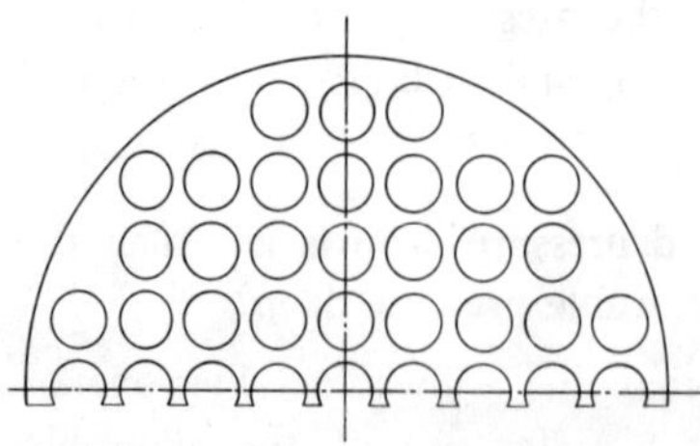

Figure 2.20 Tube support plate.

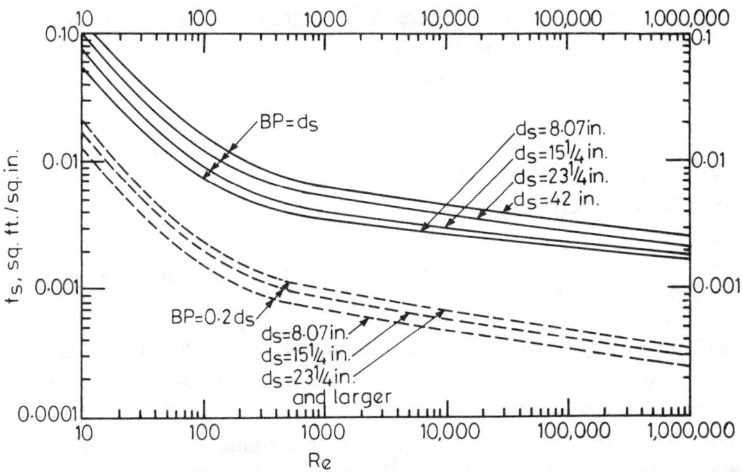

Figure 2.21 Influence of baffle spacing and shell diameter on shell-side fraction factor. *(Source unknown.)*

diameter for the shell side and the equations for the tubular flow. The other design could be the X-type shell discussed in Q. 2.98.

2.155 If there is no limit on the allowable pressure drop, what would determine the size of an exchanger?

Theoretically, the smallest shell diameter, smallest tube diameter, smallest pitch, and longest tubes should be used. In practice however, the Codes and TEMA Standards have to be satisfied, clearance for the cleaning of the tubes on the outside, available floor space (or head space for vertical units) to replace a tube or put in a tube bundle, guard against erosion and vibration damage due to very high fluid velocities, availability of the various parts and tubes in the required sizes, etc., have to be considered.

2.156 In practice, it is found that the shell-side friction factor depends on the ratio of the baffle spacing to the shell diameter as well as on the shell diameter. Why?

See Fig. 2.21. The decrease in the friction factor (and hence in the pressure drop) with decreasing shell diameter and decreasing ratio of the baffle spacing to the shell diameter (closer baffle spacing) is because of the increasing fractions of the A-, C-, and E-streams.

2.157 Why is caution suggested for the shell-side pressure drop calculations in equipment operating under vacuum?

All the correlations for the shell-side pressure drop in use have been obtained based upon experiments done at above atmospheric pressures. No significant experiments have been done under vacuum, and hence a detailed comparison between the vacuum operation and the applicability of the above-atmospheric pressure drop correlations has not been done. A unit designed for a high internal pressure may not hold under a much smaller external pressure (or internal vacuum). Hence, caution is needed in the design of the units operating under vacuum.

USES OF LIQUID–LIQUID, LIQUID–GAS, AND GAS–GAS EXCHANGERS

2.158 What are the different situations that can arise in a no-phase-change heat exchanger?

One or both of the fluids can be a gas, a liquid, or a two-phase mixture. However, situations involving a two-phase mixture, such as water–air, which does not undergo any phase change in the exchanger, are encountered less frequently. Our discussion shall, therefore, be limited to a single-phase fluid.

Thus, we can have exchangers having both fluids in the form of liquid (liquid–liquid exchangers), or in the form of gas (gas–gas exchangers), or one liquid and the other gas (liquid–gas exchangers).

2.159 Where are liquid–liquid heat exchangers used?

These are the most common variety used in almost all chemical plants, fertilizer and petrochemical complexes, petroleum refineries, and power plants. These are used to preheat the feed, cool a stream, or transfer heat from one process stream to the other.

2.160 Where are liquid–gas heat exchangers used?

These form a very large fraction of the exchangers used in industry. Among the important applications are intercoolers and aftercoolers in compressors and gas turbines, refrigeration, air conditioning, etc. The heat transfer coefficient on the gas side is very low compared to that on the liquid side, hence the former is enhanced by using finned surfaces (Chapter 4).

There is another important use of gas–liquid exchangers in condensers where whole or part of the vapor stream condenses. These are used atop the distillation towers to condense the top product and in power plants to condense the exhaust steam. However, here the gas-side heat transfer coefficient is not as low as in the noncondensing case. In fact, for pure steam condensing, it is an order of magnitude higher than that for the liquid side. Hence, the finned surfaces are not commonly used in condensers.

Air coolers, discussed in Chapter 4, form another class of liquid–gas heat exchangers.

2.161 Where are gas–gas heat exchangers used?

These are primarily used as preheaters in power and other process plants where air or other gases are needed at a high temperature for combustion or other purposes. These can be tubular exchangers with or without fins and regenerators with different kinds of matrices that give large surface area and have low pressure drop. Some gas-gas exchangers are also used as intercoolers and aftercoolers in compressors.

2.162 How does a gas-gas heat exchanger differ from a liquid-liquid one?

Since the heat transfer coefficient is less for the gaseous streams, the area required for the same amount of heat transfer is significantly higher for a gas-gas unit than for a liquid-liquid one. Moreover, very large amounts of heat are extracted from gases only at high temperatures for which regenerators are used, while the same amounts of heat can be extracted from liquids at lower temperatures because of their high volumetric heat capacity.

Often a gas-gas exchanger may be less ruggedly built than a liquid-liquid one because some interleakage may be tolerated in the former.

COOLERS AND COOLING WATER

2.163 What are coolers?

These generally use water to cool a process stream. One can find a number of such coolers in any large chemical plant.

2.164 Comment on the use of water as a heat transfer medium.

Water has a number of favorable and unfavorable points. The favorable points are the following:

1. It is available in abundance at a low cost.
2. Its heat capacity, thermal conductivity, and the latent heat of vaporization are higher than those of most other liquids. Thus, steam can be used to transfer heat to a process stream or water can be used to cool it.

The unfavorable points are the following:

1. It is corrosive to steel, particularly at high temperatures.
2. For heat transfer at temperatures above 100°C (212°F) high-pressure steam is needed, which adds to the capital cost.
3. It has dissolved salts and hence requires pretreatment for most applications, otherwise high fouling might occur.
4. It has dissolved air, which needs to be removed prior to steam generation otherwise exhaust steam condensation will be difficult and the condensate will be corrosive.
5. Above 120°F (49°C), microbial growth increases in water, which leads to excessive fouling. Hence, the operating temperature of water has to be kept below 120°F (49°C).

2.165 Comment on the use of air as a cooling medium.

Air is used under the following conditions:

1. Water is not available or is very costly.
2. Temperature of the process stream is so high that if only water is used as a coolant, its temperature will rise above 120°F (49°C) resulting in a rapid microbial growth, or else excessive quantity of water would be needed at increased costs of water and pumping power.
3. In closed loop cooling systems, cooling towers are used to cool the warm water before recirculation. Thus, the eventual heat sink, even in water cooling, is air.

The disadvantages of using air are the following:

1. Large-sized units are required since the air-side heat transfer coefficient is low and controlling.
2. Air distribution over large units does become a problem.
3. Energy taken up by the air cannot generally be reclaimed economically since the volume of air is large and its temperature rise is small.
4. The system is noisy.

- Air coolers are discussed at length in Chapter 4.

2.166 Why is the cooling water outlet temperature generally restricted to a maximum of about 120°F (49°C)?

The tendency of water to corrode the materials and the growth rate of the microbial fouling increase above 120°F (49°C). Further, there will be increased evaporation losses in the cooling tower used to cool this water if it is to be recycled. If the hot water is entering a river stream, it will adversely affect the marine life. Hence, for all these reasons, the cooling water outlet temperature is generally restricted to 120°F (49°C).

2.167 Why is the cooling water recycled?

There are multiple reasons for recycling the cooling water:

1. Since the cooling water is generally treated to reduce sedimentation and fouling, the treatment cost will be very high if fresh water is to be used always. Recycling reduces such costs to a minimum.
2. Municipal tax on fresh water is reduced since the fresh water is required only to the extent of the make up water to account for the evaporation losses in cooling towers and any leakage or purge stream.
3. In some localities, a sewerage tax is charged proportional to the fresh water usage. Hence, it also reduces if the cooling water is recycled.
4. Local regulations may require that some of the treating chemicals put in the cooling water be removed/neutralized before this water is discharged into a stream, in order to minimize the influence on marine and human lives downstream. Such treatment costs are minimal if water is recycled.

2.168 How is the heat of the cooling water removed before recycling it?

The heat is removed in cooling towers. In some cases, heat pumps or heat pipes are used to cool the cooling water and utilize the energy so released elsewhere.

2.169 What additional arrangements should be made to cool the water if it is above 120°F (49°C) so that the evaporation loss in the cooling tower is low?

It can first be cooled in an air-fin cooler (Chapter 4) to a temperature below 120°F (49°C) and then in a cooling tower.

2.170 How does the temperature of the cooling water affect the design and operation of the chemical and power plants?

The temperature of the cooling water fixes the operating pressure in the condensers atop the distillation columns and hence the operating conditions in the columns and the reboilers, too. It also affects the temperature driving potential and hence the size of the condensers and coolers used in a plant. In a power plant, it fixes the pressure in the surface condensers and hence the discharge pressure of the turbine. This influences the efficiency of the turbine. Thus, the temperature of the cooling water plays a very important role.

2.171 What caution should be exercised in designing coolers for viscous flows?

It is possible that due to the maldistribution of flow, the flow rate and temperature in some tubes will become less than those in the other tubes. This phenomena feeds on itself since at low temperatures (hence high viscosity), the fluid will slow down further, losing more heat and hence a still lower flow rate will result. The reverse will happen in the tubes having higher flow rates. Thus, the resultant cooling will be less than desired. Hence, steps should be taken, both at the design stage and during operation, to minimize the fluid maldistribution. Maintaining a minimum pressure drop helps in avoiding maldistribution (Mueller, 1974).

2.172 What are film coolers or cascade drip coolers, and where are they used?

As the name implies, the coolant flows as film over the tubes/pipes and drips from one pipe on to the other in a vertical bank of horizontal pipes through which the process fluid flows. These are very inexpensive. Cast iron ones are used in sulfuric acid cooling, and graphite impregnated ones are used in hydrochloric acid cooling. Such pipes are also placed in cooling tanks or in the basin of the cooling tower where closed cycle cooling water is used.

2.173 What is centralized or closed loop cooling?

Water, used for cooling at various locations in a plant, is brought to a central place where its heat is removed and it is recirculated to do the cooling duties (Cooper et al., 1979). The heat removal at the central place can be done by a cooling tower, air-cooled heat exchanger, or heat exchange with raw water. In the last instance only the central heat exchanger has to face the corrosion and fouling due to raw water. If it is sea water, the heat exchanger has to be of titanium. The rest of the exchangers in the plant can be of cheaper materials.

2.174 What are intercoolers and aftercoolers?

When air or other gases are compressed to high pressures, it is done in multiple stages of compression. The compression can generally be taken as adiabatic.

The temperature of the gas rises after each stage according to the relation

$$\frac{T_2}{T_1} = \left(\frac{P_2}{P_1}\right)^{(\gamma-1)/\gamma} \tag{2.35}$$

where T = temperature
P = pressure
$\gamma = C_p/C_v$
C_p = heat capacity at constant pressure
C_v = heat capacity at constant volume
1, 2 = stages 1 and 2

This temperature must be brought down to close to the ambient value or else the compressor material and seals will not withstand the increasing temperature in subsequent stages. Furthermore, cooling also reduces the power required in subsequent stages. This is achieved by intercoolers, which are shell-and-tube exchangers using water or other fluid as a coolant. The unit used after the last stage of compression is called an aftercooler.

2.175 Why is it important to have very low pressure drop in an intercooler?

Any pressure drop in the gas stream in an intercooler affects the final delivery pressure after the last stage. For example, in a four-stage compression with a compression ratio of 2.5 in each stage, 1 psi (=6895 Pa) drop in the first intercooler after the first stage will result in a loss of 1 X 2.5 X 2.5 X 2.5 = 15.6 psi (=108 kPa) at the end of the fourth stage. To this must be added the cumulative effect of the pressure drop in each of the intercoolers and in the aftercooler. Thus, for the four-stage compression with 1 psi (=6895 Pa) drop in each of the intercoolers and in the aftercooler, the total loss in the final delivery pressure would be 1 X $(2.5)^3$ + 1 X $(2.5)^2$ + 1 X 2.5 + 1 = 25375 psi (=175 kPa). Hence, the pressure drop in the intercoolers and in the aftercooler should be minimized.

CLOSED FEEDWATER HEATERS

2.176 What are closed feedwater heaters?

These are shell-and-tube heat exchangers that heat up the boiler feed water in the thermal and nuclear power plants before it enters the boiler section. This helps in increasing the thermal efficiency of the plant, since heating is usually done by extracting medium- and low-pressure condensing steam from the turbine section. These are called closed feedwater heaters because the same water comes back (with a little make up water) to the heaters after passing through the boilers, where steam is produced that generates power in the turbine section and which is then condensed in the surface condensers (Chapter 3). In open feedwater heaters, fresh water is used all through. These are generally used when steam is used up in the plant and there is no condensate available for recycling.

2.177 What are the various classes of closed feedwater heaters?

There are two classes: Low-pressure ones installed before the boiler feed pump, and high-pressure ones installed after the pump, respectively, named as l-p and h-p units. The shell-side pressure can be from vacuum to several hundred psi in the l-p units and up to over 1200 psi (8270 kPa) in the h-p units. The steam condenses in the shell, while the feedwater flows through the tubes. The tube-side pressure may exceed 5000 psi (34.5 MPa).

2.178 What are the various heat transfer zones in a feedwater heater?

Since steam is used to heat the feedwater, the steam may be superheated or saturated, and the steam condensate may be saturated or subcooled. Depending upon the conditions, therefore, the feedwater heater may have three zones (desuperheating, condensing, and subcooling), two zones (desuperheating and condensing or condensing and subcooling), or only one zone (condensing only).

2.179 What are the general tube-side construction features of the closed feedwater heaters?

In an overwhelming majority of cases, U-tubes are used on a triangular, 30° layout because the U-tubes can more easily take the differential expansion without undergoing severe thermal stresses, and there is no cleaning required on the outside since steam condenses there.

2.180 On what factors does the successful operation of a feedwater heater depend?

The most important operational parameters are the following:

1. The feedwater treatment, i.e., adjusting the pH and reducing the dissolved oxygen content to reduce corrosion
2. Venting of the noncondensable gases so that the heat transfer surface is not blocked during condensation of steam
3. Draining of the condensate to maintain a definite level in the heater to achieve the desired subcooling of the condensate without decreasing the effective heat transfer surface

DOUBLE PIPE HEAT EXCHANGERS

2.181 Describe a double pipe heat exchanger.

It has two concentric pipes of diameters d_1 and d_2 ($d_1 < d_2$), usually in the form of a hair-pin (Fig. 2.22).

2.182 Where is it used?

It is used for low heat duties requiring surface areas up to about 500 ft^2 (47 m^2). A number of double pipe heat exchangers can be connected in series or parallel as necessary. It has the advantages of flexibility since units can be added or removed as required, and it is easy to service and requires low inventory of spares because of its standardization.

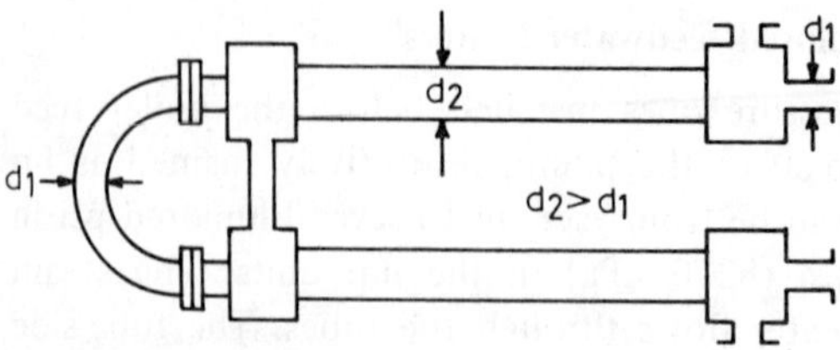

Figure 2.22 Double pipe heat exchanger.

2.183 What is the flow pattern like?

It is counterflow parallel to the pipe axis. There is no cross-flow, since there are no baffles used in it.

2.184 What factors contribute toward the total pressure drop in a double pipe exchanger?

The total pressure drop consists of the pressure drop in the exchanger, connecting piping, fittings, control regulators, and the static head of the receiver, if any.

2.185 How are they connected in series?

Both the fluids travel the total lengths of all the units (Fig. 2.23).

2.186 How are they connected in parallel?

While one fluid travels the total length of all the units, the other is distributed at different levels (Fig. 2.24). It is actually a series-parallel arrangement.

2.187 What is the difference between the equivalent diameter for the annulus of a double pipe heat exchanger for heat transfer and for pressure drop calculations?

$$\text{Equivalent diameter} = \frac{4 \times \text{cross-sectional flow area}}{\text{wetted perimeter of the surface of interest}} \tag{2.36}$$

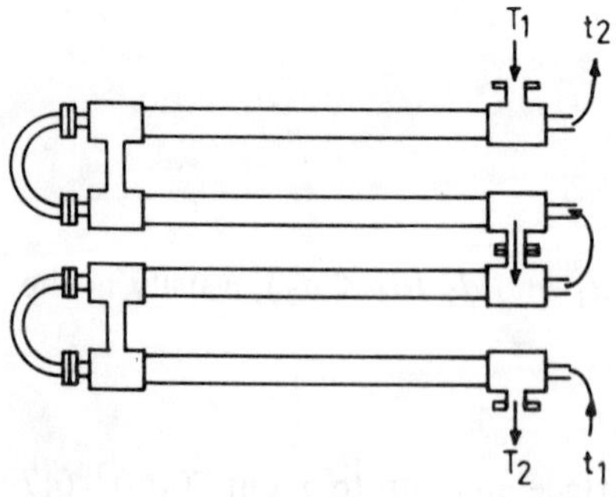

Figure 2.23 Two double pipe heat exchangers connected in series. *(From Kern, 1950.)*

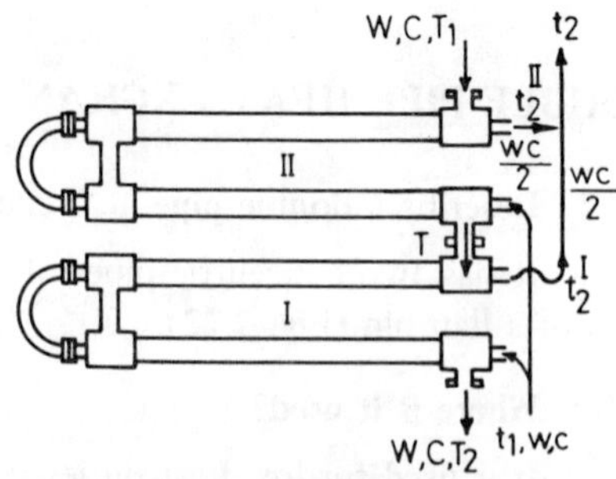

Figure 2.24 Two double pipe heat exchangers connected in series-parallel. *(From Kern, 1950.)*

While the cross-sectional area is the same for both the cases, the wetted perimeter for the heat transfer is only πd_1, while that for the pressure drop is $\pi(d_1 + d_2)$, where d_1 is the O.D. of the inner pipe and d_2 is the I.D. of the outer pipe. Thus, the equivalent diameter for the heat transfer is more than twice that for the pressure drop since $d_2 > d_1$.

2.188 Why are the double pipe heat exchangers used mostly in the form of hair-pins?

It is because of the following:

1. The differential thermal expansion between the inner and the outer pipes can be easily accommodated without producing detrimental stresses.
2. The streams can easily be manifolded on one side of a multiple hair-pin set-up.
3. All the exchangers in a multiple hair-pin set-up need not have the same dimensions.

2.189 What determines whether a series–parallel arrangement should be used in a set of double pipe heat exchangers?

If one of the two fluid streams is very large in volume compared to the other, the exchangers are set up in a series-parallel arrangement as shown in Fig. 2.24.

2.190 How does a multi-pass shell-and-tube exchanger differ from a series–parallel double pipe heat exchanger system? Draw temperature profiles for the two cases.

In a multi-pass shell-and-tube exchanger, all the fluid flows through each of the passes, unlike in the double pipe exchangers where the stream is split into substreams and is put into different hair-pins (Fig. 2.24). Further, the flow in the annulus of a double pipe exchanger is longitudinal, that in the shell of a baffled shell-and-tube exchanger is a combination of cross-flow between the baffles and longitudinal flow through the baffle windows. Thus, the flow patterns in the two systems are very different.

The F_T factor, heat duty, and the temperature ranges achievable are different in the two cases even when the inlet temperatures, surface area, mass flow rate, and physical nature (gas, liquid) of the streams used in the two cases are the same. Representative temperature profiles for the two cases are shown in Fig. 2.25.

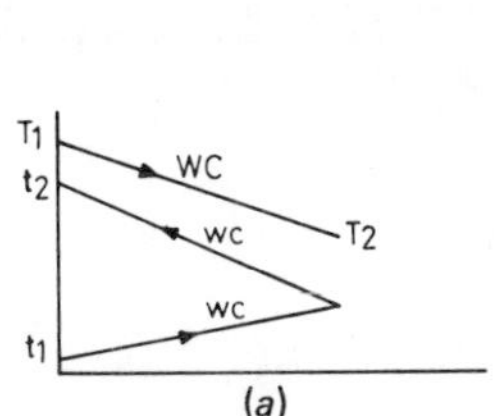

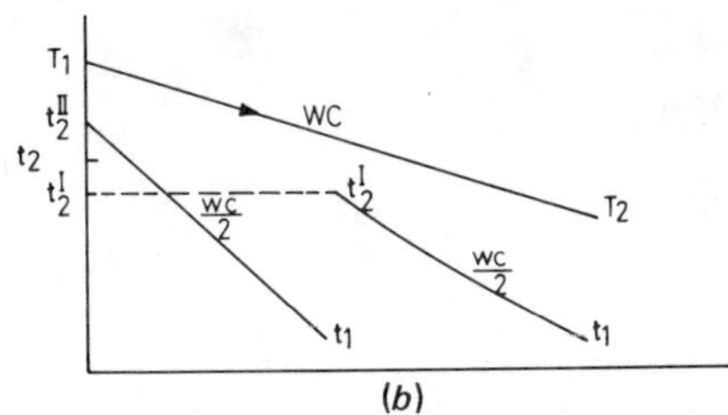

Figure 2.25 Temperature profiles. (*a*) 1-2 shell-and-tube heat exchanger. (*b*) Series–parallel arrangement in two double-pipe heat exchangers.

2.191 How is the true temperature difference calculated when the double pipes are not connected in series flow?

It is calculated by multiplying the maximum temperature span $(T_1 - t_1)$ by a correction factor based upon the inlet and outlet temperatures and the number of parallel streams (Kern, 1950).

2.192 For large heat transfer surface area, why are the double pipe heat exchangers not used?

For large surface area, a large number of double pipe exchangers will be required. These take up a lot of ground space, piping gets to be cumbersome, and the probability of leakage from the large number of connections increases. Hence, the double pipe exchangers are not used in such cases.

BAYONET TUBE HEAT EXCHANGERS

2.193 Describe a bayonet tube heat exchanger and its advantages.

Figure 2.26 shows a typical bayonet tube heat exchanger (Baum, 1979). It has a tube, capped at one end, in which is placed another concentric tube. A gap is left between the inner tube and the cap on the outer tube, thus providing a continuous passage for a fluid through the inner tube and the outer annulus. This fluid exchanges heat through the annulus with a second fluid outside.

The advantages of this exchanger are the following:

1. It can easily be fitted into any process equipment since it requires only one penetration.
2. Since the other end of the exchanger hangs freely inside the equipment, there are no bending and axially compressive stresses.

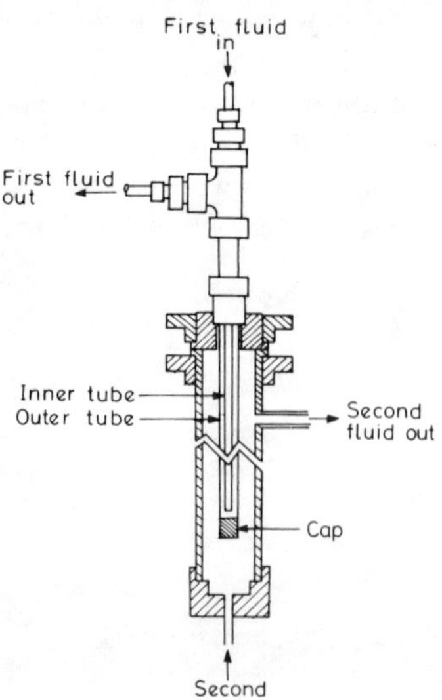

Figure 2.26 Bayonet tube heat exchanger. *(Baum, 1979.)*

3. It can be used for small heat duties as well as to study the effects of corrosion, etc., in a given environment (second fluid).

- The discussion in this chapter has been confined to the no-phase-change situations. An equally important class of the shell-and-tube heat exchangers consists of condensers, evaporators, reboilers, and vaporizers that are used with the distillation columns, thermal and nuclear power plants, and various chemical plants for producing vapors or concentrating a solution. These are discussed in the next chapter.

NOMENCLATURE

a	exponent in Eq. (2.31)
A	leakage stream; heat transfer area based on tube O.D.
A_i	inside tube surface area per unit length
A_o	outside tube surface area per unit tube length
$A_1, A_2, \ldots$	heat transfer area for zone 1, 2, . . .
b	exponent in Eq. (2.31)
B	cross-flow stream; TEMA construction class
BP	baffle pitch or baffle spacing
c	heat capacity at constant pressure
c_1	constant in Eq. (2.31)
C	heat capacity; bypass stream; TEMA construction class
C_c	total heat capacity (MC_p) for cold fluid
C_h	total heat capacity (MC_p) for hot fluid
C_{max}	maximum MC_p
C_{min}	minimum MC_p
C_p	heat capacity at constant pressure
C_v	heat capacity at constant volume
d	characteristic dimension
d_i	tube inner diameter
d_o	tube outer diameter
d_s	shell inner diameter
d_1	diameter of inner pipe in a double pipe heat exchanger
d_2	diameter of outer pipe in a double pipe heat exchanger
D	tube outer diameter
E	bypass stream; TEMA shell type (Fig. 1.1)
f	friction factor
F	bypass stream; TEMA shell type (Fig. 1.1)
F_T	ΔT_{LMTD} correction factor
g	gravitational acceleration
G	TEMA shell type (Fig. 1.1)
G	mass flux
h	heat transfer coefficient
h_i	inside heat transfer coefficient

h_{io}	inside heat transfer coefficient transferred to outside surface
h_o	outside heat transfer coefficient
h-p	high pressure
HTFS	Heat Transfer and Fluid Flow Services, United Kingdom
HTRI	Heat Transfer Research Inc., U.S.A.
j_H	Colburn factor
k	thermal conductivity
k_w	thermal conductivity of tube wall material
l-p	low pressure
lim	limit
ln	natural logarithm
L	flow path length
LMTD	log mean temperature difference
max	maximum
M	flow rate
N_{Re}	Reynolds number
N_s	number of entropy units
NTU	number of transfer units
OTL	outer tube limit
Δp	pressure drop
Δp_s	pressure drop, shell side
Δp_t	pressure drop, tube side
P	pressure
P_1	pressure at stage 1
P_2	pressure at stage 2
$q_1, q_2, \ldots$	enthalpy change in zones 1, 2, . . .
Q	total enthalpy change (or heat transferred) per unit time
$Q_{calculated}$	rate of heat transfer as calculated for a particular case
$Q_{required}$	rate of heat transfer as required in a particular case
R	TEMA construction class
Re	Reynolds number
R_f	fouling resistance
R_{fi}	inside fouling resistance
R_{fio}	inside fouling resistance transferred to outside surface
R_{fo}	outside fouling resistance
RGP	Recommended Good Practice section of the TEMA Standards
t	temperature
t_c	caloric temperature for tube-side fluid
t_{av}	average temperature
t_i	intermediate temperature
t_{in}	inlet temperature
t_{out}	outlet temperature
t_{pi}	inside wall temperature of pipe
t_1	inlet temperature of tube-side fluid
t_2	outlet temperature of tube-side fluid

t_2^{I}, t_2^{II}	outlet temperature of partial streams
T	temperature
T_1	inlet temperature of shell-side fluid; temperature at stage 1
T_2	outlet temperature of shell-side fluid; temperature at stage 2
T_{av}	average temperature
T_c	caloric temperature for shell-side fluid
T_{cond}	condensing temperature
T_{in}	inlet temperature
T_p	pipe wall temperature
T_{po}	outside wall temperature of pipe
ΔT	temperature difference
ΔT_1	temperature difference at hot end
ΔT_2	temperature difference at cold end
ΔT_{av}	average temperature difference
ΔT_B	temperature difference at exit between B-stream and tube-side fluid
ΔT_{LMTD}	log mean temperature difference
$\overline{\Delta T_{LMTD}}$	overall log mean temperature difference
$\Delta T_{LMTD 1,2 \ldots}$	log mean temperature difference in zones 1, 2, . . .
ΔT_{TMTD}	true mean temperature difference
U_c	overall clean heat transfer coefficient
$U_{c1,c2,\ldots}$	overall clean heat transfer coefficient in zones 1, 2, . . .
U_{cx}	overall clean heat transfer coefficient valid for the total heat exchanger [Eq. (2.16)]
U_d	overall design (or dirty) heat transfer coefficient
v	fluid velocity
w	total flow rate through inner pipe
W	flow stream through baffle window; flow stream through outer pipe
Δx	tube wall thickness

Greek Letters

μ	viscosity at bulk temperature
ρ	density
μ_w	viscosity at wall temperature
ϵ	defined in Eq. (2.25*a*)
β	coefficient for volumetric expansion
γ	ratio of heat capacities, C_p/C_v
π	Pi = 3.14159

REFERENCES

Baum, A. J. Optimum Design of a Bayonnet Tube Heat Exchanger. *Heat Transfer 1978: Proceedings of the Sixth International Heat Transfer Conference, Toronto, August 1978,* 4:285–289, Hemisphere: Washington, D.C. 1979.

Bejan, A. The Concept of Irreversibility in Heat Exchanger Design: Counter-flow Heat Exchangers for Gas-to-Gas Applications. *Trans. ASME, J. Heat Transfer* 99(3):374. 1977.

Bejan, A. General Criterion for Rating Heat Exchanger Performance. *Int. J. Heat Mass Transfer* 21:655–658. 1978.

Bell, K. J. Exchanger Design Based on the Delaware Research Program. *Petro/Chem Eng.* 32:C-26–C-40. 1960.

Bell, K. J. Final Report of the Cooperative, Research Program on Shell and Tube Heat Exchangers. *Bulletin No. 5.* University of Delaware Engineering Experiment Station: Newark, Delaware. 1963.

Bell, K. J. Delaware Method for Shell Side Design. In *Heat Exchangers: Thermal-Hydraulic Fundamentals and Design,* S. Kakac, A. E. Bergles, and F. Mayinger, eds., Hemisphere: Washington, D.C. 1981.

Bird, R. B., W. E. Stewart, and E. N. Lightfoot. *Transport Phenomena.* Wiley: New York. 1960.

Butterworth, D. A Calculation Method for Shell-and-Tube Heat Exchangers in which the Overall Coefficient Varies along the Length. Pp. 56–68. NEL Report No. 590. National Engineering Laboratory: East Kilbride, Glasgow. 1973.

Butterworth, D. The Development of a Model for Three-Dimensional Flow in Tube Bundles. *Int. J. Heat Mass. Transfer* 21:253–256. 1978.

Butterworth, D. A Model for Heat Transfer during Three-Dimensional Flow in Tube Bundle. *Heat Transfer 1978: Proceedings of the Sixth International Heat Transfer Conference, Toronto, August 1978.* 4:219–224. Hemisphere: Washington, D.C. 1979.

Chiou, J. P. Thermal Performance Deterioration in Cross-Flow Heat Exchanger due to Flow Non-Uniformity on Both Hot and Cold Sides. *Heat Transfer 1978: Proceedings of the Sixth International Heat Transfer Conference, Toronto, August 1978.* 4:279–284. Hemisphere: Washington, D.C. 1979.

Colburn, A. P. Mean Temperature Difference and Heat Transfer Coefficient in Liquid Heat Exchangers. *Indus. Eng. Chem.* 25:873–877. 1933.

Cooper, A., A. M. Cocks, and A. C. Henton. Improved Water Utilization with Plate Heat Exchangers. *Heat Transfer Eng.* 1(1):31–41. 1979.

Devore, A. Try This Simplified Method for Rating Baffled Exchangers. *Pet. Refiner* 40(5):221. 1961. *Errata* 40(10). 1961.

Donohue, D. A. Heat Transfer and Pressure Drop in Heat Exchanger. *Indus. Eng. Chem.* 41(11):2499–2511. 1949.

Donohue, D. A. Heat Exchanger Design. *Pet. Refiner* 34(8):94–100; (10):129–138; (11):175–184, 1955; 35(1):155–160, 1956.

Emerson, W. H. Effective Tube-Side Temperatures in Multipass Heat Exchangers with Non-Uniform Heat-Transfer Coefficients and Specific Heats. Pp. 32–55. NEL Report No. 590. National Engineering Laboratory: East Kilbride, Glasgow. 1973.

Evans, L. B., and S. W. Churchill. Economics of Using Heat Transfer Promoters. *Chem. Eng. Progr.* 58(10):55–61. 1962.

Fallows, T., C. R. Gane, R. C. Jones, and J. Lis. A Mathematical Model of the Thermal Mixing Processes in Cross-Flow Heat Exchangers. *Heat Transfer 1978: Proceedings of the Sixth International Heat Transfer Conference, Toronto, August 1978.* 4:249–254. Hemisphere: Washington, D.C. 1979.

Fischer, F. K. Mean Temperature Difference Correction in Multipass Exchangers. *Indus. Eng. Chem.* 30:377–383. 1938.

Foust, A. S., L. A. Wenzel, C. W. Clump, L. Maus, and L. B. Anderson. *Principles of Unit Operations,* 2d ed. Wiley: New York. 1980.

Frank, O. Simplified Design Procedures for Tubular Heat Exchangers. In *Practical Aspects of Heat Transfer.* AIChE: New York. 1978.

Gardner, K. A., and J. Taborek. Mean Temperature Difference: A reappraisal. *AIChE J.* 23(6):777–786. 1977.

Gay, B., N. V. Mackley, and J. D. Jenkins. Shell Side Heat Transfer in Baffled Cylindrical Shell-and-Tube Heat Exchanger—An Electrochemical Mass-Transfer Modelling Technique. *Int. J. Heat Mass Transfer* 19(9):995–1002. 1976.

Gay, B., J. D. Jenkins, and N. V. Mackley. Shell-Side Heat Transfer Coefficients in Cyındrical Heat Exchangers: The Influence of Geometrical Factors I–The Non-Leakage Case. *Letters in Heat and Mass Transfer* 4:41–52. 1977.

Gotovski, M. A., M. E. Lebedev, N. V. Mizonov, and E. V. Firsolva. Thermohydraulic Irregularities Influence on the Effectiveness of Shell-and-Tube Counter-flow Heat Exchangers. *Heat Transfer 1978: Proceedings of the Sixth International Heat Transfer Conference, Toronto, August 1978.* 4:213–217. Hemisphere: Washington, D.C. 1979.

Goyal, O. P. Speedy Analysis of Heat-Exchange Processes by Elimination of Trial and Error. *Chem. Age India* 26(6):465–470. 1975.

Goyal, O. P. Quick Determination of Temperature Profile of Multiunit Heat Exchange Systems. *Chem. Age India* 26(12):1093–1095. 1976.

Grant, I. D. R. Flow and Pressure Drop with Single-Phase and Two-Phase Flow on the Shell-Side of Segmentally Baffled Shell-and-Tube Heat Exchangers. NEL Report No. 590, pp. 1–22. National Engineering Laboratory: East Kilbride, Glasgow. 1973.

Gupta, J. P. *Heat Exchanger Design–A Practical Look,* 2d ed., C. S. Enterprises: Delhi, India. 1979.

Jones, R. C., J. Lis, and T. H. Massey. An Experimental Investigation of Thermal Mixing in Cross-Flow Tube Banks. *Heat Transfer 1978: Proceedings of the Sixth International Heat Transfer Conference, Toronto, August 1978.* 4:255–260. Hemisphere: Washington, D.C. 1979.

Kays, W. M., and A. L. London. *Compact Heat Exchangers,* 2d ed., McGraw-Hill: New York. 1964.

Kern, D. Q. *Process Heat Transfer.* McGraw-Hill: New York, 1950.

Kern, J. On the Feasibility of Heat Exchange Processes in Uninsulated Equipment. *Warme und Stoffubertragung* 9:267–272. 1976.

Macbeth, R. V. Shell-and-Tube Heat Exchanger Data Produced by an Electrochemical Mass Transfer Modelling Technique. *Heat Transfer 1978: Proceedings of the Sixth International Heat Transfer Conference, Toronto, August 1978.* 4:225–230. Hemisphere: Washington, D.C. 1979.

Machacek, S., and T. Zelenka. Design and Operation of a Large Diameter (1.67 m) Steam Line. ASME Pressure Vessel and Piping Conference, Paper No. 78-PVP-86. 1978. ASME: New York. 1979.

Massey, T. H., J. A. Hitchcock, and B. E. Launder. A Prediction Procedure for Fully Developed Turbulent Flow and Heat Transfer in Tube Banks. *Heat Transfer 1978: Proceedings of the Sixth International Heat Transfer Conference, Toronto, August 1978.* 4:261–266. 1979.

Mueller, A. C. Heat Exchangers. In *Handbook of Heat Transfer.* W. M. Rohsenow and J. P. Hartnett, eds. Chapter 18. McGraw-Hill: New York. 1973.

Mueller, A. C. Criteria for Maldistribution in Viscous Flow Coolers. *Heat Transfer–1974* 5:170–174. 1974.

Palen, J. W., and J. Taborek. Solution of Shell-Side Flow Pressure Drop and Heat Transfer by Stream Analysis Method. *Chem. Eng. Progr. Symp. Ser. No. 92* 65:53–63. 1969.

Patankar, S. V., and D. B. Spalding. A Calculation Procedure for the Transient and Steady-State Behavior of Shell-and-Tube Heat Exchangers. In *Heat Exchangers: Design and Theory Sourcebook,* pp. 155–176. N. H. Afgan and E. U. Schlünder, eds. Hemisphere: Washington, D.C. 1974.

Pipes, L. A. *Applied Mathematics for Engineers and Physicists.* 2d ed. McGraw-Hill: New York. 1958.

Roetzel, W. Heat Exchanger Design with Variable Transfer Coefficients for Cross Flow and Mixed Flow Arrangements. *Int. J. Heat Mass Transfer* 17:1037–1049. 1974.

Scaccia, C., and G. Theoclitus. Heat Exchanger-Types, Performance and Applications. *Chem. Eng.* 87(20):121–132. 1980.

Spalding, D. B. The Calculation of Convective Heat Transfer in Complex Flow System. *Heat Transfer–1974* 6:44–60.

Spencer, R. A., Jr. Predicting Heat Exchanger Performance by Successive Summation. *Chem. Eng.* 85(25):121–124. 1978.

Taborek, J. Heat Exchanger Design. *Heat Transfer 1978: Proceedings of the Sixth International Heat Transfer Conference, Toronto, August 1978.* 6:269-283. Hemisphere: Washington, D.C. 1979a.

Taborek, J. Evolution of Heat Exchanger Design Techniques. *Heat Transfer Eng.* 1(1):15–29. 1979b.

Tinker, T. Shell Side Characteristics of Segmentally Baffled Shell-and-Tube Heat Exchangers, Parts I, II, III–General Discussion on Heat Transfer. *Proc. Inst. Mech. Engrs. (Lond.)* 1:89–116. 1951.

Tinker, T. Shell Side Characteristics of Shell-and-Tube Heat Exchangers–A Simplified Rating System for Commercial Heat Exchangers. *ASME Trans.* 80:36–52. 1958.

Tubular Exchanger Manufacturers' Association, *TEMA Standards,* 6th ed. TEMA: Tarrytown, New York. 1978.

Von Riesemann, W. A. Have Engineers Been Replaced by Computers. ASME Pressure Vessel and Piping Conference, Paper No. 79-PVP-10. 1979. ASME: New York. 1979.

Whitley, D. L. Calculating Heat Exchanger Shell-Side Pressure Drop. *Chem. Eng. Progr.* 57(9):59–65. 1961.

Wolverine Truefin Engineering Data Book, 2d ed. Wolverine Tube Division, Calumet and Hecla Corp.: Detroit, Mich. 1968. (Presently located at Decatur, Alabama as a Division of UOP.)

Zukauskas, A., and R. Ulinskas. Some Aspects of the Heat Transfer for Banks of Tubes in Cross-flow for the Low Range of Reynolds Numbers. *Heat Transfer 1978: Proceedings of the Sixth International Heat Transfer Conference, Toronto, August, 1978.* 4:243–248. 1979.

Zukauskas, A., and R. Ulinskas. Surface Roughness as a Means of the Heat Transfer Augmentation for Tube Banks in Cross-flow. In *Heat Exchangers: Theory and Practice.* J. Taborek, G. Hewitt, and N. H. Afgan (eds.). Hemisphere: Washington, D.C. 1983.

CHAPTER

THREE

SHELL-AND-TUBE HEAT EXCHANGERS—III: PROCESS DESIGN WITH PHASE CHANGE

INTRODUCTION

• A very large number of heat exchangers involve a change of phase, viz. vapor to liquid (condensers) and liquid to vapor (reboilers, evaporators, vaporizers, etc.). Condensers are used on top of the distillation columns to condense vapors and in power plants to condense steam (surface condensers). These are also used in refrigeration systems to condense the refrigerant. Reboilers are generally used underneath the distillation columns to boil off part of the bottom products to aid in product enrichment and to provide the energy for distillation. Evaporators are used to concentrate an aqueous solution or to obtain pure water, while vaporizers handle nonaqueous solutions.

Thus, it is evident that all this equipment is very widely used in the chemical process industries: fertilizer, petroleum and petrochemical industries, nuclear and thermal power plants, refrigeration and air-conditioning, etc. In this chapter, we shall discuss the salient points regarding this equipment. Direct-fired boilers are not considered in this book, the emphasis being on the unfired units.

All the items discussed in Chapter 1, which covers the details of construction features, and a large number of the items in Chapter 2, which discusses the fluid flow and heat transfer aspects, also apply to the units involving a phase change. Hence, in this chapter, only the points that are specific to the phase change units, their design, and operation are mentioned. For the balance, Chapters 1 and 2 should be read.

CONDENSERS

Definitions

3.1 What is condensation? How does it occur?

It is the reduction of a vapor to a denser form, generally to liquid (e.g., steam to water), although in a few cases it could be directly to solid without any liquid forming (e.g., phthalic anhydride vapor condenses to solid under normal operating conditions).

Condensation occurs when the vapor phase is cooled below its saturation temperature, such as when it comes into contact with a cold surface. The saturation pressure (defined in Q. 3.2) of the vapor at the cold surface temperature is lower than that at the bulk vapor temperature. This difference in the saturation pressures provides the driving potential for the vapor to move toward the cold surface and get condensed there, releasing its latent heat. For example, if steam at 1 atm pressure (101.325 kPa) and 212°F (100°C) is condensing over the tubes through which cooling water is flowing at such a temperature that the temperature of the condensate film is 100°F (37.8°C), the driving potential is the difference between the saturation pressure at 212°F (100°C) of the bulk steam, 1 atm (101.325 kPa), and the saturation pressure at the temperature of the condensate film, approximately 1 psi (6.895 kPa) that is $14.7 - 1 = 13.7$ psi (94.43 kPa). Thus, the steam is being driven toward the condensing surface by a pressure differential of 13.7 psi (94.43 kPa).

3.2 What are (1) saturated vapor, (2) superheated vapor, (3) saturated liquid, (4) subcooled liquid, (5) boiling point, (6) bubble point, (7) dew point, (8) Raoult's law, (9) Dalton's law, (10) fugacity, (11) equilibrium constant, and (12) relative volatility?

The definitions are given below. Items (1) to (7) are shown in Fig. 3.1.

1. Saturated vapor is the vapor in equilibrium with its liquid at the given pressure and the corresponding boiling temperature. It has no sensible heat. On cooling, it releases only the latent heat. For example, steam in equilibrium with water at 1 atm (101.325 kPa) and 212°F (100°C) is a saturated vapor. In the case of a pure vapor that pressure is called the saturation pressure.
2. Superheated vapor is a vapor above its dew point. It is produced when a saturated vapor is removed from contact with its liquid and is heated under isobaric (constant pressure) conditions. This heat is absorbed as the sensible heat. On cooling, first the sensible heat is released, followed by the release of the latent heat at the condensing temperature. For example, steam at 1 atm pressure (101.325 kPa) and temperature >212°F (100°C) is a superheated steam. On cooling, it will first give off its sensible heat, down to 212°F (100°C), and will then condense at 212°F (100°C) releasing its latent heat.
3. Saturated liquid is a liquid at its boiling point at the given pressure. In order to produce saturated vapor from it, only the latent heat has to be provided. For

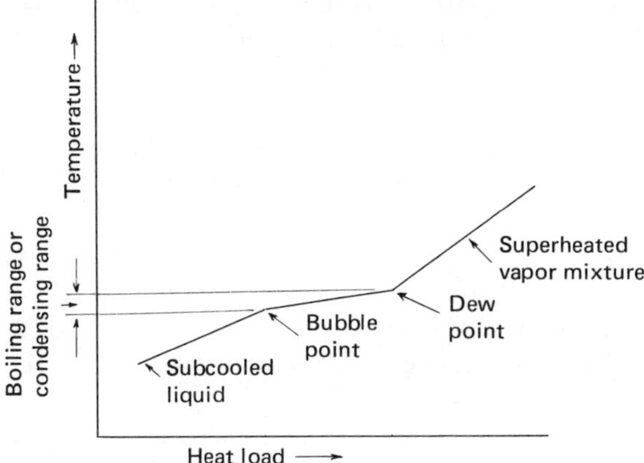

Figure 3.1 Some important terms in phase change operations.

example, water at 1 atm (101.325 kPa) pressure and 212°F (100°C) is a saturated liquid.

4. Subcooled liquid is a liquid below its boiling point. In order to produce saturated vapor at the given pressure, first the sensible heat has to be provided to bring the liquid to its boiling point and then the latent heat has to be provided to boil the liquid. For example, water at 1 atm (101.325 kPa) pressure and temperature $<$212°F (100°C) is a subcooled liquid.
5. Boiling point is the temperature at which the first vapor bubble reaches the top surface of the liquid. The vapor formed is in equilibrium with the liquid at the pressure of the system. It is only defined for a pure liquid that boils, and its vapor condenses, at one unique temperature. For example, at 1 atm pressure (101.325 kPa) pure water boils, and the steam so produced condenses, at 212°F (100°C).
6. Bubble point is the temperature at which the first vapor bubble reaches the top surface of the liquid in a liquid mixture. At that temperature, the partial pressures (or fugacities) of the components add up to the system pressure, or their mole fractions in the vapor bubble add up to 1. As the boiling continues, the more volatile (lower boiling) component keeps vaporizing faster than the rest and the boiling temperature keeps increasing. Thus, the liquid mixture boils in a temperature range that is dependent upon the components and their relative composition in the mixture. This temperature range is called the boiling range.
7. Dew point is the temperature at which the first drop of the condensate forms. The mole fractions of all the components in the drop of the condensate add up to 1. For a pure vapor, the dew point is the same as its boiling point. The difference between the dew point and the bubble point for a vapor mixture is known as the condensing range. For example, a mixture containing butane, pentane, and hexane (mole fractions 0.077, 0.613, and 0.310, respectively) has a bubble point of 102°F (38.89°C) (mole fractions in the vapor = 0.266, 0.625, and 0.109, respectively) and a dew point of 123°F (50.56°C) (mole fractions in

the liquid = 0.0167, 0.4133, and 0.570, respectively) at 1 atm pressure. Thus, the boiling (or condensing) range for the above mixture at 1 atm pressure is 21°F (11.67°C).

8. Raoult's law states that for an ideal liquid mixture where the components do not affect one another, the partial pressure of a component, p_1, is equal to the product of its vapor pressure, p_{v1}, at the given temperature and its mole fraction in the liquid, x_1:

$$p_1 = p_{v1} x_1$$

In the above mixture, the vapor pressure of butane at 100°F (37.78°C) is 3170 mm Hg (427 kPa). Hence, its partial pressure is 3170 X 0.077 = 244 mm Hg (32.5 kPa).

9. Dalton's law states that for an ideal vapor mixture (where the components do not affect one another), the partial pressure of a component, p_1, is equal to the product of the total pressure, p_t, and mole fraction of the component in the vapor phase, y_1:

$$p_1 = p_t y_1$$

and

$$p_t y_1 = p_{v1} x_1$$

In the above mixture, on complete vaporization at 1 atm (760 mm Hg, 101.325 kPa), the partial pressure of butane will be 760 X 0.077 = 58.5 mm Hg (7.8 kPa).

10. Fugacity accounts for the deviation of a mixture from ideality, which is more pronounced at higher pressures. Each component acts to lower the pressure of the other. In the Raoult's law, the vapor pressure, p_{v1} is replaced by f_{p1}, called the fugacity or corrected pressure, and in the Dalton's law, the total pressure is replaced by the total fugacity, f_t. Thus, these laws then give

$$f_1 = f_{p1} x_1 = f_t y_1$$

The values of the fugacities have to be found experimentally for each mixture at the required pressure. The units of fugacity are the same as of the pressure. In the above mixture, the boiling point of 102°F (38.89°C) was found by taking the nonideality into account while the assumption of an ideal mixture (using partial pressures instead of the fugacities) gives the boiling point to be 97°F (36.11°C). It is because the partial pressures are higher than the fugacities.

11. Equilibrium constant is the ratio of p_v/p_t for an ideal mixture and f_p/f_t for a nonideal mixture and is designated by K. Using Dalton's and Raoult's laws, it also turns out to be the ratio of the mole fraction of a component in the vapor phase to that in the liquid phase. In the above example, K for butane at 100°F (37.78°C) and 1 atm pressure (101.325 kPa) is 4.17, assuming an ideal mixture (using partial pressures), and 3.35, assuming a nonideal mixture (using fugacities).

12. Relative volatility is the ratio of the equilibrium constants for two components and is designated by

$$\alpha_{12} = \frac{K_1}{K_2}$$

It is a measure of the relative ease of separation of the two components. This is used in the condenser and distillation column design to find the composition of the mixture along the condenser path and at each tray in a distillation column. In the example cited, the relative volatilities at the bubble point are

$$\alpha_{\text{butane/pentane}} = 3.45/1.02 = 3.38$$

$$\alpha_{\text{hexane/pentane}} = 0.35/1.02 = 0.343$$

3.3 What is Gibbs' phase rule and how does it help in studying the condensation phenomenon?

The Gibbs' phase rule can be written as

$$F = C - P + 2$$

where F = degrees of freedom
C = number of components in all phases
P = number of phases

In any system, it tells the number of degrees of freedom. For example, for steam at its dew point, $C = 1$ (water), $P = 2$ (liquid and vapor). Hence $F = 1$. It means that either the pressure or the temperature may be fixed independently, which determines the value of the other. Thus, if the pressure in the system is 1 atm (101.325 kPa), the corresponding dew point temperature will be automatically fixed at 212°F (100°C).

In a multi-component system, the phase rule helps in determining which of the variables (temperature, pressure, composition) should be fixed and whether the condensation is isothermal or not. For example, while the condensation of pure steam discussed above will proceed isothermally when the pressure of the system is predetermined, this will not be the case for steam + air. Applying the phase rule, $C = 2$ and $P = 2$, gives $F = 2$. If pressure is fixed, that still leaves one degree of freedom, which means that the temperature will change. Since it is a condensing phenomenon, the temperature of the system will keep decreasing for the condensation to continue. On the other hand, for two immiscible components (steam + pentane), $C = 2$, $p = 3$ (vapor, water, liquid pentane), and $F = 1$, which is fixed by the pressure of the system since most condensers operate at one fixed pressure. This condensation proceeds isothermally. For other cases, see Kern (1950).

3.4 Can one determine the number of phases at equilibrium from the phase rule?

No. The phase rule only tells the degrees of freedom. The number of phases as well as the total number of components should be known beforehand.

Classification of Condensers

3.5 How are the condensers classified?

Condensers can be classified in numerous ways, as given in Fig. 3.2.

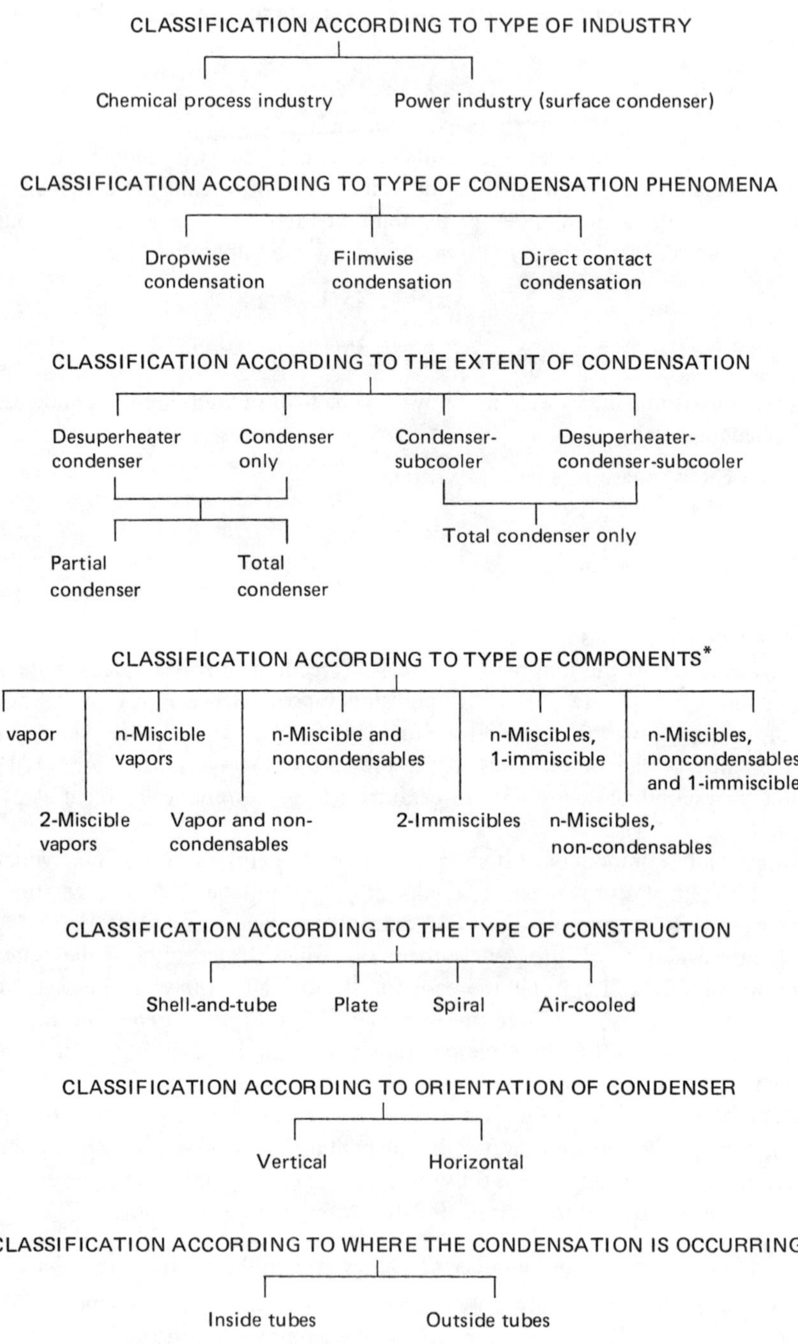

Figure 3.2 Classification of condensers. **(From Kern, 1950.)*

- Now we shall discuss this condenser classification in some detail.

3.6 What is the condenser classification according to the type of industry?

There are two broad categories used, respectively, in the chemical process industry and in the power industry.

Condensers used in the process industry are generally on top of the distillation columns and condense the vapors coming out of these columns. Part of the condensate is usually returned to the distillation column as a reflux for greater product recovery. Sometimes, only a part of the vapor mixture is condensed in a condenser (called partial condenser) for better product distribution, while the remaining is condensed in a second condenser. Usually the condensate is the desired product, although, when steam is condensed for heating a fluid, the interest is more in the large amount of latent heat released by the steam.

In the vapor coming out of the distillation column, generally more than one component is present. Actually, their number is equal to that in the liquid mixture being distilled, although some may be present in the vapor in very small amounts. Apart from this, air can be present as a noncondensable unless appropriate precautions are taken. The mechanical design of these condensers is governed by the appropriate Codes (ASME, 1980) and TEMA Standards (TEMA, 1978). Their fabrication, inspection, and testing requirements are very rigid. In most cases, the same unit can also be used for noncondensing duties (sensible heat transfer), should the process requirements or the process itself change. The rest of the classification bases in Fig. 3.2 apply to the chemical process industry condensers. These form a major part of this section on condensers.

The second category is the one used in the power industry to condense the exhaust steam from turbines and is known as the surface condenser (Fig. 3.3). It is generally a huge, unbaffled, shell-and-tube unit with a large surface area (e.g., 25,000 m^2 or 269,100 ft^2 with a 20 m or 65.6 ft length) (Butterworth, 1981a). It is usually assembled at site due to the limitations of handling and transporting large units. Smaller units are of course shop assembled and tested. These operate under vacuum with steam condensing on the outside surface of the tubes at about 50° to 60°C (122° to 140°F). This increases the efficiency of the power cycle since the enthalpy between the normal condensing temperature of 100°C (212°F) and the vacuum condensing temperature of 50° to 60°C (122° to 140°F) is also converted into work.

In order to facilitate access of steam to the lower tube rows and to reduce the vibration and erosion damage to the upper rows, the tube pitch in the upper rows is greater than that in the lower ones (Fig. 3.3), so that the steam velocity in the upper section is not too high. As the condensation proceeds, the tube pitch in the lower rows is reduced so that a moderate vapor velocity is maintained and the noncondensables are swept toward the exit. These condensers are generally two tube-pass, box-type units with a noncircular tubesheet. Their design is governed by the HEI Standards (HEI, 1978) or the British Standards (British Electrical and Allied Manufacturers' Association, 1967). The usual problem in their operation is the presence of noncondensable air, which cuts down their efficiency significantly.

Figure 3.3 Surface condenser. *(Courtesy of Ecolair–Heat Transfer.)*

These cannot be used for sensible heat transfer duties because of the tube layout, shell-side entrance and exit designs, etc. More discussion of the surface condensers is in Q. 3.64–3.69.

3.7 What is the condenser classification according to the type of condensation phenomena?

There are three types of condensation phenomena: dropwise condensation, filmwise condensation, and direct contact condensation.

1. Dropwise condensation: When condensation occurs on the outside of a tubular surface, small drops of the condensate are formed. If these drops do not wet the condensing surface, they grow and fall off leaving the surface bare for further condensation to occur. This is called dropwise condensation. An example would be the condensing of mercury vapors over steel tubes or condensing of steam over specially prepared surfaces. The surface may be gold plated (Woodruff and Westwater, 1981), thin layers of Teflon, or other organic products may be put on it. Alternately, injection into steam of small amounts of promoters like mercaptans, silicone oils, octadecylamine, etc., leads to dropwise condensation (Semeria and Hewitt, 1981). A perfect promoter (noncorrosive, nontoxic, long life, low resistance to heat transfer, and low cost) is still being sought.
2. Filmwise condensation: When the condensate drops have an affinity for the

condensing surface and wet it, they coalesce and form a film on the surface. Any further heat transfer for continuing condensation must occur through this condensate film. This is called filmwise condensation. The film introduces a high resistance to heat transfer (except with liquid metals) and reduces the condensing coefficient significantly compared to that for the dropwise condensation.

In industry, the design calculations are done using the filmwise condensation coefficient, since it is not possible yet to control and maintain the dropwise condensation phenomenon in industrial operations for long times, although controlled experiments in laboratory have been successfully done.

For authoritative reviews of dropwise and filmwise condensation, reference should be made to Tanasawa (1979) and Bell and Panchal (1979), respectively.

3. Direct contact condensation: The vapor contacts a subcooled liquid directly and gets condensed. It can be achieved by spraying liquid into vapor, such as water in steam (Fig. 3.4). In case of a dirty liquid, which will block the nozzles, tray columns are used (Fig. 3.5). Bubbling the vapor through a volume of liquid, such as steam into water is not practiced much since large volume of vapor suddenly condensing may damage the vessel. These can, however, be used as quench tanks for vents. Direct contact condensation can also be used where the condensate and the coolant are immiscible and can be easily separated. The heat transfer coefficients are very high, since in the case of pure vapors, there is little thermal resistance because of the absence of any material barrier or fouling layers between the two fluids. The pressure drop is low. Hence these are used in low pressure or vacuum operations. Both the initial and operating costs are low. However, due to the intimate mixing between the condensate and the coolant, the applications currently are few, though fast developing in the fields of power plants (Jacobs et al., 1979; Bakey and Jaszay, 1979; Efimochkin, 1983) and pressure suppression systems of water-cooled nuclear reactors (Cumo et al., 1979;

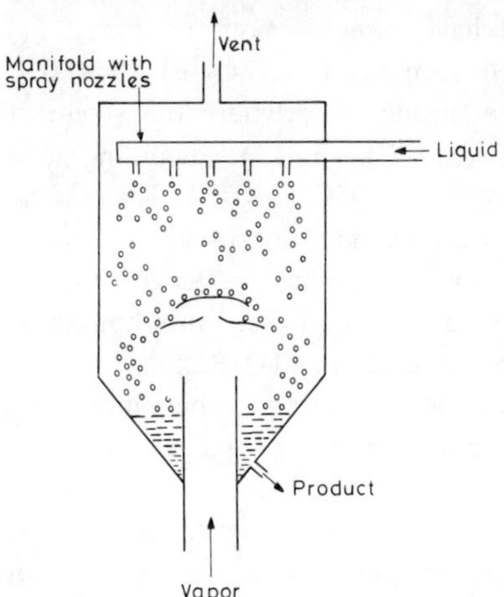

Figure 3.4 Spray tower for direct contact condensation.

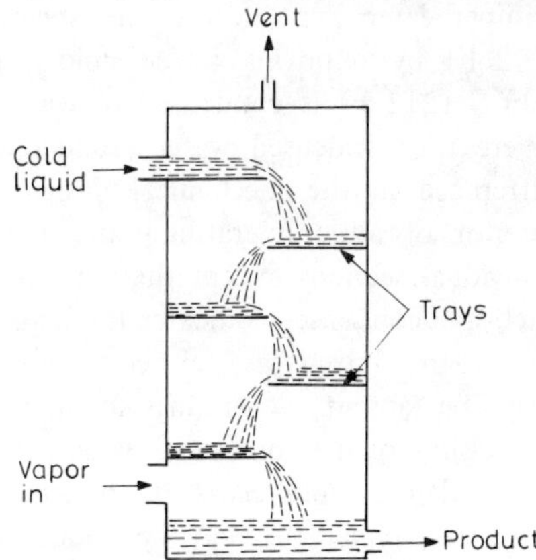

Figure 3.5 Tray tower for direct contact condensation.

Aoki et al., 1979). A big hindrance in the design of the direct contact condensers is the nonavailability of reliable design data over a wide range of variables. For good discussions on this topic see Sideman and Moalem-Maron (1982), Moalem and Sideman (1974); and Fair (1972a, b).

3.8 What is the relative order of magnitude of dropwise and filmwise condensation coefficients?

The heat transfer coefficient for the dropwise condensation is 5 to 30 times more than that for the filmwise condensation. For pure steam condensing dropwise (in controlled laboratory experiments), the heat transfer coefficient can be up to approximately 40,000 Btu/hr ft^2 °F (227,200 J/m^2 s K), while for the filmwise condensation it is approximately 2000 Btu/hr ft^2 °F (11,360 J/m^2 s K). This is because while in the former case, fresh tubular surface is intermittently exposed directly to the condensing steam, in the latter case, the steam condenses over a condensate film and the heat has to diffuse through this film, thus slowing down the process significantly.

- Further discussion on condensation in this book will imply filmwise condensation unless specified otherwise, since it is the only one used in the design of the condensers in industries.

3.9 What is the classification according to the extent of condensation?

Basically, there are two types: Total condenser and partial condenser.

Total condenser: It condenses all the vapor entering the unit, whether it is a pure vapor or a vapor mixture. It may have several sections. An example will illustrate it.

Consider superheated steam at 1 atm pressure (101.325 kPa) and 120°C (248°F) that is to be condensed to water and the condensate is desired at 70°C (158°F). The condenser handling this job will have three sections or zones: (1) desuperheater zone where the steam is cooled from 120°C (248°F) to 100°C (212°F) by removing its sensible heat, (2) condensing zone where the steam at 100°C (212°F) is condensed to water at 100°C (212°F), and (3) subcooling zone where the condensed water (condensate) is cooled to 70°C (158°F). There is no difference in the mechanical design of these sections, i.e., by looking into the interior of a nonoperating unit, it is not possible to gauge the lengths of the individual sections except that the presence of a dam baffle, if used in a horizontal shell-side condenser, indicates the height of the subcooling zone (Q. 3.25).

In a given case, either or both of the desuperheating and subcooling zones may be absent depending upon the inlet vapor conditions and whether the subcooling of the condensate is desired or not.

Partial condenser: It condenses only part of a vapor mixture entering the unit. The remainder may be condensed in another condenser or used as such or vented, depending upon the vapor composition and the process requirements. Naturally, there can be no separate subcooling zone in a partial condenser. Some of the partial condenser applications are in the removal of moisture from air in air-liquefaction plants, condensate recovery from natural gas, vapor recovery from noncondensable gas, etc.

3.10 What is the classification according to the type of components in the vapor phase?

A vapor phase may belong to any of the nine systems listed under this classification in Fig. 3.2. The design of the unit must take the exact type of system into account since it is the single most important point. Experience shows that even with pure vapors some noncondensables are generally present, and the designer should especially inquire about their concentrations (however small these may be) if not mentioned in the data sheet, because very small amounts tend to build up in the unit and adversely affect its performance (Steinmeyer and Mueller, 1974). This topic will be discussed further in this chapter.

3.11 What is the classification according to the type of construction?

1. Shell-and-tube type (Fig. 1.1): These are the most used ones, and the rest of the discussion on condensers in this chapter (after this question) is related to this variety.
2. Plate type (Fig. 6.1): These are not used much because the vapor side pressure drop is high. The vapor port is made larger than the coolant and the condensate ports. Figure 6.1 gives only a general idea of the working of a plate heat exchanger.
3. Spiral type (Fig. 6.14): These are used only when small surface area and broad flow passages are required, such as in the case of a vapor stream containing suspended particles.
4. Air-cooled type (Figs. 4.14 and 4.15): These are used very often where air is used as a coolant. Air-cooled steam condensers are discussed in Q. 3.69. A detailed section on air coolers is included in Chapter 4.

3.12 What is the classification according to the orientation of the condenser?

Orientation of the condenser refers to the direction of the overall flow passage of the vapor phase. The two extreme orientations are horizontal and vertical although inclined units are also sometimes used.

Vertical units: In these, the condensate film flows down each tube and keeps getting thicker as it flows down. The thick film hampers heat transfer (Fig. 3.6). Each tube in a tube bundle behaves independently unless the condensate layers merge and make the further tube length ineffective. Hence, for high mean condensing coefficients, the tubes should not be very long. The baffles may act to dam the condensate if the clearance between the baffle hole and the tube is very tight and the rate of condensation is greater than the amount that can flow through this clearance. These units can be only single pass, else the pressure drop requirements increase greatly.

Horizontal units: In these, the condensate film is the thinnest at the top of a tube, and it starts to thicken as the condensate flows around and drops off the

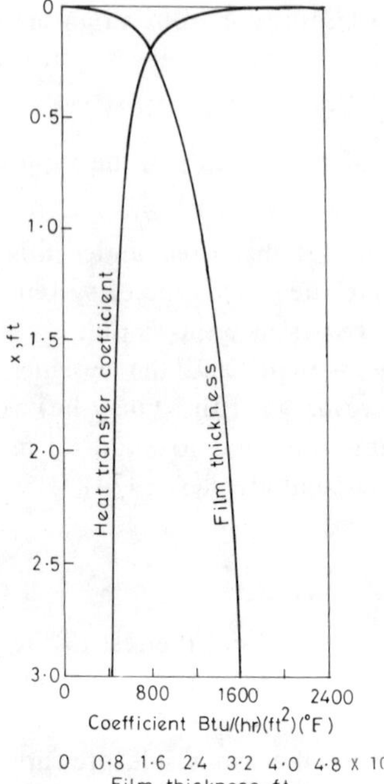

Figure 3.6 Heat transfer coefficient over a vertical tube. *(From Kern, 1950.)*

tube. The condensing coefficient is the maximum at the top and approaches zero at the bottom [the diametrically opposed location (Fig. 3.7)]. The condensing coefficient on a horizontal tube is greater than that on a vertical tube. Kern (1950) calculates for a $\frac{3}{4}$ in (19.2 mm) O.D. tube, 16 ft (4880 mm) long, the ratio of the condensing coefficients for horizontal to vertical orientations to be 3.07 for Nusselt-type condensation (Q. 3.17). Condensation inside horizontal tubes is discussed next, while condensation over a tube bundle is discussed in Q. 3.19.

3.13 What is the classification according to where the condensation is occurring, viz. inside or outside the tubes?

Inside tubes or in-tube condensation: It is used when the vapors are under high pressure or the condensate is corrosive and would damage the shell. Alloy tubings are used to avoid their corrosion in the later case. One tube pass is used in the vertical units and one or two tube passes in a horizontal one. In the latter, flow instability can result. Further, in the horizontal, two-pass case, the lower tubes of the second pass flow full of the condensate and hence are not available for condensing duties. This should be kept in mind at the design stage. The horizontal in-tube condensers are used in air-cooled condensers and for the condensation of steam to provide heat in reboilers and evaporators. If such a condenser has only one tube pass, the tubes may be slightly inclined (1°–2°) for easy draining of the condensate. The far end should be lower so that the condensate and the vapor flow in the same direction (co-current). This gives a higher heat transfer coefficient than if the inclination were reversed resulting in a countercurrent flow (Moalem-Moran and Sideman, 1982).

In a vertical in-tube condenser with downward vapor flow, the condensate film might merge at a certain distance and make further condensation difficult. These condensers are used for wide condensing range mixtures or when subcooling is required. In upward vapor flow, the condensate travels downward along the sides while the vapor flows through the central core of the tubes. Vapor shear results in larger pressure drop. Unless properly designed, part of the vapors may escape uncondensed and may also entrain some of the condensate droplets, resulting in poor function of the unit. A low vapor velocity is, therefore, required, and this results in a low heat transfer coefficient. Such units are called knock-back condensers (see also Q. 3.62).

Outside-tube condensation: It is used quite often in the chemical processing industry where low pressure drop is available or the coolant is corrosive/fouling in nature. All the water-cooled surface condensers in power plants are also of this variety. One main requirement is that the condensate should not be corrosive to the shell. Most of the discussion in the classification according to orientation (Q. 3.12) applies to this case too. Outside-tube condensation is preferred since generally water is used for cooling. The fouling due to cooling water is relatively easy to remove when it is in the tubes than if the fouling were on the outside (in-tube condensation) since that generally requires removal of the tube bundle. Furthermore, the cooling water velocity in the tubes can be higher than is possible if it were in the shell because of the vibration problems (Chapter 15). Higher tube-side velocity leads to higher overall heat transfer coefficient since, in the condenser

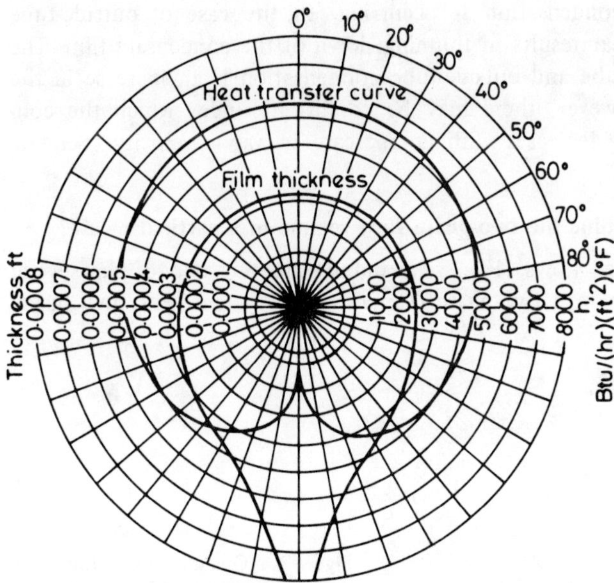

Figure 3.7 Heat transfer coefficient over a horizontal tube. *(From Kern, 1950.)*

design, the coolant heat transfer coefficient is generally the controlling one. Also, tubings with low fins on the outside (low-fin tubing) can sometimes be used with low surface tension condensates to enhance the condensing coefficient at a reduced initial cost (Q. 3.55).

3.14 What is the classification according to the controlling mechanism?

The condensation phenomenon may be gravity controlled, shear controlled, or mixed. In the first case, due to gravity, the denser phase condensate flows at the bottom and the vapor phase at the top (Fig. 3.8*a*, stratified flow). In the second case, high vapor velocity produces high shear and the effect of gravity becomes negligible. In the case of in-tube condensation, this results in an annular condensate film with the vapor flowing in the core (Fig. 3.8*b*, annular flow). The annular condensate film is wavy, and some droplets are entrained by the high-velocity vapor. Thus, annular flow occurs when the shear forces predominate and stratified flow occurs when the gravity forces predominate. Mixed flow occurs when both the forces are of the same order of magnitude.

In the stratified flow, the condensate drains down to the lower portion of the tubes resulting in a circumferential variation of the condensing coefficient (Fig. 3.8*a*). In the annular flow, on the other hand, the condensate film is all around, and hence there is no significant circumferential variation in the condensing coefficient (Fig. 3.8*b*) although its value is lower than the maximum localized value in the stratified flow.

3.15 How does vapor shear affect condensation?

As discussed in Q. 3.14, high vapor shear results in an annular flow of the condensate when in-tube condensation is occurring. In the case of outside-tube condensation, the vapor shear results in thinning down of the condensate film. The net result in both the in-tube and outside-tube condensation is an increase in the condensing coefficient. However, there may be conditions under which the condensing coefficient actually decreases with an increase in vapor velocity (Semeria and Hewitt, 1974).

3.16 How does one determine the two-phase flow regime in a particular case?

Stratified and annular flows (Fig. 3.8*a* and *b*) are but two of the several flow regimes that can exist in a two-phase (vapor–liquid) system. For flow inside

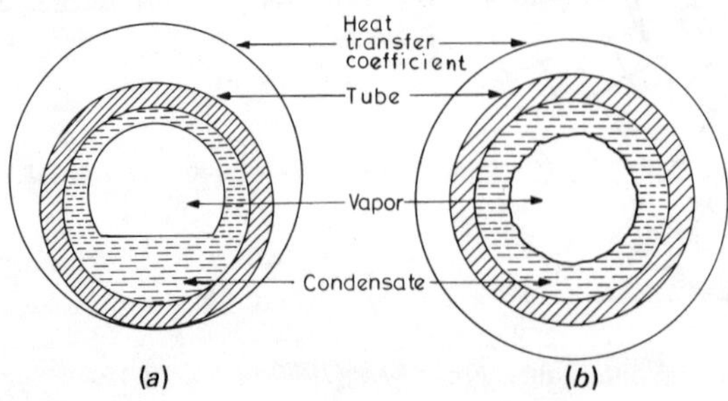

Figure 3.8 Qualitative heat transfer coefficient. (*a*) Stratified. (*b*) Annular flow.

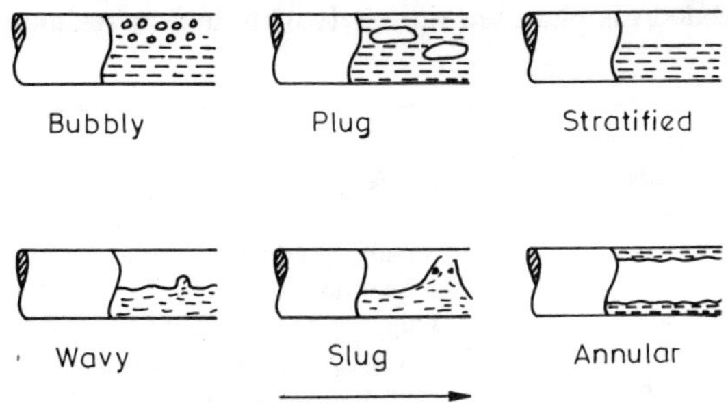

Figure 3.9 Two-phase flow regimes in tubes. *(From Collier, 1972.)*

tubes, the possible flow regimes are shown in Fig. 3.9 while, for the flow in shells, the flow regimes are shown in Fig. 3.10. Relative liquid and vapor velocities for flow inside tubes are given in Table 3.1. Since both the heat transfer and pressure drop are different in each regime, one needs to find out the flow regime that exists in a particular case. The factors involved are the mass fluxes of vapor and liquid and their density, viscosity, surface tension, etc. Baker (1954) gave a two-phase flow map for flow inside tubes showing various flow regimes (Fig. 3.11). Modifications to this plot have been suggested by Mandhane et al. (1974) and by Taitel and Dukler (1976). Palen et al. (1979) have suggested a flow regime plot based upon a force balance parameter (gravity force/shear force) versus liquid volume fraction (Fig. 3.12).

For the two-phase flow on the shell side, separate plots for the vertical and horizontal units were obtained by Grant (1975) as reported by Butterworth (1977). These are shown in Fig. 3.13. Two phase flow patterns have been dealt with in detail in Bergles et al. (1981).

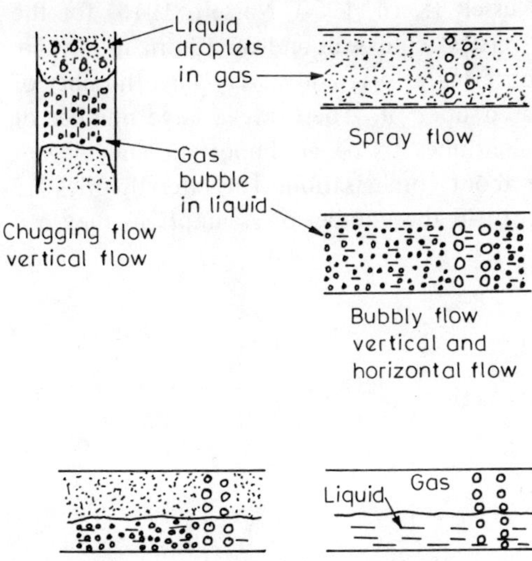

Figure 3.10 Two-phase flow regimes in shells. *(From Butterworth, 1977.)*

Table 3.1 Relative velocities of liquid and gas phases in various two-phase flow regimes in a horizontal tube

Flow regime	Liquid velocity, m/s	Gas velocity, m/s	Remarks
Bubble	1.5–5	0.3–3	Small gas bubbles dispersed in bulk liquid
Plug	0.6	<1	Plug of gas in liquid phase
Stratified	<0.15	0.6–3	Layer of gas over liquid layer
Wavy	<0.3	>5	High velocities in gas
Slug	Low	High	Slug of liquid in gas phase
Annular	Low	>6	Liquid in annular flow, gas in central core
Spray or dispersed	Low	>60	Liquid droplets dispersed in bulk gas

Source: From Coulson and Richardson (1977).

In a condenser, there may be more than one flow regime due to a relative change in the mass fluxes of vapor and liquid phases. For example, in horizontal tubes, an initially annular flow regime (high vapor shear) may change finally to stratified flow (gravity controlled) as the condensation proceeds since the vapor mass flux decreases further down the tubes. One should keep this in mind when designing a unit.

Pure Vapor Condensation: Nusselt's Theory

3.17 What is Nusselt's theory of condensation?

It is a formulation derived by Nusselt (Kern, 1950; Nusselt, 1916) for the condensation of a pure vapor on a flat vertical cold wall and on a horizontal tube. At the time it was derived, it was a pioneering effort, and a very large number of condensers have since been designed based upon it. There are a large number of assumptions (listed next). However, it continues to be an important and elegant theory and is still taught to all learning about condensation. The later theories are usually compared with the Nusselt's theory in the number of assumptions they can

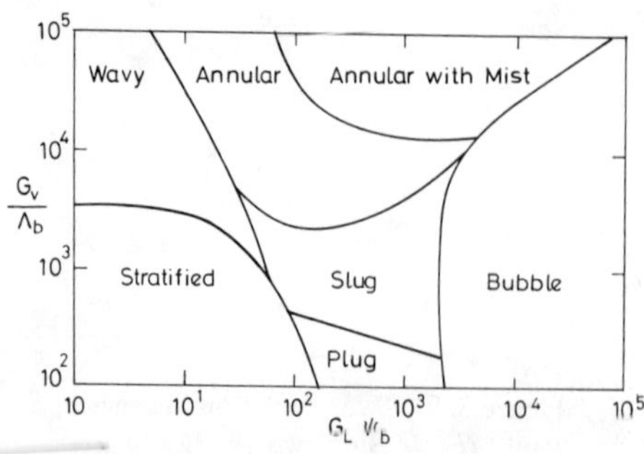

Figure 3.11 Baker's flow regime map for tubular flow. *(From Baker, 1954.)*

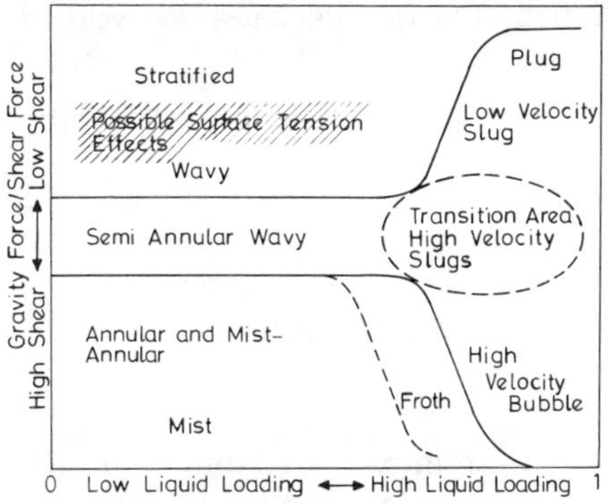

Figure 3.12 Modified flow regime map for tubular flow. *(From Palen et al., 1979.)*

relax. It has also been extended to tubes of elliptical cross section (Moalem and Sideman, 1975) and helically coiled tubes (Karimi, 1977). Improved models for film condensation continually appear in the literature (e.g., Kutateladze, 1982).

3.18 What are the assumptions in Nusselt's derivation of the condensing coefficient?

Nusselt's derivation is for a single tube only. The important assumptions are the following:

1. Vapor is saturated. On condensing, it delivers only the latent heat of condensation to the condensate film.
2. Heat transfer through the film to the condensing surface is by conduction through the condensate film, which flows down due to the gravity in a laminar

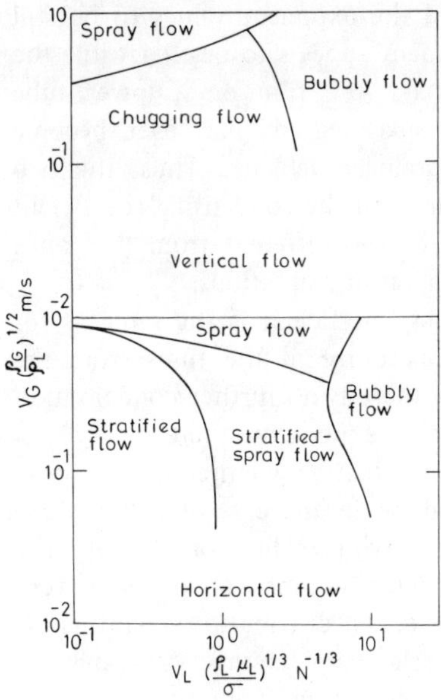

Figure 3.13 Grant's flow regime map for shell-side flow. *(From Butterworth, 1977.)*

fashion. Also, there are no ripples in the film. The film thickness and velocity are determined by a momentum balance.
3. Temperature gradient through the film is linear, i.e., the heat capacity of the condensate is neglected.
4. Temperature of the condensing surface is constant, and its curvature is neglected. The surface is smooth and clean.
5. Physical properties of the condensate are taken at the mean film temperature.
6. Longitudinal conduction through the condensate film is neglected.
7. There is no vapor shear on the condensate surface.

• We shall now discuss the effects of removal of some of the assumptions or how the deviations are taken into account. Related matters are also discussed as and when they arise. Assumptions 3–6 usually do not have much effect and hence would not be discussed.

3.19 How does the condensation over a tube bundle differ from that over a single tube?

Condensation over a single tube has been discussed in Q. 3.12. Now we will discuss the condensation over a tube bundle. The situation is different for horizontal and vertical tube bundles.

Nusselt (1916) had given a relation for the average condensing coefficient over N horizontal tubes, $\bar{h}_N$, in a vertical row to be $\bar{h}_N = \bar{h}_1 N^{-1/4}$, where $\bar{h}_1$ is the condensing coefficient for one tube and $N^{-1/4}$ accounts for the flow of the condensate from an upper tube on to a lower one, assuming the film on each tube to be laminar and without splashing and associated disturbances when the condensate drops on it from an upper tube. This is too conservative. Kern (1950) suggested $\bar{h}_N = \bar{h}_1 N^{-1/6}$. Some people have found the exponent values to be still closer to zero (Bell and Panchal, 1979). The argument appears to be that while the condensate falling from an upper tube would make the film on a lower tube thicker, this film will be highly disturbed due to splashing and may even become turbulent, increasing the heat transfer and the drainage velocity. Thus, the two effects would counter one another to a large extent, and the condensing coefficient over a vertical row of horizontal tubes may not be too different from that on a single tube. More data are needed to confirm this in various situations.

In a bundle of vertical tubes, the condensing coefficient is the same as for the condensation outside a single vertical tube. Of course, if the films from the adjacent tubes merge at any point down the tube, there is no further condensation over the remaining length of the tubes since the vapor cannot flow there.

For the inside-tube condensation, it does not matter whether it is a single tube or a tube bundle. The only difference could be in the case of a two-phase mixture entering a condenser or second pass of a two-pass horizontal unit. The equal distribution of a two-phase mixture in a large number of tubes is very difficult (Steinmeyer and Mueller, 1974). An unequal distribution could cause different condensing coefficients in different tubes, and the average value may be different and lower than the one obtained in experiments using single tube.

3.20 Latent heat of vaporization of a pure substance is the same as that of condensation at a given pressure. Does it mean that the rates of vaporization and condensation are the same?

No. The latent heat has nothing to do with the rates of transfer of heat. The rates of vaporization and condensation of a pure substance are dependent upon the driving potential, the turbulence or agitation in the liquid film, the resistance to heat transfer, the superheat of vapor, etc. The rate is usually more for vaporization than for condensation since, in the former, the formation and release of bubbles causes the liquid to agitate and thus increase the heat transfer. Furthermore, the resistance to heat transfer by the condensate film is greater than that encountered during vaporization by nucleate boiling (Q. 3.72).

Condensation of Superheated Vapor

3.21 How is the superheat in a vapor accounted for in the design of a condenser?

Generally, the superheated vapor is taken to be at its dew point while designing a condenser. While it results in a higher calculated heat transfer due to the neglect of the vapor desuperheating zone, this is compensated by the lower value of ΔT_{LMTD} based upon the dew point, neglecting the higher temperature of the entering superheated vapor. The designs done in this manner have been found to be satisfactory, especially for steam where the superheat is a very small fraction of the latent heat (see Chapter 2 for a discussion on ΔT_{LMTD}).

3.22 When does the above not hold? What method is adopted in such a case?

The above assumption of taking a superheated vapor at its dew point for the condenser design does not hold when the superheat is a sizable fraction (say $\geqslant$25 percent) of the latent heat. In such cases, a large desuperheating zone does exist and has to be calculated separately. This can be done by finding the separate heat load and temperature profiles of the fluids in the desuperheating and condensing zones (and in the subcooling zone, if one exists), and calculating individual LMTD for each zone. Then

$$\Delta T_{\mathrm{LMTD}_{\mathrm{overall}}} = \frac{Q'}{\Sigma_i \, (q_i'/\Delta T_{\mathrm{LMTD}\,i})}$$

In such cases, each zone is designed separately and $U_{\mathrm{d}i}$ and A_i for each zone are obtained. Then, the overall U_{d} is calculated by

$$U_{\mathrm{d}_{\mathrm{overall}}} = \frac{\Sigma_i \, U_{\mathrm{d}i} A_i}{\Sigma_i \, A_i}$$

The total heat transfer area is $A = \Sigma_i A_i$ (see also Q. 2.29). A good discussion on the condensation of superheated vapor is given by Foxall and Chappell (1980).

Subcooling of Condensate

3.23 When is it desirable to subcool a condensate?

If the bubble point of the condensate is close to or lower than the ambient temperature, a significant amount of the condensate is lost in transporting it from the condenser to a storage tank if it is pumped without subcooling. Such a condensate should be subcooled significantly before it is pumped so that its vapor pressure is rather low and losses in transporting are negligible. It is also desirable to subcool the condensate in power and refrigeration industries.

3.24 In a condenser–subcooler, how is the subcooling taken into account in the calculation of the surface area?

The subcooling in all the cases, whether inside or outside the tubes and whether in a vertical or horizontal unit, is primarily by natural convection, since the flow velocity of the condensate is usually rather small. A conservative natural convection heat transfer coefficient is assumed, and the area so calculated is added to the area for condensation to obtain the total condenser–subcooler area. A loop seal is provided to keep the required condensate level in the unit to effect the subcooling. For large subcooling when the surface area for subcooling is comparable to that for condensing, a separate cooler may be provided. Good discussion and a solved example have been given by Rubin (1982).

3.25 How does a vertical condenser–subcooler differ from a horizontal condenser–subcooler?

The two are shown in Figs. 3.14 and 3.15. A vertical unit has two distinct zones in series for condensing and subcooling. These can be calculated separately since the subcooling zone is like an ordinary liquid–liquid heat exchanger. In a horiziontal unit, on the other hand, both the condensing and subcooling zones are parallel to one another. A few rows of tubes are completely submerged in the condensate. The condensate level in the shell is maintained by a loop seal or dam baffle. The design has to be done based upon the principle of balanced pressure drop, which implies that the pressure drop for the vapor phase and for the condensate traveling the length of the condenser has to be equal. Furthermore, the

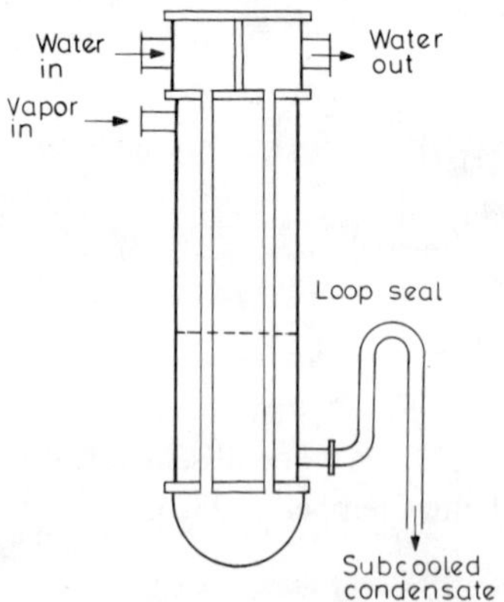

Figure 3.14 Vertical condenser–subcooler.

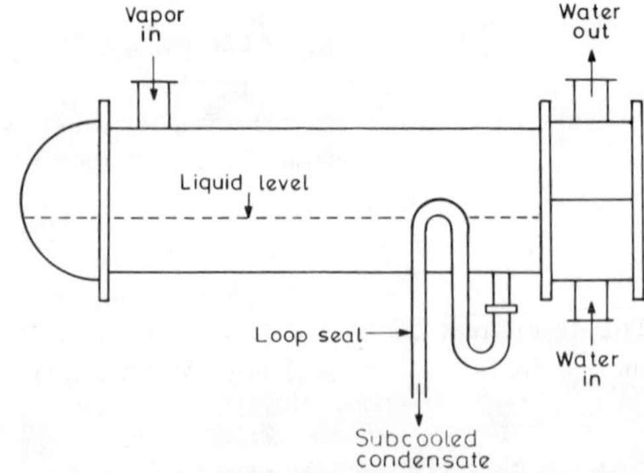

Figure 3.15 Horizontal condenser–subcooler.

assumed submergence of the tube rows has to perform the required subcooling, while the remainder of the tube rows perform the condensing duty. The two zones are not very well defined. Hence, the design is difficult. However, the overall heat transfer coefficient is higher in a horizontal unit than in a vertical unit because the condensing coefficient in the former is higher (Q. 3.12). Furthermore, the horizontal unit is easily mounted above a tall distillation column, whereas a vertical unit requires a large head room that may not always be available. Hence, a larger number of horizontal units are in use compared to the number of vertical units.

Multi-Component Condensation

- Under this heading will be included vapor mixtures containing two or more components, including noncondensables and immiscibles.

3.26 What is the controlling resistance to heat transfer in a condenser?

In the case of condensation of a pure vapor, it is generally the resistance on the coolant side. If, however, noncondensables are also present that form a layer over the condensate film, the diffusion of the vapors through this film also contributes significantly to the total resistance and, in some cases, may even be the controlling one.

3.27 Is the condensing coefficient constant in such a case?

No, it varies a great deal from the inlet to the outlet of the condenser. The condensing coefficient is very high at the inlet. As the vapor condenses, its flow rate, temperature, and thermal and physical properties vary along the condenser length. Its partial pressure in the bulk gaseous stream also reduces and so does the driving potential for the vapor to diffuse through the noncondensable gas film toward the condensate film. Thus, the rate of condensation and the condensing coefficient keep getting lower along the condenser path. For example, steam containing 1 percent air may have a condensing coefficient of 1000 Btu/hr ft^2 °F (5678 J/m^2 s K) at the inlet and 50 Btu/hr ft^2 °F (284 J/m^2 s K) at the outlet.

3.28 Is the heat released per unit condenser length uniform in the above case?

The heat is released both due to sensible cooling and the latent heat of condensation. Since the condensing coefficient varies as the composition and hence the properties of the gaseous mixture, both the rate of condensation and sensible cooling will vary along the condenser length. Therefore, the heat released per unit length is not uniform.

3.29 In what important aspects does the condensation of vapors in the presence of noncondensable gas differ from the case of pure vapor condensation?

As condensation proceeds, the composition and hence the thermophysical properties of the gas–vapor mixture keep changing. The mass flow rate of the vapor to the condensate film also changes due to the decreasing vapor pressure and the

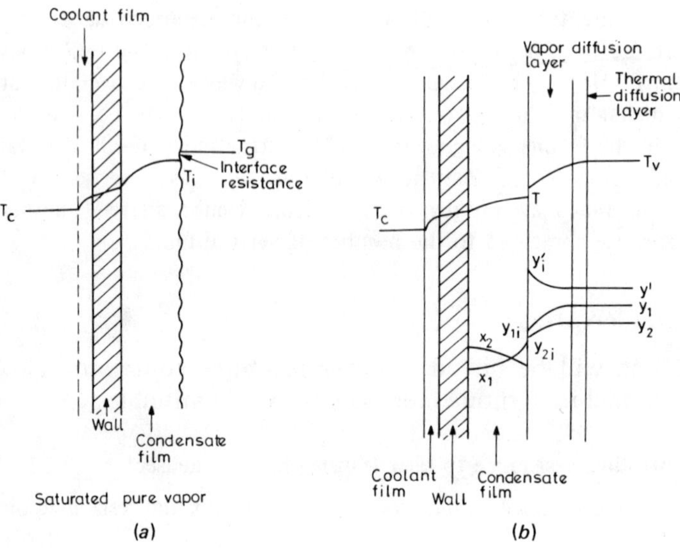

Figure 3.16 (*a*) Pure vapor diffusion. (*b*) Multi-component diffusion.

increasing thickness of the noncondensable gas film on the condensate surface. The condensing coefficient and the heat released per unit length are also not uniform as discussed earlier. These phenomena are not encountered in the condensation of a pure vapor.

The two cases are shown schematically in Fig. 3.16*a* and *b*. The multi-component mixture is taken to contain two condensable and one noncondensable components. (The various terms used are explained in the Nomenclature.) The difference between the two are evident from the figures. Furthermore, the calculation procedures are also different, since the noncondensable film offers high resistance to mass transfer (diffusion) of the condensables to the condensate film. Even when all the components are condensable, the calculation procedure is different than that for a pure vapor, since the mixture condenses over a range of temperature (except a binary mixture of immiscible or an azeotrope), while the pure vapor condenses at a fixed temperature for a given operating pressure (see also Q. 3.39).

3.30 What is the principal difference in the condensation of a binary mixture and a multi-component mixture?

A binary mixture generally condenses over a small temperature range, whereas a multi-component mixture condenses over a wide temperature range unless it is a mixture of components of a homologous series (e.g., butane, pentane, hexane, etc.). In a wide temperature range, the heat released per degree drop in temperature is not uniform. Hence, ΔT_{LMTD} cannot be used in the calculation of a multi-component condenser. Instead a condensing curve is used for the purpose. Generally, it gives a greater temperature difference than the ΔT_{LMTD}. Hence, the use of ΔT_{LMTD} will be too conservative in the case of a multi-component condenser design.

3.31 When a binary mixture condenses, is the composition of the condensate at each point uniform? How is the rate of condensation of each component related to one another?

The composition of the condensate is not uniform at each point in the condenser. The higher boiling (low volatility) component condenses in greater proportion near the condenser inlet, while the lower boiling (high volatility) component condenses in greater proportion near the condenser outlet.

Similar to the driving potential for a pure vapor discussed in Q. 3.1, the driving potential for each component is the difference in its partial pressure in the vapor and its vapor pressure at the temperature of the condensate film. This also gives the relation between the condensation of the two components, since their partial pressures (or fugacities) in the vapor add to give the total pressure (or total fugacity).

3.32 Since the composition of the condensate formed from a binary mixture of vapors varies along the condenser path (Q. 3.31), what would be the average composition of the condensate at the outlet of a total condenser or a partial condenser?

In a total condenser, since all the vapor condenses, the average composition of the condensate at the outlet will be the same as the vapor composition at the inlet, assuming all the condensate is well mixed.

In a partial condenser, only a part of the vapor condenses, while the remainder leaves in the vapor form itself. Since the composition of the condensate varies along the condenser path, so does the composition of the remaining vapor. Hence, it is not possible to predict what the average condensate composition will be except that it will not be the same as either the inlet or the outlet vapor composition.

3.33 How is a condensing curve drawn for a multi-component system?

The condensing curve is a plot of vapor heat content versus vapor temperature for the condensing range. To plot it, point-to-point compositions of the liquid and vapor phases have to be determined along the condensing range and the heat content of the vapor evaluated. For this purpose, various temperature intervals are chosen in the condensing range and the composition and heat content determined for each interval. The condensing curve is based upon two assumptions (Bell and Panchal, 1979):

1. The entire vapor-liquid mixture at any cross section in the condenser is isothermal.
2. The vapor and liquid phases are in thermodynamic equilibrium at each cross section.

Both these assumptions do not quite hold but lead to acceptable designs. A good example has been given by Kern (1950), wherein the calculations have been

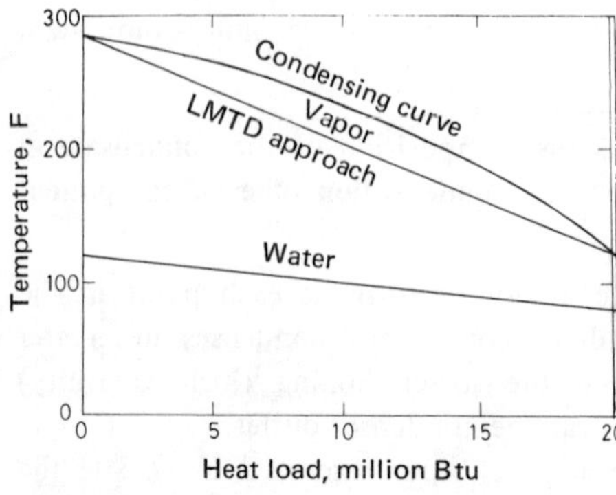

Figure 3.17 Multi-component condensing curve versus LMTD approach. *(From Kern, 1950.)*

done by both a condensing curve and the ΔT_{LMTD} approach (Fig. 3.17). The latter came out to be approximately 13 percent more conservative.

3.34 What is meant by fronts and tails in a multi-component condensation?

Fronts refer to the presence of small amounts of lower boiling components in a multi-component vapor mixture that reduce the bubble point significantly even though the heat removed between the old and new bubble points will not be a significant fraction of the total heat removed. For example, the presence of a small amount of propane in a mixture of pentane, hexane, and heptane constitutes a front.

Similarly, tails refer to the presence of small amounts of higher boiling components that raise the dew point considerably, although only a small amount of heat is removed till the main components start to condense at a much lower temperature. For example, the presence of a small amount of decane in a mixture of pentane, hexane, and heptane constitutes a tail.

Fogging

3.35 What is fog in a condenser?

It is the homogeneous condensation on minute nuclei in the condenser shell (away from the tubes) when the vapor is supercooled (or supersaturated) below its dew point. This situation comes about in the following cases (Steinmeyer, 1972):

1. Mixing of two saturated vapor streams at different temperatures produces a supersaturated mixture at the mixed temperature.
2. Heat transfer (cooling) in the vapor is faster than the vapor mass transfer to the cold condensate film, and it is more likely to occur when a noncondensable film of lower molecular weight gas is present.

The nuclei on which the vapor condenses to form the fog are foreign nuclei, such as particles and ions in the air or entrained liquid droplets. If enough nuclei are present, fog will form. Fogging is undesirable since it reduces the amount of condensate formed and some of the condensate on the suspended nuclei is carried away by the exiting gaseous stream and is lost.

3.36 How can fogging be minimized?

A large temperature difference (ΔT) between the bulk vapor and the cold wall in the presence of a low molecular weight noncondensable or a wide range of molecular weights of the condensables results in a faster rate of heat transfer compared to the rate of mass transfer and hence in fogging. Therefore, in the above situations, a large ΔT should be avoided.

Colburn and Edison (1941) gave an equation for predicting fogging. Branan (1976) gave a simple procedure to predict fogging by considering heat and mass transfers as separate processes and taking the ratio of the two, which, for a given case, is proportional to the respective driving potentials. Hence, a design should be checked for possible fogging. If fogging is predicted, ΔT should be lowered, noncondensables should not be allowed to accumulate, and larger surface area should be used. Prediction of fog formation along with an example is also discussed by LoPinto (1982).

If fogging does occur, special mist eliminators are necessary to coalesce the fog particles into droplets that will settle down.

Design

3.37 Why is the design of a condenser more complicated than that of a heat exchanger without any phase change?

In a heat exchanger without any phase change one has to account for the sensible heat transfer and the related pressure drop in the unit along with the effects of the bypass and leakage streams.

In a condenser, on the other hand, one has to account for the following:

1. Sensible heat transfer in the vapor and liquid phases and the latent heat of condensation. Sensible heat transfer coefficients are rather low
2. The associated pressure drop that is different for the vapor, two-phase flows, and the condensate streams because of their widely different flow rates and physical properties
3. The mass transfer through a film of noncondensables. This is very important and also difficult to calculate since data do not exist for all the possible systems
4. The decreasing driving potential for mass transfer due to the depletion of the condensable vapors
5. The nonlinear heat release that requires piecewise calculations instead of an overall calculation as is generally done for a no-phase change unit
6. The two-phase flow regimes
7. The change in the vapor velocity that may change the Reynolds number (from turbulent to laminar)
8. The presence of the immiscibles that affects the operation of the condenser
9. The effects of the bypass and leakage streams as in the no-phase-change units
10. The inside- or outside-tube condensation and the vertical or horizontal orientation of the unit

11. The changes in the physical properties of the mixture at each step since the vapor and liquid compositions change as the condensation proceeds
12. The change in the overall heat transfer coefficient in multiple coolant passes

Thus, it is evident that the process design of a condenser is significantly more involved than that of a no-phase-change heat exchanger. Consequently, there is a strong incentive to computerize the condenser design methods.

3.38 What factors, both process- and geometry-related, affect the condensing coefficient and the condenser design?

The process-related factors are the following:

1. Number and amounts and molecular weights of the components in the vapor phase; presence of noncondensables, miscible, and immiscible components
2. Superheated or at saturation temperature
3. Dew point and bubble point
4. Condensing at a unique temperature or in a temperature range
5. Partial or complete condensation
6. Temperature, pressure, and flow rates of the vapors and the coolant
7. Overall heat transfer coefficient or condensing coefficient
8. Physical properties of the vapor, condensate, and the coolant and their dependence on temperature, pressure, and composition: The properties include density, viscosity, mass diffusivity, heat capacity, thermal conductivity, and surface tension, and also, any polymerizing or fouling tendencies of the condensate at the operating conditions.
9. Latent heat of condensation of vapors
10. Extent of subcooling desired
11. Gas blanketing of the condenser surface
12. Gravity-controlled or shear-controlled condensation
13. Vibration of the tubes due to fluid flow or acoustic vibrations in shell-side condensation
14. Erosion of tubes due to suspended droplets or particles

The geometry-related factors are the following:

1. Condensation inside tubes or outside tubes
2. Vertical or horizontal tubes
3. Number of passes in the condenser
4. For horizontal, outside-tube condensation, the number of tube rows crossed
5. Total number of tubes in the shell
6. Plain or finned tubes
7. Liquid seal at the bottom for subcooling of the condensate

All this information should be known to the designer. The rest of the information needed is given in the heat exchanger specification sheet (Fig. 1.37) and discussed in Q. 1.115–1.121, see also Q. 3.50.

3.39 Briefly, what are the various methods for calculating the condensing coefficients?

There are several methods proposed in the open literature, both for specific situations and general condensing phenomena, and newer ones are also being evolved. The salient features of the more popular ones are briefly mentioned below. For details, the references cited should be consulted.

1. Nusselt's method (Kern, 1950; Nusselt, 1916): Its application and assumptions have already been discussed in Q. 3.17 and 3.18. It was the first method proposed and has been used a lot, even when it should not have been applied because of its limitations.
2. Colburn's methods: There are several. One accounts for the turbulent motion of the condensate film and has been used in some designs (Colburn, 1934). The others (Colburn and Edison, 1941; Colburn and Hougen, 1934) are for the condensation of a vapor from a binary mixture containing one vapor and another noncondensable in a single-pass unit.
3. Kern's method (Kern, 1950): It assumes no resistance to heat transfer in the vapor phase for fully condensable mixtures, i.e., the vapor is assumed to be at its dew point. Then, the area so calculated is multiplied by the ratio of the total heat load to the condensing load. For the condensation of a vapor from moncondensables, Kern recommends the Colburn-Hougen (Colburn and Hougen, 1934) method and also gives some rough estimating techniques.
4. Dukler's method (Dukler, 1959): This is valid for vertical condensers. The condensing coefficients are much higher than predicted by either the Nusselt or the Colburn method. It accounts for the vapor shear, Prandtl and Reynolds numbers, and holds both in the laminar and turbulent regimes.
5. Ward's method (Ward, 1960): It neglects the sensible heat of cooling of the condensate film. It is otherwise a good design method for multi-component systems and multi-coolant passes.
6. Bell and Ghaly's method (Bell and Ghaly, 1972): It accounts for the sensible heat of cooling of the condensate film. The overall heat transfer coefficient is assumed constant in all the passes, as is also the case in all the above methods. This method is applicable to condensing inside or outside the tubes, with or without vapor superheat, and is widely used in the industry for multi-component/partial condensers with one, two, or any even number of coolant passes. It is similar to Silver's method (Silver, 1947), although it was proposed independently.
7. Butterworth's method (Butterworth, 1975): It is a slight generalization of the Bell and Ghaly's method. It accounts for the variation in the overall heat transfer coefficient along the length of the shell with two coolant passes and one shell pass (E-type shell).
8. Emerson's method (Emerson, 1975): It is similar to Butterworth's method except that it can be used for any number of coolant passes.

Besides the above rigorous methods, other simplified methods also appear in the literature from time to time for some specific situations. These use nomographs and/or simplified equations (Starczewski, 1981). These may be good for quick estimates.

It is difficult for any method to account for all kinds of situations that arise in the design of condensers. Hence, before using any method, its assumptions, limitations and range of applications should be thoroughly examined.

3.40 How does the turbulent flow influence condensation?

In a condenser, four situations can occur with regard to the flow regimes in gravity driven condensate films:

1. Vapor and condensate both in laminar flow
2. Vapor in laminar flow, condensate in turbulent flow
3. Vapor in turbulent flow, condensate in laminar flow
4. Vapor and condensate both in turbulent flow

It should be pointed out that all condensing equipment does have the condensate in laminar flow for at least part of the length until the condensate film becomes turbulent.

The net effect of the turbulence in the condensate film is to increase the condensing coefficient. The heat transfer to the cold wall in that case is not only by conduction but also by the eddies, and hence the rate of heat transfer is increased. The Reynolds number is also high, which in turn increases the heat transfer coefficient. The turbulent in the vapor phase causes greater shear on the condensate film and reduces its thickness and hence also the resistance to heat transfer to the cold wall. The net effect is an increase in the condensing coefficient.

Case (1) is the classic Nusselt's theory. Case (2) was first dealt with by Colburn (1934), and both cases (2) and (4) have been dealt with by Dukler (1959) who accounts for the Prandtl number and the vapor shear. Case (3) has been dealt with by Blangetti and Schlünder (1979). It is found that the condensing coefficient is higher in cases (2)-(4) compared to case (1), except at very low Prandtl number (Dukler, 1959).

3.41 What are the various methods for calculating the pressure drop in a condenser?

Pressure drop varies with the kind of the two-phase flow regime. In a stratified flow, a conservative estimate is to take the arithmetic average of the pressure drop calculated at the inlet and exit vapor conditions. It neglects the pressure drop due to condensate flow, which is actually quite small. In the case of a total condenser, the exit pressure drop is close to zero since only the condensate leaves the condenser. Hence, in such a case, it is half the pressure drop based upon the inlet conditions.

In a more general case, for inside-tube condensation, standard methods as given, for example, by Collier (1980) can be used. An early method was given by Lockhart and Martinelli (1949) and Rouet (1983), which has been extensively used. It considers each phase separately and then a combined effect is formulated. For other methods, refer to Collier (1980).

For the outside-tube (shell-side) condensation, the situation is more difficult. The various flow regimes are shown in Fig. 3.10. A review of various methods up to 1977 for predicting the pressure drop has been given by Ishihara et al. (1980). More relations have appeared since, including one by Grant and Chisholm (1979). An extensive discussion on the pressure drop in tubes and shell and also in singularities (bends, junctions, sudden area changes, manifolds, etc.) is given in Bergles et al. (1981). A systematic presentation of the applicable relations for various situations has been given by Butterworth (1981b).

3.42 Is the pressure drop per unit length constant in a condenser?

No. It is maximum near the inlet and minimum near the outlet because the velocity decreases due to the vapor condensing from the inlet to the outlet. The pressure drop being proportional to the velocity squared decreases at a faster rate than does the velocity. The condensate does not contribute much to the pressure drop, since its volumetric flow rate is very small. Furthermore, there is some pressure recovery due to the slowing of the vapor. However, in many methods, the pressure recovery is neglected as a safety measure to make the calculations more conservative.

3.43 How should the baffles be spaced in a condenser?

For good vapor distribution over the condensing surface it is necessary that the vapor travels fast through the condenser. As condensation occurs, the amount of vapor decreases. In order to keep nearly the same vapor velocity, one should stagger the baffles, spacing them more closely toward the outlet compared to the spacing near the inlet (Fig. 3.18). However, this will rule out the effective utilization of this unit in any noncondensing duty should a change in the process or in the demand so require. Hence, the staggered baffle spacing, although advantageous, is not generally used.

3.44 What factors influence the choice of velocity for gaseous flow on the shell side?

The factors are the molecular weights and composition of the components in the gaseous phase, pressure drop, vibrations of the tubes due to cross-flow, acoustic

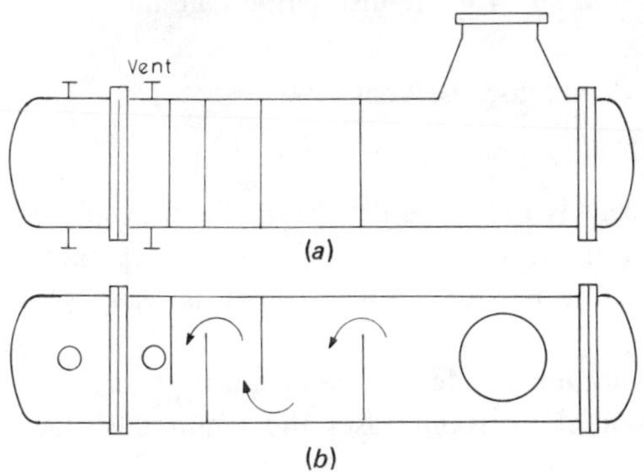

Figure 3.18 Staggered baffle spacing in a condenser. (*a*) Side view. (*b*) Top view. (*From* The Process Engineer's Pocket Handbook *by Carl Branan. Copyright 1976 by Gulf Publishing Company, Houston, Texas. Used with permission. All rights reserved.)*

vibrations, presence of suspended droplets or particles, partial or complete condensation inside the shell, etc.

3.45 How is the pressure drop for steam condensing on the shell side calculated and what is its range of values?

The pressure drop is calculated by taking one-half of the pressure drop for the inlet conditions since the contribution of the condensate to the pressure drop is negligible. The value of the pressure drop generally lies in the range of 0.1 to 1 psi (0.6895 to 6.895 kPa). This is a rather conservative way since the pressure drop is proportional to the square of the velocity. As the velocity decreases, the pressure drop decreases faster.

3.46 How is the pressure drop calculated in a partial condenser?

An average of pressure drops calculated on the basis of the inlet and the outlet vapor flow conditions gives a good estimate of the pressure drop in the partial condenser both for the horizontal and the vertical orientations of the unit.

In a horizontal condenser, more precise calculations may be done based upon the principle of the balanced pressure drop. It states that the pressure drop suffered by the vapor and the condensate should be the same since the two are flowing in the same unit parallel to one another. The pressure drop for the two phases should be individually computed based upon the average flow area, hydraulic radius, Reynolds number, and the physical properties. If the individual pressure drops do not match, then the assumed flow area should be changed and calculations repeated until the two match.

3.47 If the operating pressure of a distillation column and hence of the condenser is increased, what effects is it likely to have?

At the higher operating pressure, both the dew point and the bubble point of the mixture increase. For a given cooling water temperature, the ΔT_{LMTD} will increase, and hence the surface area required for condensation will decrease, resulting in a lower initial cost. However, the reboiler will have to supply more heat to maintain the high temperature in the system. The mechanical design will require thicker shells for the condenser, reboiler, and also for the distillation column.

3.48 Why should the pressure drop in a condenser be kept low?

There are two important reasons:

1. Usually the pressure of the vapor stream is only slightly above the atmospheric pressure. Hence, in a condenser operating at 1 atm, there is not much pressure drop available. In vacuum condensers, the pressure drop available is really very small.
2. A high pressure drop lowers the equilibrium condensing temperature and hence reduces the temperature driving potential. This increases the required surface area.

3.49 Qualitatively what are the effects of various properties, and operating and design parameters on the condensing coefficient?

The effects of some of the properties and the parameters are discussed below:

1. A high value of the thermal conductivity of the condensate results in a higher condensing coefficient since it increases the rate of heat conduction through the condensate film.
2. A high value of the viscosity of the condensate results in a lower condensing coefficient since it gives a thicker condensate film, which results in a greater resistance to heat transfer.
3. A high value of the density of the condensate results in a higher condensing coefficient since it increases the drainage velocity of the film whose thickness therefore decreases, resulting in lesser resistance to heat transfer.
4. A high value of the surface tension of the condensate results in a lower condensing coefficient since the film does not drain easily. The effect is more pronounced in condensation over the finned tubes.
5. A high value of the latent heat of condensation of the vapor results in a higher condensing coefficient since it releases proportionately more heat per unit mass of vapor condensed.
6. Reynolds number, lying in the turbulent zone, results in higher condensing coefficient due to transfer of heat by the eddies and the higher forced convection heat transfer.
7. A low amount of superheat results in higher average condensing coefficient since the heat transfer coefficient for sensible heat is far less than for pure vapor condensation.
8. Presence of noncondensables cuts down the condensing coefficient significantly.
9. Condensing coefficient is generally higher in a horizontal unit than in a vertical one (Q. 3.12).

3.50 In Fig. 1.37, a heat exchanger specification sheet is given for a heat exchanger. What additional information is required for designing a condenser?

The additional information required is the condensing curve, dew point, bubble point, latent heat of condensation, molecular weights and vapor pressures of the components, operating pressure, amounts of condensables and noncondensables, if subcooling is also required then the amount of subcooling, orientation (vertical or horizontal), inside- or outside-tube condensation, etc. For a pure vapor condensing some of the above items will not apply.

3.51 What are the problems in the formulation of more accurate design procedures?

There are several problems, and the important ones are the following:

1. Lack of data or good correlations for mass diffusivity at various temperatures, pressures, and concentrations

2. Lack of data or good correlations for vapor–liquid equilibrium
3. Lack of good data or good correlations for two-phase flow correlations when amounts of vapor and liquid are changing
4. Uncertainty about condensing of a vapor mixture containing an immiscible component. The condensation could result in droplets of various components, or each one flowing in separate channels, or one forming a film over the surface while the other one condenses as drops over this film. The condensing coefficient in each of these three cases would be different and is difficult to calculate.
5. Lack of good data or good correlations for the effects of various geometrical features of the condenser.

Items 1–3 are very difficult to obtain for the wide range of mixtures and operating conditions. Presence of noncondensables and immiscibles complicates the picture still further.

Apart from the above, the industries tend to use simpler design procedures even though the large overdesign is costly both initially and during the operation. The reason is that the industries want to be very certain that their condensers will work. Furthermore, some of the more complex design methods require a lot of computation time, and many designers are also reluctant to use them. The need of simple computer programs (Hayde, 1981) continues, although the cooperative research groups (e.g., HTRI, HTFS) are bringing out computer packages using complex design procedures.

3.52 What are some of the design recommendations for condensers?

For preliminary design the following are recommended:

1. When pure steam is condensing, the condensing coefficient can conservatively be taken to be 1500 Btu/hr ft^2 °F (8520 J/m^2 s K), regardless of whether it is condensing on the inside or outside of a tube. Since it is not the controlling resistance, there is no need to calculate it very accurately.
2. The following overall heat transfer coefficients should be used for fresh-water-cooled, single-component condensers. Values for sea-water-cooled condensers are given in the parentheses
   a. Organic vapors: 120–160 Btu/hr ft^2 °F or 681–909 J/m^2 s K (100–120 Btu/hr ft^2 °F or 570–680 J/m^2 s K)
   b. Organic vapors with small acmount of noncondensables: 90–120 Btu/hr ft^2 °F or 510–680 J/m^2 s K (70–110 Btu/hr ft^2 °F or 400–625 J/m^2 s K)
   c. Aqueous vapors: 200–250 Btu/hr ft^2 °F or 1130–1420 J/m^2 s K (180–220 Btu/hr ft^2 °F or 1020–1250 J/m^2 s K)
   d. Vacuum condensers with small amounts of noncondensables: 50–90 Btu/hr ft^2 °F or 285–510 J/m^2 s K (40–75 Btu/hr ft^2 °F or 230–425 J/m^2 s K)
3. Desuperheating should be neglected in the design unless it contributes about 25 percent or more of the total heat load.
4. When the heat flux is small, use low-fin tubing (Q. 4.16). However, aqueous vapors should not be condensed on the low-fin tubes since the high surface tension of water bridges the gap between the fins, thus blanketing out a large

fraction of the available surface. This may make a finned tube worse than a plain tube for the high surface tension condensates. Use of finned tubes is discussed in Q. 3.55–3.58.

5. For most applications, horizontal outside-tube (shell-side) and vertical inside-tube condensers are generally suitable and should be evaluated before doing any special design. In horizontal shell-side condensers, the baffle cut should be vertical so that the condensate can easily flow without damming behind the baffles, which would otherwise make a larger number of tubes ineffective.
6. In tube-side condensation, the maldistribution of vapors should be prevented. In such horizontal units with one tube pass, a 1° to 2° tilt should be provided for easy flow of the condensate.
7. In shell-side condensation, impingement plates should be provided to prevent tube erosion from suspended condensate droplets and to prevent tube vibrations due to high velocity of the entering vapor.
8. In shell-side condensers, the vapor must be made to flow over the tube bundle by judiciously using the baffles and seal strips (if required) and minimizing manufacturing tolerances. Direct flow of the vapor from the inlet to the outlet should be prevented. A proper location of the inlet and outlet nozzles and vent lines also helps a great deal in it. The vent should be near the exit, and no condensate should flow out from it.
9. To achieve high vapor velocity throughout, staggered baffle spacing should be used. This helps in maintaining a high condensing coefficient, greater mass transfer through the film of noncondensables, greater vapor shear on the condensate film whose thickness will be reduced, and pushing the noncondensables toward the exit without letting them build up too much. Overall heat transfer area required reduces significantly compared to the case of equal baffle spacing. Semeria and Hewitt (1974) found a reduction of 36 percent in the area required for a condenser set-up with unequal baffle spacing compared to the one with equal baffle spacing. The pressure drop in both the cases was small (see Q. 3.43 also).
10. Baffles should be sized and arranged such that there is no dead region for the accumulation of the condensate or the noncondensables.
11. Dirty, corrosive, and/or high-pressure stream should preferably be put on the tube side.
12. The exit nozzle should be properly sized for the drainage of the condensate, which should not accumulate since it will submerge some of the tube rows and adversely affect the performance.
13. Since maximum uncertainty exists in horizontal tube-side condenser designs, extra surface should be provided as a safety factor.
14. Possible change in the two-phase flow regime due to reduction in the vapor velocity as the condensation proceeds should be kept in mind.

3.53 Why cannot the usual design procedures be applied to the condensation of mercury vapors?

Mercury does not wet the condenser surface and hence condenses dropwise instead of filmwise. It does not also follow the no-slip condition (or zero velocity at the liquid–solid contact surface), which is implicitly assumed in the momentum balance. The actual heat transfer coefficient in the dropwise condensation is significantly higher than the one predicted by the filmwise condensation. Therefore, a unit designed by the usual procedures of filmwise condensation for the mercury vapors will be grossly overdesigned.

3.54 What are the types of shell and tube bundles generally used in the condensers?

The shells are generally E-, J-, and X-types (Fig. 1.1), depending upon the pressure drop available (minimum for the X-type). The tube bundles can be fixed, floating-head, or the U-tube type (Fig. 1.1) depending upon the severity of differential thermal expansion and the shell-side fouling. The tubes may be plain or finned, depending upon the vapor-side coefficient and the surface tension of the condensate. (Finned tubes are discussed next.) The baffle spacing can be single- or multiple-segmental, or no-tube-in-window design may be used for very low pressure drops (Figs. 1.14–1.17). The units may be horizontal, vertical, or inclined, with condensation on the outside of the tubes (shell side) or inside the tubes. In the latter case with a vertical unit, the vapor flow may be down or up.

There are some advantages and some disadvantages related to each type of construction. These relate to the initial costs, cleaning the fouling deposits, subcooling of the condensate, accumulation of the noncondensables, blanketing of the tubular surface, bypass and leakage streams, pressure drop, vapor distribution, flexibility in operation (change in throughputs), possible alternate usage, temperature driving potential, number of tube passes possible, condensing range handled, vapor velocity, mounting of the unit, instabilities in condensation, etc. These have to be evaluated for each case.

Use of Finned Tubes

3.55 What are the advantages of low-fin tubes in condensation?

The condensing coefficient per unit length is higher, and hence a shorter condenser can be used with consequent reduction in the pressure drop and initial and operating costs. It is used when the gas side condensing coefficient is low, e.g., in the condensation of refrigerants.

3.56 Why is the condensing coefficient higher for a finned tube than for a plain tube?

The reasons for this are the following:

1. The surface area of the finned tube is greater than that of the same length of the plain tube.
2. The condensate formed on the fins drains into the space between two fins thus exposing a large part of the fins to the vapor. This offers very little resistance to further condensation. The thick film formed in the trough drains down due to gravity. Surface tension plays a role in the retention of the condensate (Fig.

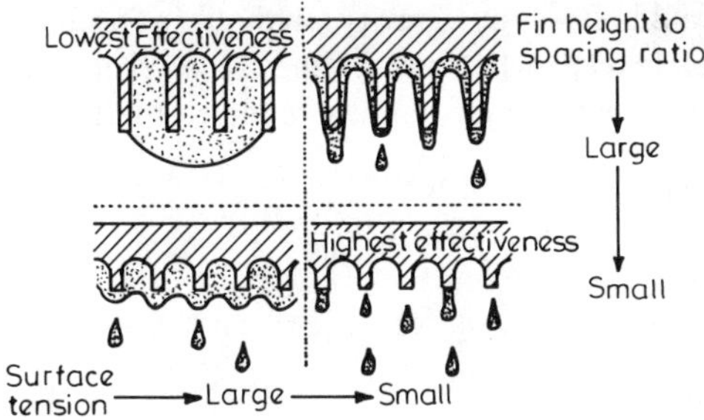

Figure 3.19 Condensate retention on finned tubes. *(From Taborek, 1974.)*

3.19). Significant work on it has been done by Beatty and Katz (1948). For later developments, see Tanasawa (1979) and Webb (1982).

3.57 What factors influence the effectiveness of the fins in a finned tube condenser?

The factors are: flow rate of vapor, vapor composition, surface tension of the condensate, bonding of the fins to the tube, the dimensions of the fins and their number per unit length.

3.58 How do the fluted tubes work in condensers?

In these tubes, although the condensation is filmwise, the condensation coefficients produced are in the dropwise range. See Q. 1.21 and Carnavos (1974) for the details of operation of the fluted tubes. Finned and fluted tubes are discussed at length in Chapter 4.

Operating Problems

3.59 When does the condenser operation become unstable?

If the rate of condensation is much faster than the rate at which the condensate can be drained, the tubes start to get filled up. This is more likely to happen in a gravity-controlled condensation case. When, due to the condensate thus filling the tubes, there is not enough surface available for condensation, the incoming vapor will blow out the condensate from the tubes and the process is repeated. This leads to both flow and pressure instabilities and should be avoided.

Another kind of instability noticed by Wild (1969) has been reported by Semeria and Hewitt (1974). It relates to the subcooled condensate level fluctuating due to the condensation of a pure vapor which alternately releases and builds up pressure in the vapor space causing the condensate level to fluctuate. Addition of small amounts of noncondensables reduced the rate of condensation and hence the fluctuations were not encountered anymore.

3.60 What are the problems in condenser operation?

These are the following:

1. Maintaining cleanliness: The fouling due to cooling water and the condensate should be minimized, both by proper water treatment and by avoiding dead zones.

2. Draining of the condensate: It should flow by gravity without a need to maintain a head within the condenser (except when subcooling is also required). Otherwise, the lower portion of the unit gets flooded and is rendered ineffective for condensation duty.
3. Efficient venting of the inerts
4. Avoiding fog formation

Problems in formulating more effective design procedures have been discussed in Q. 3.51.

Reflux, Knock-Back, and Vacuum Condensers

3.61 What is a reflux condenser?

It is a total condenser used to return all the condensate to the unit from which the vapor stream is originating (Fig. 3.20). It is used in distillation experiments to gather data, in situations where the process requirements are to keep a liquid at its boiling point, and in the storage of volatile liquids where the vapor is condensed and returned to the storage vessel. The condensate can get subcooled in this type of condenser since the vapor enters at the top and flows down through the tubes.

3.62 What is a knock-back condenser?

In these also the condensate is returned to the unit from which the vapor steam is originating. These differ from the reflux-type condensers in the sense that the vapor enters the tubes from the bottom (Fig. 3.21) instead of from the top. As the condensate forms on the tube wall, it drains down flowing counter to the hot

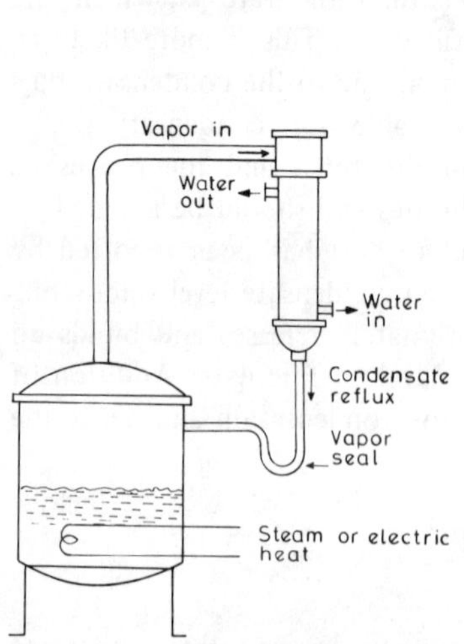

Figure 3.20 Reflux condenser.

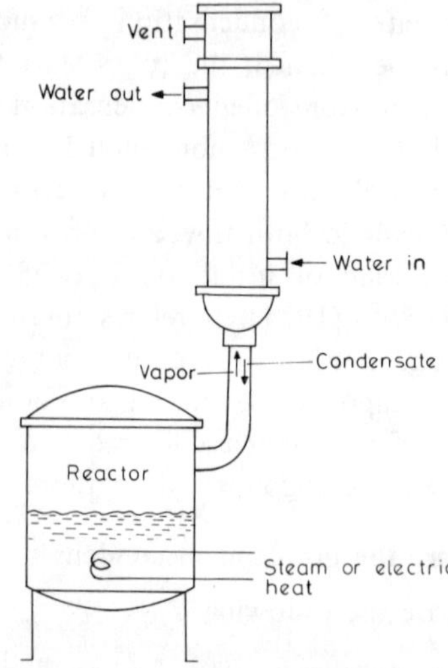

Figure 3.21 Knock-back condenser.

vapor stream. Thus, the condensate cannot get subcooled. Furthermore, if the vapor velocity is very high, it can result in the flooding of the tubes with liquid unable to flow down. This should be avoided. See also the discussion in Q. 3.13 on vertical condensers with upward flow of vapor.

Both the reflux and the knock-back condensers can be mounted on the units generating the vapor stream to reduce the piping length.

3.63 What is a vacuum condenser?

It is a condenser operating under vacuum. It is used on top of the distillation columns that operate under vacuum, such as in vacuum distillation in a petroleum refinery. Surface condensers in power plants also operate under vacuum (see Q. 3.64–3.69).

The important points to keep in mind are to prevent the leakage of air into the vacuum condenser, to have a system to vent out the accumulated noncondensables, and to reduce the pressure drop in the system. The pressure drop is minimized by (1) mounting the condensers very close to the distillation column to reduce the frictional losses through long pipings, (2) an enlarged inlet to the shell to reduce the entrance pressure loss (Fig. 1.22), and (3) using staggered baffle spacing, more at the entrance and less toward the exit to reduce the pressure loss when the vapor flow is high at the entrance (Fig. 3.18).

Surface Condensers

- For general description see Q. 3.6 and Fig. 3.2.

3.64 What are the functions and operation of a surface condenser?

The functions of a surface condenser are the following (Baker, 1981):

1. To condense the turbine exhaust steam
2. To maintain a specified vacuum at the turbine exhaust in order to keep the power cycle efficiency high
3. To recover the clean condensate for return to the boiler

As far as the operation is concerned, on the shell side, the steam velocity is kept at 100-200 ft/sec (30.5-61 m/s) at a low vacuum of 26 in Hg (13.5 kPa) or 200–400 ft/sec (61-122 m/s) at a high vacuum of 29 in Hg (3.38 kPa). The pressure drop is kept at below 0.5 in Hg (1.69 kPa). Noncondensable gases and air accumulate near the exit from which they are sucked out. On the tube side, the cooling water velocity is kept at 6–8 ft/sec (1.8-2.5 m/s) to increase the heat transfer coefficient and reduce the fouling deposit. The cooling water is sometimes chlorinated and otherwise also treated with appropriate chemicals to reduce algae, microbial, and other types of fouling (Chapter 13). Proper water treatment is an important factor in the satisfactory operation of the surface condenser.

In the mechanical design, the shell, operating under vacuum, has to withstand the external atmospheric pressure. The pressure differential across the tubes is quite high too, with the cooling water at 5–25 psig (135–275 kPa) and the steam in the shell at a vacuum of approximately 14 psi (4.83 kPa).

3.65 What factors govern the design of a surface condenser?

A surface condenser is designed based upon the HEI Standards (1978) or the British Standards (1967). The important factors involved in the design are the heat load, extent of noncondensables, vacuum on the shell side, cooling water inlet temperature and velocity, rise permitted in the cooling water temperature, and the fouling factors. A data specification sheet is given in Fig. 3.22 and a detailed section on specifications is given in HEI Standards (HEI, 1978).

3.66 Why is steam velocity maintained high in surface condensers?

High steam velocity (Q. 3.64) shears off the condensate film, which becomes very thin and the resulting heat transfer coefficient is high. Furthermore, it helps steam to penetrate deeper into the tube bundle laid out on a radial pitch and thus use the available surface area more effectively. It also helps in driving the noncondensables out of the condenser.

3.67 What are the major problems in the operation of a surface condenser?

The major problems relate to the following:

1. The removal of air as completely as possible from the boiler feed water
2. The maintenance of vacuum on the shell side and prevention of any air leakage into the shell
3. The tube vibrations due to high velocity of the entering steam and inadequate design of the impingement baffles
4. The fouling and corrosion of the tubes due to cooling water
5. The erosion–corrosion of the tube outside surface and tube-to-tubesheet joints (Impagliazzo and Bell, 1978) due to high steam velocity containing suspended droplets of the condensate

3.68 How does the presence of air influence the condensing coefficient of steam?

The presence of only 1 percent air causes the condensing coefficient of steam to drop by about 50 percent (Othmer, 1929). This is because the steam has to diffuse through a film of air to reach the condensing surface. The resistance offered by the air film is very high. The condensing coefficient drops to almost one third when only 2 percent air is present. The problem of air in steam condensing is encountered in the surface condensers operated under vacuum power plants, where, despite best efforts, air from the boiler feed water or leakages tends to accumulate.

Furthermore, a temperature gradient also exists due to the noncondensable air film. This means that the actual steam temperature has to be higher than without the noncondensables in order to overcome this temperature gradient and also to provide the requisite ΔT for the condensation to occur. This also raises the pressure in the condenser shell and adversely affects the power cycle efficiency.

Air dissolved in the condensate (water) also leads to corrosion of the unit.

SURFACE CONDENSER SPECIFICATION SHEET

No.	Item	Unit	Shell side		Tube side	
1	CUSTOMER	*	EQUIPMENT TAG NO. *			
2	PLANT LOCATION	*	MANUFACTURER			
3	SERVICE OF UNIT	*				
4	SIZE	TYPE	SURFACE ft^2			
5	**CONDENSER PERFORMANCE**					
6	FLUID ALLOCATION		SHELL SIDE		TUBE SIDE	
7	FLUID CIRCULATED		*		*	
8	FLUID QUANTITY TOTAL	lb/h	*		*	
9	STEAM	lb/h	*			
10	WATER	lb/h	*		*	
11	NONCONDENSABLE	lb/h	*			
12	TEMPERATURE (IN/OUT)	°F	*		*	
13	INLET PRESSURE		*	in. Hg.	*	psig
14	VELOCITY	ft/s				
15	PRESS. DROP ALLOW./CALC.	psi	* /		* /	
16	FOULING RESISTANCE (MIN.) or C.F.		*		*	
17	HEAT EXCHANGED Btu/h		*		MTD CORRECTED °F	
18	TRANSFER RATE, Btu/h-ft^2 °F SERVICE				CLEAN	
19	**CONSTRUCTION OF SURFACE CONDENSER**					

			SHELL SIDE	TUBE SIDE	
20					DIVIDED WATER BOX: YES/NO
21	DESIGN/TEST PRESSURE, psig		* /	/	SHELL EXPANSION JOINT: YES/NO
22	DESIGN TEMPERATURE, °F		*		TYPE:
23	NO. PASSES PER SHELL				HOTWELL: (VERT. CYL./BATHTUB)
24	CORROSION ALLOWANCE, in.		*		TUBE/TUBE SHEET JOINT:
25	CONNECTIONS	IN			GASKETS:
26	SIZE &	OUT			
27	RATING				

28	TUBES: NO. OD THKNS (MIN./AVG) (BWG) LENGTH PITCH	
29	TUBE: ASTM *	CHANNEL: *
30	SHELL: MATL. * ID OD	CHANNEL COVER: *
31	TUBE SHEETS: MATL. *	BAFFLE SUPPORTS: * NO.
32	**AUXILIARIES**	
33	UTILITIES MOTIVE STEAM, * psig AT * °F	
34	COOLING WATER, INLET TEMP. * °F, MAX. OUTLET TEMP. * °F	
35	HOGGING JET: FOR AIR EVACUATION AT START-UP	
36	SYSTEM PRESSURE mm Hg (abs): INITIAL * ;FINAL * ;AIR TEMP. * °F	
37	SYSTEM VOLUME ft^3; EVACUATION TIME * minutes, AIR LEAKAGE RATE lb/h	

Figure 3.22 Surface condenser specification sheet. *(From Baker, 1981.)*

SURFACE CONDENSER SPECIFICATION SHEET (*Continued*)

38 JETS:	FIRST STAGE	SECOND STAGE
39 SIZE-NOZZLES		
40 MATERIAL		
41 STEAM, lb/h		
42 CONDENSERS:	INTERCONDENSER	AFTERCONDENSER
43 SIZE & TYPE		
44 SURFACE, ft^2		
45 MATERIAL		
46 C.W. FLOW, lb/h		
47 C.W. TEMP, °F	/	/
48 CODE REQUIREMENTS:	CODE STAMP: YES/NO	HEI SPEC.
49 WEIGHTS, lb.-SURFACE COND.:SHIPPING	OPERATING	FILLED WITH WATER
50 AUXILIARIES:SHIPPING	OPERATING	FILLED WITH WATER
51 REMARKS: * = INFORMATION BY BUYER		
52		

Figure 3.22 Surface condenser specification sheet. *(From Baker, 1981.)* (*Continued*)

3.69 Besides water-cooled surface condensers what other type can be used to condense the exhaust steam?

Air-cooled steam condensers can also be used. Their advantages are that cooling water treatment, pumping, recycle system, etc., are not necessary (Larinoff et al., 1978; Christman, 1980). However, such condensers have steam in the tubes and air is blown over them on the outside. Hence, steam-side pressure drop is higher. Further, the operation is carried out under vacuum for higher efficiency of the turbine. So, air leakage into the tubes reduces the condenser performance. Removal of the noncondensables from the tubes also poses problems. For details about the operating characteristics and the data needed for the design of the air-cooled steam condensers, Larinoff et al. (1978) should be consulted.

A commendable history of the surface condenser has been provided by Sebald (1979).

- Now we shall discuss reboilers.

REBOILERS

On Bubbles

3.70 How does a bubble form, grow, and detach from a surface?

Bubbles form at nucleation sites on hot surfaces. A nucleation site on a clean surface may be any crevice or indentation. The liquid in it gets superheated and a vapor nucleus forms. Both the wettability of the surface (contact angle) and the geometrical shape of the cavity influence the formation of the bubbles. If the cavity is large, it will get completely filled with the liquid and no bubbles will form. The approximate maximum size for an active nucleation site on a smooth metallic surface is 5 μm (0.2 mil) for water, 0.5 μm (0.02 mil) for organics and refrigerants, and 0.1-0.3 μm (0.004-0.012 mil) for cryogenic fluids (Collier, 1981a). These sizes increase considerably on fouled, rough, or especially prepared porous surfaces.

The bubble growth is limited in the initial stages by the inertia of the surrounding liquid and in the later stages by the rate at which the latent heat of vaporization can diffuse to the vapor–liquid interface. In between, the growth is spontaneous and fast. The growth rate is faster for a bubble formed from a small cavity than from a large one.

There are several forces acting on the bubbles. These are buoyancy and hydrodynamic drag forces, which try to detach the bubble from the surface, and surface tension and liquid inertia forces, which try to prevent it from detaching. When the surface tension forces dominate, the departing bubbles tend to be spherical; when the inertial forces dominate, the bubbles tend to be hemispherical; and when both the forces are significant, the bubbles have an oblate shape (Johnson et al., 1966). The frequency of bubble formation and release varies with the nucleation site. To an approximation, the product of the bubble release frequency and departure diameter is a constant. For water at atmospheric pressure, bubble diameters are 1 to 2.5 mm (40 to 100 mil) and frequencies are 20–40 s^{-1}.

Pool Boiling

3.71 What is pool boiling?

It is the boiling of a stagnant liquid contacting a hot surface. The heat transfer to the liquid is by natural convection and by agitation produced by the release and movement of the vapor bubbles through it.

If the liquid is already at its boiling point, it requires only the latent heat to produce vapors. This is called saturated pool boiling.

If the liquid is below its boiling point, it first has to be heated (sensible heat) to its boiling point and then the latent heat is required to produce the vapor. This is called subcooled pool boiling.

The heat flux, q, to the liquid depends upon the difference in the temperature of the heated surface, T_w, and the bulk liquid boiling temperature, T_{sat}. $(T_w - T_{sat})$ is also known as the driving potential or the wall superheat (Fig. 3.23).

3.72 What are the different regions and specific points in the pool boiling curve (Fig. 3.23)?

The various regions are the following:

1. Natural convection region: Sensible heat transfer occurs from the hot surface to the liquid. Temperature gradients are set up. Liquid loses its heat by evaporation

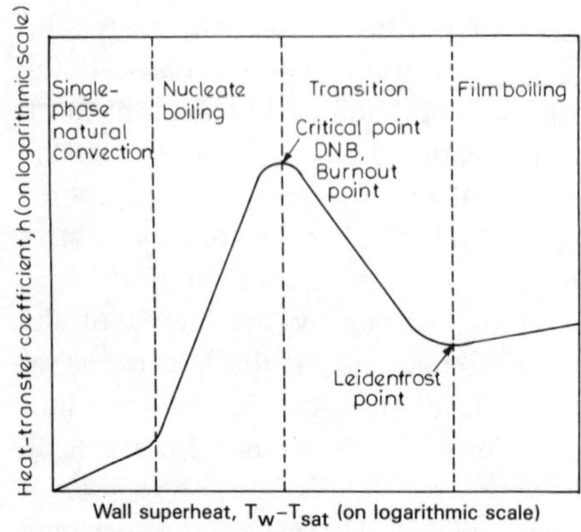

Figure 3.23 Heat transfer mechanisms in pool boiling. *(With permission from Butterworth, 1978.)*

at the free surface. Even though the hot surface temperature is greater than the boiling point of liquid, no boiling occurs since the heat flux and hence the driving potential $(T_w - T_{sat})$ is not great enough to produce vapor bubbles.

2. Nucleate boiling region: Vapor bubbles are formed at the hot surface at specific sites called the nucleation sites. As $(T_w - T_{sat})$ increases, the number of active nucleation sites increases and more bubbles are formed and released. The surface temperature increases slowly for relatively large changes in the heat flux because a large fraction of the heat goes as the latent heat of vaporization. This is a very important region for the design of reboilers, evaporators, etc. A large number of factors influence the heat transfer in nucleate boiling. Some of the important ones are (Collier, 1981a) (a) the system pressure, (b) the surface condition, (c) the size and orientation of the surface, (d) the wettability of the surface, (e) the subcooling of the liquid, (f) the hysteresis in the boiling curve, (g) the presence of noncondensable gases, and (h) the gravitational acceleration. A state-of-the-art review has been presented by Cooper (1979).
3. Critical heat flux: As vapor forms on the hot surface during the nucleate boiling, liquid rushes to replenish the surface. However, as the heat flux is increased, the rate of vaporization keeps increasing until a point is reached when the liquid cannot keep the whole surface wet. This is the highest heat flux that can be supplied to the liquid and is called the critical heat flux (CHF). Patches of vapor form on the surface, and heat transfer at these locations occurs through the vapor film. It is no more nucleate boiling and, therefore, this point is also known as the point of departure from the nucleate boiling (DNB) or burnout. The CHF depends on a number of variables, some of which are (Collier, 1981a) (a) the subcooling of the liquid, (b) the liquid viscosity, (c) the surface condition, and (d) the geometry and orientation of the surface. In most cases, DNB causes failure of the equipment to do the desired job.
4. Transition boiling region: It is also called the partial film boiling region. Since the heat transfer through the vapor surrounding the surface in patches is primarily by radiation, the net heat flux reduces while the surface temperature, T_w, increases. Vapor is released in spurts, which also causes instabilities in

boiling. Gradually, more and more of the surface is covered by the vapor and less and less is in contact with the liquid, resulting in a continuous fall in the heat flux as $(T_w - T_{sat})$ increases. For water at 1 atm, this normally occurs at a temperature difference of 40°–125°C (100°–260°F). Since, in this region, $(T_w - T_{sat})$ changes very fast, not too many studies are available. Some of the factors which affect the transition boiling are (Bankoff and Mehra, 1962) (a) the surface condition (density of nucleation sites), (b) the temperature difference $(T_w - T_{sat})$, (c) the thermophysical properties of the liquid, (d) the thermophysical properties of the heat transfer surface, and (e) the frequency of contact between the liquid and tube wall at any given location.

Some work has recently been published (Witte and Lienhard, 1982) indicating the existence of two separate transition boiling curves that are extensions of the nucleate and film boiling curves. These explain some of the anomalous behavior noticed in the past. More work is needed on this topic.

5. Film boiling region: Vapor film covers the entire hot surface. Heat transfer is primarily by radiation at high surface temperature, although there is also some conduction and convection through the vapor film. The vapor bubbles are released from the top of the film rather than from the hot surface. As is apparent from Fig. 3.23, the heat flux in the early regions of the film boiling is less than that in the nucleate boiling, although the driving potential $(T_w - T_{sat})$ is significantly higher. Hence, this is not a desired region for operating the process equipment. The point at which the film boiling starts is called the Leidenfrost point. Some of the factors that affect the film boiling are (a) the surface condition, (b) the surface geometry and orientation, (c) the subcooling of the liquid, (d) the properties of the vapor, and (e) the temperature difference $(T_w - T_{sat})$.

The relations for the heat transfer coefficient in various regions are given by Collier (1981a). An earlier review with further research needs is given by Afgan (1979).

3.73 How does pool boiling outside of the tubes in a horizontal tube bundle differ from that outside of a single tube?

Discussion in Q. 3.72 relates to pool boiling outside a single tube or on an isolated surface in a large body of water. When a large number of tubes are present in a horizontal tube bundle, both the overall size of the bundle and the tube density (i.e., tube pitch and layout angle) matter. In close-packed, fat tube bundles, liquid circulation can become limiting because not enough space is available for sufficient liquid to flow around and enter the bundle from the bottom to maintain an adequate amount of water in the bundle. In such a case, the liquid can only enter from the top but is hindered in doing so by the vapor released from the tube bundle and therefore the liquid cannot reach all the tubes. A large number of tubes are thus starved of the liquid. The critical heat flux (CHF) in such a situation is less than that for a single tube. Palen and Small (1964) have accounted for this phenomena in their design formulations.

Convective or Flow Boiling

3.74 What is convective or flow boiling?

It is the boiling of a flowing liquid over a heated surface. Pumps are used to make the liquid flow. The heat transfer is by forced convection. Unless the liquid is at its boiling point, a part of the unit is used to heat it up to the boiling point. Generally, boiling on the tube surface is avoided by maintaining a hydrostatic head and keeping a high recirculation ratio, which is defined as the ratio of the amount circulated to the amount vaporized per unit time.

3.75 What are the different regions in convective or flow boiling inside a vertical tube?

These are shown in Fig. 3.24. It is a composite figure showing the wall and fluid tempeatures, flow patterns, and heat transfer regions for a constant heat flux at the tube wall. As the heat flux is increased, the various regions will shift and change. A detailed discussion on this by the use of a three-dimensional heat transfer surface has been given by Collier (1981b). In Fig. 3.24, the local heat transfer coefficient is inversely proportional to the temperature difference between the wall temperature and the fluid temperature for the constant flux case.

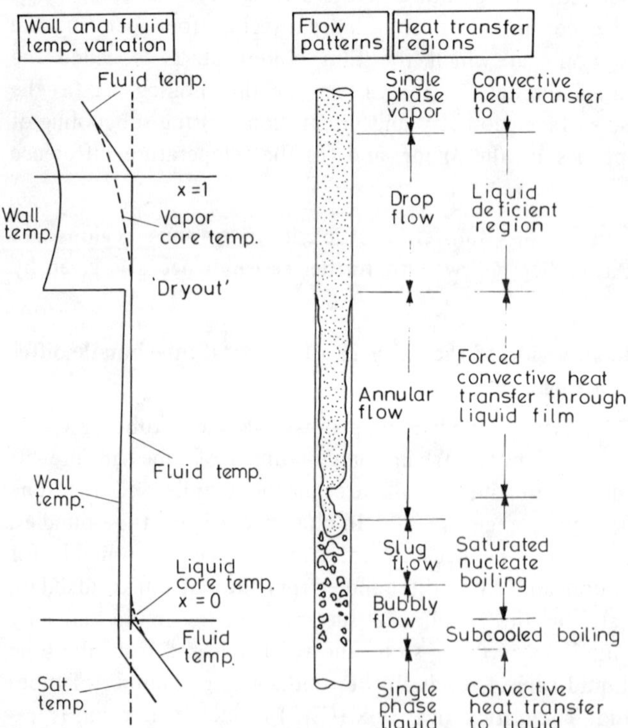

Figure 3.24 Temperature profiles, flow regimes, and boiling mechanisms in a vertical tube. *(From Collier, 1981b.)*

The various two-phase flow patterns have already been explained in Q. 3.16. The various heat transfer regions in Fig. 3.24 are the following:

1. Convective heat transfer to liquid: Subcooled liquid entering the tube is heated in this region. It is a single-phase heat transfer process.
2. Subcooled boiling region: The heat transfer coefficient is greater than that in the earlier region. Vapor bubbles are formed on the surface, since the liquid temperature there exceeds the saturation temperature while the bulk is still subcooled. Most of these bubbles then condense in the bulk subcooled liquid.
3. Saturated nucleate boiling region: The bulk of the liquid is at the saturation temperature. Net generation of vapor occurs. Liquid and vapor flow together giving rise to the various flow regimes. The liquid film moves along the wall giving rise to forced convective heat transfer region.
4. Liquid-deficient region: After a critical heat flux is reached, the liquid film at the surface breaks down and the wall becomes dry. This point is known as departure from nucleate boiling (DNB) and the region is also known as dry-wall, dry-out, or post-burnout region. The heat transfer coefficient is low. Liquid droplets are present in the core. As the heat flux increases, even these droplets evaporate, and only dry vapor is obtained. The heat transfer in this region is by convection and radiation and by the impacting of the liquid droplets on the hot surface.

A review of critical heat flux in flow boiling has been done by Hewitt (1979).

Correlations for the heat transfer coefficient in various regions above are given by Collier (1981b) and Polley (1981).

3.76 How do the flow regimes differ in vertical and horizontal tubes where flow boiling is occurring?

These are shown in Figs. 3.24 and 3.25. While most of the regimes are similar, the differences lie in the effect of gravity in the horizontal tubes (stratification) and boiling suppression due to the hydrostatic head in the vertical tubes. The flow stratification can lead to a significant temperature difference between the bottom and the top of a horizontal tube.

The transition from one flow regime to the other depends on several factors, such as the flow rate, subcooled liquid temperature at the entrance, thermophysical properties of the liquid and the vapor, latent heat of vaporization, and heat flux at the tube wall.

The heat transfer coefficient keeps changing as the temperature and the liquid-to-vapor ratio in the flow stream change. The relations for the pool and flow boiling have been summarized by Collier (1981a, b) and Polley (1979). Corrections have to be made for the following cases (Perry and Chilton, 1973):

1. When the vapor fraction is so high that the remaining liquid does not wet the total tube wall
2. When the velocity of the mixture at the exit approaches sonic velocity
3. When the surface tension of the liquid is very low

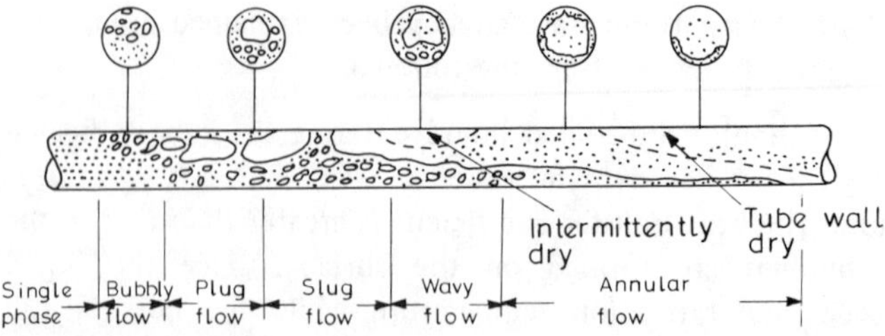

Figure 3.25 Flow regimes in horizontal tubes.

While (1) and (2) result in a decrease in the boiling coefficient, a decrease in the surface tension increases the boiling coefficient. The increase has been found to vary by an exponent of –1.4 to –2 of the surface tension.

3.77 What are the limitations of the methods available in the literature to evaluate the boiling heat transfer coefficients?

These methods generally relate to single (pure) boiling liquids or mixtures with a short boiling range (<5°C or 9°F). They are not directly applicable to mixtures or solutions with a wide boiling range.

Reboilers and Their Classification

3.78 What are reboilers?

These provide the vapor reflux at the bottom of a distillation column. The energy for distillation is supplied in the form of latent heat of vaporization.

3.79 What are the heating mediums used in the reboilers?

These are the following (Fair, 1960):

1. Condensing of steam or other vapors
2. Hot water
3. Hot oil
4. Patented heat transfer media (e.g., Dowtherm)
5. Liquid metals and molten salt
6. Flue gases

Of these, the condensing steam is the one used most often.

3.80 What are the advantages of using condensing steam?

The advantages are the following (Mayinger and Schult, 1981):

1. Even with small temperature difference between the two sides of the evaporator surface, an economic heat transport is possible.
2. Due to the large latent heat released on condensation, the mass flow rate of the steam required for a given heat duty is small compared to that of the other heating mediums.

3. Since the condensing coefficient is very high, the surface area required is low.

Such units where the condensation of one vapor results in the vaporization of a liquid are also called condenser-reboilers or double-phase-change heat exchangers. A typical case with Freon-12 vaporizing in a tube due to the condensation of steam in a concentric annulus is shown in Fig. 3.26.

3.81 How are the reboilers classified?

The reboilers can be classified in numerous ways as given in Fig. 3.27.

3.82 How are reboilers classified according to the type of circulation?

There are two types of reboilers in this classification: natural circulation and forced circulation.

1. Natural circulation (Figs. 3.28 and 3.29): These are also known as thermosiphon reboilers. The circulation or flow is due to the hydrostatic pressure difference caused by the density difference between the liquid in the downcomer from the bottom of the distillation column and the two-phase mixture in the line leading to the distillation column. The two-phase mixture is lighter than the pure liquid. The design of piping is very important so that the pressure drop is minimized. These units can be either horizontal or vertical.
2. Forced circulation: The circulation or flow is due to a pump in the downcomer line. The design of piping is not as important as in the natural circulation type. The hydrostatic head prevents boiling at the surface. Heat is transferred from the surface to the liquid as in the single phase. These are used where the hydrostatic head is insufficient to provide the required circulation in a thermosiphon

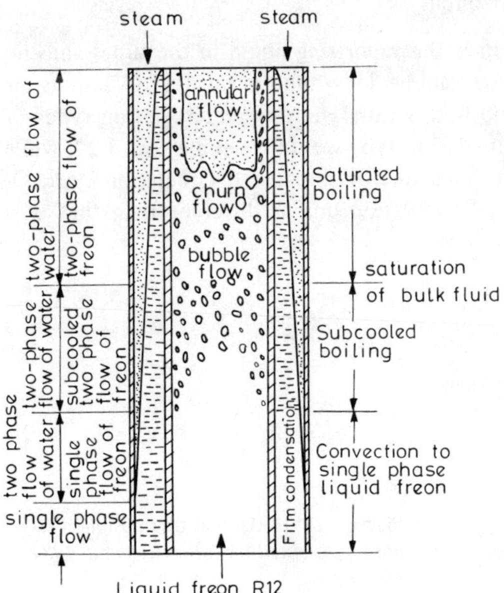

Figure. 3.26 Double-phase-change heat exchanger. *(From Mayinger and Schult, 1981.)*

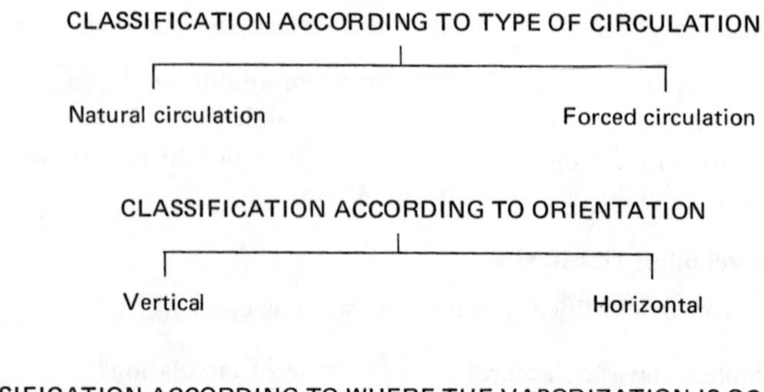

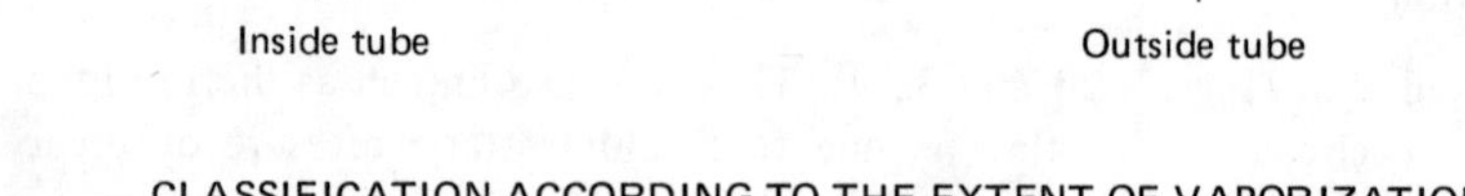

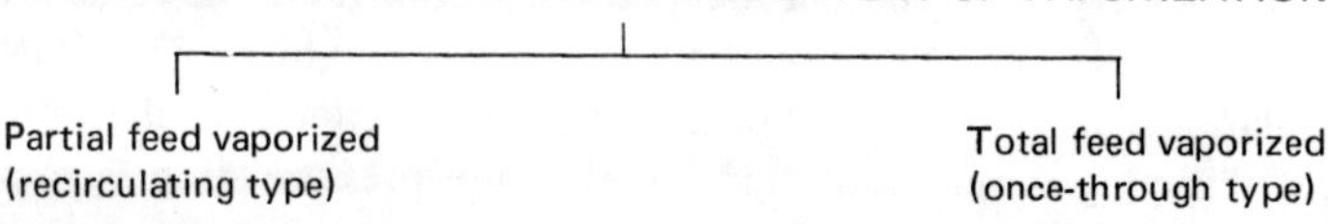

Figure 3.27 Classification of reboilers.

reboiler. The general set-up resembles Fig. 3.28 except for the inclusion of a pump in the liquid feed line to the reboiler.

3.83 How are the reboilers classified according to orientation?

Reboilers can be vertical or horizontal.

1. Vertical reboilers: These generally have the vaporizing liquid in the tubes and the heating medium in the shell. These can be long tube vertical (LTV) units or short tube vertical (STV) units, and both natural and forced circulation types. A general schematic of the natural circulation type would resemble Fig. 3.29, while a pump would be included in the feed line of the forced circulation type. Of course, the total design of a forced circulation unit is different than that of a

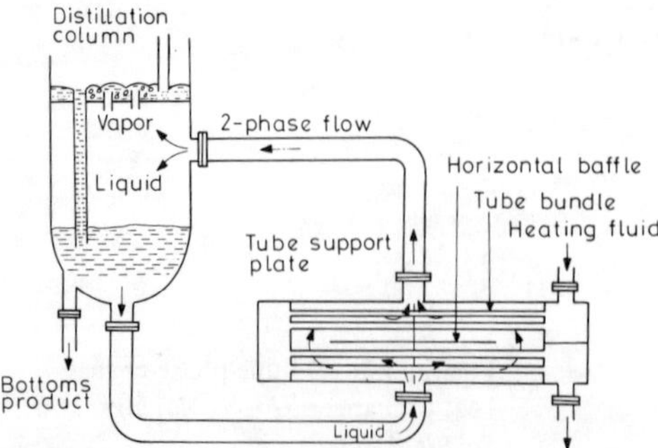

Figure 3.28 Horizontal thermosiphon reboiler. *(Adapted from Bergles et al., 1981, p. 512.)*

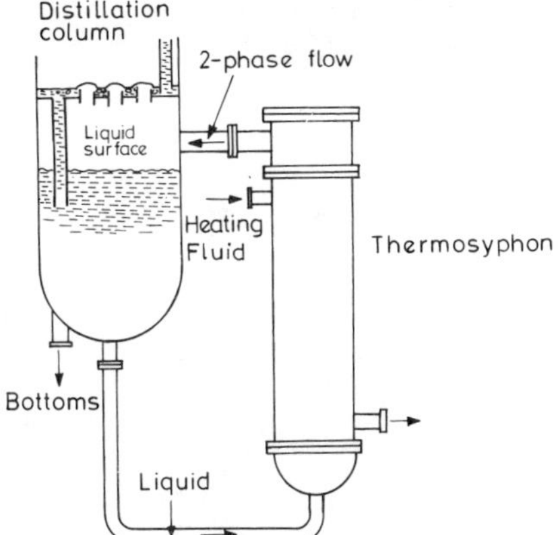

Figure 3.29 Vertical thermosiphon reboiler. *(Adapted from Bergles et al., 1981, p. 512.)*

natural circulation one. Furthermore, the liquid can be flowing up or down the tubes in a forced circulation unit. The flow regimes in an LTV are shown in Fig. 3.24. The heat transfer coefficient is different in different zones.

2. Horizontal reboilers: These generally have the vaporizing fluid in the shell and the heating medium in the tubes. These can be both the natural circulation or forced circulation type. A special type, called the kettle reboiler (Fig. 3.30) where only the vapor leaves the reboiler instead of the two-phase mixture, is also a horizontal unit (see Q. 3.86 and 3.87 also). Another variation is the internal reboiler where the tube bundle is inserted at the bottom of the distillation column itself (Fig. 3.31) (see Q. 3.88 also).

3.84 How are reboilers classified according to where the vaporization is occurring?

The vaporization can be inside the tubes or outside the tubes.

1. Inside-tube vaporization: As discussed in Q. 3.83, inside-tube vaporization is generally practiced in the vertical units. Both natural and forced circulation are used.

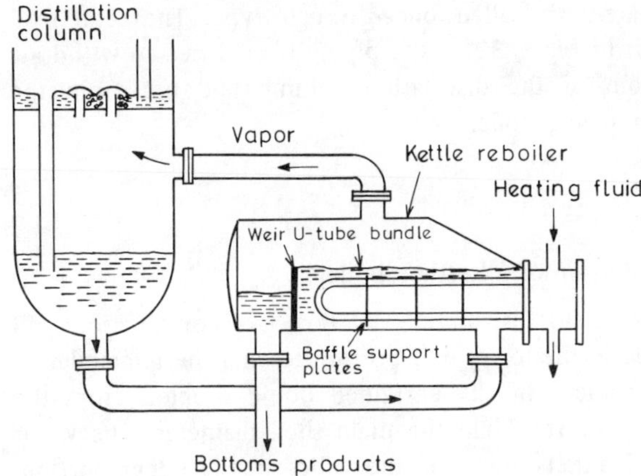

Figure 3.30 Kettle reboiler. *(Adapted from Bergles et al., 1981, p. 511.)*

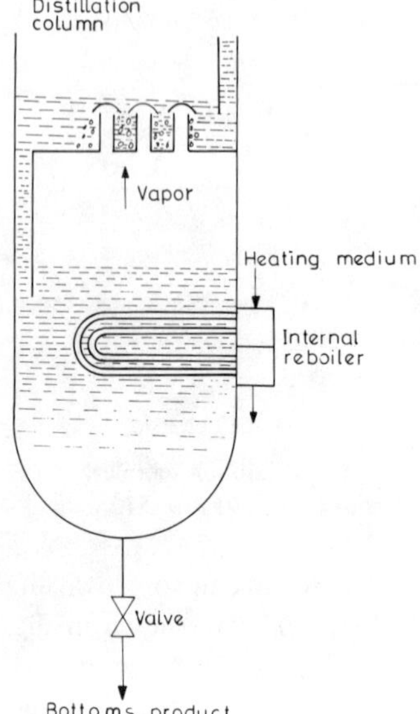

Figure 3.31 Internal or bundle-in-column reboiler.

2. Outside-tube vaporization: As discussed in Q. 3.83, outside-tube vaporization is generally practiced in the horizontal units. Both natural and forced circulation are used.

3.85 How are reboilers classified according to the extent of vaporization?

The feed to the reboiler can be either partly or fully vaporized.

1. Partial-feed reboilers: These are also called recirculating type. Only part of the feed (up to approximately 30 percent) is vaporized, while the rest is recirculated as liquid. This reduces the fouling deposit as well as the pressure drop due to acceleration and flow surges encountered when most or all of the liquid is vaporized. Kettle reboiler (Fig. 3.30) is an example of a partial-feed reboiler, so are the thermosiphon reboilers in Figs. 3.28 and 3.29.
2. Total-feed reboilers: These are also called once-through type. The total feed entering the unit is vaporized. In Fig. 3.30, if the piping is modified to withdraw the product from the bottom of the distillation column and not from the reboiler, it will become a total-feed reboiler.

Kettle Reboilers

3.86 What are the uses and operation of a kettle reboiler?

The kettle reboiler (Fig. 3.30) is unique in both its construction and operation. It has a U-tube bundle in the lower half portion leaving the remainder of the shell space for the disengagement of the entrained liquid droplets from the vapor. The tubesheet diameter is shorter than the main shell diameter. The vapor produced in the lower portion blankets some of the tubes in the upper portion, thus reducing the heat flux there. However, the movement of the vapor also results in circulation within the liquid phase. The net effect is a variable heat flux. This variation is both from bottom to top, and from the edge of the bundle to inside. A typical case is shown in Fig. 3.32. Thus, for better analysis, a two-dimensional approach is required. In order that the vapor can move up easily, the tube pitch-to-tube diameter ratio is generally larger (1.5–2) than in a single-phase heat exchanger (1.1–1.5). The tubes are kept immersed in the liquid by means of a weir. Excess liquid drains out. Recently, work has been reported on the increase in the level due to vapor generation. This work will help in the proper sizing of the disengagement space (Singh and Gupta, 1981; Tandon and Gupta, 1981; Sharma et al., 1982). This is more important in the case of a kettle reboiler used as an emergency steam condenser in nuclear power plants where, in an emergency, the high-pressure steam condenses in the tubes and the water stored in the shell boils (Singh and Gupta, 1981). However, the change in the liquid level due to vapor generation should be considered in all applications of the kettle reboilers.

The pressure drop in these units is low. The construction of the tube bundle should allow distribution of the liquid over its total length. Two inlets may be used for better liquid distribution.

Mesh pads are generally used below the exit nozzle to remove the entrained droplets. This reboiler is not suitable for liquids that foam, since the two-phase level rises very high.

Residence time is high since natural circulation is used for fluid flow. Therefore, it is not suitable for heat-sensitive materials. It has low flow velocity compared to the forced circulation units. It has higher fouling deposits, but is easier to clean and maintain. Proper blowdown is required to prevent residue buildup. Since the shell size is larger than the tube bundle, it is not used for high-pressure operations because it will require a rather thick shell.

If vapor blanketing occurs, nucleate boiling will give way to film boiling and the heat transfer will come down. A long, thin bundle is preferable to a short, fat one to avoid vapor blanketing of the inner tubes in the bundle. Generally, only a part of the feed is vaporized. The concentrated bottom products are withdrawn from the other side of the weir.

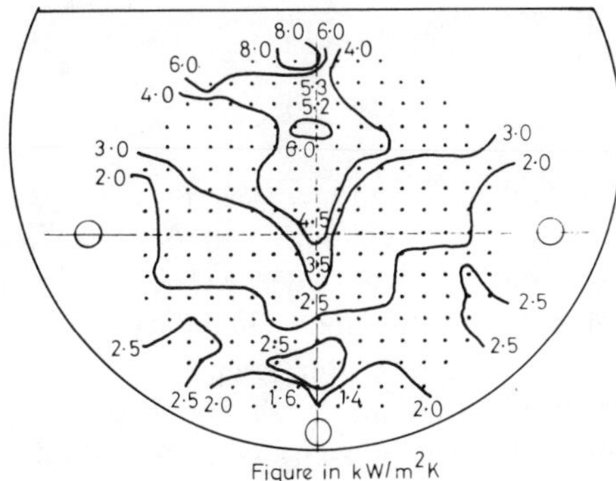

Figure 3.32 Variable heat flux in a kettle reboiler. *(From Leong and Cornwell, 1979.)*

3.87 What are some of the design recommendations for a kettle reboiler?

Generally accepted design recommendations are the following:

1. The heat transfer coefficient and the maximum heat flux for various liquids to avoid film boiling are shown in the tabulation below.

	Heat transfer coefficient Btu/hr ft² °F (J/m² s K)	Maximum heat flux Btu/hr ft² (J/m² s K)
Organic	200–400 (1140–2270)	800–1500 (2520–4730)
Aqueous solutions	20,000 (114,000)	28,000–35,000 (88,300–110,000)

2. The pitch to tube diameter ratio should be 1.5-2. This will avoid the vapor generated at a tube from getting entrapped in between the tubes above, which otherwise will cut down the heat transfer and hence also the rate of vapor generation. This is called the vapor binding of the tubular surface.
3. A long thin tube bundle is preferable to a short fat bundle to avoid the vapor blanketing of the inner tubes in the bundle.
4. Enough space should be left in the shell above the tube bundle to allow for the increase in the two-phase level and for the disengagement of the entrained liquid from the vapor. The amount of disengagement space in terms of the ratio of the shell diameter to the tube bundle diameter increases with increasing heat flux.

Internal Reboilers

3.88 What are the uses and operation of internal reboilers?

These are also called bundle-in-column reboilers (Fig. 3.31) and are used when the heat duty required is not very large and the liquid is nonfouling. One or more U-tube bundles are inserted at the bottom of the distillation column to provide the necessary boil-off rate. The advantages are the following:

1. No separate shell is needed.
2. No piping is needed.

The disadvantages are the following:

1. Use is possible only in low or nonfouling cases.
2. Tube bundle has to be short and thick and can be used for only small heat duties.
3. Since generally U-tubes are used, they overhang from one side. In order to accommodate the resulting bending moments, the lower portion of the distillation column has to be made thicker and/or a special flange is needed.
4. Every time the reboiler has to be cleaned, the distillation column has to be shut down and drained. Furthermore, before restarting, the column has to be tested for leakage, etc. On the other hand, if an outside reboiler is used, a stand-by unit can be used while the main reboiler is being cleaned without shutting down the column.

5. The height of the column increases due to the tube bundle at the base and due to the need to provide disengagement space above the bundle.

Hence, internal reboilers are not used very often.

Natural Circulation and Thermosiphon Reboilers

3.89 What are the uses and operation of natural circulation (thermosiphon) reboilers?

These can be vertical or horizontal units.

Vertical units: These are usually E-type exchangers with boiling in the tubes. Only part of the feed is vaporized. Hence, the circulation rate is high (low residence time). Therefore, fouling is less and heat-sensitive materials can be processed. It occupies less ground space, but the distillation tower has to be elevated to provide enough head for the flow.

To make the operation stable, valves (restrictions) are incorporated in the inlet pipe. Furthermore, the outlet pipe cross section is made equal to the total cross section of all the tubes. Design of the exit nozzle is important so that the pressure drop is minimal. Liquid level is maintained equal to the top tubesheet. ΔT should be $\leqslant 50°C$ ($90°F$), or else film boiling might result.

These can be falling film or rising film type. The heat transfer coefficient for the falling film type is higher than that for the rising film type (Fig. 3.33). Their performance can be predicted better than that of the horizontal units and these are better controlled too. These are also easier to clean if the liquid is a fouling type since the fouling is on the tube side. The flow distribution is more uniform, and the pressure drop is lower compared to that in the horizontal units. These also provide more hydraulic head and higher circulation rates. Capital and operating costs are lower too.

Horizontal units: These are usually G-type exchangers with boiling in the shell. Only part of the feed is vaporized, hence the circulation rate is high (low residence time). Vapor binding of the tubes can happen. A horizontal unit is preferred when

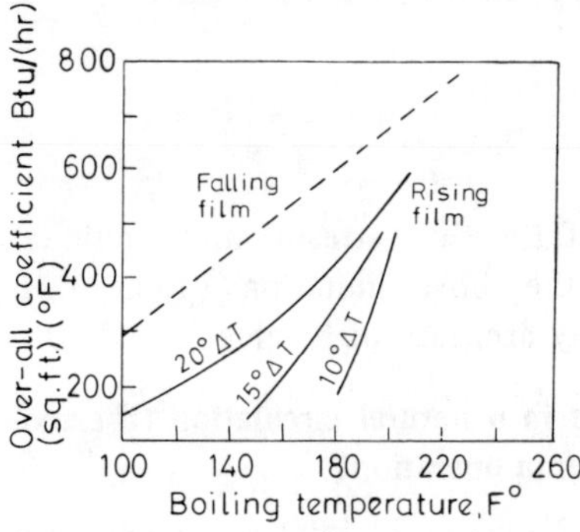

Figure 3.33 Heat transfer coefficients for falling and rising film thermosiphon reboilers for a typical case. *(With permission from Perry and Chilton,* Chemical Engineers' Handbook, *5th ed. 1973. Copyright McGraw-Hill Book Co., New York, N.Y.)*

1. The area requirement is large.
2. Low boiling point elevation due to hydrostatic head is desired, such as in vacuum operation.
3. The tower cannot be elevated to provide enough head.

In the horizontal units, seal strips should be installed in order to reduce by-passing and phase separation within the unit (Q. 1.87). More than one inlet and outlet are provided for good distribution and high velocities. Aim should be to have bubbly flow within the reboiler.

Experiments on a full-scale thermosiphon reboiler have been reported by von Bockh (1983). These gave data on the limits of stable operation such as dry-out and flooding. It was concluded that with small temperature difference, short tubes should be used, while with large temperature difference, long tubes give better heat transfer.

Pressure loss characteristics of the pipework have significant influence on both the horizontal and vertical units.

3.90 What are the advantages of a thermosiphon reboiler?

These can be used for general services. They do not require any circulation pumps and the piping arrangement is very simple.

3.91 What are the recommended heat transfer coefficient values for the thermosiphon reboilers?

These are given in the tabulation below for different liquid types.

	Btu/hr ft² °F	J/m² s K
Heavy organics	100–160	560–910
Light hydrocarbons	150–220	850–1250
Aqueous solutions	200–350	1140–1990

3.92 What factors limit the heat flux in a vertical, in-tube thermosiphon reboiler?

Palen et al. (1974) conducted tests on such thermosiphon reboilers and concluded that there are three types of limitations to heat flux in such cases:

1. Film boiling
2. Mist flow
3. Flow instability

A combination of vapor fraction (or quality of the flow stream) and frictional pressure drop may produce one or the other of the above limitations (Table 3.2). Hence, the heat flux in such a case cannot have any arbitrary high value.

3.93 What is the usual pressure drop permitted in a natural circulation (thermosiphon) reboiler? How does it affect the column operation?

A low pressure drop of approximately 0.25 psi (1.72 kPa) is allowed in a natural circulation reboiler. This pressure drop has to be provided by the hydro-

Table 3.2 Heat flux limitations in vertical in-tube thermosiphon reboilers

Condition	Type of limitation	Effect of increasing heat flux
High exit vapor fraction	Mist flow	Destruction of liquid annular film
Low exit vapor fraction, low reduced pressure	Flow instability	Choking and density wave oscillations
High reduced pressure	Film boiling	Bubble layer retardation. Wall temperature exceeds Leidenfrost point

Source: From Palen et al. (1974).

static head below the bottom plate of the column. The higher the pressure drop, the higher the hydrostatic head has to be. This will require elevating the entire distillation column to provide the necessary head, and this can be rather costly.

3.94 Why is the piping layout important in thermosiphon reboilers?

The circulation in these is due to the hydrostatic pressure difference between the inlet and the outlet lines. In order to have a high circulation rate and better control, the pressure drop in the piping should be minimized. This requires a judicious choice to be made regarding the pipe material, diameter, layout, and use of bends, valves, and other pipe fittings.

3.95 How does a horizontal thermosiphon reboiler differ from a kettle reboiler?

It differs from the kettle reboiler in the sense that the disengagement of the entrapped droplets occurs in the distillation column and not in the reboiler shell, as in the case of the kettle reboiler. However, the heat transfer coefficients in the two cases are of the same order of magnitude. Also, in both, the circulation of the liquid occurs by natural circulation. For the same heat load and vapor generation rate, however, the thermosiphon reboiler is smaller in diameter since no disengagement space is provided in its shell.

Forced Circulation Reboilers

3.96 What are the uses and operation of forced circulation reboilers?

In these, pumps are used to circulate the feed through the reboiler. Higher circulation rates can be maintained so that the fouling is low. These are also used with viscous and solid-bearing liquids. The boiling may occur either inside or outside the tubes.

3.97 When is an inside-tube forced circulation reboiler preferred?

It is preferred for fouling services, viscous liquids, liquids containing suspended solids, and low pressure systems. In most other systems, boiling on the outside of the tube bundle is preferred.

3.98 Why is a proper selection of pump important in the forced circulation units?

The pump should develop sufficient head to meet the pressure drop requirements in the system. If the head developed is more, the operating point will change. This will require more NPSH (net positive suction head) and may result in cavitation, which will damage the pump. If the head developed is insufficient, the required flow velocity may not be achieved, thus increasing the fouling deposit and adversely affecting the heat transfer rate (Lord et al., 1970a).

Choice and Performance of Reboiler Systems

3.99 How does one choose the type of reboiler system for a specific application?

It depends upon a number of factors, including the flow rates, properties of the liquid and vapor, composition of the liquid, any suspended material in the liquid, fouling and corrosion tendencies, amount of vapor required, overall heat transfer coefficient, space and piping requirements, heat-sensitive nature of the liquid, controllability of the unit, elevation of the distillation column required because of the hydrostatic head for the reboiler, cost of pump and pumping, etc. The characteristics of different types are given in Table 3.3.

3.100 How does the quality of feed to the distillation column affect the reboiler duty?

Quality of the feed is a measure of the state of the stream entering the distillation column. It is generally designated by the symbol q. $q = 1$ is a liquid at its boiling point, $q < 1$ is a two-phase feed, $q = 0$ is a vapor at its dew point, $q < 0$ is a superheated vapor, and $q > 1$ is a liquid below its boiling point.

The reboiler duty decreases as the q value of the feed decreases.

3.101 What is the difference between the pressure drop in a two-phase system compared to that in a single-phase system?

The pressure drop in the two-phase system is made up of (1) frictional losses, (2) gravitational losses, and (3) acceleration losses, while in the single-phase system, the acceleration losses are absent. These relate to the fact that the pressure drop, both for the liquid and the vapor, has to be the same in a system (the principle of balanced pressure drop). Since the vapor density is very low compared to the liquid density, the vapor velocity has to be very high to obtain the same pressure drop as for the liquid stream. This results in a shear exerted by the vapor on the liquid surface, which accelerates the liquid layer. The pressure drop associated with it is called the acceleration losses.

3.102 In a distillation column, the condenser is usually the cleanest part and the reboiler the dirtiest. Why?

The reboiler accumulates dirt from the liquid stream. Furthermore, the temperature being the highest in the reboiler, the reverse solubility salts might deposit and any coking, etc., might occur too. On the other hand, the condenser

Table 3.3 Characteristics of reboilers

Type	Advantages	Disadvantages
Kettle type (Fig. 3.30)	1. Easy maintenance and cleaning 2. Equivalent to a theoretical plate in the distillation column 3. Contains vapor disengagement space 4. Easy to control and operate	1. Lower heat transfer rate and large area 2. Easily fouled 3. Higher residence time 4. Vapor blanketing of tubes in fat bundles 5. Extra shell volume makes it expensive 6. Requires large floor space and piping
Vertical thermosiphon (Fig. 3.29)	1. Very high heat transfer rates 2. Compact, simple piping 3. Low residence time 4. Not easily fouled 5. Requires less floor space 6. Low cost 7. Useful for vacuum or low pressure 8. Only tube-side exposed to corrosive fluids	1. Distillation column may need to be raised to provide enough head 2. Maintenance and cleaning are awkward 3. Equivalent to theoretical plate only at high recycle 4. Not used where excessive and sudden process fluctuations are expected 5. Higher boiling point elevation results in lower vapor generation
Horizontal thermosiphon (Fig. 3.28)	1. Moderately high heat transfer rates 2. Easy maintenance and cleaning 3. Good controllability 4. Large area can be obtained 5. Low residence time 6. Not easily fouled	1. Extra piping and space required 2. Design formulations are less certain than for other units 3. Equivalent to theoretical plate only at high recycle 4. Not used when excessive and sudden process fluctuations are expected
Internal or bundle-in-column (Fig. 3.31)	1. Cheap since no shell or piping is needed 2. Equivalent to a theoretical plate in the distillation column 3. Contains vapor disengagement space	1. Tube bundle length is limited by column diameter 2. Column has to be shut down and drained every time the reboiler is cleaned 3. Attachment of tube bundle to column poses mechanical difficulties 4. Lower heat transfer rate and large area 5. Easily fouled 6. Higher residence time 7. Vapor blanketing of tubes in fat bundles
Forced circulation	1. Viscous fluids can be used 2. Liquids with suspended particles can be used 3. Low fouling 4. Circulation can be controlled 5. Higher heat transfer coefficient	1. Pump and its operation add to the total cost

receives only vapor from the distillation column, and any fouling is confined only to the cooling water side. To minimize the fouling deposit, the liquid velocity in the reboilers and cooling water velocity in the condensers should be kept high if sufficient pressure drop is available.

3.103 What are the specially prepared surfaces and how do they affect the heat transfer coefficient?

This refers to the following:

1. The plain tubes with a specially prepared, porous, thin layer bonded onto the tubular surface. The layer has small pores that provide a large number of nucleation sites. Their mechanism of operation is shown in Fig. 3.34. Boiling is achieved with reduced superheat compared to a smooth surface (Fig. 3.35). Their use is limited to nonfouling liquids of low molecular weights.
2. Contoured or profiled tubes, also known as enhanced surfaces (Fig. 1.6), help both in the condensing and evaporative heat transfer. On vertical fluted tubes, up to 300 percent increase in overall heat transfer coefficients has been observed in simultaneous evaporation and condensation. The enhancement is greater with finer grooves compared to the coarser ones (Sideman, 1981a). It is reported that straight, square-edged grooves (Fig. 3.36) are most advantageous for Reynolds number (Re) > 500, circular bottom square-groove for $250 < \text{Re} < 500$, and the triangular grooves for $\text{Re} < 500$. The circular grooves were not found advantageous at all. The heat transfer rates increase as the contact angle between the liquid and the surface decreases.

It is claimed that the tube material does affect the heat transfer rate in enhanced tubes, since in these tubes one has a two-dimensional temperature field compared to a one-dimensional (radial) field in the smooth tubes (Sideman, 1981b). This temperature field is greatly affected by the thermal conductivity of the tube material. Furthermore, the resistance offered by the tube wall should not be neglected, since this is of the same order of magnitude as that offered by a thin film of liquid. Film thickness on an enhanced surface varies greatly.

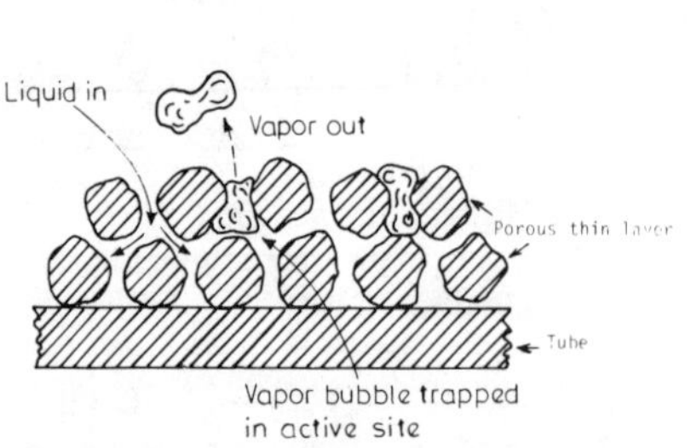

Figure 3.34 Porous thin layer bonded on tubular surface.

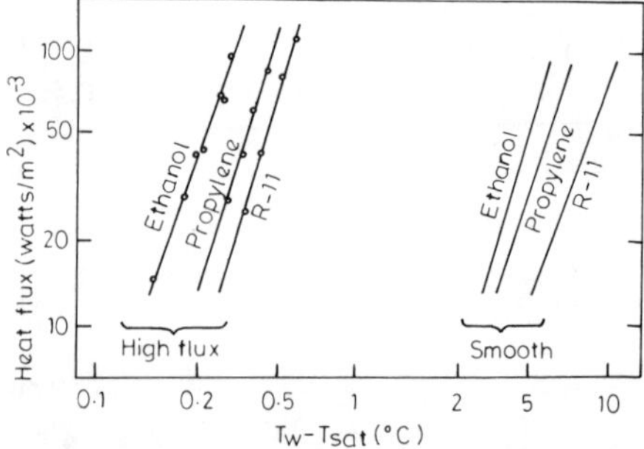

Figure 3.35 Comparison of high flux and smooth tubing.

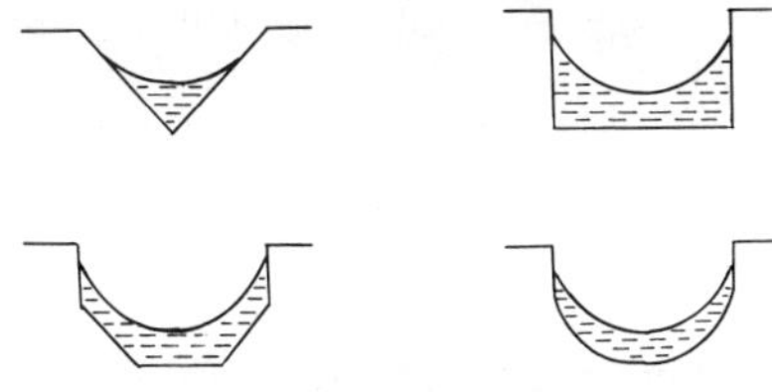

Figure 3.36 Some typical grooves used in enhanced surfaces. *(From Sideman, 1981b.)*

3.104 What are the effects of the manufacturing clearances and bypass lanes on the performance of a unit?

Manufacturing clearances and pass partition lanes result in the bypass of a large fraction of the liquid. The effect of such clearances is significant and should be accounted for in the design, since most of the thermal design data has been obtained using ideal bundles. Due to the bypass channels, a marked maldistribution of the liquid and vapor occurs with a lot of liquid not contacting the tubes. There is a twofold damaging effect:

1. Liquid vaporization is reduced since a large fraction of the liquid does not contact the tubes. Some of the tubes may not be completely wet.
2. The vapor formed passes through the tube bundle and results in vapor binding of the tubes.

A model, based on equal pressure drop throughout the liquid and vapor, showed that with the bypass area equal to 20 percent of the total flow area, 75 percent or more of the liquid bypassed the tube bundle, thus adversely affecting its performance (Polley and Grant, 1979). To reduce the effects of bypassing, seal strips and seal rods should be used (see Q. 1.87–1.89).

3.105 When does a reboiler operation become unstable?

There are several situations:

1. If the heat flux is high while the liquid distribution is nonuniform and inadequate to keep the tubes covered, local hot spots develop, and vapor is formed in a spurt as soon as any liquid reaches those hot spots. This results in an unstable operation.
2. Wavy and slug flow patterns in the two-phase flow regimes (Fig. 3.9) are inherently unstable and should be avoided.
3. Changes in the inlet flow rates, composition, and temperature of the evaporating and/or heating medium make a system depart from its current steady operation and try to seek another steady state. It is likely that the new state is not a stable one, and minor fluctuations would tend to throw the operation out of gear.
4. If the circulation rate is low, the boil-off of the available volatile components will result in an increase in the boiling temperature of the remaining liquid, thus reducing the vaporization rate further. Some liquid from the trays will be dumped on the base, which will replenish the volatile components in the circulating stream, and the rate of vaporization will increase until the volatile

components decrease again. Thus, surges in the vaporization rate will occur periodically resulting in an unsteady operation.

The dynamic behavior of an evaporator–condenser system has been dealt with by Mayinger and Schult (1981).

3.106 What factors should be kept in mind to minimize the operational problems in a reboiler system?

These are the following:

1. The start-up should be gradual or else thermal stresses are likely to be set up (see Chapter 9 for thermal stresses).
2. If steam is used for heating, inert gases must be vented by suitably located vents. Also, the condensate formed should be drained effectively. If either of these is not done adequately, it will block part of the heat transfer surface, thus reducing the total amount of heat transferred.
3. Proper liquid level should be maintained. A low liquid level will cause loss of thermosiphon. Film boiling might then set in, which sharply reduces vapor generation rate. Too high a liquid level will flood the reboiler outlet and will damage the distillation column by hydraulic hammer.
4. Temperature difference ($T_w - T_{sat}$) should not be too high. A high-temperature driving potential (>50°-55°C or 90°-100°F) will cause film boiling with a reduced heat flux and reduced vapor generation (Fig. 3.23).
5. Piping layout is important. Improper piping layout severely cuts down the thermosiphon reboiler efficiency (Q. 3.94).

A number of other operational problems, such as pressure surging, low heat transfer, or leakages, in reboiler and other systems, along with their probable causes and suggested cures are given by Lord et al. (1970b).

3.107 What are the controls needed on a reboiler or evaporator?

The controls may be needed for all or some of the following in a given case:

1. The pressure
2. The liquid level in the column
3. The liquid level above the tubes in an in-shell reboiler
4. The temperature of the heating medium
5. The circulation rate
6. The blowdown, if any

3.108 What are the unresolved problems in boiling/evaporation/condensation on which further studies are warranted?

These relate to the following:

1. Pool and convective boiling of pure organic liquids and binary and multi-component mixtures, in particular with regard to finding the critical heat flux (CHF)

2. Studies mentioned in (1) should be carried out in single tubes and over tube bundles of various sizes and layouts, and also including pass partition lanes.
3. Flow distribution in two-phase flow from a header into a number of tubes or from one tube pass to the other. The distribution is usually nonuniform and unpredictable.
4. Minimum liquid flow rates are required to maintain a surface wet in the falling film evaporators at a given local heat flux. If the surface film breaks down, there will be dry patches, and the heat transfer will be adversely affected. The effects of the surface tension and concentrations need also to be determined.
5. Condensation studies of vapor mixtures with noncondensables/immiscibles
6. Studies of direct contact condensation
7. Studies related to the industrial feasibility of dropwise condensation
8. Better enhanced heat transfer surfaces

EVAPORATORS AND VAPORIZERS

• Distinctions among reboilers, evaporators, and vaporizers have been pointed out at the beginning of this chapter. However, basically, all these generate vapors, and hence some factors, upon which their classifications are based, are common as is evident from Figs. 3.27 and 3.37. Nevertheless, we have discussed them again in this section for the sake of completeness, and more importantly, because further subclassification and usage are different than is the case with the reboilers.

For ease in writing and reading, generally only the term evaporator has been used, although most of the discussion applies to the vaporizer too. The situations where it is not so are self-evident as, for example, in the power plant evaporators, there being no vaporizers in the conventional power plants.

General

3.109 What are evaporators and vaporizers?

These are equipment used to concentrate a solution by evaporating the solvent, as in caustic soda production, or to produce pure water, as in the power plants. If the solvent is water, these are customarily called evaporators, otherwise these are called vaporizers. These consume a lot of energy.

3.110 How do evaporators differ from dryers and crystallizers?

In the evaporators, a portion of solvent is vaporized, and the product is a concentrated solution without any solid product. The solvent can be vaporized either by supplying heat of vaporization or by reducing the pressure. In a majority of cases, heat is supplied, while in some a combination of the two is used.

In the dryers, the total solvent is vaporized and the product is a dry powder or solid.

In the crystallizers, even though only a portion of the solvent may be vaporized or none at all, the desired product is the solid crystals. Evaporators used to produce common salt, potassium nitrate, etc., can also be classified as crystallizers.

3.111 Where are evaporators largely used?

These are used in the following ways:

1. In power plants to produce pure water for boiler feed
2. In chemical plants in a variety of industries, such as sugar, caustic soda, paper and pulp, dye, and fruit juice, to concentrate the liquor

Classification

3.112 How are the evaporators classified?

The evaporators can be classified in numerous ways, as given in Fig. 3.37.

• Now we shall discuss the various types of evaporators. Other items related to each type will also be discussed as the need arises.

3.113 What are the power plant evaporators?

These are primarily used to produce pure water from raw or treated water for boiler feed in a power plant or wherever large amount of steam is regularly required. Only a fraction of the water fed to the evaporator is vaporized, while a continuous bleed or blowdown maintains the dissolved salt content in water within the specified limits.

3.114 What are chemical plant evaporators?

These are used to concentrate solutions. The interest in these is in the residue left behind after the evaporation of water, while the interest in the power plant evaporators is in the vapors formed (and recondensed). There is no blowdown in the chemical evaporators since the interest is in the residue. Hence, greater heat recovery is possible compared to the power plant evaporators since no heat is lost due to blowdown in the former.

3.115 What is the evaporator classification according to the type of circulation?

There are two broad categories: natural circulation and forced circulation evaporators. Both are discussed below.

Natural circulation evaporators are used for general services where low viscosity and low fouling liquids are used. No pump is required to circulate the feed. Circulation is maintained due to the hydrostatic pressure difference. A high head room is required to obtain the required circulation rate. Control is not easily possible. There are four types of natural circulation evaporators (Kern, 1950):

1. Horizontal-tube type where the feed is in the shell: These are not used often these days.

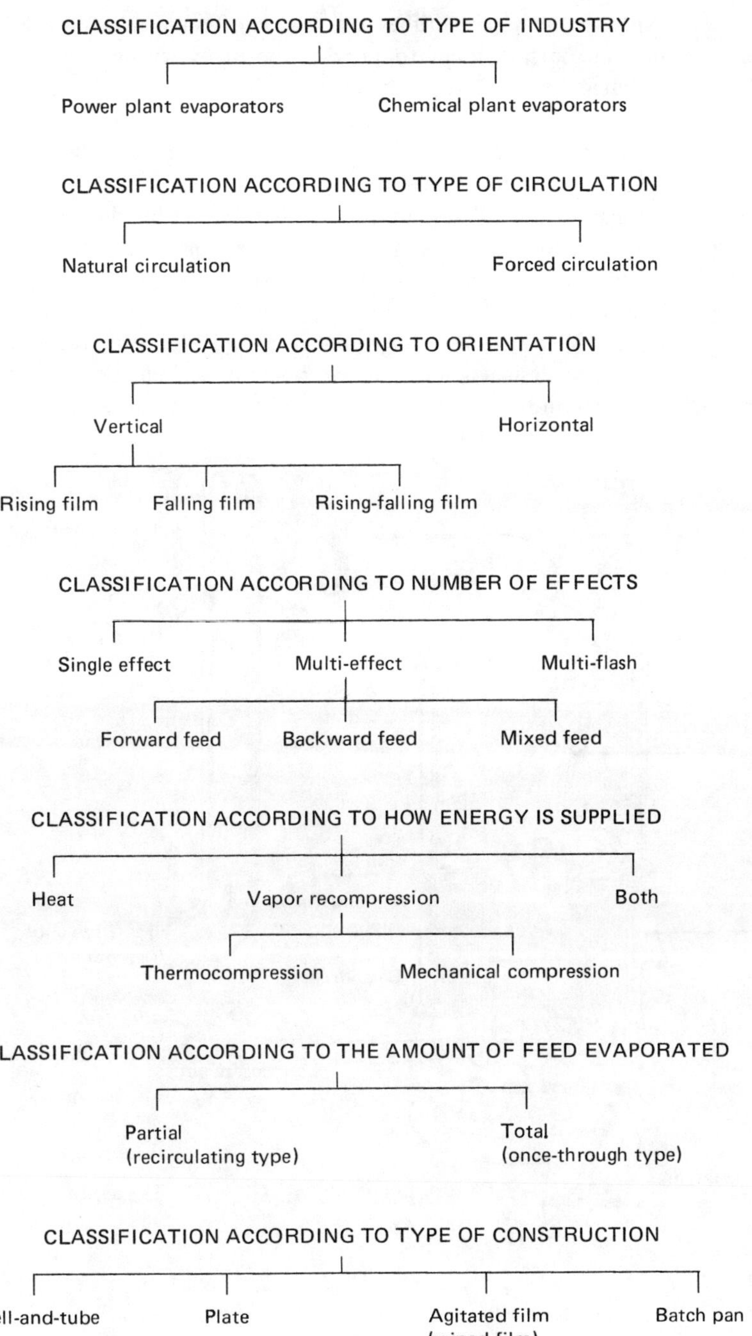

Figure 3.37 Classification of evaporators.

2. Calandria type where short vertical tubes are used with evaporation in the tubes and steam condensing outside them to provide the necessary heat (Fig. 3.38): The tubes are attached to two fixed tubesheets. A large cylindrical space is left in the middle to permit recirculation of the feed. Large diameter tubes, up to 3 in (76.2 mm) O.D., are used to reduce the pressure drop. These are widely used and can be modified to suit the application.
3. Basket type is similar to the calandria-type, but has a tube bundle that can be removed for cleaning, and hence it can be used for scaling and fouling liquids. Furthermore, it can take differential thermal expansion easily without producing unnecessary stresses.
4. Long-tube vertical (LTV) type where long tubes are used with evaporation inside the tubes: These can be designed to be once-through or recirculation type (Figs. 3.39 and 3.40). These are used for foamy, frothy liquors, since the foam is

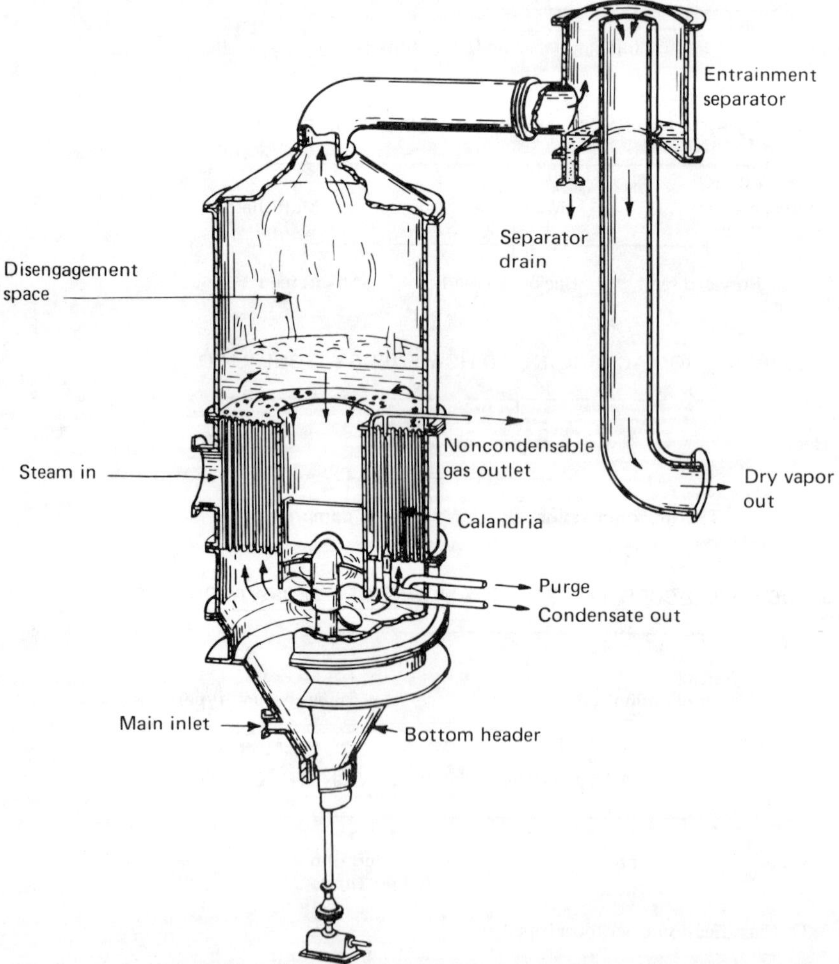

Figure 3.38 Calandria evaporator. *(From Smith, 1983.)*

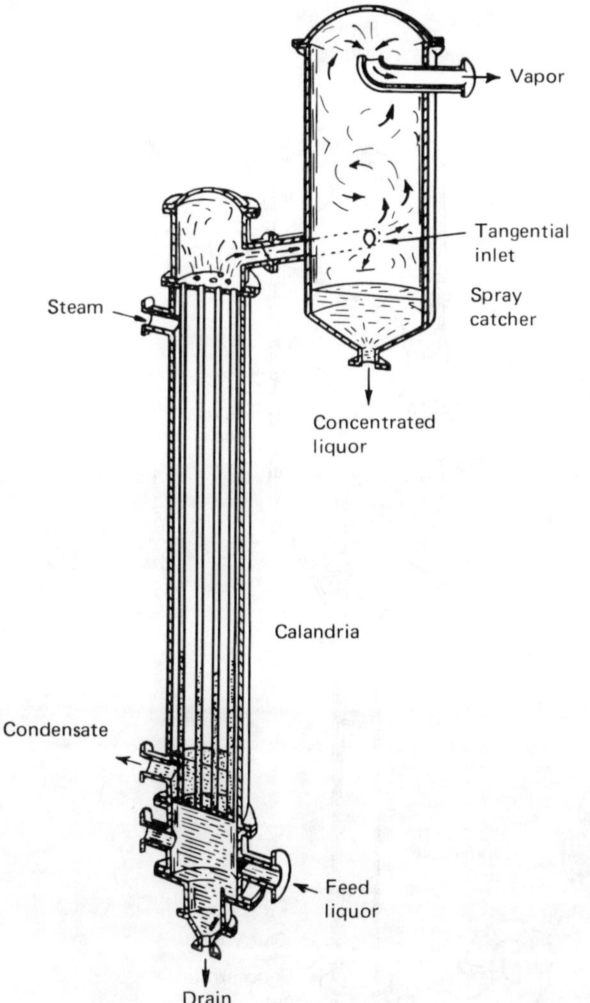

Figure 3.39 Long-tube vertical (LTV) evaporator: once-through type. *(From Smith, 1983.)*

broken when the liquid–vapor mixture hits a baffle at the vapor head (see also Q. 3.116). The vapor moves through the central core of the tubes, forcing the liquid against the tube wall and dragging it along due to the high vapor shear. These are not used for very viscous and highly fouling liquids.

Forced circulation evaporators are used for very viscous and highly scaling liquids, since very viscous liquids have poor natural circulation, and a high velocity of forced circulation keeps the scaling within limits. The tube bundle can be vertical or horizontal (Figs. 3.41 and 3.42). Boiling in the tubes is prevented by a large hydrostatic head. Vapor disengagement occurs in a separate drum or vessel operated at a slightly lower pressure wherein the liquid flashes and vapor is produced. Flow control and low residence time are easily achieved, if desired. The circulation rate is

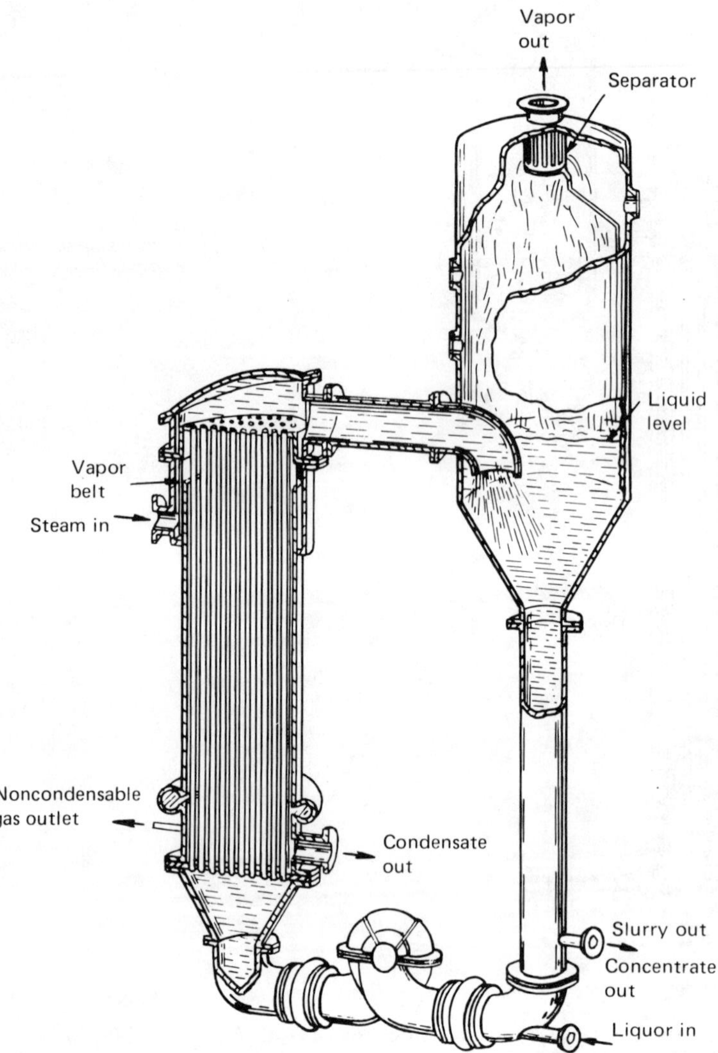

Figure 3.40 Long-tube vertical (LTV) evaporator: recirculation type. *(From Smith, 1983.)*

very high, about 200 to 300 times the evaporation rate. This permits only a small rise in the temperature and hence very little scaling. The equipment operating costs are rather high due to the high cost of pumping large amounts of very viscous liquids. It can be designed to fit into a restricted space, unlike the natural circulation units which generally require large space.

3.116 What is the evaporator classification according to their orientation?

The evaporators can be vertical or horizontal:

Vertical units can be both natural and forced circulation type as discussed in

Q. 3.115. There are three subdivisions with regard to the direction the liquid film can take.

1. Rising film type: Feed level is maintained up to the bottom tubesheet. By condensing steam around the tubes, the feed is made to boil. Vapors so produced rise up through the central core of the tubes, forcing the liquid against the tube wall. Due to the high velocity of the vapor, shear force is exerted on the liquid. This and the surface tension forces result in the movement upward of the liquid film up the tube wall, against gravity. These require high heat flux and high-temperature driving potential, which may not be possible with a heat-sensitive liquid.
2. Falling film type: Feed is input at the top of the unit and is distributed in each tube by means of a perforated distributor plate or by nozzles. The liquid film flows down due to gravity. Therefore, it is thin and highly turbulent, giving a high heat transfer coefficient. The vapor produced also travels down and is separated from the liquid at the bottom. The temperature driving potential required is low, and also the residence time is less than that in the rising film type. Hence, it can be used for heat-sensitive materials, such as fruit juices. Usually, no recirculation is allowed.

   Inert gases may sometimes be injected into a falling film evaporator to reduce the partial pressure required to vaporize the volatile components. This often eliminates the need for vacuum operation. A minimum liquid flow rate is required to form the film. After the film is formed, the flow rate may then be reduced.
3. Rising-falling film type: It has the advantages of both the rising and falling film types. Liquid distribution is easy as in the rising film type, and the heat transfer coefficient is high as in the falling film type. The height of this unit is less than that of either of the rising or the falling film type.

Horizontal units can be both natural and forced circulation types as discussed in Q. 3.115. These are not used to the same extent as the vertical units.

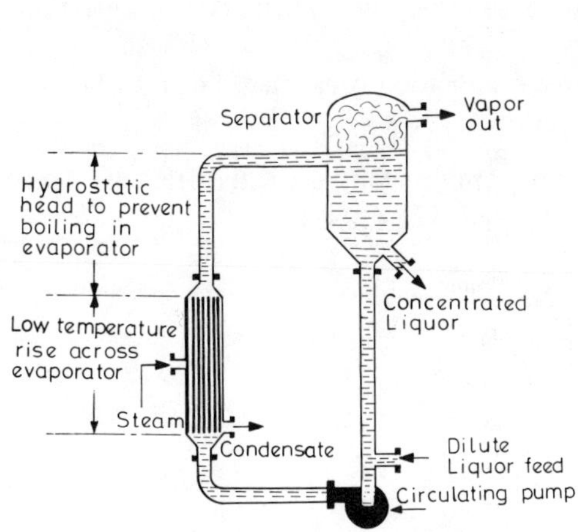

Figure 3.41 Vertical forced circulation evaporator.

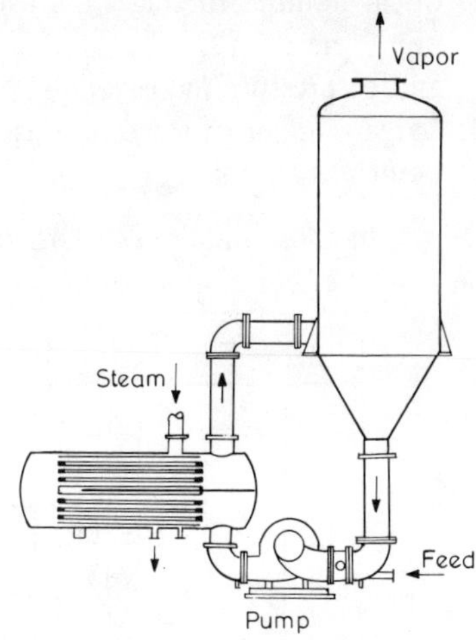

Figure 3.42 Horizontal forced circulation evaporator.

3.117 What is the evaporator classification according to the number of effects?

Effects means the number of units in series. One may have single- or multi-effect evaporators.

1. Single-effect evaporators: These are the simplest to design and use and are preferred when small evaporation rates are required. Figures 3.38–3.42 show single-effect evaporators.
2. Multi-effect evaporators: These have a number or a cascade of evaporators connected together and are used where large evaporation rates are required as is the case in most large industries. Vapor from one effect is used to supply heat to the next effect, thus economizing on the steam usage (Fig. 3.43). Each effect behaves as a single-effect unit. Fresh steam is supplied only to the first effect. The pressure is the highest in the first effect and gradually reduces in each subsequent effect. There is a temperature difference across the heat transfer surface in each effect. The vapor produced in the last effect is withdrawn by a condenser and ejector system. Multi-effect units are used both in the power plants and in the chemical industries.
3. Multi-flash evaporators: These are a variation of the multi-effect units. The vapor produced in each effect is used to preheat the feed, which enters the first effect and no heat transfer is done within the effects (Fig. 3.44). Vapor in each effect is formed due to flashing of the hot solution to a lower pressure that keeps decreasing in each effect. This is preferred in solutions with low boiling point elevations and large throughputs.

3.118 How is the number of effects or stages decided in an evaporator system?

It is based upon the initial and operating costs for the optimum plant economics since the amount of steam produced in successive evaporators keeps decreasing. As a general rule:

1. Small number of stages are used for small evaporation capacities and high boiling point elevations. The boiling point of a solution rises when its concentration and/or pressure increases.
2. Large number of stages are used for large evaporation capacities and small boiling point elevations.

In most industries, 3 to 6 evaporators are generally used. Experiments done on a three-effect evaporator are reported by Karagil et al. (1983).

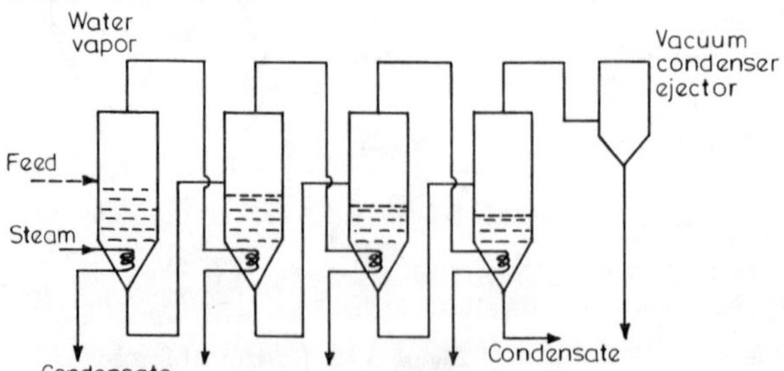

Figure 3.43 Multi-effect evaporator arrangement.

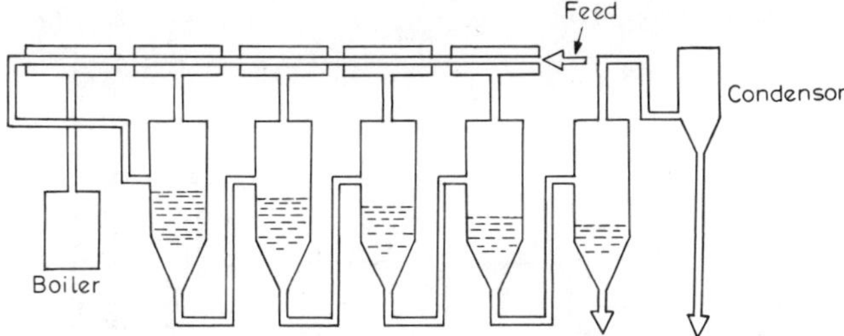

Figure 3.44 Multi-flash evaporator arrangement. *(With permission from Escher Wyss, Limited.)*

3.119 How are the multi-effect evaporators operated?

Figure 3.45 shows the various flow patterns for the feed possible in a multi-effect evaporator.

a. Forward feed: Both steam and liquor flow in the same direction. Since the pressure in subsequent effects is lower, no feed pump is necessary, except for the dilute feed input into the first stage and for removing the concentrated liquor from the last stage. The overall coefficient is the lowest in the last effect since the feed is the most concentrated there. The low pressure in each stage also results in the flashing of the feed, which produces vapor, and the steam economy is favorably affected. These are simplest to use and should be evaluated first before other systems are considered. One example is the use in the sugar industry.

b. Backward feed: The dilute feed is input into the last effect, while fresh steam is fed to the first effect. Thus, the feed and steam move counter to one another. Since the feed encounters higher pressure in each effect, there is no flashing, and sensible heat is needed in each effect. Furthermore, a pump is needed for each effect besides the ones needed in the forward feed arrangement discussed earlier. Backward feed gives higher capacity when the liquor is very viscous and when the feed is cold, since the heat transfer coefficient is maintained high due to higher temperature in each effect as the feed travels backward. One example is the use in the caustic soda industry.

c. Mixed feed: The feed enters an intermediate effect, flows without the aid of pumps until the last effect, and is then pumped to the first effects for final concentration. This combines the advantages of both the forward and backward feeds, since the number of pumps is reduced and the final evaporation is still done at the highest temperature in the first effect where the fresh steam is introduced.

d. Parallel feed: The feed is input to each effect separately, while the fresh steam is input only to the first effect, and the vapor generated in each effect is fed to the next effect to supply the heat of vaporization. Thus, it behaves as a multi-effect unit as far as the steam is concerned and as a single-effect unit as far as the feed is concerned. This is practiced in some crystallizing evaporators. One example is the common salt industry (Kern, 1950).

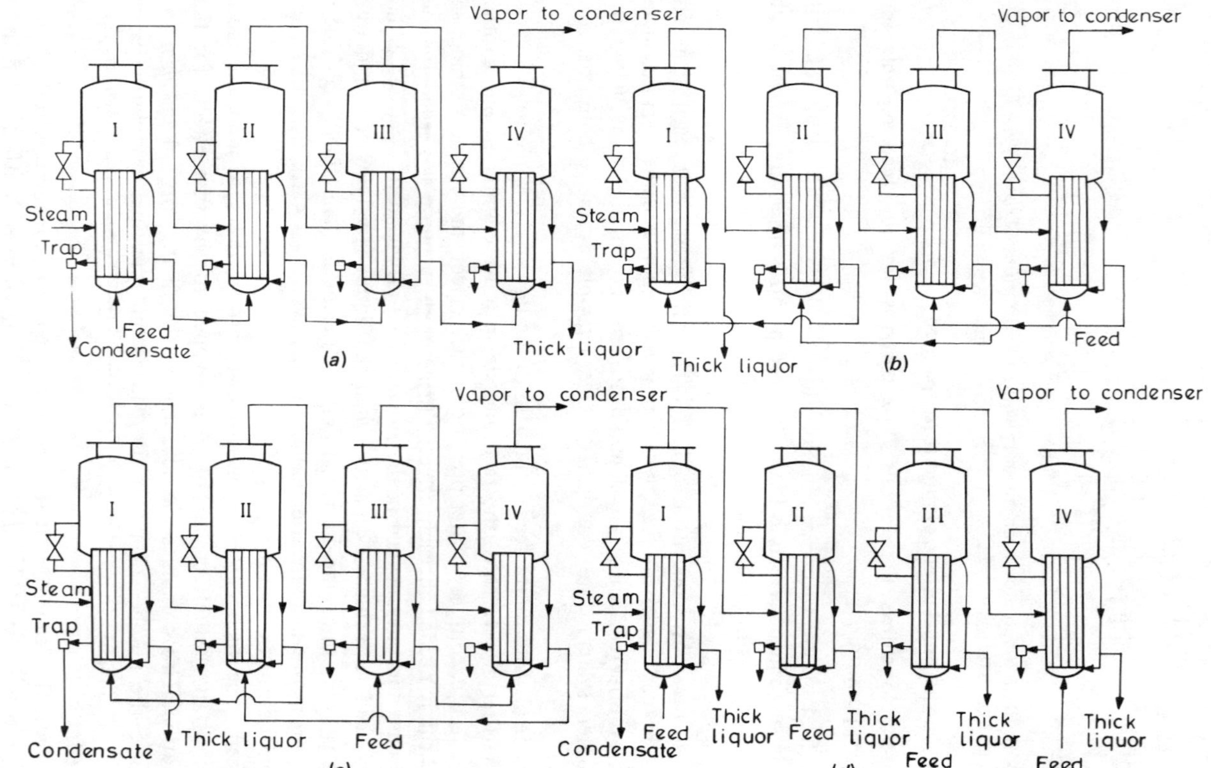

Figure 3.45 Feed arrangement in multi-effect evaporators. (*a*) Forward feed. (*b*) Backward feed. (*c*) Mixed feed. (*d*) Parallel feed. *(From* Unit Operations of Chemical Engineering *by McCabe and Smith. Copyright 1976. Used with the permission of McGraw-Hill Book Co.)*

3.120 Is equal amount of steam produced in each effect?

No. The amount of steam produced decreases with the number of effects, since a finite temperature difference is required between the condensing steam and the boiling point of the solution in the evaporator. Furthermore, in the power plant usage, part of the water is discharged as blowdown to control the buildup of undesired solutes, and this carries with it a certain amount of sensible heat. Thus, all the heat given out by the condensing steam is not used to produce an equivalent amount of steam in the evaporator. As a rough estimate, 1 kg (2.2 lbm) of fresh steam supplied to the first evaporator will produce about 0.85 kg (1.87 lbm) of steam in that evaporator, which in turn would produce about 0.75 kg (1.65 lbm) of steam in the second evaporator, giving a total of 1.6 kg (3.53 lbm) of distilled water for each 1 kg (2.2 lbm) of steam input in a two-effect evaporator.

3.121 When are the areas of different effects in a multi-effect evaporator different?

This happens when the heat transfer coefficients in different effects differ appreciably, as when flashing occurs in some of the effects and not in the rest (e.g., in mixed feed units).

3.122 How do the temperature (*T*), pressure (*P*), concentration (*C*), and viscosity (μ), vary in a four-effect evaporator with different feed arrangements shown in Fig. 3.45?

These are given in Table 3.4 (Gupta, 1979) in decreasing order of magnitude. Fresh steam is supplied to the first effect. These are relative values. The highest values achieved in different feed arrangements are not necessarily equal.

3.123 What is the classification according to how the energy is supplied for evaporation?

Energy can be supplied as heat, by vapor recompression, or by a combination of both heat and vapor recompression.

1. Heat: All the evaporator systems discussed so far had energy supplied as heat in the form of steam to the first effect from a boiler or a steam generator.
2. Vapor recompression: The vapor coming off of an evaporator is compressed to a higher (saturation) pressure (Fig. 3.46). Consequently, its temperature also rises. It is then sent to the service side of the evaporator to supply the heat. The cycle is repeated. The cost of recompression is very small compared to the latent heat available from the recompressed vapor.
3. In most of the multi-effect evaporator installations, a combination of (1) and (2) is used to supply the energy (Fig. 3.47). This turns out to be more effective and optimal, costwise.

3.124 How is vapor recompressed?

Vapor can be recompressed in two ways:

1. Thermo-compression: High-pressure steam is used in an ejector system (Fig. 3.46) to recompress the vapors. In this case, a mixture of live steam and compressed

Table 3.4 Relative values of temperature (T), pressure (P), concentration (C), and viscosity (μ) in different feed arrangements in a four-effect evaporator set-up[a,b]

	Highest	Second highest	Third highest	Lowest
Forward feed	T_1	T_2	T_3	T_4
	P_1	P_2	P_3	P_4
	C_4	C_3	C_2	C_1
	μ_4	μ_3	μ_2	μ_1
Backward feed	T_1	T_2	T_3	T_4
	P_1	P_2	P_3	P_4
	C_1	C_2	C_3	C_4
	μ_1	μ_2	μ_3	μ_4
Mixed feed	T_1	T_2	T_3	T_4
	P_1	P_2	P_3	P_4
	C_1	C_2	C_4	C_3
	μ_1	μ_2	μ_4	μ_3
Parallel feed	T_1	T_2	T_3	T_4
	P_1	P_2	P_3	P_4
	C_1	C_2	C_3	C_4
	μ_1	μ_2	μ_3	μ_4

Source: From Gupta (1979).
[a]Subscripts denote the effect number.
[b]The feed arrangements are shown in Fig. 6.45.

recycled vapor are supplied to the service side. This is used when live or fresh steam is available at a pressure higher than that required in the evaporator.

2. Mechanical compression: When fuel is costly, which makes the steam generation costly, the vapor compression is done by using a centrifugal compressor (Fig. 3.47). This is also practiced widely in heat pumps to save energy.

3.125 What are the advantages and disadvantages of vapor recompression?

The advantages are that no large boiler set-up is needed and that no condenser is needed after the last effect, since the vapor produced is recompressed and recycled.

It has been reported (APV, 1978) that while a double-effect evaporator requires approximately 0.55 kg steam for every kg (0.55 lbm steam/lbm) of water evaporated, the steam demand is reduced to 0.39 kg/kg (0.39 lbm/lbm) water evaporated by adding first effect recompression, thus economizing on the steam. The actual amount of live steam saved depends upon the pressure of the live steam, the pressure of the vapor generated, and the pressure of the saturated steam required on the service side for heat transfer by condensation. Table 3.5 gives the savings when live steam is available at 150 psig (1140 kPa) and vapor is available at 15 psig (204.8 kPa). It also lists the amount of vapor entrained in each case. The lower the required service steam pressure, the higher is the live steam savings.

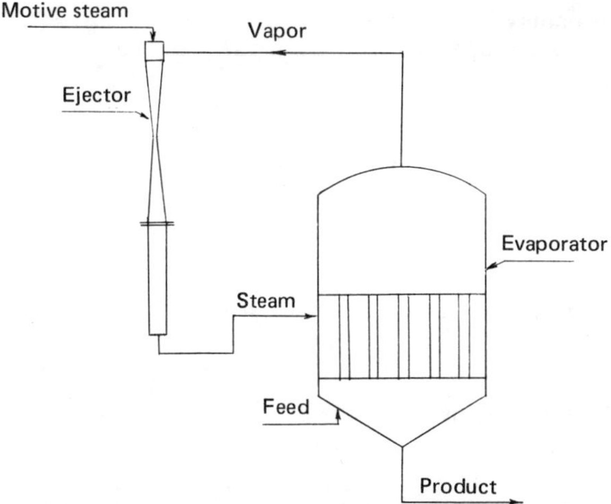

Figure 3.46 Single effect evaporator with vapor recompression.

The primary disadvantage of vapor recompression is the higher initial cost. Various parameters, such as the capacity, costs, and interest payment, govern the choice of whether vapor recompression should be used or not. This has been discussed at length using industrial examples by Walter and Kondorosy (1980). On the whole, in most situations, vapor recompression saves energy (Beesley and Rhinesmith, 1980). A knowledge of thermodynamic aspects helps in the design (Weiner et al., 1980).

3.126 What are the evaporator classifications according to the amount of feed evaporated?

In an evaporator, either partial or full feed may be evaporated. Most of the evaporators discussed thus far can be designed for both the cases. Usually, heat-sensitive liquids are not recycled, while highly scaling liquids are only partially evaporated by using a large recycle ratio.

3.127 What are the evaporator classifications according to the type of construction?

Evaporators can be shell-and-tube, plate, agitated (wiped) film, or batch pan type.

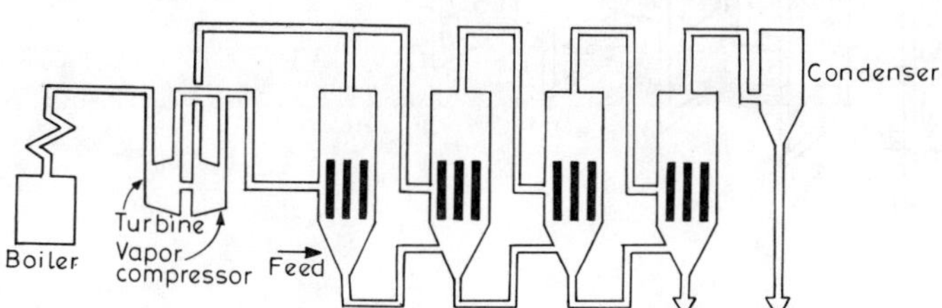

Figure 3.47 Combination steam and vapor compression evaporator set-up. *(With permission from Escher Wyss, Limited.)*

Table 3.5 Thermo-compression economy[a]

Steam pressure required on service side, psig	18	20	22	24	26
Entrainment ratio, lbm entrained vapor/lbm live steam	3.33	2	1.43	1.08	0.87
Amount of live steam saved, percent	77	67	59	52	46

Source: From Kern (1950).
[a]Live steam pressure equals 150 psig; vapor coming off the evaporator equals 15 psig.

1. Shell-and-tube type: These are the most used. All the evaporators discussed so far are shell-and-tube type, although some can be made in other types too.
2. Plate type: A typical set-up is shown in Fig. 3.48 and a line sketch is given in Fig. 3.49. These can be rising, falling, or rising–falling film types. These are used for low to moderate heat duties.

   The advantages are low residence time as required for the heat-sensitive liquids (pharmaceutical, fruit, and dairy products); low liquid hold-up, which gives rapid start-up and shutdown with a minimum wastage; easy and fast cleaning; and flexible capacity that can be decreased or increased.

   The disadvantages are that these cannot be used with very viscous liquids since the liquid distribution over the plates may not be uniform and that these cannot be used with liquids containing large suspended particles or where crystallization might occur due to evaporation.

   A good discussion of plate evaporators has been given by Worrall (1978). Plate heat exchangers are discussed at length in Chapter 6.
3. Agitated film type: These are also called wiped-film type (Fig. 3.50). Liquid is introduced at the top, and a very thin film of liquid is formed by means of an agitator in a jacketed vessel. The clearance between the agitator blades and the vessel may be fixed or adjustable. The film flows down in turbulent flow, and

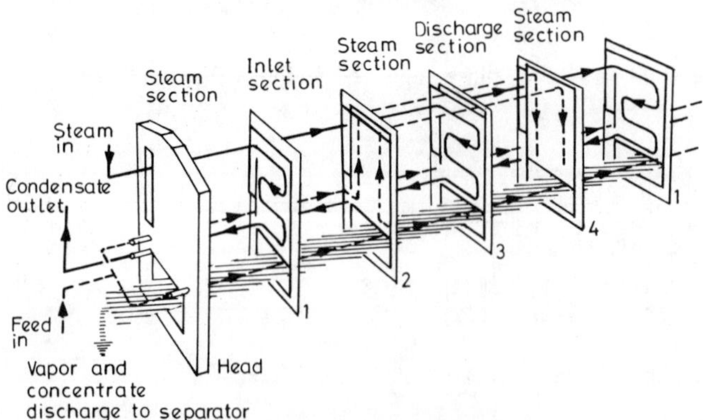

Figure 3.48 Plate-type evaporator. *(With permission from APV Company, Ltd., 1979.)*

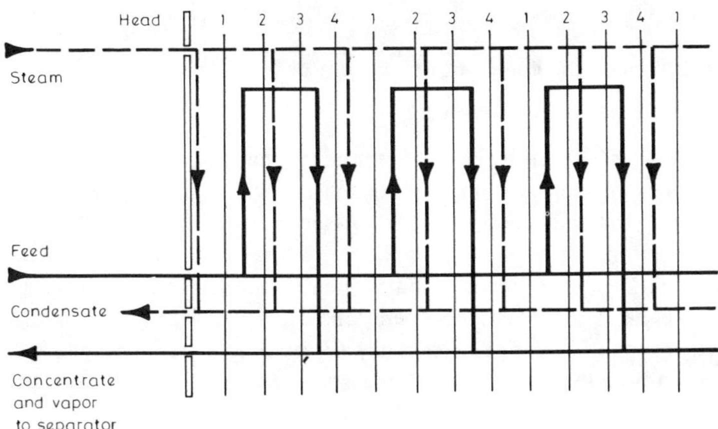

Figure 3.49 Line sketch of plate-type evaporator. *(With permission from APV Company, Ltd.,, 1979.)*

the heat transfer coefficient is very high. These are used for very viscous liquids, highly foaming liquids, and for heat-sensitive liquids requiring low residence time. The equipment cost is high.

4. Batch pan: These are open pans used for evaporation. These work in a batch mode, and several may be used in parallel. The pan is either jacketed or has internal coils for heat transfer. The heat transfer coefficient is low, and the residence time is large. These are of limited utility now because of larger plant sizes favoring continuous production instead of batch operation.

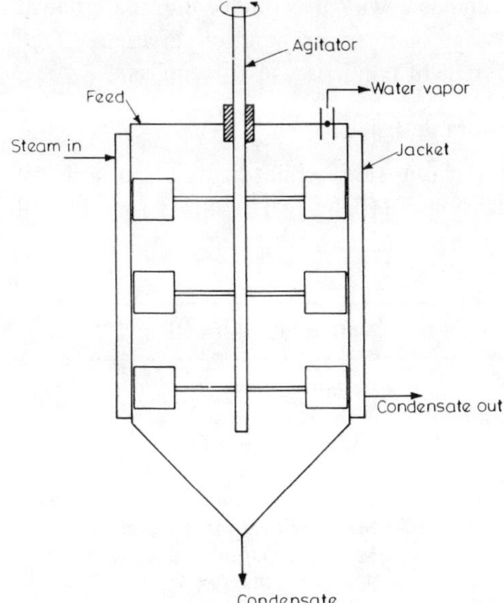

Figure 3.50 Agitated or wiped-film type evaporator.

Choice, Design, and Operation

3.128 What factors influence the choice of an evaporator?

The important factors are the following:

1. Liquid characteristics: These relate to the viscosity, density, surface tension, concentration, changes in viscosity and surface tension with temperature, scaling and corrosive nature, foaming tendency, pH, presence of solids, crystallization tendency on evaporation, latent heat of vaporization, rise in boiling point with increase in concentration, gases liberated on boiling (if any), explosion hazards, heat sensitivity, the necessity of sterile operations as in the dairy and pharmaceutical industries, droplet entrainment tendency of vapor as in organics, etc.
2. Desired product: Amount of liquid to be evaporated, concentration of the desired product
3. Material of construction: It has to be compatible with the corrosive nature of the liquid. The costlier the material, the higher the design heat transfer coefficient should be so as to reduce the evaporator size and hence the cost. The materials of construction used for different products are listed in Table 3.6.

3.129 In Fig. 1.37, a heat exchanger specification sheet is given for a heat exchanger. What additional information is required for designing an evaporator?

The additional information required is the vaporizing temperature, latent heat of vaporization, molecular weight, operating pressure, quantity of liquid to be vaporized, tendency to crystallize, total solids content, boiling point elevation, type of unit desired, space restrictions, etc. It would further aid the design engineer if the details about the available steam supply and pumping requirements are also provided. From all this, the design engineer will determine the material of construction, size and number of effects, amount of evaporation in each effect, temperature and pressure in each effect, design of condenser at the end, etc.

3.130 How are the shell-and-tube evaporators designed?

In an evaporator, steam condenses on one side, while the solution boils off on the other side of the tube. The condensing coefficient of steam is very high if

Table 3.6 Materials of construction of evaporators

Products	Material of construction
General use	Carbon steel
Food, dairy, pharmaceuticals	304/316 stainless steel
Sulfuric acid < 50%, hydrochloric acid, phosphoric acid	Carbon steel or rubber lined carbon steel
Sulfuric acid < 72%	Lead lined carbon steel/or cast lead
Caustic soda < 40%	Stress relieved carbon steel
Caustic soda (with NaCl)	Monel/nickel/nickel lined carbon steel
Ammonium nitrate	304/304 L stainless steel
Ammonium sulfate	316 stainless steel

Table 3.7 Typical overall coefficients in evaporators

	Overall coefficient, U	
Type	Btu/ft^2 hr °F	W/m^2 °C
Long-tube vertical evaporators		
Natural circulation	200–600	1000–3000
Forced circulation	400–2000	2000–10,000
Short-tube evaporators		
Horizontal type	200–400	1000–2000
Calandria type	150–500	800–2500
Coil evaporators	200–400	1000–2000
Agitated-film evaporators, Newtonian liquids, viscosity:		
1 cP (0.001 Pa s)	400	2000
1 P (0.1 Pa s)	300	1500
100 P (10 Pa s)	120	600

Source: From McCabe and Smith (1976).

inerts are not allowed to accumulate. The primary resistance to heat transfer is due to the boiling liquid. In forced circulation units, the heat transfer coefficient can be calculated by using the general relations for the sensible heat transfer. In natural circulation units, the value of the heat transfer coefficient is significantly lower and less amenable to exact calculations. Usually, the units are designed based upon an overall coefficient, which accounts for the type of evaporator and whether natural or forced circulation is occurring (Table 3.7). These values are based upon wide experience and practices in industry that have stood the test of time. It is likely that the units are significantly overdesigned on this basis. A higher value may result in vapor blanketing of the tubes and in film boiling with consequent reduction in the evaporator duty.

3.131 Is the tube pitch the same in the heat exchangers and in the evaporators?

No. Usually the tubes are spaced farther apart in the evaporators compared to the no-phase-change heat exchangers, since the vapor generated in the lower section of the evaporator tube bundle should not blank out the tubular surface in the upper section of the tube bundle.

3.132 How does one choose the operating temperature?

It is based upon the product stability or the heat sensitivity of the liquid, the material of construction of the evaporator, the ready availability of a heat source to supply the required heat so as to maintain the operating temperature, the number of effects that determines the pressure (and temperature), the progression from the first to the last effect, etc.

3.133 What are the important factors that affect the rate of evaporation?

These are the following:

1. The temperature difference between the hot surface and the saturation temperature or boiling temperature. A very high value may lead to film boiling and reduced evaporation (Q. 3.72).
2. The nature of the heating surface as far as promoting greater nucleation is concerned.
3. The properties of the liquid, such as viscosity, surface tension, heat capacity, density, and their variation with temperature. These affect the rate of circulation and the sensible heat required to bring the liquid to its boiling point.
4. The thermal condition of the feed–subcooled, saturated, or two-phase mixture. This greatly affects the energy requirements.
5. The pressure in the evaporator, since this affects the amount of vapor produced due to flashing and also the boiling point and hence the temperature driving potential.
6. The increase in concentration, since it results in an increase in the boiling point of the solution and decreases the temperature driving potential.
7. The hydrostatic head above the heat transfer surface, since it increases the pressure and hence the boiling temperature of the solution at the heat transfer surface. This has the same effect as in (6) above.

3.134 What is the effect of the liquid level on the evaporator performance?

Liquid level exerts a hydrostatic pressure, thus increasing the boiling point of the solution. This decreases the driving potential between the condensing steam and the boiling liquid, resulting in a reduced amount of heat transferred to the boiling liquid.

3.135 How does the loss of water due to evaporation affect the evaporator performance?

The loss of water increases the concentration of the solution. This results in the following:

1. An increase in the viscosity which affects the circulation within that evaporator as well as in the remaining effects downstream. It also requires higher power to pump the concentrated liquor.
2. A decrease in the vapor pressure of water in solution compared to that for pure water at the same temperature. Hence, the solution has to be heated to a higher temperature so that boiling occurs. This additional temperature rise is called the boiling point rise (BPR) or boiling point elevation. It cuts down the temperature driving potential between the hot surface and the boiling solution, and thus reduces the rate of evaporation and hence the performance of an evaporator. This is more pronounced in the case of a multi-effect evaporator and is an important factor in limiting the number of effects.
3. Dry-out of part of the tubular surface in extreme cases. This may lead to corrosion failure of the tubes (Fisher, 1979).

3.136 Whereas the disengagement space for the entrained liquid droplets is provided within the evaporator shell for the power plant evaporator, the disengagement is usually done in a separate drum for the chemical plant evaporator. Why?

The upper 50–60 percent space in a power plant evaporator is used for the disengagement of the entrained water droplets. The pressure in these evaporators is usually low, and hence the thickness of the shell, tubesheets, and the flanges is not much. On the other hand, the chemical plant evaporators usually operate at high pressures. Since the thickness of the shell, flanges, etc., varies with the pressure and the shell diameter, the cost of the chemical plant evaporator becomes very high when the disengagement space is included in the same shell. Hence, to reduce the cost, the disengagement is done in a separate drum. The thickness of both the evaporator and the drum is significantly less than would be the case if the disengagement is done in the evaporator itself.

3.137 How can energy be conserved in an evaporator?

The following steps will help:

1. Preheating the feed to its boiling point
2. Using multiple effects, if feasible
3. Using the low-temperature waste heat from the evaporator in an alternate process
4. Keeping heat losses from the equipment to a minimum

• This completes our discussion on heat exchangers involving phase change. A short review of the topic has been done by Bell (1983). We will consider extended and enhanced surfaces in the next chapter.

NOMENCLATURE

A	total heat transfer surface area
BPR	boiling point rise
C	number of components; concentration
CHF	critical heat flux
DNB	departure from nucleate boiling
E	TEMA type shell (Fig. 1.1)
f_1	partial fugacity of component 1 in a mixture
f_p	fugacity
f_{p_1}	fugacity of a pure component
f_t	total fugacity of a vapor mixture
F	degrees of freedom
G_L	mass velocity of liquid
G_v	mass velocity of vapor
$\bar{h}_1$	average condensing coefficient for the first row of horizontal tubes

$\bar{h}_N$	average condensing coefficient for *N* rows of horizontal tubes
J	TEMA type shell (Fig. 1.1)
K	equilibrium constant
LTV	long tube vertical unit
N	number of rows of horizontal tubes in a condenser
p_1	partial pressure of a component in a vapor mixture
p_t	total pressure of a vapor mixture
p_v	vapor pressure of pure component
P	number of phases; pressure
q'	quality of feed to a distillation column; heat flux
q'_i	heat load in zone i
Q	total heat load
STV	short tube vertical unit
T	temperature
T_c	coolant temperature
T_{sat}	saturation temperature or bulk liquid boiling temperature
T_v	bulk vapor phase temperature
T_w	temperature of heated surface
ΔT	temperature difference
ΔT_{LMTD}	log mean temperature difference
ΔT_{LMTD_i}	log mean temperature difference in zone i
$\Delta T_{LMTD_{overall}}$	overall log mean temperature difference
U	overall heat transfer coefficient
U_{di}	design (or dirty) heat transfer coefficient in zone i
$U_{d_{overall}}$	overall design (or dirty) heat transfer coefficient
V	velocity
x	mole fraction in liquid phase
X	TEMA type shell (Fig. 1.1)
y	mole fraction in vapor phase

Greek Letters

α_{12}	relative volatility of component 1 with respect to component 2
$\alpha_{butane/pentane}$	relative volatility of butane with respect to pentane
$\alpha_{hexane/pentane}$	relative volatility of hexane with respect to pentane
μ	viscosity
ψ_b	Baker parameter (Fig. 3.11)
Λ_b	Baker parameter (Fig. 3.11)
ρ	density
σ	surface tension

Subscripts

i	interface; zones in a combined desuperheater, condenser and subcooler

1	component 1
2	component 2
g, G	gas
L	liquid

Superscript

noncondensable component

REFERENCES

Afgan, N. H. Intermittent Phenomena in Pool Boiling. In *Heat Transfer 1978: Proceedings of the Sixth International Heat Transfer Conference, Toronto, August 1978.* 6:407–415. Hemisphere: Washington, D.C. 1979.

Aoki, S., A. Inoue, Y. Kozawa, and H. Akimoto. Direct Contact Condensation of Flowing Steam into Injected Water. In *Heat Transfer 1978: Proceedings of the Sixth International Heat Transfer Conference, Toronto, August 1978.* 5:107–112. Hemisphere: Washington, D.C. 1979.

APV Company. *Falling Film Plate Evaporators,* Publication No. A458c. APV Co., Ltd.: Crawley, England. 1978.

APV Company. *APV Plate Evaporator Climbing and Falling Film Type,* Publication No. A3351. APV Co., Ltd.: Crawley, England. 1979.

ASME. *ASME Boiler and Pressure Vessel Code,* Section VIII, Division 1 and 2. ASME: New York. 1980.

Baker, O. Simultaneous Flow of Oil and Gas. *Oil Gas J.* 53:185–195. 1954.

Baker. W. J. How to Specify and Select Surface Condensers, *Hydrocarbon Process.* 60(9):239–245. 1981.

Bakey, A., and T. Jaszay. High Performance Jet Condensers for Steam Turbines. In *Heat Transfer 1978: Proceedings of the Sixth International Heat Transfer Conference, Toronto, August 1978.* 2:61–65. Hemisphere: Washington, D.C. 1979.

Bankoff, S. G., and V. S. Mehra. A Quenching Theory for Transition Boiling, *Ind. Eng. Chem. Fundamentals* 1:38–40. 1962.

Beatty, K. O., Jr., and D. L. Katz. Condensation of Vapors on Outside of Finned Tubes, *Chem. Eng. Prog.* 44(1):55–70. 1948.

Beesley, H. A., and R. D. Rhinesmith. Energy Conservation by Vapor Compression Evaporators, *Chem. Eng. Prog.* 78(8):37–41. 1980.

Bell, K. J. Heat Exchangers with Phase Change. In *Heat Exchangers: Theory and Practice.* J. Taborek et al. (eds.). Hemisphere: Washington, D.C. 1983.

Bell, K. J., and M. A. Ghaly. An Approximate Generalized Design Method for Multicomponent/Partial Condensers, *AIChE Symp. Ser. No. 131* 69:72–79. 1972.

Bell, K. J., and C. B. Panchal. Condensation. In *Heat Transfer 1978: Proceedings of the Sixth International Heat Transfer Conference, Toronto, August 1978.* 6:361–375. Hemisphere: Washington, D.C. 1979.

Bergles, A. E., J. G. Collier, J. M. Delhaye, G. F. Hewitt, and F. Mayinger. *Two-Phase Flow and Heat Transfer in the Power and Process Industries.* Hemisphere: Washington, D.C. 1981. (Highly recommended reading.)

Blangetti, F., and E. U. Schlünder. Local Heat Transfer Coefficients on Condensation in a Vertical Tube. In *Heat Transfer 1978: Proceedings of the Sixth International Heat Transfer Conference, Toronto, August 1978.* 2:437–442. Hemisphere: Washington, D.C. 1979.

Branan, C. *The Process Engineer's Pocket Handbook.* Gulf Publishing: Houston, Texas. 1976.

British Electrical and Allied Manufacturers' Association. *Recommended Practice for the Design of Surface Type Steam Condensing Plant.* British Electrical and Allied Manufacturers' Association: London. 1967.

Butterworth, D. *A Calculation Method for Shellside Heat Exchangers in which the Overall Coefficient Varies along the Length,* NEL Report No. 590, pp. 56–71. National Engineering Laboratory: East Kilbride, Glasgow. 1975.

Butterworth, D. Developments in the Design of Shell-and-Tube Condensers, Paper 77-WA/HT-24. ASME Winter Annual Meeting, Atlanta, Georgia. 1977.

Butterworth, D. *Introduction to Heat Transfer.* Oxford University Press: Oxford. 1978.

Butterworth, D. Condensers: Thermohydraulic Design. In *Heat Exchangers: Thermal-Hydraulic Fundamentals and Design.* S. Kakac, A. E. Bergles, and F. Mayinger (eds.). Hemisphere: Washington, D.C. 1981a.

Butterworth, D. Condensers: Basic Heat Transfer and Fluid Flow. In *Heat Exchangers: Thermal-Hydraulic Fundamentals and Design.* S. Kakac, A. E. Bergles, and F. Mayinger (eds.). Hemisphere: Washington, D.C. 1981b.

Carnavos, T. C. Some Recent Developments in Augmented Heat Exchange Elements. In *Heat Exchangers: Design and Theory Sourcebook.* N. Afgan and E. U. Schlünder (eds.). Hemisphere: Washington, D.C. 1974.

Christman, J. Vacuum Steam Condensers: Air vs. Water Cooling. *Hydrocarbon Process.* 59(9):257–259. 1980.

Colburn, A. P. Note on the Condensation When a Portion of the Condensate Layer Is in Turbulent Motion. *Trans. AIChE* 30:187–193. 1934.

Colburn, A. P., and A. G. Edison. Prevention of Fog in Cooler-Condensers, *Ind. Eng. Chem.* 33:457–458. 1941.

Colburn, A. P., and O. A. Hougen. Design of Cooler Condensers for Mixtures of Vapors with Noncondensing Gases. *Ind. Eng. Chem.* 26:1176–1182. 1934.

Collier, J. G. *Convective Boiling and Condensation.* McGraw-Hill: New York. 1972.

Collier, J. G. *Convective Boiling and Condensation,* 2d ed. McGraw-Hill: New York. 1980.

Collier, J. G. Boiling and Evaporation–I. In *Heat Exchangers: Thermal-Hydraulic Fundamentals and Design.* S. Kakac, A. E. Bergles, and F. Mayinger (eds.). Hemisphere: Washington, D.C. 1981a.

Collier, J. G. Boiling and Evaporation–II. In *Heat Exchangers: Thermal-Hydraulic Fundamentals and Design.* S. Kakac, A. E. Bergles, and F. Mayinger (eds.). Hemisphere: Washington, D.C. 1981b.

Cooper, M. G. Nucleate Boiling. In *Heat Transfer 1978: Proceedings of the Sixth International Heat Transfer Conference, Toronto, August 1978.* 6:463–469. Hemisphere: Washington, D.C. 1979.

Coulson, J. M., and J. F. Richardson. *Chemical Engineering,* 3d ed., vol. 1. Pergamon: Oxford. 1977.

Cumo, M., G. E. Farello, and G. Ferrari. Direct Heat Transfer in Pressure-Suppression System. In *Heat Transfer 1978: Proceedings of the Sixth International Heat Transfer Conference, Toronto, August 1978.* 5:101–106. Hemisphere: Washington, D.C. 1979.

Dukler, A. E. Dynamics of Vertical Falling Film Systems. *Chem. Eng. Prog.* 55(10):62–67. 1959.

Efimochkin, G. I. Direct-Contact Heat Exchangers for Large Steam Turbine Installations. In *Heat Exchangers: Theory and Practice.* J. Taborek et al. (eds.). Hemisphere: Washington, D.C. 1983.

Emerson, W. H. *Effective Tube-side Temperature Differences in Multipass Heat Exchangers with Nonuniform Heat Transfer Coefficients and Specific Heats.* NEL Report No. 590, pp. 32–55. National Engineering Laboratory: East Kilbride, Glasgow. 1975.

Escher Wyss Ltd. *Plants for Evaporation, Crystallization, Concentration,* Bulletin C22.16.31 ZBge 50. Escher Wyss: Zurich, Switzerland. 1980.

Fair, J. R. What You Need to Design Thermosiphon Reboilers. *Pet. Refiner* 39(2):105–123. 1960.

Fair, J. R. Process Heat Transfer by Direct Fluid-Phase Contact. *AIChE Symp. Ser. No. 118* 68:1–11. 1972a.

Fair, J. R. Designing Direct-contact Coolers/Condensers, *Chem. Eng.* 79:91–100. 1972b.

Fisher, S. A. Premature Dryout in Conventional and Nuclear Power Station Evaporators. In *Heat Transfer 1978: Proceedings of the Sixth International Heat Transfer Conference, Toronto, August 1978.* 2:49–54. Hemisphere: Washington, D.C. 1979.

Foxall, D. H., and H. R. Chappell. Superheated Vapor Condensation in Heat Exchanger Design. *Chem. Eng.* 87(26):41–50. 1980.

Grant, I. D. R. *Flow and Pressure Drop with Single Phase and Two-Phase Flow in the Shell-Side of Segmentally Baffled Shell and Tube Heat Exchangers.* NEL Report No. 590, pp. 1–22. National Engineering Laboratory: East Kilbride, Glasgow. 1975.

Grant, I. D. R., and D. Chisholm. Two-Phase Flow on the Shellside of a Segmentally Baffled Shell-and-Tube Heat Exchanger. *Trans. ASME, J. Heat Trans.* 101:36–42. 1979.

Gupta, J. P. *Heat Exchanger Design–A Practical Look,* 2d ed., C. S. Enterprises: Delhi, India. 1979.

Hayde, L. J. Calculator Program for a Steam Condenser, *Chem. Eng.* 88(3):99–102. 1981.

Heat Exchange Institute. *HEI Standards for Steam Surface Condensers,* 7th ed. HEI: Cleveland, Ohio. 1978.

Hewitt, G. F. Critical Heat Flux in Flow Boiling. In *Heat Transfer 1978: Proceedings of the Sixth International Heat Transfer Conference, Toronto, August 1978.* 6:143–171. Hemisphere: Washington, D.C. 1979.

Impagliazzo, A. M., and K. J. Bell. Steam Surface Condenser Operating and Design Considerations, paper presented at the Joint ASME/IEEE/ASCE Power Generation Conference, Dallas, Texas. September 1978.

Ishihara, K., J. W. Palen, and J. Taborek. Critical Review of Correlations for Predicting Two-Phase Flow Pressure Drop across Tube Banks. *Heat Transfer Eng.* 1(3):23–32. 1980.

Jacobs, H. R., K. D. Thomas, and R. F. Boehm. Direct Contact Condensation of Immiscible Fluids in Packed Bed. In *Condensation Heat Transfer,* pp. 103–110. ASME: New York. 1979.

Johnson, M. A., J. de la Pera, and R. B. Mesler. Bubble Shapes in Nucleate Boiling. *Chem. Eng. Progr. Symp. Ser. No. 64* 62:1–5. 1966.

Karagil, G., A. Kabaalioglu, E. Ekinci, I. Cataltas, and F. Civan. Modelling and Experimental Investigation of a Multieffect Evaporator. In *Heat Exchangers: Theory and Practice.* J. Taborek et al. (eds.). Hemisphere: Washington, D.C. 1983.

Karimi, A. Laminar Film Condensation on Helical Reflux Condensers and Related Configurations. *Int. J. Heat Mass Transfer* 20:1137–1144. 1977.

Kern, D. Q. *Process Heat Transfer.* McGraw-Hill: New York. 1950.

Kutateladze, S. S. Semi-empirical Theory of Film Condensation of Pure Vapors. *Int. J. Heat Mass Transfer* 25:653–660. 1982.

Larinoff, M. W., W. E. Moles, and R. Reichhelm. Design and Specification of Air-Cooled Steam Condensers. *Chem. Eng.* 85(12):86–94. 1978.

Leong, L. S., and K. Cornwell. Flow Boiling Heat Transfer Coefficients in a Kettle Reboiler Tube Bundle, *The Chem. Engr. (Lond.)* 57:219. 1979.

Lockhart, R. W., and R. C. Martinelli. Proposed Correlation of Data for Isothermal Two-Phase, Two-Component Flow in Pipes. *Chem. Eng. Prog.* 45(1):39–48. 1949.

LoPinto, L. Fog Formation in Low-Temperature Condensers. *Chem. Eng.* 89(10):111–113. 1982.

Lord, R. C., P. E. Minton, and R. P. Slusser. Design Parameters for Condensers and Reboilers, *Chem. Eng.* 77(6):127–134. 1970a.

Lord, R. C., P. E. Minton, and R. P. Slusser. Guide to Trouble-Free Heat Exchangers. *Chem. Eng.* 77(11):153–160. 1970b.

Mandhane, J. M., G. A. Gregory, and K. Aziz. A Flow Pattern Map for Gas-Liquid Flow in Horizontal Pipes. *Int. J. Multiphase Flow* 1:537–553. 1974.

Mayinger, F., and M. Schult. Dynamic Behavior of Double-Phase-Change Heat-Exchangers. In *Heat Exchangers: Thermal-Hydraulic Fundamentals and Design.* S. Kakac, A. E. Bergles, and F. Mayinger (eds.). Hemisphere: Washington, D.C. 1981.

McCabe, W. L., and J. C. Smith. *Unit Operations of Chemical Engineering,* 3d ed. McGraw-Hill: New York. 1976.

Moalem-Maron, D., and S. Sideman. Analysis of Direct Contact Condensers: Single and Two-Phase Systems. In *Heat Exchangers: Design and Theory Sourcebook.* N. Afgan and E. U. Schlünder (eds.). Hemisphere: Washington, D.C. 1974.

Moalem, D., and S. Sideman. Theoretical Analysis of a Horizontal Condenser Evaporator Elliptical Tube. *Trans. ASME, J. Heat Transfer* 97(3):352–359. 1975.

Moalem-Maron, D., and S. Sideman. Condensation Inside Near Horizontal Tubes in Co-Current and Counter-Current Flow. *Int. J. Heat Mass Transfer* 25(9):1439–1443. 1982.

Nusselt, W. Surface Condensation of Water Vapor. Parts I and II. *Zeitschr. Ver. deutsch. Ing. (VDI Z).* 60(27):541–546 and 60(28):569–575. 1916.

Othmer, D. F. The Condensation of Steam, *Ind. Eng. Chem.* 21:576–583. 1929.

Palen, J. W., and S. M. Small. A New Way to Design Kettle and Internal Reboilers. *Hydrocarbon Process.* 43(11):199–208. 1964.

Palen, J. W., C. C. Shih, A. Yarden, and J. Taborek. Performance Limitations in a Large Scale Thermosiphon Reboiler. In *Heat Transfer 1974. Proceedings of the Fifth International Heat Transfer Conference, Tokyo, September 1974.* Hemisphere: Washington, D.C. 1974.

Palen, J. W., G. Breber, and J. Taborek. Prediction of Flow Regimes in Horizontal Tube-Side Condensation. *Heat Transfer Eng.* 1(2):47–57. 1979.

Perry, R. H., and C. H. Chilton (eds.). Heat Transmission. In *Chemical Engineers' Handbook*, 5th ed. McGraw-Hill: New York. 1973.

Polley, G. T. Reboilers. In *Developments in Heat Exchanger Technology–I.* D. Chisholm (ed.). Applied Science: London. 1981.

Polley, G. T., and I. D. R. Grant. Pressure Drop Prediction for Two-Phase upward Flow through a Tube Bundle with Bypassing, Paper A10. *Proceedings of the International Meeting on Industrial Heat Exchangers,* Liege, Assoc. Eng. Grad. Univ. Liege. November 1979.

Rouet, A. A Note on the Lockhart–Martinelli Correlation in the Turbulent–Turbulent Case. *Int. J. Heat Mass Transfer* 26:145–146. 1983.

Rubin, F. L. Multizone Condensers: Tabulate Design Data to Prevent Errors. *Hydrocarbon Process.* 61(6):133–135. 1982.

Sebald, J. F. A History of Steam Surface Condensers for the Electric Utility Industry. *Heat Transfer Eng.* 1(3):80–87. 1979.

Semeria, R., and G. F. Hewitt. Aspects of Heat Transfer in Two-Phase Gas-Liquid Flow. In *Heat Exchanger: Design and Theory Sourcebook,* N. Afgan and E. U. Schlünder (eds.). Hemisphere: Washington, D.C. 1974.

Sharma, A., D. Garg, and J. P. Gupta. Further Experiments on Transient Swelling during Pool Boiling. *Letters in Heat and Mass Transfer* 9:221–226. 1982.

Sideman, S. Some Design Aspects of Thin Film Evaporators/Condensers in Water Desalination. In *Heat Exchangers: Thermal-Hydraulic Fundamentals and Design.* S. Kakac, A. E. Bergles, and F. Mayinger (eds.). Hemisphere: Washington, D.C. 1981a.

Sideman, S. Film Evaporation and Condensation in Desalination. In *Heat Exchangers: Thermal-Hydraulic Fundamentals and Design.* S. Kakac, A. E. Bergles, and F. Mayinger (eds.). Hemisphere: Washington, D.C. 1981b.

Sideman, S., and D. Moalem-Maron. Direct Contact Condensation. *Advan. Heat Transfer* 15:227–281. 1982.

Silver, L. Gas Cooling with Aqueous Condensation. *Trans. Inst. Chem. Eng.* 25:30–42. 1947.

Singh, K. P., and J. P. Gupta. Transient Swelling of Liquid Level during Pool Boiling in Emergency Condenser. *Letters in Heat and Mass Transfer* 8(1):25–33. 1981.

Smith, R. A. Types of Evaporators. In *Heat Exchanger Design Handbook.* Vol. 3, *Thermal and Hydraulic Design of Heat Exchangers.* E. U. Schlünder et al. (eds.). Hemisphere: Washington, D.C. 1983.

Starczewski, J. Simplify Design of Partial Condensers. *Hydrocarbon Process.* 60(3):131–136. 1981.

Steinmeyer, D. E. Fog Formation in Partial Condensers. *Chem. Eng. Prog.* 68(7):64–68. 1972.

Steinmeyer, D. E., and A. C. Mueller. Why Condensers Don't Operate as They Are Supposed to. *Chem. Eng. Prog.* 70(7):78–82. 1974.

Taborek, J. Design Methods for Heat Transfer Equipment: A Critical Survey of the State-of-the-Art. In *Heat Exchangers: Design and Theory Sourcebook.* N. Afgan and E. U. Schlünder (eds.). Hemisphere: Washington, D.C. 1974.

Taitel, Y., and A. E. Dukler. A Model for Predicting Flow Regime Transitions in Horizontal and Near Horizontal Gas Liquid Flow. *AIChE J.* 22(1):47–55. 1976.

Tanasawa, I. Dropwise Condensation–The Way to Practical Applications. In *Heat Transfer 1978: Proceedings of the Sixth International Heat Transfer Conference, Toronto, August 1978.* 6:393–405. Hemisphere: Washington, D.C. 1979.

Tandon, N., and J. P. Gupta. Experiments on Transient Swelling during Pool Boiling, *Letters in Heat and Mass Transfer* 8(3):247–251. 1981.

Tubular Exchanger Manufacturers' Association. *Standards of Tubular Exchanger Manufacturers Association,* 6th ed. TEMA Inc.: Tarrytown, New York. 1978.

von Bockh, P. Experimental Investigation on a Full Scale Thermosiphon Reboiler. In *Heat Exchangers: Theory and Practice.* J. Taborek et al. (eds.). Hemisphere: Washington, D.C. 1983.

Ward, D. J. How to Design a Multicomponent Partial Condenser. *Petro/Chem. Eng.* 32(10):C-42–C-48. 1960.

Walter, R. O., and P. Kondorosy. *Fundamental Economic Principals in Choosing the System for Evaporation Processes.* Bulletin e22.17.31ZBgi 5. Escher Wyss: Zurich, Switzerland. 1980.

Webb, R. L. High-Performance Surfaces for Non-Boiling Heat Transfer. In *Heat Transfer in Energy Problems.* T. Mizushina and W. J. Yang (eds.). Hemisphere: Washington, D.C. 1982.

Weiner, L. D., H. R. Dolf, and D. A. Austin. A Systems Engineering Approach to Vapor Recompression in Evaporators. *Chem. Eng. Prog.* 76(11):70–77. 1980.

Wild, N. H. Non-condensable Gas Eliminates Hammering in Heat Exchangers. *Chem. Eng.* 76(9):132–134. 1969.

Witte, L. C., and J. H. Lienhard. On the Existence of Two Transition Boiling Curves. *Int. J. Heat Mass Transfer* 25:771–779. 1982.

Woodruff, D. W., and J. W. Westwater. Steam Condensation on Various Gold Surfaces. *Trans. ASME, J. Heat Transfer* 103:685–692. 1981.

Worrall, P. Selecting and Designing Evaporators. In *Practical Aspects of Heat Transfer.* CEP Technical Manual. AIChE: New York. 1978.

CHAPTER

FOUR

AUGMENTED SURFACE HEAT EXCHANGERS

INTRODUCTION

• The desire to reduce the size, weight, and/or volume of a heat exchanger has lead to the development of various kinds of augmentation techniques to increase the heat transfer. In a number of situations these also lead to a saving in the capital and/or operating cost compared to a shell-and-plain-tube unit. With ever increasing plant sizes and energy costs, the augmentation techniques are being used in an increasing number of situations. In this chapter we shall discuss some of these techniques. Some specific applications, such as air-fin coolers and compact heat exchangers, are also discussed.

4.1 What are the ways by which heat transfer may be improved?

There are numerous ways to improve the heat transfer. Some are in practical use, while others are awaiting commercial applications. Some increase the heat transfer coefficient, h, while others increase the surface area per unit length, A, and still others decrease the temperature driving potential, ΔT, needed for heat transfer (i.e., they increase both h and A).

The various ways of increasing the heat transfer can be classified into two broad categories (Bergles et al., 1981; Larson and Kreith, 1981):

1. Passive techniques: These do not require direct application of any external power other than the pumping power. These include (a) treated or specially prepared

surfaces, (b) rough surfaces to promote turbulence, (c) extended surfaces, (d) enhanced surfaces, (e) turbulators or tube inserts that create rotating and/or secondary flows, (f) coiled tubes, and (g) additives to liquids and gases.

2. Active techniques: These require external power and include (a) heat transfer surface rotation, (b) heat transfer surface wiping, (c) heat transfer surface vibration at low frequency, (d) fluid vibration by pulsations introduced into the fluid, (e) electrostatic fields (d.c. or a.c.) applied to dielectric fluids, and (f) suction of vapor or condensate film through a porous heat transfer surface to decrease the resistance to heat transfer.

In a given situation one may use two or more techniques simultaneously for greater enhancement of heat transfer. Most of the active techniques are not presently being used due to the increased power and equipment cost and the reduced reliability of the total system. Bergles et al. (1979) have tabulated the present and potential commercial use of various augmentation techniques (Table 4.1). This shows that only some of the techniques are in common use.

An exhaustive survey of enhancement of six modes of heat transfer (single-phase free convection, single-phase forced convection, pool boiling, convective boiling, vapor space condensation without vapor shear effects, and forced in-tube condensation) using the augmentation techniques noted earlier has been done by Bergles (1979).

One should be careful and ensure that the use of the heat transfer augmentation techniques does not violate any code and standards, will not increase corrosion and/or fouling, will continue to operate effectively and reliably even after long times, and will be cost effective.

• In this chapter, we shall discuss the following passive augmentation techniques: (1) extended (or finned) surface heat exchangers, (2) turbulators, (3) air-fin coolers, (4) enhanced surfaces, and (5) compact heat exchangers. Although the air-fin coolers are a special case of the extended surfaces, these are discussed as a separate section in this chapter because of their important applications (Q. 4.41–4.56). Plate heat exchangers, which augment heat transfer due to corrugations, and spiral plate heat exchangers, which enhance heat transfer due to ever changing flow direction, are discussed in Chapter 6.

EXTENDED (OR FINNED) SURFACE HEAT EXCHANGERS

4.2 What are extended or finned surface heat exchangers?

These are units that have a heat transfer surface density (area per unit volume) greater than that of a plain-tube unit with the same O.D., layout, and pitch. In the discussion here, this will imply individually finned tubes with fins on

Table 4.1 Present and potential commercial use of various augmentation techniques[a, b]

	Commercial availability	Mode: Forced convection	Mode: Boiling	Mode: Condensation	Typical material	Performance
Inside tubes						
Metal coatings	Yes	–	2	–	Al,Cu,St	High
Integral fins	Yes	2	3	4	Al,Cu	High
Flutes	Yes	4	4	4	Al,Cu	Moderate
Integral roughness	Yes	2	3	4	Cu,St	High
Wire coil inserts	Yes	3	4	4	Any	Moderate
Displaced promoters	Yes	2	4	4	Any	Moderate (laminar)
Twisted tape inserts	Yes	2	3	4	Any	Moderate
Outside circular tubes						
Coatings						
Metal	Yes	–	2	4	Al,Cu,St	High (boiling)
Nonmetal	No	–	4	4	Teflon	Moderate
Roughness (integral)	Yes	3	2	4	Al,Cu	High (boiling)
Roughness (attached)	Yes	3	4	–	Any	Moderate (for convection)
Axial fins	Yes	1	4	4	Al,St	High (for convection)
Transverse fins						
Gases	Yes	1	–	–	Al,Cu,St	High
Liquid/two-phase	Yes	1	1	1	Any	High
Flutes						
Integral	Yes	–	–	2	Al,Cu	High
Nonintegral	Yes	–	–	4	Any	High
Plate-fin heat exchanger						
Metal coatings	Yes	–	3	–	Al	High
Surface roughness	Yes	4	3	4	Al	High (boiling)
Configured or interrupted fins	Yes	1	2	2	Al,St	High
Flutes	No	–	–	4	Al	Moderate
Plate type heat exchanger						
Metal coatings	No	–	4	–	St	Low
Surface roughness	No	4	4	4	St	Low
Configured channel	Yes	1	3	3	St	High (for convection)

Source: Bergles et al. (1979). With permission.
[a]1, common use; 2, limited use; 3, some special gases; 4, essentially no use.
[b]Al, aluminum; Cu, copper; St, steel.

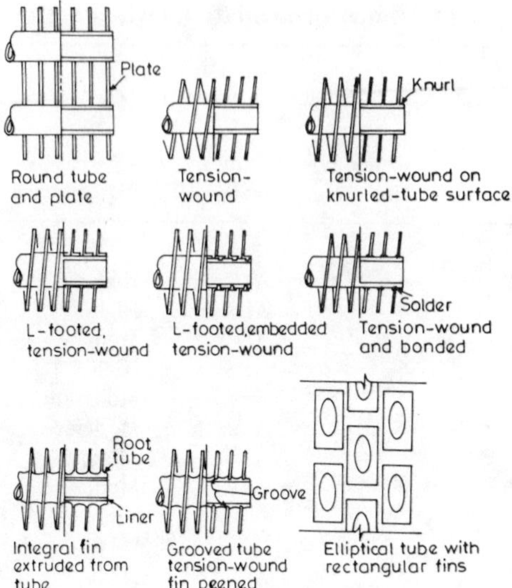

Figure 4.1 Externally finned tubes. *(From Kern and Kraus.* Extended Surface Heat Transfer. *McGraw-Hill. 1972. By permission.)*

the outside and/or inside (Figs. 4.1 and 4.2) (see also Fig. 4.5). Units with continuous fins (tube-fin units) will be dealt with in the discussion of compact heat exchangers (Q. 4.63–4.72).

4.3 Where are these used?

These are generally used when one of the two fluids is a gas, since the gas side heat transfer coefficient is very low. The fins increase the heat transfer area per

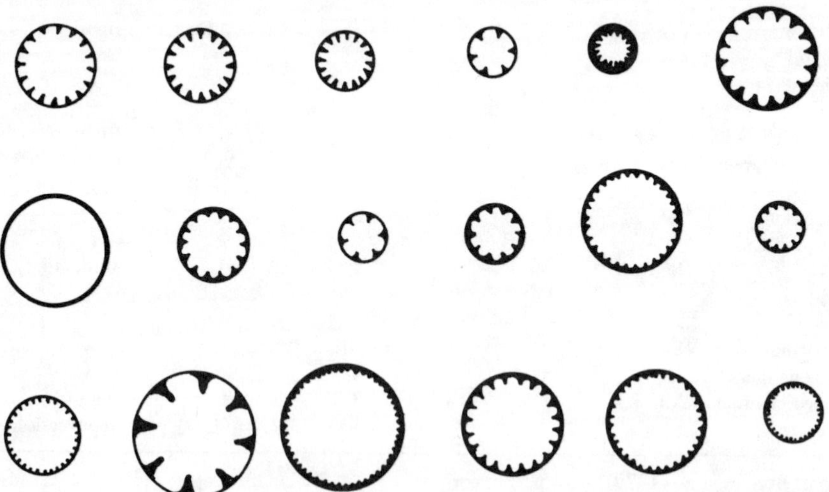

Figure 4.2 Internally finned tubes. *(From N. H. Afgan and E. U. Schlünder.* Heat Exchangers: Design and Theory. *McGraw-Hill/Hemisphere. 1974. By permission.)*

unit tube length and thus help increase the total heat transfer. Some of the typical applications are the following:

1. Cooling and heating of gases
2. Cooling of process fluids in air-fin coolers
3. Condensation of organic vapors
4. Refrigeration and air conditioning
5. Cryogenic industries
6. Automobiles
7. Electronic components and equipment

There are other applications, as described below:

Some units have one fluid at a very high pressure compared to the other fluid. The high-pressure fluid is placed in the tubes (Q. 2.59). In order to decrease the tube length and/or the number of tubes and hence the weight and the cost of the unit, fins are sometimes placed on the outside of the tubes. The fins do not have to withstand the high pressure (since the low-pressure fluid is on the shell side) and hence can be thin. However, they increase the heat transfer area 3 to 10 times, thus decreasing the total tube length required and hence the cost.

If a very costly tube material is required due to the corrosive nature of the tube-side fluid, fins of cheaper material can be used on the outside with the same net effect as above.

4.4 Since fins increase the heat transfer area, why are they not used in all the cases?

The important situations in which the fins are not used and the reasons for that are the following:

1. If both the fluids are gases, the heat transfer coefficient for both will be low. Hence, in order to increase the overall heat transfer in such cases, fins will have to be used on both the inside and the outside of the tubes. This will make the unit rather costly. The same is the case when one side is gas and the other side is a very viscous oil in laminar flow whose film coefficient is also low. However, if the cost of the unit is not a consideration but the weight and/or the space occupied are of greater interest (e.g., in mobile situations), such units may prove advantageous over a plain-tube unit. Some tubes with enhancement on both side have fins on the outside and spiral ridges on the insdie (UOP, 1981) or fins on the outside and wound wire inserts on the inside (Spiral Tube, 1980).
2. While the fins are used in the condensation of organic vapors, their use in steam condensation is not usually feasible because the high surface tension of the condensate (water) results in its retention within the fins (Fig. 3.19), greatly decreasing the utility of the fins.
3. These are not used on the cooling water side of a condenser since water will bridge the space between the fins, rendering them ineffective. Furthermore, the film heat transfer coefficient for most liquids is rather high, and this would result in a very low fin efficiency.
4. In water-tube boilers, fins are generally not used on the outside because the fin

tip would be hotter than the base due to radiation from the boiler furnace. This would require high chromium stainless steel fins with an increase in cost and decrease in their thermal conductivity (Kern and Kraus, 1972). Besides, the gases are dirty and might plug the space between the fins. Also, the gas velocity is low, and hence a thick boundary layer will build up reducing the advantage of the fin. Obtaining a reliable joint between a thin fin and the tube for operation at high temperatures of several hundred degrees is difficult. However, electric resistance welding is claimed to provide such a joint with high efficiency and lower costs (Tolubinsky and Zozulya, 1974). Finned tubes are used in some economizers where the temperature is not very high.

5. If the fluid is corrosive or of the heavily fouling type, the fins are not used, since mechanical/chemical cleaning of the fouling deposit from in between the fins is difficult.

4.5 How can one quickly know whether the externally finned tube would be advantageous before attempting a complete design?

The externally finned tubes are used when the outside heat transfer coefficient, h_o, is significantly less than the inside one, h_i. The finned tubes increase the outside heat transfer coefficient per unit length by an order of 3 to 4, while the cost of the tube per unit length also increases by an order of 1.5 to 2.0. Hence, for the finned tubes to be more economical, the reduction in the number of finned tubes over the plain tubes should be significant so that the shell diameter decreases with a resultant decrease in the sizes of the tubesheets and the flanges. The percent reduction in the number of tubes when finned tubes are used can be estimated from Fig. 4.3. The abcissa is based upon the film coefficients and the fouling factors for the plain tubes.

4.6 What are the types of fins that are used in heat exchangers?

There are several types of fins that are used depending upon the fluids, operating conditions, availability of the specific fins, and past experience in using them. They fall into two broad categories:

1. Longitudinal fins: These may be straight, spiral (large pitch), channel type, etc. Some of these are made discontinuous to break the boundary layer and increase the heat transfer.
2. Transverse fins: These may be radial, helical, or spiral (short pitch), wire-wound, etc. Some newer varieties have saw-tooth cuts on the edge to increase the condensing coefficient (Fig. 4.4). Besides these, spines or studs are also used in some cases. An example is their use in large steam boilers where studded tubes are embedded in the refractory brick walls to keep the latter at a temperature below the combustion temperature (below 3400°F or 1900°C).

Generally, the longitudinal fins on the outside of a tube are used with axial flow, while the transverse fins are used with cross-flow. However, some studies (Obermeir and Schaber, 1979) claim that the tubes with transverse fins used in a shell with axial flow in the shell, give larger heat transfer and lower pressure drop

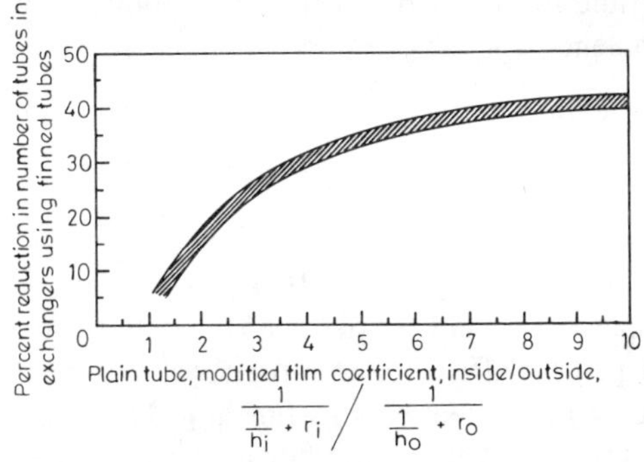

Figure 4.3 Percent reduction in number of tubes when finned tubes are used. *(With permission from Wolverine Division, UOP, Inc., Decatur, Alabama.)*

up to a certain small ratio of fin pitch to fin height. The tests were conducted using only one tube, and further investigation is required.

4.7 How does one decide whether longitudinal or transverse fins should be used?

This depends upon the type of the heat exchanger, pressure drop, surface heat transfer coefficient,* corrosive/fouling nature of the fluid, ease in cleaning, whether the flow is required to be parallel to the tubes or in cross-flow, availability of fins, and experience in using them.

4.8 How can better mixing (hence higher heat transfer) be achieved in an externally finned tube?

It can be done by one or more of the following:

1. In longitudinally finned tubes: by interrupting the fin, i.e., by removing small sections of it at certain intervals to cause circumferential mixing, by making slots in the fin face to cause circumferential mixing, by serrating the fin edges to increase turbulence, or by turning the fin at an angle so that swirling motion is created and turbulence increased.
2. In transverse finned tubes: by cutting the fin edges to form a starlike

*Longitudinal fin efficiency goes down significantly when $1/(r_{fo} + 1/h_o)$ is greater than 150 Btu/hr ft² °F (850 W/m² K) while the transverse fins can be used up to 1000 Btu/hr ft² °F (5680 W/m² K). See Q. 4.18 for definitions of fin efficiency.

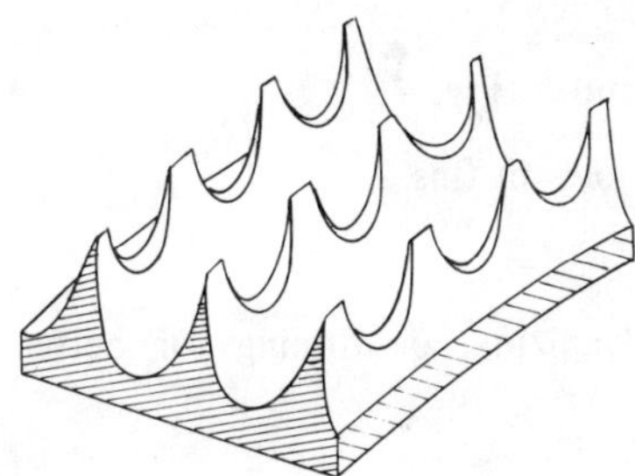

Figure 4.4 Fins with saw-tooth cuts.

(saw-toothed type) shape so that the fluid can enter better up to the root, or by crumpling the fins to increase the turbulence.

Such methods are difficult to use in the internally finned tubes to promote mixing.

4.9 How is the fin material chosen?

Several factors affect this choice. One is the range of the fin-side heat transfer coefficient and the fin height. For example (Fraas and Ozisik, 1965), a comparison between steel and copper fins for an 80 percent efficiency in a particular case of 3 in O.D. (76.2 mm) fins over 1 in O.D. (25.4 mm) tube with 10 fins/in (394 fins/m) gives two different values of the heat transfer coefficient, shown in the tabulation below. Thus, for the lower h value, steel fin is better since the copper

	k, Btu/hr ft °F (W/m K)	h, Btu/hr ft² °F (W/m² K)	Fin thickness, in (mm)
Steel	34 (58.8)	6 (34.07)	0.0694 (1.76)
		30 (170.35)	0.347 (8.8)
Copper	215 (372.1)	6 (34.07)	0.011 (0.28)
		30 (170.35)	0.055 (1.4)

one would be very thin and will get damaged in handling. For the higher h value, steel fin is very thick, and 10 of these cannot be put in 1 in (or 394 in 1 m). On the other hand, copper fins would do well in this case.

Other important factors to be considered in choosing the fin material are the differential thermal expansion between the tubes and the fins, the corrosive nature of the fin-side fluid, the availability of material in proper sizes, the ease of attaching the fins on the tube, the possibility of galvanic corrosion between the fins and the tube, the cost of finned tube, etc.

4.10 Are the base tube and the fins of the same material?

Not necessarily so. Except in the case of integral or extruded fins, the tube and fins are generally of dissimilar materials. Table 4.2 lists some such pairs. It does not exclude other combinations.

4.11 How are the fins attached to the tubes?

There are several ways for the outside-finned tubes (Fig. 4.1):

1. Snug fitting of tubes through tight (undersized) holes in fins
2. Tension wound on plain tubes
3. Tension wound and embedded on to the plain tube
4. Tension wound and bonded by soldering, galvanizing, or tinning for better thermal contact and strength
5. Integral fin made from thick-walled plain tube

Table 4.2 Tube and fin materials pairs

Tube/fin	Carbon steel	Stainless steel	Copper	Aluminum	Cupro-nickel	Brass
Carbon steel	X		X	X		
Stainless steel	X	X		X		
Copper			X	X	X	
Aluminum			X	X	X	
Cupro-nickel			X		X	
Brass			X		X	X

Source: Hunt Heat Exchanger (1975). With permission.
Note: This does not exclude other pairs.

6. Tension wound into grooves on the plain tube
7. Extruded fins drawn from melt
8. Electric resistance welded (ERW)

For the inside-finned tubes of small diameter, only integral fins can be formed (Fig. 4.2). These can be spiral or straight (longitudinal) (Carnavos, 1979). In large diameter tubes, longitudinal fins can be attached by soldering or by other techniques.

The type of attachment method used depends upon the materials and size of the fin and the tube and also the operating temperature.

● The discussion above is restricted to individually finned tubes. The plate fin and tube geometry will be dealt with in the discussion of compact heat exchangers later in this chapter (Q. 4.66).

4.12 Does the contact between the fins and the tubes affect heat transfer?

Yes, indeed it does. In most design calculations, the contact is assumed to be perfect without any contact or bond resistance. If the contact is not perfect, the heat transfer is reduced significantly. In some cases, due to improper methodology adopted or due to corrosion during use or differential thermal expansion of the tubes and the tension-wound fins at high temperatures, a gap may form between the tubes and the fins. In such cases, the resistance to heat transfer is greatly increased and the overall heat transfer coefficient of such a finned tube might even be less than that of the bare tube. Hence, proper attachment of the fins to the tubes is a very important factor.

4.13 How close can the consecutive fins be placed on a tube?

In cases where fins are advantageous, the heat transfer is increased by placing them as near to one another as is possible. However, if the fins are very close together, the fluid in between the fins will move at a lower velocity (for constant pumping power) giving more time for fouling to occur. It is also very difficult and costly to put fins very close together. Thus, fouling, manufacturing, and cost considerations limit the minimum spacing between the fins.

4.14 How are the finned tubes joined to the tubesheet?

The fins terminate slightly short of the tubesheets, leaving a small leading portion of the tube bare, which is then joined to the tubesheet in the usual manner (Chapter 10). The situation as it arises with the radial and longitudinal finned tubes is shown in Fig. 4.5.

4.15 How are the outside finned tubes supported by the baffles?

In the longitudinally finned tubes, rings are welded or shrunk-fit around each tube at different locations so that the fins do not get interlocked. Then, large doughnut-type baffles (Fig. 1.13) are used to enclose the tube bundle. The ring on one tube touches the fins of the surrounding tubes. This provides support in the horizontal units, reduces vibrations, and also cuts down the flow that would ensue through this space, bypassing the fins. If conventional cross-baffles are used, a lot of fluid will bypass the tubes, and the heat transfer will be reduced since the turbulence is cushioned by the longitudinal fins.

In the radial low-fin tubes (discussed in Q. 4.16 and 4.17), conventional cross-baffles are used as in the plain tube units. The leakage stream A (Q. 2.3) is larger than with the plain tubes. Radial high-fin tubes are generally used in air-fin

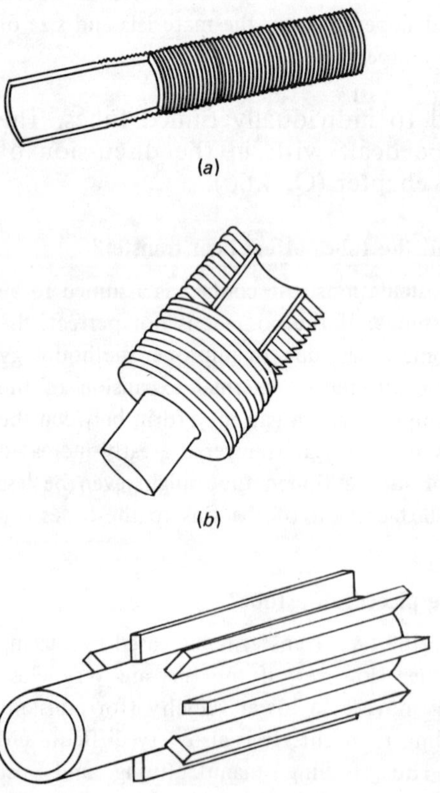

Figure 4.5 Finned-tube ends for joining to tubesheet. (*a*) Radial low-finned tube. (*b*) Radial high-finned tube. (*c*) Longitudinal finned tube.

coolers that have a single-pass cross-flow of air and need only support plates instead of the baffles. Air-fin coolers are discussed later in this chapter (Q. 4.37–4.52).

4.16 What is the surface area ratio of finned to bare tube for the low-fin tubes used in general condensing and boiling services and for the high-finned tubes used in the air-fin coolers?

The low-fin tubes are generally $\frac{3}{4}$-1 in (19.2-25.4 mm) O.D. with fins of $\frac{1}{16}$ or $\frac{1}{8}$ in (1.6 or 3.2 mm) height, and 11-19 fins/in (433-748 fins/m). Some newer versions have 19–40 fins/in (748–1575 fins/m) with fin height from 0.03 to 0.061 in (0.76 to 1.55 mm). The O.D. of the tubes is the same as the O.D. of the finned tubes so that these can easily replace the plain tubes in a tube bundle. The surface area ratio of the low-fin tubes to the plain tubes is 3.5 to 5.0. The high-fin tubes are 1-1$\frac{1}{2}$ in (25.4 to 38.1 mm) O.D. with $\frac{1}{2}$-1 in (12.7-25.4 mm) high fins and 7-11 fins/in (276 to 433 fins/m). The surface area ratio of the high-fin tubes to the plain tubes can be very high, of the order of 7–20.

4.17 Compare a 19 fins/in (748 fins/m) low-fin tube with a plain tube.

Let us take a $\frac{3}{4}$ in (19.2 mm), 16 BWG (1.65 mm) plain tube. Its wall thickness is 0.065 in (1.65 mm), and its I.D. is 0.63 in (16 mm).

A 19 fins/in low-fin tube (748 fins/m) with a fin height of $\frac{1}{16}$ in (1.6 mm) will be made out of the $\frac{3}{4}$ in (19.2 mm) O.D. tube by machines having two or three stacks of external discs that form double or triple helixes and displace the I.D. inside while leaving the O.D. relatively unchanged. Tube diameter under the fin is called root diameter. The tube wall becomes 2 BWG thinner than the original one. Thus, the dimensions of the 19 fins/in (748 fins/m) low-fin tube will be O.D. = $\frac{3}{4}$ in (19.2 mm); root diameter = $\frac{5}{8}$ in (15.9 mm); I.D. = 0.527 in (13.4 mm), same as for a $\frac{5}{8}$ in (15.9 mm) O.D., 18 BWG (1.24 mm thick) tube. This tube will be called a $\frac{3}{4}$ in (19.2 mm), 18 BWG (1.24 mm), 19 fins/in (748 fins/m) low-fin tube (Fig. 4.6). Data on surface area, fin dimensions, etc., for various types of fins are available in Kern and Kraus (1972) or from the manufacturers, such as UOP (1981).

4.18 What is fin efficiency? What does it depend upon?

Fin efficiency is the ratio of the actual heat dissipation from a fin to that which would occur if the entire fin surface were at the same temperature as its base. Thus, it is the ratio of the heat conducted through the fin with finite thermal conductivity to that what would be conducted with an infinite thermal conductivity. In actual practice, a temperature gradient must exist in the fin for heat conduction to occur from the base to the tip. It also points out the fact that after a certain height, further increase in the fin height may not be economical since the incremental efficiency would become very low.

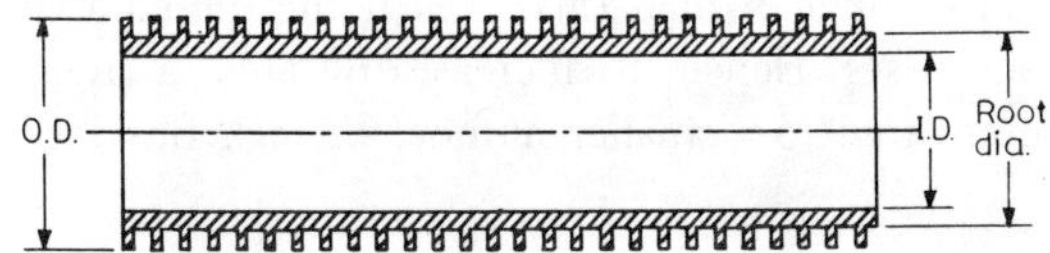

Figure 4.6 Finned-tube dimensions.

Fin efficiency depends upon the properties of the fin material, its shape, height, and surface heat transfer coefficient. For a longitudinal fin of rectangular cross section with insulated top, the efficiency, η, is given by

$$\eta = \frac{\tanh(mb)}{mb} \tag{4.1}$$

where

$$m = \left(\frac{hC}{kA}\right)^{1/2} \tag{4.2}$$

The value of tanh $(x)/x$ increases with decreasing x. Thus, the efficiency increases with increasing thermal conductivity, k, which reduces m. Hence, in increasing order of fin efficiency, some materials are admirality brass, aluminum, and copper. On the other hand, with increasing heat transfer coefficient, h, the efficiency decreases, since it increases m.

The relation for fin efficiency of radial or transverse fins is more complicated involving the modified Bessel functions. It is available in several references, such as Kern and Kraus (1972).

4.19 How does the transfer of latent heat in a finned-tube cooler–condenser affect the fin efficiency?

The fin efficiency in a cooler–condenser is smaller than that in a sensible heat transfer case, and it varies from inlet to the exit of the unit (Mizushina, 1974) as more and more vapor is condensed.

4.20 Would the efficiency of a copper fin bonded on to a copper tube be different than if it were bonded on to an aluminum tube? Assume that the bond resistance is zero.

No. The efficiency of a fin does not depend upon the base tube characteristics but only on the properties of the fin material, its dimensions, and the heat transfer coefficient (Q. 4.18).

4.21 How should the finned side area be selected?

It would be ideal if the fins would help bring the gas side heat transfer factor (heat transfer coefficient × finned tube surface area) to equal that on the liquid side per unit tube length. But since the fin efficiency decreases with increasing height, the attempt should be to make the gas side factor at least half that on the liquid side. Since the heat transfer coefficient ratio of liquid to gas side runs 5 to 20 times, the area ratio of gas to liquid side may be selected to be 3 to 10 times.

4.22 Why cannot the fins be made very high to increase the area?

The fin efficiency goes down with increasing fin height. A larger number of low fins transfer more heat than a small number of high fins, even if the finned area per unit length is the same in both the cases. Hence, to increase the area, a larger number of low fins should be used compared to a smaller number of high fins [see Eq. (4.1)].

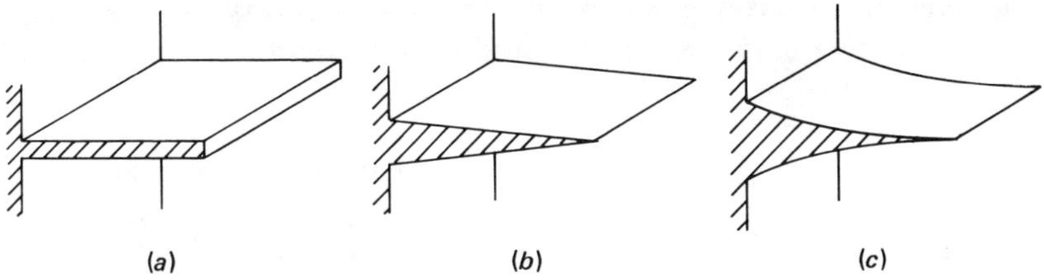

Figure 4.7 Three types of longitudinal fins. (*a*) Rectangular. (*b*) Triangular. (*c*) Inverse parabolic.

4.23 Compare longitudinal fins of different sections.

The three sections that have been analyzed are (Fig. 4.7) (a) rectangular, (b) triangular, and (c) inverse parabolic.

It is found that the inverse parabolic is a fin that requires the least material to obtain a given heat dissipation rate. Further results obtained are [Kern and Kraus (1972)]:

1. For the same conditions, triangular fins require 69 percent and inverse parabolic 65 percent material compared to the rectangular fins.
2. The relative weight for any one type of fins made out of different materials is directly proportional to the material density and inversely proportional to its thermal conductivity. Thus, for fins made out of steel, copper, and aluminum, the relative weights are 1.0:0.137:0.08.
3. The fin profile area and the fin volume are directly proportional to the cube of the heat load at the base. Hence, to double the heat load in a given case, one may either double the number of fins or keep the same number of fins but make them eight times as large. Hence, a larger number of smaller fins are preferred.
4. Rectangular fins are easiest to manufacture, while the inverse parabolic fins are too costly to make and not easy to handle. It is unlikely that these are made commercially.

4.24 How is the tube wall temperature for a finned tube calculated?

There is no easy way to calculate the tube wall temperature, and a successive approximation method is adopted. The six resistances to be accounted for are the following:

1. The outside film resistance
2. The outside fouling resistance
3. The resistance equivalent to fin efficiency
4. The plain-tube wall resistance
5. The tube-side fouling resistance
6. The tube-side film resistance

A wall temperature is assumed, on the basis of which a viscosity correction factor is obtained. This is used to calculate the outside film coefficient (or the outside film resistance). Besides this, there are five more resistances noted above.

Since very different heat fluxes pass through these resistances because area keeps changing, the temperature drop across each resistance is obtained by multiplying the resistance by the heat flux corrected to the outside gross surface. The wall temperature thus calculated is compared with the assumed value, and the sum of the component temperature differences is compared with the actual temperature difference between the inside and the outside bulk fluids. If these match, the assumed value of the wall temperature is accepted, or else a new one is assumed and the process repeated. Detailed procedure and an example have been given by Kern and Kraus (1972).

If the viscosity correction factor can be neglected in a given case, the wall temperature can be directly calculated without trial-and-error, since all the resistances are then known.

4.25 On what factors do the heat transfer coefficient (Nusselt number) and the friction factor (Euler number) on an extended surface heat exchanger depend?

The Nusselt number depends upon (Mirkovic, 1974)

$$\mathrm{Nu} = \mathrm{Nu}\,(\mathrm{Re}, \mathrm{Pr}, h/d, \delta/d, n/d, S_{\mathrm{T}}/d, S_{\mathrm{L}}/d, n_{\mathrm{R}}) \tag{4.3}$$

where Nu = Nusselt number
Re = Reynolds number
Pr = Prandtl number
d = bare tube outer diameter
h = height of fins
δ = thickness of fins
n = number of fins per unit length
n_{R} = number of finned-tube rows
S_{T} = transverse pitch of tube bundle
S_{L} = longitudinal pitch of tube bundle

In some correlations, the ratio of viscosity at the bulk temperature to that at the tube wall temperature is also used as a correction factor to account for the effect of temperature on physical properties. Since its exponent is low, the value is generally close to one.

The Euler number, Eu, depends upon (Mirkovic, 1974)

$$\mathrm{Eu} = \mathrm{Eu}\,(\mathrm{Re}, h/d, \delta/d, n/d, S_{\mathrm{T}}/d, S_{\mathrm{L}}/d, n_{\mathrm{R}}) \tag{4.4}$$

Empirical correlations obtained from experimental results are also given by Mirkovic.

4.26 What are the assumptions in the simplified thermal analysis of the finned surfaces?

The assumptions are the following (Kern and Kraus, 1972):

1. Steady state operation
2. The thermal conductivity of the fin does not vary with temperature or direction (homogeneous material).

3. The heat transfer coefficient of the fin to the surroundings is constant and uniform over the entire surface of the fin.
4. The temperature of the medium surrounding the fin is constant and uniform.
5. The fin is very thin. Hence the temperature difference from the fin surface to the center line is negligible compared to that between the fin and the outside fluid.
6. The temperature at the base of the fin is uniform and constant.
7. There is no contact resistance between the fin and the tube.
8. There is no heat source or heat sink within the fin.
9. The heat transferred through the edge of the fin is negligible compared to that dissipated through its lateral surfaces.
10. Heat transfer to or from the fin is proportional to the difference in temperature between the fin and the surrounding medium.

Among these assumptions, (9) may be important in low fins of rectangular cross section. This can be relaxed if need be. Furthermore, one can assume the heat transfer coefficient to be different for the edge and the lateral surface, thus disregarding assumption (3) to some extent.

In applications, such as space vehicles, where the heat loss is by radiation, assumption (10) may not hold. This can also be taken into account.

In actual practice, there is a two-dimensional temperature distribution in the tube wall, and hence the temperature at the base of the fin is not uniform (Heggs and Stones, 1983), violating assumption (6). For rectangular fins, neglecting this may lead to an error of up to +16 and −10 percent in the prediction of the heat transfer coefficient in extreme cases of very thin and very thick fins (Heggs and Stones, 1983). In most designs though, the two-dimensional distribution is neglected, and assumption (6) is used in the calculations.

4.27 When is an unsteady state analysis required?

It is required for the start-up and shutdown and during emergency and upset conditions in critical usage. The calculations are quite involved and are done numerically.

4.28 In what ways do the calculations for extended surface heat exchangers differ from those for the plain-tube ones?

Some of the differences are the following:

1. The calculations for the extended surface heat exchanger take into account the fin dimensions as well as their number.
2. The surface area on the finned side is more, hence the equivalent diameter (or hydraulic radius) is less than that in a plain-tube unit with the same O.D.
3. In longitudinally finned tubes, the flow gets "channeled," and hence the level of turbulence is reduced, which reduces the heat transfer coefficient. Ways to counter it include wavy longitudinal fins or interrupted fins (Q. 4.8).

4.29 Is the heat flux beneath the base of a fin the same as that at the bare tube surface between two consecutive fins?

No, it is significantly greater beneath the fin base because the heat loss from a fin is more than that through a segment of bare tube equal in area to the fin base. If it were not so, there would be no advantage of using the finned tubes.

4.30 How are the various resistances added in determining the overall heat transfer coefficient, U_d, for a finned tube?

Let us assume that the tube has fins on the outside only. Then, U_d is given by

$$\frac{1}{U_d} = \frac{1}{h_o} + R_{fo} + \frac{\Delta x}{k} + \frac{\eta A_o}{A_i}\left(\frac{1}{h_i} + R_{fi}\right) \quad (4.5)$$

The first three terms on the right-hand side are common with the plain-tube case (Q. 2.9). The fourth term on the right-hand side transfers the in-tube resistances to outside tube surface by multiplying by the area ratio, A_o/A_i, and the fin efficiency, η.

4.31 Conservative fouling resistance specification can be very costly in finned-tube units compared to the plain-tube one. Why?

Consider a tube with fins on the outside. If it is a 1 in (25.4 mm) 16 BWG (1.65 mm thick) tube, the area ratios (outside area/inside area) are 1.15:1.0 for a plain tube and 3.3:1.0 for 19 fins/in (748 fins/m) low-fin tube. If a fouling factor of 0.002 ft^2 hr °F/Btu (0.0003522 m^2 K/W) is assigned to the outside (finned side), it makes no difference whether it is a finned or a plain tube. However, if the same fouling resistance is assigned to the inside (unfinned side), it gets transformed to the gross outside surface (Q. 4.30). With an assumed efficiency of 90 percent, the fouling resistance value becomes 0.00594 ft^2 hr °F/Btu (0.001046 m^2 K/W), which is unrealistic and cuts down severely the advantages of using the finned tubes. Of course, the net effect also depends upon the outside film resistance. The lower that resistance (higher film coefficient), the higher is the effect of the fouling resistance.

The detrimental effect on fin efficiency by a uniform fouling deposit has been treated by Epstein and Sandhu by means of two mathematical models (Epstein and Sandhu, 1979). In one, the deposit is treated as a resistance in series with the resistance offered by the surrounding fluid. In the other, the deposit is treated as a thermal conductor in parallel with the fin. It is found that the fin efficiency actually increases (due to increase in the fin temperature). However, the total heat transfer is reduced in the case of the series model due to increased resistance and is increased in the case of the parallel model because of an increase in conductance. (The series model seems more realistic.)

4.32 How is the heat transfer increased in the internally finned tubes?

Some internal fins are shown in Fig. 4.2. These are straight or spiral type. These provide increased heat transfer area, and the spiral type also increase the turbulence and/or mixing and hence are superior to the straight type, specially at lower Reynolds number ($\leqslant$10,000). When compared to the plain tubes, the sensible

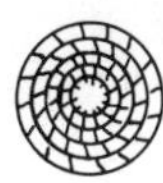
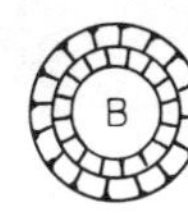

Figure 4.8 Internally finned tubes for multi-passages. *(From Carnavos, 1978. By permission of ASME.)*

heat transfer rate increased by up to 90 percent for the same pumping power (Carnavos, 1974). A significant increase was also found in the case of laminar flow (Marner and Bergles, 1979; Trupp and Soliman, 1983). Analytical studies have shown that the Nusselt number increases with the number of fins up to a certain critical fin number, and thereafter it decreases (Soliman and Feingold, 1979).

For condensation of steam in internally finned tubes, Royal and Bergles (1979) found a decrease both in the condenser length and the pressure drop.

Multi-passage internally finned tubes, formed by mechanically coupling parallel individual internally finned tubes of different diameters so that the fin tips of an outer tube contacted the next innermost tube's outside surface (Fig. 4.8), were tested for heat transfer characteristics (Carnavos, 1978). It was found that such tubes could reduce the required plain-tube length by a factor of 2–10.

An annotated bibliography of augmentation of heat transfer in laminar flow through tubes has been given by Bergles and Joshi (1983). The techniques used for augmentation, test arrangement, parameter ranges studied, and conclusions are presented.

TURBULATORS AND STATIC MIXERS

4.33 What are turbulators?

These are devices inserted in the tubes to increase the turbulence and improve the heat transfer (Figs. 4.9–4.12). These are also known as tube inserts and turbulence promoters.

Figure 4.9 Twisted tape.

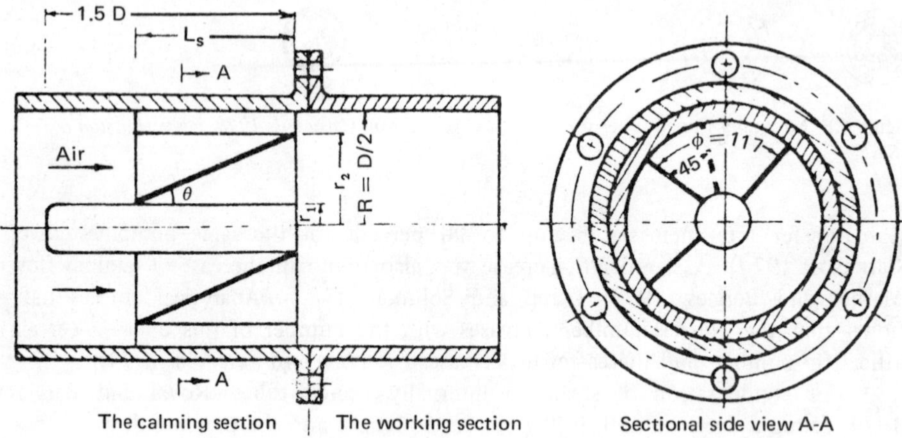

Figure 4.10 Position and dimensions of van swirlers. *(From Abdel-Salam et al., 1983.)*

4.34 How do the turbulators differ from internally finned tubes?

While the purpose of the turbulators and the internal fins is the same, there are certain differences between the two as given in Table 4.3.

4.35 What are static mixers?

These are devices placed inside the tubes for mixing of very viscous fluids in laminar flow by dividing the flow in two or more parts and then a little distance downstream combining these with other parts, dividing once more, and so on, depending upon the number of mixing elements. There are a variety of them. A typical mixing element is shown in Fig. 4.13 (McCabe and Smith, 1976). Others are like corrugated sheets called Sulzer static mixers (van der Meer and Hoogendoorn, 1983).

Besides producing mixing, the static mixers also renew the liquid film at the tube wall and increase the velocity gradient there, thus increasing the laminar flow heat transfer by several hundred percent (Marner and Bergles, 1979). Both the velocity and temperature profiles are close to a plug flow, giving nearly uniform values across the cross section. Thus, viscous liquids, which are easily thermally degraded, can also be heated by using static mixers since the thermal gradient within the flow stream is very little.

The difference between the static mixers and the turbulators lies primarily in the use pattern. While the former are used with highly viscous liquids in laminar flow, the latter are used with less viscous fluids in laminar and/or turbulent flow.

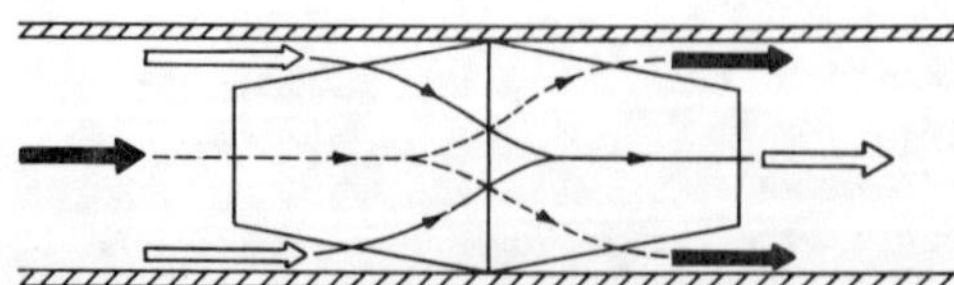

Figure 4.11 Everters. *(From Maezawa and Lock, 1979. By permission.)*

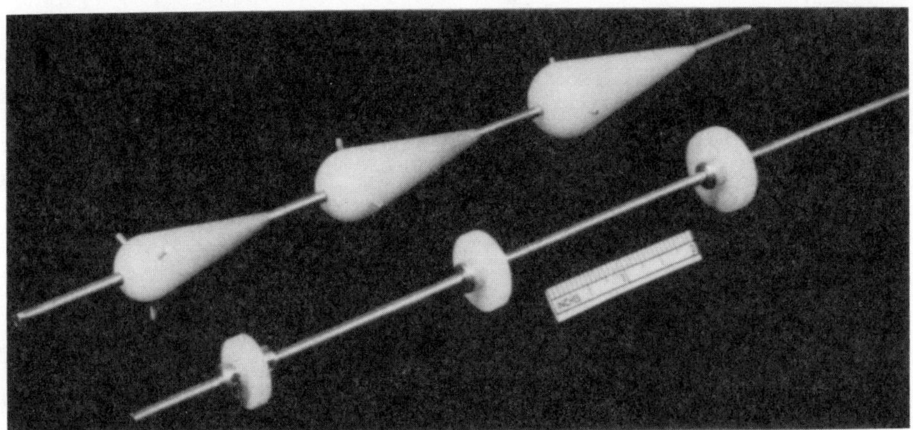

Figure 4.12 Disk and streamline inserts. *(From L. B. Evans and S. W. Churchill.* Chemical Engineering Progress *58(10):55, October 1962. By permission.)*

Table 4.3 Comparison between turbulators and internally finned tubes

	Turbulators	Internal fins
Attachment to tubes	Generally held by friction or spring action	Generally extruded or resistance welded (ERW), thus held rigidly
Removable	Yes, in most cases	No
In situ installation in existing plain tubes	Yes	No, tubes have to be taken out for ERW of the fins or for grooving the tubes
Variety	Great many types, e.g., twisted tapes (Fig. 4.9), vane swirlers (Fig. 4.10), inlet vortex generators, metal wool packing, helically wound spring temper wire, spiral brush, everters (Fig. 4.11), ring inserts with various spacings, streamline shapes, and disk inserts (Fig. 4.12)	Relatively limited types (Fig. 4.2)
Cleaning	Very difficult since turbulators are volume elements and generally cover the entire cross section in a twisted fashion with no straight passage all through	Relatively easy since these are attached to the surface while the bulk core is free
Uses	Both with liquids and gases. No restriction on surface tension since retention is little, if any. However, the fluid should be clean since cleaning is difficult. These can be used in evaporators, steam generators, boiling of liquid metals, condensation,[a] curing a hot spot, etc.	Generally with gases or low surface tension condensate

[a]Bergles (1979).

A survey of various types of turbulators and internal fins in laminar flow has been done by Bergles and Joshi (1983).

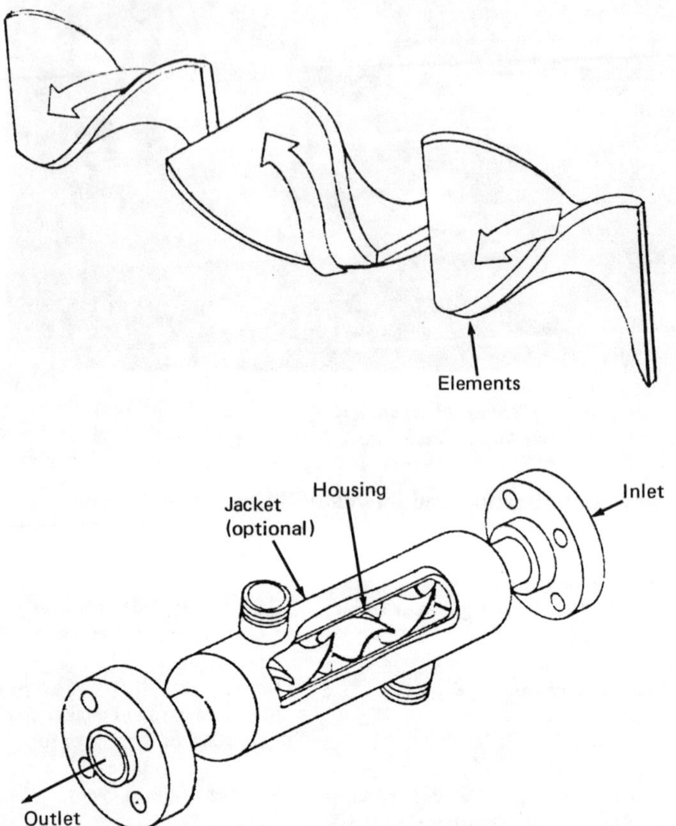

Figure 4.13 Static mixers. *(From* Unit Operations of Chemical Engineering *by W. L. McCabe and J. C. Smith. Copyright © 1976. Used with permission of McGraw-Hill Book Company.)*

4.36 What does the performance of turbulators and static mixers depend upon?

Since the turbulators and static mixers are of widely different kinds, it is difficult to characterize the factors that affect their performance. These will need to be determined in each case. In general though, the factors will include the following:

1. The size of the turbulator
2. The hydraulic diameter of the tube with the turbulator in place
3. The properties of the fluid
4. The mass flow rate of the fluid

- Most of the references cited in this section discuss the performance characteristics of individual turbulators and static mixers and should be consulted for details.

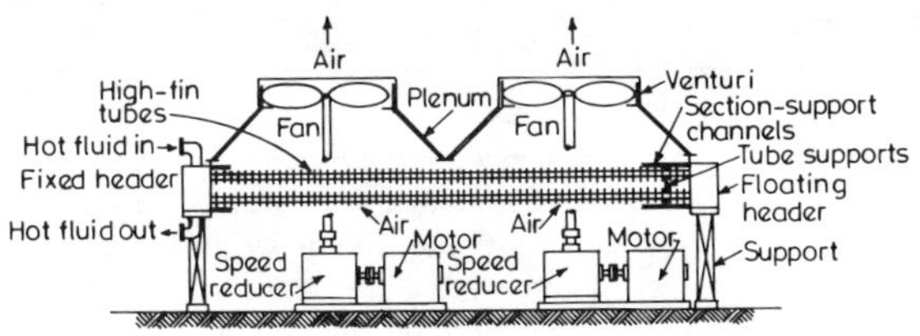

Figure 4.14 Induced draft air-fin cooler.

AIR-FIN COOLERS

4.37 What are air-fin coolers?

These are units of tube bundles with a hot **process fluid flowing inside** the tubes and air flowing in cross-flow over the tubes to cool the **process fluid** (Figs. 4.14 and 4.15). There are high transverse fins on the outside of the tubes to increase the heat transfer area. Such units are also used for condensing steam, petroleum, and high boiling organic vapors, ammonia, etc., and in cooling the cooling water for reuse. In the last case, a combination of air-fin coolers and cooling tower may be used to reduce the fan noise as well as the cooling tower height (Q. 4.52).

4.38 Compare air cooling versus water cooling.

Earlier, air cooling was used only when water was scarce. However, with lower limits now set on thermal pollution of certain water sources, air cooling is economical in other circumstances too. The two are compared in Table 4.4. Air-fin cooler is used with air cooling, while a shell-and-tube unit is used with water cooling.

4.39 What are the sizes of air-fin coolers?

These can be very small, such as automobile radiators, or very large, such as used in the process industries with a long bay having multiple fans in a row. The tubes may be arranged in several bundles of 4–8 rows. If the heat duty of one process fluid is not very high, more than one fluid may be passed through different bundles to make an effective use of the fan.

Units, such as used in automobile radiators, are known as compact heat exchangers and are dealt with separately in this chapter (Q. 4.63–4.72). Hence, we

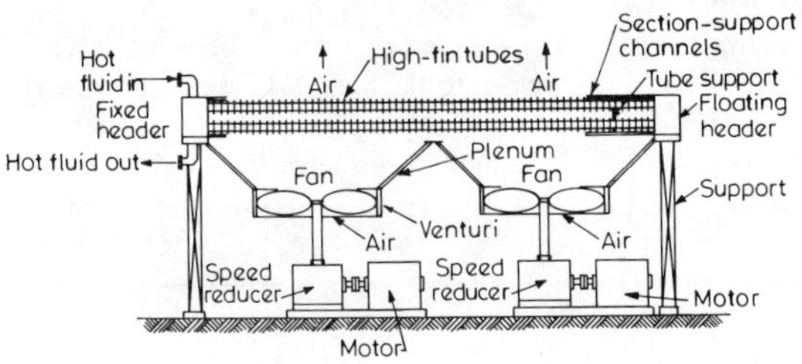

Figure 4.15 Forced draft air-fin cooler.

Table 4.4 Air cooling versus water cooling

	Air cooling	Water cooling
Cost of coolant	Free	One has to pay for the water. Besides it is brought to the site by pipes and pumps that add to the cost
Site	Unrestricted	Restricted to the vicinity of source of water if large volume of water is needed. The land is costlier, too
Pretreatment of coolant	None	Required to reduce fouling and corrosion
Allowance for fouling and corrosion	Very little, if any. Hence cheaper materials can be used	Significant both in terms of resistance to heat transfer due to fouling and loss of material due to corrosion. Hence costlier materials may be required in cases where sea water is used
Danger due to contamination if a tube leaks	Very little. Besides leaking tube is easily detected	More danger. Leaking tube is detected with greater efforts
Running cost	Low due to all the above cited reasons	High due to all the above cited reasons
Maintenance cost	Low since no shell is to be opened and very little fouling deposit on the outside	High since shell is to be opened and tubes cleaned on the outside. Then the unit has to be reassembled and tested
Standards on thermal pollution	Easily met	Often requires cooling towers, etc., to meet the thermal pollution standards in water streams. These add to the cost
Size	Large since heat transfer coefficient is low	Relatively small since heat transfer coefficient is high
Seasonal variation	Highly susceptible to seasonal and diurnal variations. Rain and sunshine affect too. Freezing might occur in very cold weather	Less susceptible to these variations
Noise	Very high. Noise control proves costly	Very low
Effect of nearby structures	Air recirculation can set in which will affect efficiency	No effect
Use in alternate service	Not easily possible	Easily possible
Floor space	Very little since it can be mounted above other equipment	Installed on ground and hence requires floor space for equipment as well as for servicing
Gaseous process stream	Not suitable since the overall heat transfer coefficient is very low	Suitable by using finned tubes on the gas side
Extent of cooling	Limited. Preferable when process liquid is at least 50°C (90°F) higher than the air	The temperature can be brought down close to the web bulb temperature
Waste heat recovery	Not possible because of large volume and low temperature rise	Possible but not always economical

will concentrate on large units employing high-fin circular tubes and used in the process industries.

4.40 List the parts of an air-fin cooler.

These are the following:

1. Finned-tube bundle and manifold for the process fluid
2. Fan and drive
3. Structure supporting the tube bundle and the fans
4. Louvres to direct and control the air flow and hence the heat transfer

4.41 What are the kinds of drives used for an air-fin cooler fan?

These are the following:

1. Direct drive for small fans
2. Vee-belt transmission for intermediate duties
3. Gear-box drive for very heavy duties

The choice of the drive is decided by the duty as well as by the API Standard 661 (API, 1978). Generally, electricity is used as the motive force, although steam turbines and hydraulic motor drives also are used sometimes.

4.42 How is air made to flow over the tubes?

There are two ways:

1. By using a fan: Induced draft or forced draft
   a. Induced draft (Fig. 4.14): The fan is mounted above the tube bundle. It throws air away from the tube bundle and as a result, creates a slight vacuum underneath the fan. Air from the surroundings rushes in and passes over the tube bundle.
   b. Forced draft (Fig. 4.15): The fan is mounted below the tube bundle. It forces air over the bundle due to the positive pressure created by the fan.
2. By free-convection effects: This is also called fanless air cooler (Doyle and Benkly, 1973). Its size is larger and is not used often.

4.43 Compare induced and forced draft air-fin coolers.

The comparison is given in Table 4.5 and the two set-ups are shown in Figs. 4.14 and 4.15.

4.44 In air-fin cooler fans why are an even number of blades ($\geqslant 4$) preferred?

These make less noise than if the blades number less than 4. Furthermore, if one blade goes bad or gets damaged and has to be taken off, the corresponding opposite blade can also be removed to keep the fan balanced and without significantly reducing the fan performance.

4.45 What factors affect the fan noise?

Fan noise is an important factor in the selection of a fan for the air-fin cooler service. Noise within the fan is created by the interrupted air flows at high velocities

Table 4.5 Induced draft versus forced draft air-fin coolers

	Induced draft	Forced draft
Heat transfer coefficient	Lower due to the effect of the number of rows	Higher because of turbulence caused by the fan
Power consumption	Higher since the fan is in high temperature air	Lower since fan blows cooler air
Recirculation of air	Negligible since the air is thrown vertically up at a high velocity	Significant due to lower velocity of the exit air from the tube bundle
Air distribution	Good since it draws air from a wider area	Poor unless steps are taken to improve the air distribution
Environmental effects	Negligible since high air velocity prevents rain from entering	Significant unless protected otherwise
Mounting and structural support	High level mounting and hence heavy structural support required	Low level mounting and hence nominal structural support required
Maintenance	Due to high mounting it is a bit difficult	Easy access
Drive and lubrication	Since air temperature is high, these should be designed to withstand it	Low temperature. Hence design is easy and cost is low
Tube bundle	Horizontal or V shape	Horizontal or A shape

and by air turbulence. A reduction in the noise reduces vortex shedding and hence improves its efficiency also. About two-thirds of the sound is emitted by the fan and one-third by the passage of the air through the tube bundle. The noise produced is dependent upon the following:

1. Fan-tip speed
2. Type of drive
3. Plenum configuration
4. Inlet configuration

The fan-tip speed is the most important. The noise can be reduced by using a lower tip speed, lower fan power, and larger surface area. Such fans usually have wider blades and large hubs to attain the same volumetric flow rates. Other means of reducing the sound level include the use of sound-absorbing screens and sound baffles to attenuate the sound level (Paikert and Ruff, 1983). If the sound level is to be guaranteed for long distances, moisture in the air should also be considered since it affects sound transmission. Variable pitch fans may also be used for sound control.

An equation relating the level of noise to different variables is available (Farrant, 1983). Standard testing procedures for air-fin coolers are available to carry out field tests and to interpret the data regarding its performance (AIChE, 1979).

4.46 What are important design parameters of an air-fin cooler?

The parameters are the following:

1. Flow rate of the process fluid
2. Inlet and outlet temperatures of the process fluid
3. Sensible and/or latent heat load
4. Available tube-side pressure drop
5. Air temperature, including daily and seasonal variation (Q. 4.47)
6. Induced or forced draft (Q. 4.43)
7. Design of louvres

4.47 What factors related to air affect the design?

These are the following (API, 1978; Briggs and Young, 1963):

1. The maximum, minimum, and the design air temperature for the plant location
2. The sizes of the nearby structures and their distances from the coolers. This affects air recirculation, which may in fact be desirable in the winter to avoid freezing (North, 1980).
3. The presence of heat sources, such as furnaces and low stacks in the vicinity
4. Induced draft or forced draft design (Q. 4.43)
5. The direction and magnitude of the prevailing winds to determine their effects on air recirculation and support structure design
6. The height of the fan mountings from the ground to decide the structural support needed

4.48 How are the air-fin coolers designed?

Because of the large number of variables that influence their design, a successive approximation procedure is generally required to size up the unit. Initially, only the chosen air flow, the inlet air temperature, and the inlet and outlet process fluid temperatures are known. Account has to be taken of (1) the fin efficiency, (2) the effect of plugging of fins by dust-laden air, and (3) the bond resistance in tension-wound fins. A 10 percent penalty in the heat transfer coefficient has been recommended for tension-wound aluminum L-fin on a carbon-steel tube for a temperature rise of 100°C (180°F) above the finning temperature (North, 1980); and (4) correction for cross-flow. The heat transfer coefficient for the finned side, h_o, is calculated by the Briggs and Young equation (Ganapathy, 1980), which accounts for the fin dimensions and pitch. For further details of the design procedure, Kern and Kraus (1972), North (1980), Ganapathy (1980), Mueller (1973), and Urbicain and Paloschi (1981) may be consulted.

4.49 How does the air temperature vary in an air-fin cooler compared to that on the shell side of a baffled shell-and-tube unit?

The air temperature in an air-fin cooler is different at different tube rows in a vertical plane or cross section. On the other hand, in a shell-and-tube unit, the shell-side fluid temperature is taken to be the same over an entire cross section in a shell pass.

4.50 What are the important factors in the proper performance of an air-fin cooler?

The proper distribution of the process fluid in the tubes and of the air over the tube bundle are two important factors, with the latter being more dominant. In order to ensure proper air distribution, its leakage and bypass should be minimized by adequate sealing and use of louvers. The required hydrostatic head for air flow should account for the pressure drop through the tube bundle and that due to the louvers, etc.

Effects of a sudden shutdown of the fan should be evaluated. It might lead to severe stresses in the system due to high-temperature process fluid in the tubes and the resultant expansion and bowing of the tubes with fixed ends.

4.51 How does water spray affect the air-fin cooler performance?

The effect of water spray on the performance of air-fin coolers has been studied by Tree et al. (1979), among others. The heat transfer is enhanced due to the evaporation of water on the cooler surface, which draws the required latent heat of evaporation from the process fluid being cooled. If, however, the drop size in the spray is very small, these might evaporate within the air, thus cooling the air, but the net enhancement would be minimal in that case.

It should be kept in mind that the use of water spray may bring about corrosion and fouling deposit on the outside of the tubes and the fins.

4.52 How does a combination of air-fin cooler and cooling tower work?

The incoming air is cooled by water spray in the cooling tower during the summer months. This cool air is blown over the air-fin cooler tubes to cool the process fluid or cooling water below the ambient air temperature. It reduces the size of both the cooling tower and the air-fin cooler compared to what they would be if only one of these were used. The fan used in the air-fin cooler is of a lesser power and the fan noise generated is lower too compared to the case when only the air-fin cooler is used. Furthermore, no pretreatment of the water used in the cooling tower is necessary. During the winter months, the cooling tower is not used, since only the air-cooler suffices to achieve the required cooling. This is one of the several types of arrangements possible.

ENHANCED SURFACES

4.53 What are enhanced surfaces?

These are surfaces that enhance the heat transfer coefficient. In the discussion here, these will be taken to mean:

1. Surfaces with ribs or flutes in the form of spirals that form peaks and valleys both on the inside and the outside (Q. 1.21 and 3.103 and Figs. 1.6, 1.7, and 3.36)
2. Surfaces with a special porous coating on the outside used primarily in boiling heat transfer (Q. 3.103 and Figs. 3.34 and 3.35)

4.54 Where are the enhanced surfaces used?

These are primarily used in phase change situations, such as in condensation and evaporation, wherein enhancements of up to 300 percent have been reported. Both experimental and theoretical work have been done as briefly surveyed by Sideman (1981), who has also discussed the use of various kinds of grooves (Fig. 3.36) as enhanced surfaces.

4.55 What are the advantages of the enhanced surfaces?

The advantages are the following (Kalinin, 1974):

1. A decrease in the heat transfer surface, volume, and weight of an exchanger for a given heat duty, flow rate, and pressure drop
2. An increase in the heat transfer for a given size unit, flow rate, and pressure drop. This will be achieved when the plain tubes are replaced by the enhanced tubes.
3. A decrease in the pressure drop for a given size unit and heat duty
4. An appropriate combination of the above

Some of these results have been reported by Kalinin (1974) for circumferentially rolled tubes. Carnavos (1974) has reported similar data on a number of fluted tubes, while Palen et al. (1971) have reported extensive work on spirally grooved tubes and compared them with plain tubes in a shell-and-tube exchanger. A detailed survey of the principles, status, and applications of the enhanced surfaces has been given by Bergles (1981a,b).

Yampolsky (1983), using spirally fluted tubes, found that while the heat transfer was enhanced, the friction factor actually decreased compared to a smooth tube at an equivalent Reynolds number. The explanation offered is that the rotation of the fluid caused by the spiral flutes stabilizes the core flow, which reduces the pressure drop. However, the turbulence between the core flow and flute flow is increased, which enhances heat transfer near the wall. This hypothesis would benefit from further work.

4.56 On what factors does the enhancement depend?

The factors that affect enhancement are the pitch of rolling, the depth and type of the grooves, number of groove starts, Reynolds number, type of heat transfer (sensible and/or phase change), magnitudes of latent heat and surface tension, horizontal or vertical orientation, tube materials, size of the tube bundle, layout angle, etc. For condensing services, the heat transfer coefficient is enhanced because only a thin film of the liquid condensate is left on the crests, while most of it drains into the troughs due to the surface tension and gravity effects. The thin film offers very little resistance, and hence the condensing coefficient is very high.

4.57 How does the surface tension of the condensate affect enhancement?

A lower surface tension liquid, such as acetic acid, spreads out faster than a higher surface tension one, such as water. Since the enhancement is greater with a thin film of liquid on the crests, a faster spreading or draining liquid will naturally produce a thinner film. Hence, for a given enhanced surface tube, the enhancement is higher for a lower surface tension condensate.

4.58 How does the latent heat of the condensing vapor affect enhancement?

A vapor with a low latent heat, such as Freon 11, produces a larger amount of condensate that has to be drained compared to the condensation of steam or ammonia whose latent heat is higher. Thus, the Freon 11 will fill the troughs to a higher level, leaving less surface exposed on which the condensation will occur. Therefore, the enhancement in this case will be lower than with condensing steam or ammonia on the same tube as far as the effect of the latent heat is concerned.

4.59 How does the Reynolds number of the condensing vapor affect enhancement?

High Reynolds number flow of the vapor results in a high vapor shear on the condensate film, making it drain faster into the troughs and leaving only a very thin film on the crests. This thin film offers very little resistance to heat transfer, and hence the enhancement is high. The vapor shear also results in a faster drainage of the condensate from the troughs, thus reducing the flooded level and increasing the heat transfer.

4.60 How does the size of tube bundle affect the enhancement achieved?

As discussed in Chapter 3, the heat transfer coefficient over a tube bundle is less than that over a single tube due to the vapor binding and reduced circulation of the liquid in boiling or dripping of the condensate from one tube row to the one below in condensation, which increases the thickness of the condensate film. The same holds for enhancement in a bundle. It is less than that over a single tube and decreases with the increasing number of tubes in a bundle.

4.61 What factors influence the performance of the enhanced surface tubings having a porous coating? How can maximum enhancement be achieved?

Such tubes are generally used in the boiling duties. Their mechanism of action and the effect produced in a specific case are shown in Figs. 3.34 and 3.35. Basically, these provide a large number of re-entrant nucleation sites that help produce boiling with a low temperature differential between the surface and the bulk. Hence, for the enhancement to be maximum for a given tubing, one should have the following (Palen et al., 1983):

1. Low heat flux or else vapor binding might occur within the nucleation sites drastically decreasing their effectiveness
2. Low system pressure and low hydrostatic head or else boiling will occur at higher temperature, reducing the effectiveness of the surface
3. Low boiling range mixture or else the higher boiling components will tend to concentrate in the nucleation sites. Enhancement for pure substances is two or three times that for the mixtures (Antonelli and O'Neill, 1983).

Table 4.6 Comparison between enhanced and extended surfaces

	Enhanced surfaces	Extended surfaces
Material of construction	Made of one metal or alloy	Except for extruded type, generally different materials used for the base tube and the fins
Extent of protrusions compared to a plain tube	Small	Large
Enhancement on one or both sides	Generally used on both sides. Hence improves heat transfer coefficient on both sides	Generally used on one side where gas is flowing. Improves heat transfer coefficient on that side only
Efficiency	Not defined for it	It is less than one and decreases with increasing height of the fin
Ratio of increase in heat transfer to that in surface area	High (>1). Hence the heat transfer coefficient per unit area is increased	Low (<1). Hence the heat transfer coefficient per unit fin area is decreased
Bond resistance	None	Possible if fins are not attached properly or get loosened due to differential thermal expansion
Effect of corrosion	Uniform	May damage the fin bond and introduce bond resistance

4. Less number of tubes in a bundle and larger pitch so that the vapor binding of the upper tubes is minimized
5. Highest possible nucleate boiling or pool boiling and lowest possible convective boiling, or else the effect of the nucleation sites would be lost
6. None or very little of the surface should be used for sensible heating since sensible heat transfer enhancement is minimal (Antonelli and O'Neill, 1983).

4.62 Compare enhanced surfaces with extended (finned) surfaces.

Both kinds of surfaces are used to increase the heat transfer per unit length of the tube compared to a plain tube. There are, however, several differences between the two kinds. These are given in Table 4.6.

COMPACT HEAT EXCHANGERS

4.63 What characterizes a unit to be compact?

It is the surface area per unit volume or the surface area density. It is shown in Fig. 4.16 for different types of exchangers. An arbitrary value $\geqslant 700\ m^2/m^3$ (213 ft^2/ft^3) has been suggested for the compact units. The overall heat transfer coefficient is also generally higher. Thus, both the high surface density and high heat transfer coefficient contribute toward reducing the volume. Furthermore, since compact surfaces can achieve structural stability and strength with thinner sections, the exchanger mass is lowered significantly.

4.64 Where are the compact heat exchangers used?

These are used in a variety of situations under extremes of conditions, such as in cryogenic industry; in spacecrafts, aircrafts, and missiles; and in rotary regenerators for waste heat recovery, where weight and/or volume are to be kept to a minimum. There are several other applications between these extremes, such as automobile radiators, air conditioning and refrigeration, and chemical process industries. With an increasing interest in energy savings, compact heat exchangers are becoming more popular.

While one fluid has to be gaseous, the other can be a gas, a liquid, two-phase flow, a boiling liquid, or a condensing vapor (Robertson, 1980).

4.65 What are the advantages and disadvantages of a compact heat exchanger?

The advantages are the following (Shah and Webb, 1983):

1. Savings in weight and volume
2. Sometimes the total cost is also lower than that of the other types
3. A variety of surfaces are available that can be used to custom-make a unit as per the requirements. The surface areas on the hot and the cold side can be made significantly different if desired.

The disadvantages are the following:

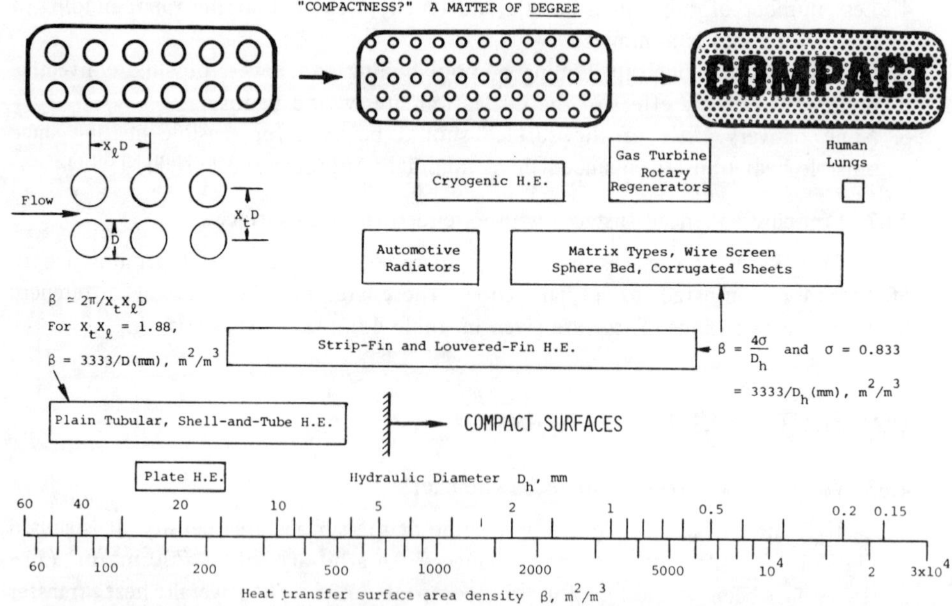

Figure 4.16 Heat transfer surface area density spectrum of exchanger surfaces. *(From Shah, 1981a. With permission.)*

1. Their applications are somewhat limited since at least one of the two fluids must be a gas.
2. Fluids should not have suspended particles and should be noncorrosive. If suspended particles are present, as encountered by automotive and industrial radiators, it must be accounted for in the design by having an overdesign and/or larger spacing between fins, or else it will be difficult to remove a deposit from in between the fins. Furthermore, a corroded part cannot generally be repaired; it instead requires the whole unit to be replaced.
3. Pressure drop and hence the pumping cost is high.
4. Due to large frontal area, the header design for fluid distribution is very important.
5. The operating temperature and pressure are lower than those in a shell-and-tube unit due to the construction features, and lower strength of brazing at high temperature, etc.
6. Each new design has to be evaluated separately for its heat transfer and pressure drop characteristics and operating limits on temperature and pressure. This entails a large amount of cost, time, and effort.

4.66 What are the different types of compact heat exchangers?

There are several types but the two broad types generally used are plate-fin exchangers and tube-fin exchangers. Each can take a variety of shapes as discussed below (Shah and Webb, 1983; Robertson, 1983):

1. Plate-fin exchangers (Fig. 4.17): These are primarily used with gases and two-phase fluids. The fins are sandwiched between parallel plates and are attached to the plates by brazing, soldering, welding, mechanical fit or extrusion. The fins serve several purposes: (1) These impart strength to the whole assembly, enabling the exchanger to withstand high pressure. (b) These enhance heat transfer as do finned tubes. (c) These permit heat transfer across the exchanger between nonadjacent streams. An analysis incorporating this factor has been done by Haseler (1983).

   The types of fins used are (a) plain and straight fins, such as triangular and rectangular shapes; (b) plain and wavy fins; (c) interrupted fins, such as offset or serrated strips, louvre, perforated, and pin fins. One can obtain a wide variety of surface density by varying the basic geometric variables. Typical fin densities are 120 to 700 fins/m (3 to 17 fins/in), fin thickness 0.05–0.25 mm (2–10 mil), fin heights 2–20 mm (0.08 to 0.8 in) and surface area density up to about 1500 m^2/m^3 (460 ft^2/ft^3). Some of these are also used in rotary regenerators to recover waste heat.
2. Tube-fin exchanger (Fig. 4.18): These can be used when both fluids are gaseous (one may be at a high pressure) or when one fluid is a liquid or two-phase. Round, rectangular, or elliptical tubes with fins on the outside are used. Individual tubes with fins on the outside and/or inside have already been discussed (Q. 4.2–4.32). Their surface area density is only 100-200 m^2/m^3 (30.5–61 ft^2/ft^3), much below the criterion used for the compact units

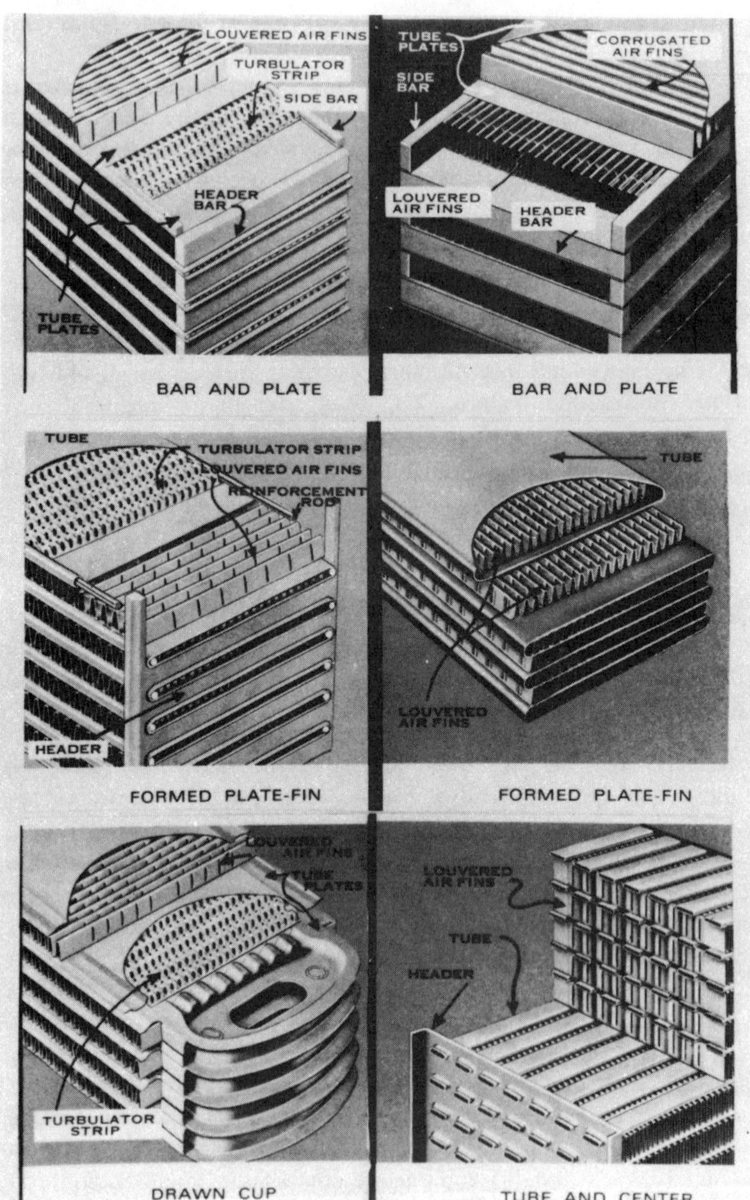

Figure 4.17 Plate-fin compact heat exchangers. *(Courtesy of Harrison Radiator Division, GMC, Lockport, New York.)*

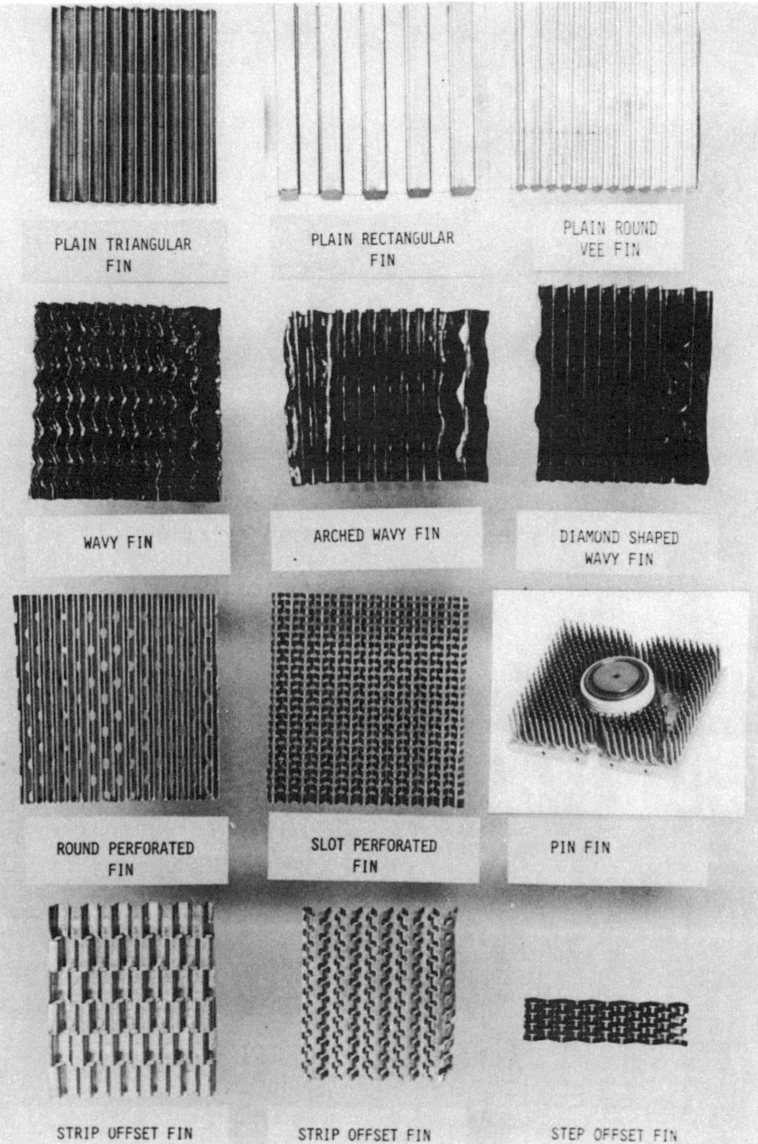

Figure 4.17 Plate-fin compact heat exchangers. *(Courtesy of Harrison Radiator Division, GMC, Lockport, New York.) (Continued)*

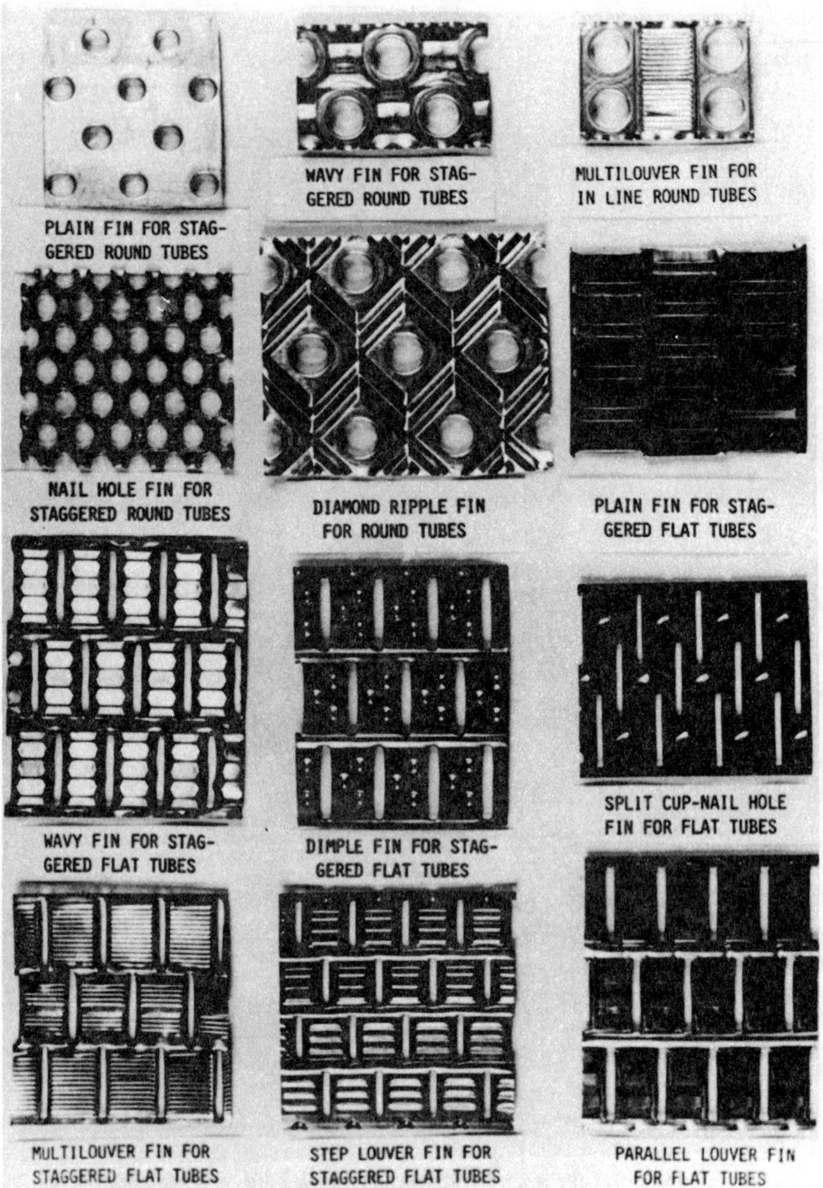

Figure 4.18 Tube-fin compact heat exchangers. *(Courtesy of Harrison Radiator Division, GMC, Lockport, New York.)*

(Q. 4.63). The tube-fin units falling in the range of the compact units have continuous external fins that may be plain, wavy (corrugated), or interrupted fins around circular, rectangular, or elliptical tubes. The typical fin densities are 300-600 fins/m (7.5-15 fins/in), fin thickness 0.1-0.25 mm (4-10 mil), fin flow length 25-250 mm (1-10 in), surface area density up to about 1000 m^2/m^3 (305 ft^2/ft^3). The type and the density of the fins distinguishes these units from the air-fin coolers discussed earlier (Q. 4.26-4.37).

4.67 How do the different kinds of compact surfaces differ from one another?

These differ in the following:

1. Shapes and sizes are the principal causes of differences between different fins
2. Number of fins per unit length
3. Height of fins in plate-fin units and tube pitch in tube-fin units
4. Surface area density
5. Kinds of tubes used in a tube-fin unit, viz. circular, rectangular, or elliptical
6. Thermal conductivity of the fin material
7. Thickness of the material used
8. Methods of formation of the fin
9. Experience in using them since some are used more often than the others
10. Heat transfer characteristics as a function of the Reynolds number
11. Friction factor characteristics as a function of the Reynolds number
12. Cost, which comes down with large volume production

4.68 What factors affect the choice between different types of compact units?

These are the following:

1. Operating pressure and temperature
2. Whether both the fluids are gaseous or one is liquid or two-phase
3. Fouling characteristics and suspended impurities
4. Allowed pressure drop
5. Mechanical strength (ruggedness) desired
6. Restrictions on size, volume, and/or weight
7. Cost
8. Experience in using a specific variety
9. Extent of acceptable fluid intermixing as in regenerators

4.69 What kinds of flow arrangements are possible in a compact heat exchanger?

Cross-flow is the most common type of flow used since the header design is simplified. Multi-passes are also possible for one or more fluids. More than two fluids can also be used for exchanging heat as is done in the cryogenic industry.

Counter-flow is not used often since, in the plate-fin units, it would require separating the fluids at each end, thus complicating the header design, and in the tube-fin units, the continuous fins preclude counter-flow except in the interrupted variety where an excessive pressure drop would be encountered to force the fluid through the slits. Parallel flow is not used since it has all the disadvantages of counter-flow and also has a lower efficiency than the cross-flow.

The flow is generally laminar because of a low hydraulic diameter and the desire to restrict the pressure drop. With plain fins, the flow may be fully developed laminar, while with the interrupted fins, it is generally developing laminar flow over each interruption even if the bulk flow is turbulent. This holds up to as high a Reynolds number as 10,000.

The compact surfaces in the laminar region yield a higher heat transfer coefficient than a plain tube with a highly turbulent flow. For example, an 800 fins/m (20 fins/in) plain rectangular fin geometry in laminar flow would provide the same heat transfer coefficient as a 19 mm ($\frac{3}{4}$ in) diameter plain tube at a Reynolds number of 25,000. If the plain fin is replaced by an offset strip fin, the heat transfer coefficient is increased further by 2.5 times (Shah and Webb, 1983).

4.70 How are the compact heat exchangers designed?

For the design, a large number of variables have to be accounted for (Q. 4.68). It is obvious that no one design correlation for heat transfer and for pressure drop will suffice. Besides, analytical correlations can be obtained only for relatively simple shapes. In most others, empirical correlations have to be obtained by experiments (steady state, transient, or periodic) covering wide ranges of the variables. The experiments may involve either heat transfer or mass transfer since the two phenomena are analogous. An extensive survey of laminar heat transfer and friction factor in 40 different shapes with a number of original analytical contributions has been presented by Shah and London (1978). These include solutions for the developed and developing velocity and temperature profiles and the entry length required for the fully developed flow. These show that for the constant temperature and constant wall flux boundary conditions, the Nusselt number is least for the triangular ducts and maximum for two infinite parallel plates, with intermediate values in an increasing order for the square, hexagonal, circular, and rectangular ducts of decreasing height to width ratio.

For the plate-fin units, Kays and London (1964) provided the first extensive set of j- and f-factor correlations for many different plate-fin geometries using air flow. The existence or otherwise of j and f correlations for various types of plate fin has been discussed at length by Shah and Webb (1983). The interrupted fins give the j-factor to be 2 to 4 times higher, and the f-factor to be 3 to 5 times higher than the equivalent plain-fin surfaces with the same surface density and hydraulic diameter. This is because at each interruption the boundary layer is disrupted and a new one starts to form. The developing boundary layer is thinner than a fully developed one, and hence the heat transfer coefficient is higher in the former case. Snider and Kraus (1981) have given a general algorithm using graph theory for the analysis of any arbitrary configuration of plate fins.

For a review of the tube-fin units with plain, wavy, or interrupted fins, one should consult Shah and Webb (1983) and Webb (1980). The tube fins behave very differently than the plate fins. A low velocity wake exists between the tubes with rather low heat transfer coefficients as demonstrated by mass transfer techniques

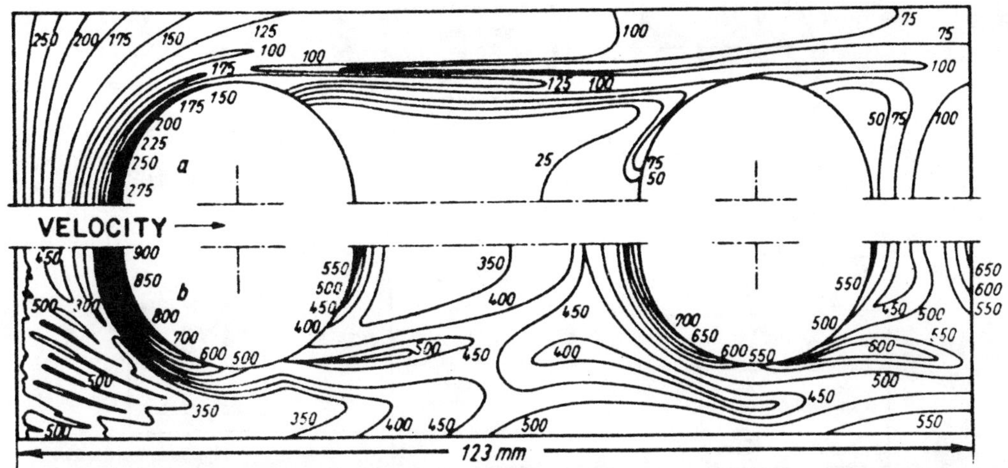

Figure 4.19 Distribution of mass transfer coefficients on a two-row tube plain-fin unit. *(From Kruckels and Kottke, 1970.)*

(Fig. 4.19). A 10:1 variation of the local mass transfer coefficient occurs over the fin surface. Other geometrical parameters, such as the tube pitch, layout, number of tube rows, and number of fins per unit length, also influence the performance. In a three tube-row test section, it was found (Saboya and Sparrow, 1976) that at a low Reynolds number ($\simeq$200), the fins associated with the first row transferred about 50 percent of the total heat, the fins associated with the second row transferred about 28 percent, and the fins associated with the third row transferred the balance. These differences almost disappeared at high Reynolds number ($\simeq$1000–1200) since vortex-induced heat transfer becomes more important compared to the boundary-layer-induced one, which is important at low Reynolds numbers. With wavy fins, the heat transfer was 45 percent higher compared to a plain fin at a Reynolds number of 1000, while interrupted fins increased it still further by 30 to 50 percent.

Flat or rectangular tubes give higher heat transfer coefficient for in-tube flow at low Reynolds number and lower form friction for flow on the outside. Thus, the flat tubes turn out to be better both in heat transfer and pressure drop, but their use is restricted to low pressures (Shah and Webb, 1983).

Compact heat exchanger design is a very vast field. The discussion here is of necessity only peripheral. The references cited should be consulted for the details of the design procedures. Generally, the ϵ-NTU method is used (Chapter 2). Shah (1981b) and London (1983) have discussed and illustrated with examples both the rating and the sizing of compact heat exchangers using this method. A flow sheet for the overall design procedure is shown in Fig. 4.20.

4.71 What can be the problems in the operation of a compact heat exchanger?

The following problems can occur:

1. Flow maldistribution due to (a) an inadequate header design; (b) bad soldering or brazing that obstructs flow passages and/or offers contact resistance to heat

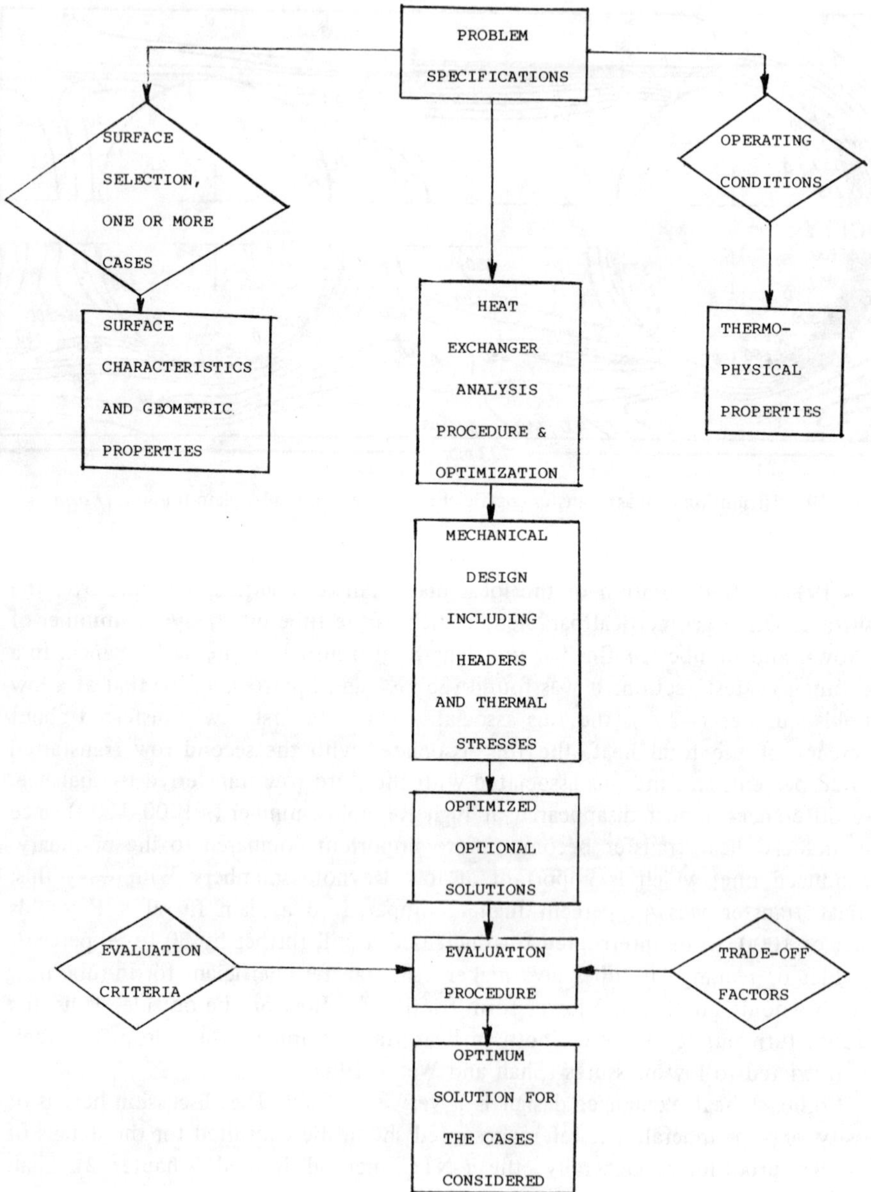

Figure 4.20 Methodology of compact heat exchanger design. *(From London, 1983.)*

transfer, or (c) malformation of flow passages, some of which cannot be avoided due to their small sizes and manufacturing tolerances.

2. Interfluid mixing due to leakages between the flow passages. Its detrimental effects should be evaluated while choosing a particular design.
3. Operating conditions being different than the design conditions

4. Fouling
5. Longitudinal conduction

Of these, the flow maldistribution is the single most serious problem. It results in higher pressure drop and lower heat transfer, thus reducing the effectiveness of the unit, which is generally calculated on the assumption of a uniform flow distribution. Relations for the reduction in effectiveness for several flow arrangements due to maldistribution have been derived by Shah (1981c). Effects of longitudinal conduction have been discussed by London (1983). These decrease the effectiveness of a unit, and the decrease is higher with higher effectiveness designs. In a specific case, London found an 84 percent penalty in surface area when the unit was designed for 98 percent effectiveness, and 10 percent penalty at 90 percent effectiveness design.

4.72 What are the differences in the process design and the fabrication methods for shell-and-tube and compact heat exchangers?

The process design of any heat exchanger requires the design of a unit that will meet both the heat transfer requirements and the pressure drop limitations

Table 4.7 Comparison of shell-and-tube and compact units

	Shell-and-tube unit	Compact heat exchanger
Leakage and bypass streams	Present due to manufacturing tolerances (Q. 2.3)	Absent
Design correlations	Same for different sizes	Vary with sizes and shapes
Physical size and weight	Can be very large	Not very large
Effects of flow maldistribution and longitudinal conduction	Negligible for tube side. Leakages and bypass on shell side can be accounted for (Chapter 2)	These can be significant (Q. 4.74)
Handling of fouling fluids, two-phase systems, high pressure, and temperature	Easily possible	Difficult, if not impossible
Flow patterns	Shell-side flow pattern can be varied by using different kinds of baffles and nozzle arrangements. Tube-side flow is linear in all cases	Both fluid flow patterns can be varied by choosing appropriate passage geometries
Design codes	Several codes exist worldwide that are supplemented by TEMA Standards (Chapter 8). Hence a very large number of manufacturers is available	Only certain parts conform to codes. The core is generally a patented design of each company. No codes exist for it. Number of manufacturers is relatively less
Fabrication technique	Welding	Brazing or soldering
Material thickness	Rather thick	Relatively thin since fins provide rigidity and pressure is relatively low

imposed by the specific usage. However, there are subtle differences between the design and fabrication of a shell-and-tube unit and of a compact heat exchanger as described in Table 4.7.

• As is evident from the discussion in this chapter and the list of references, the augmented surface heat exchangers are already in substantial use, and their use will continue to increase because of their several advantages in cost, size, and/or weight.

NOMENCLATURE

A	surface area per unit length; cross-sectional area of a rectangular fin
A_i	inside tube surface area per unit length
A_o	outside tube surface area per unit length
API	American Petroleum Institute
b	length of fin
BWG	British wire gauge, a measure of thickness
C	fin circumference
d	bare tube outer diameter
ERW	electric resistance welded
Eu	Euler number
f	friction factor
fpi	fins per inch
h	heat transfer coefficient; height of fin
h_i	inside heat transfer coefficient
h_o	outside heat transfer coefficient; finned-side heat transfer coefficient
j	heat transfer factor, Colburn factor
k	thermal conductivity
m	defined by Eq. (4.2)
n	number of fins per unit length
n_R	number of finned tube rows
NTU	number of transfer units
Nu	Nusselt number
Pr	Prandtl number
Re	Reynolds number
R_{fi}	inside fouling resistance
R_{fo}	outside fouling resistance
S_L	longitudinal pitch of tube bundle
S_T	transverse pitch of tube bundle
ΔT	temperature driving potential
U_d	design (or dirty) heat transfer coefficient
x	argument in Q. 4.18
Δx	tube wall thickness

Greek Letters

η efficiency
δ fin thickness
ϵ-NTU a design method (Q. 4.70)

REFERENCES

Abdel-Salam, M. S., M. M. Hilal, E. E. Khalil, and A. M. A. Mostafa. Effect of Swirl Angle and Geometry on Heat Transfer in Turbulent Pipe Flow. In *Heat Exchangers: Theory and Practice*. J. Taborek, G. Hewitt, and N. H. Afgan (eds.). Hemisphere: Washington, D.C. 1983.

American Institute of Chemical Engineers. *AIChE Equipment Testing Procedure for Air Cooled Heat Exchangers.* AIChE: New York. 1979.

American Petroleum Institute. *API Standard 661: Air-Cooled Heat Exchanger for General Refinery Services,* 2d ed. API: Washington, D.C. 1978.

Antonelli, R., and P. S. O'Neill. Design and Application Considerations for Heat Exchangers with Enhanced Boiling Surfaces. In *Heat Exchangers: Theory and Practice.* J. Taborek, G. Hewitt, and N. H. Afgan (eds.). Hemisphere: Washington, D.C. 1983.

Bergles, A. E. Enhancement of Heat Transfer. *Heat Transfer 1978: Proceedings of the Sixth International Heat Transfer Conference, Toronto, August 1978.* 6:89–108. Hemisphere: Washington, D.C. 1979.

Bergles, A. E. Principles of Heat Transfer Augmentation, I: Single-Phase Heat Transfer, II: Two-Phase Heat Transfer. In *Heat Exchangers: Thermal–Hydraulic Fundamentals and Design.* S. Kakac, A. E. Bergles, and F. Mayinger (eds.). Hemisphere: Washington, D.C. 1981a.

Bergles, A. E. Applications of Heat Transfer Augmentation. In *Heat Exchangers: Thermal-Hydraulic Fundamentals and Design.* S. Kakac, A. E. Bergles, and F. Mayinger (eds.). Hemisphere: Washington, D.C. 1981b.

Bergles, A. E., and S. D. Joshi. Augmentation Techniques for Low Reynolds Number In-Tube Flow. In *Low Reynolds Number Flow Heat Exchanger.* S. Kakac, R. K. Shah, and A. E. Bergles (eds.). Hemisphere: Washington, D.C. 1983.

Bergles, A. E., R. L. Webb, and G. H. Junkhan. Energy Conservation via Heat Transfer Enhancement, *Energy* 4:193–200. 1979. [Here noted from Bergles et al. (1981).]

Bergles, A. E., J. G. Collier, J. M. Delhaye, G. F. Hewitt, and F. Mayinger. *Two-Phase Flow and Heat Transfer in the Power and Process Industries.* Hemisphere: Washington, D.C. 1981.

Briggs, D. E., and E. N. Young. Convection Heat Transfer and Pressure Drop of Air Flowing Across Triangular Pitch Banks of Finned Tubes, *Chem. Eng. Progr. Symp. Ser. No. 59* 41:1–10. 1963.

Carnavos, T. C. Some Recent Developments in Augmented Heat Exchange Elements. In *Heat Exchangers: Design and Theory Sourcebook.* N. Afgan and E. U. Schlünder (eds.). McGraw-Hill: New York. 1974.

Carnavos, T. C. Cooling Air in Turbulent Flow with Multi-Passage Internally Finned Tubes, Paper No. 78-WA/HT-52, presented at the ASME Winter Annual Meeting, San Francisco. December 1978.

Carnavos, T. C. Cooling Air in Turbulent Flow with Internally Finned Tubes, *Heat Trans. Eng.* 1(2):41–46. 1979.

Doyle, P. T., and G. J. Benkly. Use Fanless Air Coolers, *Hydrocarbon Process.* 52(7):81–86. 1973.

Epstein, N., and Sandhu, K. Effect of Uniform Fouling Deposits on Total Efficiency of Extended Heat Transfer Surfaces. *Heat Transfer 1978: Proceedings of the Sixth International Heat Transfer Conference, Toronto, August 1978.* 4:397–402. Hemisphere: Washington, D.C. 1979.

Evans, L. B., and S. W. Churchill. Economics of Using Heat Transfer Promoters, *Chem. Eng. Progr.* 58(10):55–62. 1962.

Farrant, P. E. Noise and Its Influence on Air-Cooled Heat Exchanger Design. In *Heat Exchangers: Theory and Practice.* J. Taborek, G. Hewitt, and N. H. Afgan (eds.). Hemisphere: Washington, D.C. 1983.

Fraas, A. P., and M. N. Ozisik. *Heat Exchanger Design.* Wiley: New York. 1965.

Ganapathy, V. Design of Air-Cooled Exchangers: Process Design Criteria. In *Process Heat Exchange.* V. Cavaseno (ed.). McGraw-Hill: New York. 1980.

Haseler, L. E. Performance Calculation Methods for Multi-Stream Plate-Fin Heat Exchangers. In *Heat Exchangers: Theory and Practice.* J. Taborek, G. Hewitt, and N. H. Afgan (eds.). Hemisphere: Washington, D.C. 1983.

Heggs, P. J., and P. R. Stones. Two-Dimensional Heat Flow through Fin Assemblies. In *Heat Exchangers: Theory and Practice.* J. Taborek, G. Hewitt, and N. H. Afgan (eds.). Hemisphere: Washington, D.C. 1983.

Hunt Heat Exchanger. Publication No. 142, *Extended Surface Heat Exchangers.* Hunt Heat Exchanger Ltd.: Middleton, Manchester, England. 1975.

Kalinin, E. K. Tubular Heat Exchangers with Bilateral Heat Transfer Augmentation and Calculation of a Heat Exchanger under Unsteady Operating Conditions. In *Heat Exchangers: Design and Theory Sourcebook.* N. Afgan and E. U. Schlünder (eds.). McGraw-Hill: New York. 1974.

Kays, W. M., and A. L. London. *Compact Heat Exchangers,* 2d ed. McGraw-Hill: New York. 1964.

Kern, D. Q., and A. D. Kraus. *Extended Surface Heat Transfer.* McGraw-Hill: New York. 1972.

Kruckels, S. W., and V. Kottke. Investigation of the Distribution of Heat Transfer on Fins and Finned Tube Models. *Chemie-Ingenieur-Tech* 42:355–362. 1970. [Here noted from Robertson (1981).]

Larson, R., and F. Kreith. Workshop on the Research Needs of Heat Transfer Processes and Equipment. Vail, Colorado. June 1981. SERI: Colo. 1981.

London, A. L. Compact Heat Exchanger-Design Methods. In *Low Reynolds Number Flow Heat Exchangers.* S. Kakac, R. K. Shah, and A. E. Bergles (eds.). Hemisphere: Washington, D.C. 1983.

Maezawa, S., and G. H. S. Lock. Heat Transfer inside a Tube with a Novel Promoter. *Heat Transfer 1978: Proceedings of the Sixth International Heat Transfer Conference, Toronto, August 1978.* 2:595–600. Hemisphere: Washington, D.C. 1979.

Marner, W. J., and A. E. Bergles. Augmentation of Tubeside Laminar Flow Heat Transfer by Means of Twisted-Tape Inserts, Static-Mixer Inserts and Internally Finned Tubes. *Heat Transfer 1978: Proceedings of the Sixth International Heat Transfer Conference, Toronto, August 1978.* 2:583–588. Hemisphere: Washington, D.C. 1979.

McCabe, W. L., and J. C. Smith. *Unit Operations of Chemical Engineering,* 3d ed. McGraw-Hill: New York. 1976.

Mirkovic, Z. Heat Transfer and Flow Resistance Correlations for Helically Finned and Staggered Tube Banks in Crossflow. In *Heat Exchangers: Design and Theory Sourcebook.* N. Afgan and E. U. Schlünder (eds.). McGraw-Hill: New York. 1974.

Mizushina, T. Design of Cooler Condensers and Evaporative Coolers. In *Heat Exchangers: Design and Theory Sourcebook.* N. Afgan and E. U. Schlünder (eds.). McGraw-Hill: New York. 1974.

Mueller, A. C. Heat Exchangers. In *Handbook of Heat Transfer.* W. M. Rohsenow and J. P. Hartnett (eds.). McGraw-Hill: New York. 1973.

North, C. D. R. Air Coolers. In *Developments in Heat Exchangers.* D. Chisholm (ed.). Applied Science: London. 1980.

Obermeir, E., and A. Schaber. Experimental Investigation of Heat Transfer from Transverse Finned Tubes with Longitudinal Flow. In *Heat Transfer 1978: Proceedings of the Sixth International Heat Transfer Conference, Toronto, August 1978.* 2:613–617. Hemisphere: Washington, D.C. 1979.

Paikert, P., and K. Ruff. State of Art for Design of Air Cooled Heat Exchanger with Noise Level Limitation. In *Heat Exchangers: Theory and Practice.* J. Taborek, G. Hewitt, and N. H. Afgan (eds.). Hemisphere: Washington, D.C. 1983.

Palen, J. W., B. Cham, and J. Taborek. *Comparison of Condensation of Steam on Plain and Turbotec Spirally Grooved Tubes in a Baffled Shell-and-Tube Condenser,* HTRI Report 2439-300/6. Alhambra, Calif. 1971. [Here noted from Carnavos (1974).]

Palen, J. W., J. Taborek, and S. Yilmaz. Comments on the Application of Enhanced Boiling Surfaces in Tube Bundles. In *Heat Exchangers: Theory and Practice.* J. Taborek, G. Hewitt, and N. H. Afgan (eds.). Hemisphere: Washington, D.C. 1983.

Robertson, J. M. Review of Boiling, Condensing and Other Aspects of Two-Phase Flow in Plate-Fin Heat Exchangers. In *Compact Heat Exchangers: History, Technological Advancement and Mechanical Design Problems.* R. K. Shah, C. F. McDonald, C. P. Howard (eds.). Book No. G00183. ASME: New York. 1980.

Robertson, J. M. Heat Exchange Equipment for the Cryogenic Process Industry. In *Heat Exchangers: Theory and Practice.* J. Taborek, G. Hewitt, and N. H. Afgan (eds.). Hemisphere: Washington, D.C. 1983.

Royal, J. H., and A. E. Bergles. Pressure Drop and Performance Evaluation of Augmented In-Tube Condensation. In *Heat Transfer 1978: Proceedings of the Sixth International Heat Transfer Conference, Toronto, August 1978.* 2:459–464. Hemisphere: Washington, D.C. 1979.

Saboya, F. E. M., and E. M. Sparrow. Experiments on a Three-Row Fin and Tube Heat Exchanger. *Trans. ASME, J. Heat Trans.* 98:520–522. 1976. [Here noted from Robertson (1983).]

Shah, R. K. Classification of Heat Exchangers. In *Heat Exchangers: Thermal-Hydraulic Fundamentals and Design.* S. Kakac, A. E. Bergles, and F. Mayinger (eds.). Hemisphere: Washington, D.C. 1981a.

Shah, R. K. Compact Heat Exchanger Design Procedures. In *Heat Exchangers: Thermal-Hydraulic Fundamentals and Design.* S. Kakac, A. E. Bergles, and F. Mayinger (eds.). Hemisphere: Washington, D.C. 1981b.

Shah, R. K. Compact Heat Exchangers. In *Heat Exchangers: Thermal-Hydraulic Fundamentals and Design.* S. Kakac, A. E. Bergles, and F. Mayinger (eds.). Hemisphere: Washington, D.C. 1981c.

Shah, R. K., and A. L. London. Laminar Flow Forced Convection in Ducts. *Adv. Heat Trans.* Suppl. 1. 1978.

Shah, R. K., and R. L. Webb. Compact and Enhanced Heat Exchangers. In *Heat Exchangers: Theory and Practice.* J. Taborek, G. Hewitt, and N. H. Afgan (eds.). Hemisphere: Washington, D.C. 1983.

Sideman, S. Film Evaporation and Condensation in Desalination. In *Heat Exchangers: Thermal-Hydraulic Fundamentals and Design.* S. Kakac, A. E. Bergles, and F. Mayinger (eds.). Hemisphere: Washington, D.C. 1981.

Snider, A. D., and A. D. Kraus. A General Extended Surface Analysis Method, *Trans. ASME, J. Heat Trans.* 103:699–704. 1981.

Soliman, H. M., and A. Feingold. Analysis of Heat Transfer in Internally Finned Tubes Under Laminer Flow Conditions. In *Heat Transfer 1978: Proceedings of the Sixth International Heat Transfer Conference, Toronto, August 1978.* 2:571–576. Hemisphere: Washington, D.C. 1979.

Spiral Tube, Bulletin No. 177, McC/11/80, *Spiral Tube Heat Transfer Data.* Spiral Tube (Heat Transfer) Ltd.: Derby, England. 1980.

Tolubinsky, V. I., and N. V. Zozulya. Application of Heat Transfer Finned Surfaces in High Temperature Heat Exchangers. In *Heat Exchangers: Design and Theory Sourcebook.* N. Afgan and E. U. Schlünder (eds.). McGraw-Hill/Hemisphere: New York. 1974.

Tree, D. R., V. W. Goldschmidt, R. W. Garrett, and E. Kach. Effect of Water Sprays on Heat Transfer of a Fin and Tube Heat Exchanger. In *Heat Transfer 1978: Proceedings of the Sixth International Heat Transfer Conference, Toronto, August 1978.* 4:339–344. Hemisphere: Washington, D.C. 1979.

Trupp, A. C., and H. M. Soliman. Performance Optimization of Internally Finned Tubes for Laminar Flow Heat Exchangers. In *Heat Exchangers: Theory and Practice.* J. Taborek, G. Hewitt, and N. H. Afgan (eds.). Hemisphere: Washington, D.C. 1983.

UOP, *Engineering Data Book.* Wolverine Division, UOP Inc.: Decatur, Alabama. 1981.

Urbican, M. J., and J. Paloschi. Simulation of Air-Cooled Heat Exchangers. *Computers Chem. Eng.* 5(2):75–81. 1981.

van der Meer, T. H., and C. J. Hoogendoorn. Heat Transfer Enhancement by Static Mixers for Very Viscous Fluids. In *Heat Exchangers: Theory and Practice.* J. Taborek, G. Hewitt, and N. H. Afgan (eds.). Hemisphere: Washington, D.C. 1983.

Webb. R.L. Air-side Heat Transfer in Finned Tube Heat Exchangers. *Heat Trans. Eng.* 1(3):33–49. 1980.

Yampolsky, J. S. Spirally Fluted Tubing for Enhanced Heat Transfer. In *Heat Exchangers: Theory and Practice.* J. Taborek, G. Hewitt, and N. H. Afgan (eds.). Hemisphere: Washington, D.C. 1983.

CHAPTER

FIVE

HEAT EXCHANGERS FOR SPECIAL SERVICES

INTRODUCTION

● In this chapter, we will discuss the heat exchangers for (1) high-temperature services for which ceramic units are being developed so that the heat recovery from very high-temperature flue gases is increased; (2) cryogenic services for which special tests are required for material selection since many materials become brittle; (3) corrosive services for which nonmetallic units, such as impervious graphite, glass, Teflon, and composites are used; (4) nuclear power plants for which different codes and classifications exist and more stringent quality control, stress analysis, and documentation systems are required; and (5) liquid metals and molten salt heat exchangers for use at high temperatures with their attendant advantages and disadvantages compared to steam. These topics supplement the other chapters on shell-and-tube heat exchangers in this book.

HIGH-TEMPERATURE HEAT EXCHANGERS

5.1 What are high-temperature heat exchangers?

These are units having one stream at a temperature above 1500°F (800°C, arbitrary limit). This stream is usually flue or waste gases coming out of combustion chambers and furnaces as in metallurgical industries, and generally carry dirt particles and contaminants, such as metallic salts, sulfates, SO_2, and SO_3. It is estimated that over 40 percent of the exhaust gas in the United States is at 1500°F (800°C) or higher (Larson and Kreith, 1981). Of this, about 100 trillion Btu/yr (3340 MW) is recoverable.

5.2 How do these differ from low-temperature heat exchangers?

The differences are the following:

1. Radiation heat transfer plays an important role in the high-temperature units.
2. Tube diameters and pitch are larger in high-temperature units so that the pressure drop is kept low. The cost of adding a fan or blower to work at high temperatures can be prohibitive.
3. Even though the gases have low heat transfer coefficient, fins are generally not used in high-temperature units because (Scaccia and Theoclitus, 1980) (a) the gaseous stream usually carries suspended dirt particles that will foul or fill up the space between the fins and make a finned tube worse than a plain tube and (b) the gas velocity is low anyway, due to low pressure drop available, hence the advantage of a fin is negligible.
4. The materials for construction are different in the two cases. In high-temperature units, one uses ceramics, or high alloy and costly tubings, while low alloy tubings are usually used in the low-temperature units.
5. The selection of material and their thickness and the mechanical design is governed by the thermal stresses in the high-temperature units. Other factors to be considered are the extent of oxidation of the material, thermal shock bearing capability, and erosion due to suspended dirt particles. Fouling and corrosion due to metallic salts, sulfates, etc., also have to be kept in mind. Stress analysis has to be carried out for a safe and reliable design.
6. A variety of designs are used in high-temperature units. Some are (a) stationary regenerators made of refractory brickwork where cyclic gas flows and more than one unit is used, (b) rotary regenerators or recuperators made of compact metal surfaces, (c) shell-and-tube heat exchangers made of metals and ceramics, and (d) compact ceramic units.
7. Differential expansion is an important factor in the high-temperature units and should be accounted for either by using expansion bellows or by using bayonnet-type units. Floating tubesheets cannot generally be used, since sealing gasket or packing materials do not work effectively at such high temperatures.
8. Heat losses from the outside surface to the environment have to be prevented by using proper insulation, and the weight of the insulation should be considered in the mechanical design of the unit and in the foundation design.
9. For high-temperature heat transfer, gases, air, or liquid metals and molten salts are preferred over steam, since the latter requires very thick shell and tubes to contain its high pressure. Liquid metals and molten salt heat exchangers are discussed later in this chapter.

• In this chapter, we shall consider only the fixed type of units (not requiring cyclic gaseous flow) for high-temperature services. Thus, regenerators and rotary recuperators will not be considered since these form a big discipline in themselves.

5.3 What materials are used in the fabrication of high-temperature heat exchangers?

These are the following:

1. High alloy steels (primarily ferritic and austenitic): These have a temperature limit of 850°-950°C (1560°-1740°F) above which they lose strength and also get oxidized. Hence, high-temperature waste gases have either to be cooled to the limiting temperature of the material by diluting with air or the heat flux to the tubes by radiation is limited. Either of these methods reduces the efficiency of heat recovery to a rather low value. See Q. 10.13-10.15 regarding steel alloys for high-temperature usage. Other materials, such as tungsten, molybdenum, and tantalum, have high melting points compared to that of steel, but they get easily oxidized at high temperature and lose strength and are also prohibitively expensive.
2. Ceramics: These have not yet reached a mass usage stage, but a lot of developmental effort is underway. These can take the flue gases directly without air dilution and can, in turn, preheat the furnace combustion air to a temperature of 1000 °C (1800 °F) or above compared to 500 °-600 °C (900-1100 °F) possible in the metallic units. Since more furnace fuel is saved with higher preheated combustion air temperature, a further saving of 10 to 15 percent in the furnace fuel will result with the combustion air at 1000 °C (1800 °F) compared to that at 550 °C (1000 °F) (Tennery, 1983). Furthermore, in most cases, ceramics do not get oxidized at high temperatures encountered nor do they lose strength. They have low coefficients of thermal expansion and hence can take thermal shocks.
3. Coated surfaces: Compact units can be made from low cost materials and then coated using aluminum or aluminum compounds in the gaseous phase. While the compact units provide the necessary strength, the coating prevents oxidation. Other coatings that can be used are siliconizing, chromizing, and multi-element combinations with aluminum or silicon as the principal element (Larson and Kreith, 1981). Coatings do pose some problems related to thermal expansion mismatch of the metal and the coating, grain growth, etc. These affect their ability to sustain cycles of heating and cooling, thermal shock, and mechanical damage.

5.4 What are the types of ceramic heat exchangers?

These are compact honeycomb type and tube-and-header type.

1. Compact honeycomb type (Fig. 5.1): These are compact units with small flow passages (e.g., 6 mm high and 16 mm wide) alternating in a cross-flow configuration. Larger surface area per unit volume is obtained compared to the

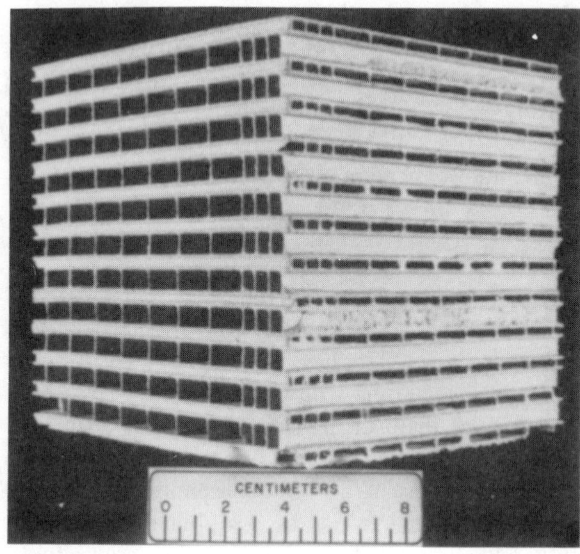

Figure 5.1 Ceramic compact honeycomb heat exchanger. *(Courtesy of Oak Ridge National Laboratory.)*

shell-and-tube units. The ceramics used are magnesium aluminosilicates (MAS) and lithium aluminosilicates (LAS), with the former being preferred due to its properties and lower cost. Air leakage toward flue gas side is about 5 percent of the air mass flow, since the air is at a higher pressure than the flue gas.

2. Tube-and-header type: These have ceramic tubes with headers at the two ends. One of the ways to connect the tubes to the headers is to compression load the tubes individually on the air inlet header side. One of the two streams (air or the flue gases) flows through the tubes, while the other flows in cross-flow over the tube bundle (Fig. 5.2). The mass and size of the tube bundle are very large. Hence, fins are used to increase the heat transfer coefficient and to reduce both the mass and size of the units. Since, on both the sides we have gases (flue gases and air), the enhancement in some designs is done on both the sides. Circular fins are used on the outside, while straight longitudinal fins are used on the inside. This cuts down the mass by 40 percent or more as shown by conceptual designs of large units. One can also use circumferential dimples on the inside. The energy saved by using these ceramic units is very large, and the usual payback period is less than 2 years. Tubes are made of silicon carbide, while for the high-pressure usage siliconized and sintered alpha silicon carbide may also be used. Due to the currently available tube-to-header joining techniques, a minimum leakage of 2–3 percent air into the flue gases cannot be avoided, although extensive developmental efforts are underway to reduce this. Ceramic and other nonmetallic heat exchangers are compared in Table 5.1. These units are used in exchanging heat from flue gases. Other uses envisaged include the following:
   a. In turbine systems where dirty or contaminated fuel or coal is burnt (Tennery, 1983; Coombs, 1979). In some such cases, the contaminated combustion gases are not allowed to contact the turbine directly. Instead, they heat air in a ceramic unit and the hot air, in turn, drives the turbine.

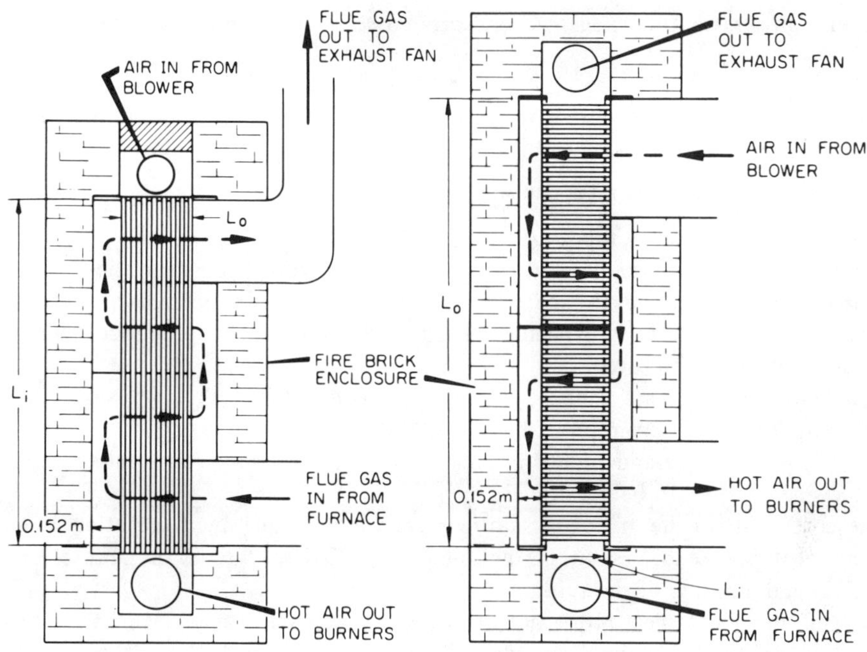

Figure 5.2 Two examples of tube-and-header type units. *(Courtesy of Oak Ridge National Laboratory.)*

Table 5.1 Comparison of nonmetallic heat exchanger materials

	Graphite	Glass	Teflon	Ceramic
Usable for highly corrosive fluids	Yes	Yes	Yes	–
Pressure limit	High	Medium	Low	Medium
Temperature limit, °C (°F)	180 (356) (due to binder limitations)	175 (347)	180 (356)	Very high
Heat duty	High	Medium	Low to medium	High
Surface area	High	Medium	Low to medium	Low to medium
Ability to absorb pressure shocks	Moderate	Nil	Moderate	Low
Brittleness	Low	High	Low	High
Thermal conductivity	High	Low	Low	Low
Tube wall thickness	High	Low	Low	High
Scale deposition	Low	Low	Low	Moderate
Shapes of units possible (i)	←———	Shell-and-tube	———→	
(ii)	Modular	Coil	Coil	Compact
Weight/unit area	High	Low	Low	High
Tube diameters	Up to 1 in (25.4 mm)	Any size	0.1, 0.125, and 0.25 in (2.5, 3.18, and 6.35 mm)	Any size

b. In high temperature usage of concentrated solar energy produced in heliostat fields

5.5 What are the major unresolved problems in the use of the ceramic heat exchangers?

These are the following (Tennery, 1983):

1. High cost of the ceramic tubes
2. Long tubes of uniform properties (devoid of microflaw concentrations) are difficult to produce by the currently available methods. These may produce fracture when it is least expected. Hence, the ceramic tube production techniques need to be improved.
3. Tube-to-tubesheet joints cannot yet be made completely leak tight, since the differential thermal expansion of the tubes has to be accommodated.
4. Durability of the ceramic tubes in an environment of highly contaminated hot process gases is yet to be confirmed.
5. The coefficient of thermal expansion may change with time due to reaction with some of the species present in the flue gases. This may result in unforeseen stresses and cracking of the tubes.
6. Most of the mechanical properties known were evaluated using small bars. The exact behavior of the same material in the form of long, hollow tubes is not known well.
7. Flue gases from coal or residual oil result, after long time, in corrosion reactions with siliconized and sintered alpha silicon carbide, changing their fast fracture strength. The exact reasons for this and also the preventive steps need to be worked out.
8. Ceramic-metal composites are also being considered, but fabrication would pose greater problems.

CRYOGENIC HEAT EXCHANGERS

5.6 Where are cryogenic temperatures encountered?

These are encountered in the following:

1. Air separation plants
2. Liquefaction of air and other gases (hydrogen, helium, natural gas, etc.)
3. Ethylene plants
4. Hydrogen purification
5. Research facilities, such as bubble chambers containing liquid hydrogen for particle physics, superconducting magnet systems, ultra-high vacuum, and purification and separation of deuterium and tritium in fusion research.

5.7 In what important aspects do the heat exchangers used in cryogenic services differ from those used in general services?

The primary difference is the operating temperature. In the cryogenic services, it is generally less than –100°C (–150°F). This puts a lot of importance on material selection, fabrication, and design (see Chapter 10). The change in dimensions (contraction) due to the low temperature introduces stresses in the system and in the welded joints. This requires the use of expansion joints and/or bellows. Furthermore, complete moisture removal is required, or else it will freeze at low temperatures and may plug the tubes. Therefore, these units need to be dried before use by circulating warm air. The pressure drop available is very small. On the other hand, high heat transfer coefficients are necessary. Since the temperature difference between the heat exchanger and the environment is rather large, insulation is necessary to minimize heat leakage into the system. Sometimes, heat is exchanged between multiple streams in the same unit.

5.8 For use in cryogenic services, what important property must the materials possess?

The materials must be able safely to resist high stress changes and shock loads. For this, notch tough materials should be used. Impact tests should be done if required by the Code.

5.9 What is the major change produced by the low temperatures in the materials of construction?

At temperatures below –20°F (–28.9°C), most materials show a marked decrease in impact resistance. They lose their ductility and become brittle. They are liable to brittle fracture at full stress.

5.10 How can brittle fracture be avoided in a unit at low temperature?

A number of precautionary steps are recommended with regard to the selection of material, welding, etc. They are the following:

1. The material selected should not become brittle at low temperature. Such materials are some stainless steels, wrought aluminum, and nonferrous metals such as copper and copper-nickel alloys. Should these materials fail, they do not suddenly undergo a brittle fracture but give a warning by bulging, leaking, etc. Corrective measures can then be taken in time to avoid a complete failure.
2. Welding electrodes and welding procedures should be carefully selected for the intended service.
3. All corners, notches, and sharp changes of sections should be avoided since they act as stress raisers.
4. Localized yielding should be avoided.

5.11 What are the different types of heat exchangers used in cryogenic services?

These are listed below in chronological order of development (T. K. Poddar, 1978, unpublished):

1. Linde exchanger (1895)
2. Hampson coiled exchanger

3. Coiled tubular exchanger
4. Switching regenerator
5. Reversing type
6. Compact units
7. Heat pipes

A number of these are coiled tubes. The first two, viz. Linde and Hampson ones, are not used anymore because of their low efficiency compared to the heat exchangers used in modern plants for the cryogenic services. The most used are (3)–(6), which are discussed below. Item (7) is used in selected cases only.

5.12 Discuss the design and advantages of coiled-tube exchangers.

In this, different groups of tube bundles carrying different liquids are coiled around a core cylinder in a number of layers and collected at each end of the exchanger shell in headers (Fig. 5.3). Successive layers of tubes are wound in opposite directions and separated by spacing bars so that the shell-side fluid can contact all the tubes. The core cylinder gives mechanical stability, especially during the manufacture of the exchanger. Its minimum diameter is given by the maximum allowable curvature of the coiled tubes, for which no flattening of the tubes occurs. High-pressure gases flow through these tubes, while low-pressure cold gas flows on the outside over them (through the shell).

The advantages are the following:

1. Heat exchange is possible with more than two fluids.
2. It can take high pressure and large temperature ranges.
3. Large heat transfer area (up to 25,000 m^2 or 270,000 ft^2) can be provided. Large units are made by coiling machines. The size is limited only by the handling and transport capabilities. The geometry can be modified to suit the requirements.
4. Thermal expansion and contraction can occur without inducing undue stresses.
5. No vibration problem is encountered.
6. The design is very compact.

5.13 Discuss the design and advantages of switching regenerators.

These consist of towers containing high heat capacity and high surface area packings, such as thin, corrugated strips of aluminum. Two or more towers are used. The warm and cold streams are switched periodically as in the regenerators used at high temperatures.

The advantages are the following:

1. Very large sizes possible for industrial production of liquified gases
2. Low pressure drop
3. Impurities (e.g., moisture and carbon dioxide) can be removed from the feed stream by condensation and subsequently evaporated into the outgoing waste stream by carefully controlling the operating conditions.

5.14 Discuss the design and advantages of reversing heat exchangers.

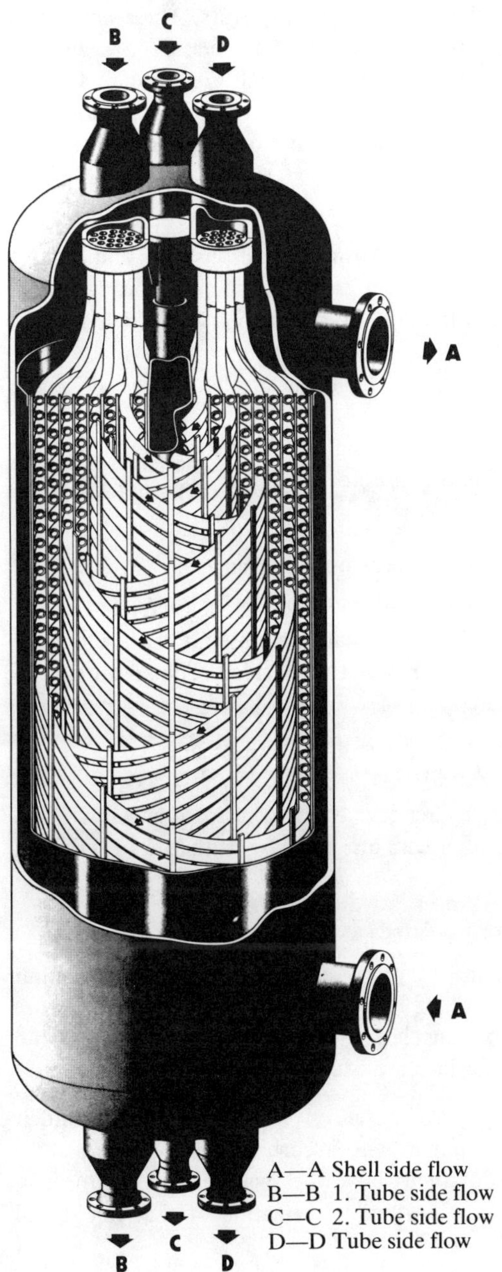

Figure 5.3 Coiled-tube cryogenic heat exchanger. *(Courtesy of Linde Aktiengesellschaft, 1977.)*

These provide an alternative to the switching regenerators. These consist of several, usually four, coaxial finned tubes. Feed air, oxygen, waste nitrogen, and unbalanced flow (a portion of waste nitrogen) pass through different annuli and the central tube. The flow direction of air and waste nitrogen are reversed periodically for purification purposes. The advantages of reversing heat exchangers over switching regenerators are the following:

1. Complete removal of impurities is possible.
2. Oxygen produced is so pure that it can be used for medical purposes and in steel making by L.D. (Litz-Donawitz) process. On the other hand, the switching regenerators give tonnage oxygen, containing various impurities, which is good only for industrial use.
3. The size of these exchangers is only about one-third of that of the switching regenerators.
4. Although the initial cost is high, the low operating costs and high product purity make this unit a welcome choice. These consume less energy than the switching regenerators (Wolff et al., 1979).

5.15 Discuss the design and advantages of compact heat exchangers.

Compact heat exchangers have been discussed at length in Chapter 4. Their use in cryogenic services started in the 1940's. Primarily aluminum or its alloys are used. The heat exchanger is generally brazed rather than welded. The advantages of compact heat exchangers are the following:

1. High surface/volume ratio resulting in compact and lightweight units
2. Very wide design flexibility
3. High efficiency in heat transfer, approaching over 95 percent in some cases
4. Yield strength and tensile strength of aluminum increase with the lowering of temperature

5.16 How is a cryogenic heat exchanger insulated?

Since the cryogenic temperatures are very low compared to the environment temperature, there will be a large amount of heat gained by the low temperature unit unless it is properly insulated. No one insulation will do the job adequately and at a minimum cost. Hence, multiple insulations are used:

1. Vacuum layer around the heat exchanger, i.e., an envelope is formed around it and is evacuated. Activated charcoal is put in it to absorb any gas leakage.
2. A layer of higher boiling liquid is maintained above the vacuum layer so that any heat from outside is used as the latent heat of vaporization of that liquid and does not enter the heat exchanger.
3. A layer of low conductivity packing powder is put over the liquid layer.
4. Finally, a radiant insulating layer (e.g., shining aluminum foil) is wrapped around the packing powder layer.

The thicknesses and types of insulation can vary depending upon the requirements, availability, experience in handling, and the cost. In fact, the cost of insulation can be a significant fraction of the total cost of a cryogenic heat exchanger.

The multiple insulations listed above would suffice for very low temperatures (–455°F or –270°C) such as encountered in hydrogen and helium liquefaction plants. In air separation plants where the temperature is not so low (around –390°F or –200°C), a single layer insulation of low thermal conductivity may be used.

HEAT EXCHANGERS FOR CORROSIVE SERVICES

● Most chemicals and water are corrosive in nature and attack the materials of construction, reducing their effective thickness and degrading their mechanical properties. For some of the fluids, the normal materials, such as mild steel, can be used by increasing the thickness to compensate for loss due to corrosion. For some others, special alloys, such as stainless steel, nickel, chromium, copper, and their alloys, are used. However, there are other chemicals, such as strong acids and alkalies, that corrode most materials at a fast rate unless exotic materials and alloys, such as titanium and Hastelloy are used at a prohibitive cost. For such fluids, nonmetallic units are preferred. These are made of impervious graphite, glass, Teflon*, or composite materials. Their salient features are discussed below, and a comparison between them has been made in Table 5.1.

Impervious Graphite Heat Exchangers

5.17 What are the principal advantages and disadvantages of impervious graphite heat exchangers?

The principal advantages are the following:

1. Resistance to highly corrosive environments
2. High thermal conductivity that is more than three times that of steel
3. No metal contamination of the fluids occurs in these
4. Do not scale or foul easily
5. Can withstand rapid and uneven changes in the temperature since the coefficient of thermal expansion is very low
6. Low friction factor since the flow passages are very smooth
7. Properties of graphite can be varied to suit the requirements. These can be made more dense and compact or light and porous. Properties of graphite are listed in Perry and Chilton (1973).
8. Can easily be machined to very fine tolerances.

The principal disadvantages are the following:

*Registered trademark E. I. duPont de Nemours, Wilmington, Delaware.

1. Very high cost per unit area
2. Tubes are rather thick to impart sufficient strength. The fracture strength of graphite is rather low, therefore, they occupy more space.
3. Require resin or other chemicals to impregnate pores to make graphite impervious. These chemicals affect resistance to corrosion and the high temperature limit.
4. These are brittle units and hence require careful handling in manufacture, transport, installation, and operation.
5. Low operational limits on pressure (75 psig or 620 kPa with liquids, 50 psig or 450 kPa with steam), and temperature (180°C or 355°F).
6. Not all oxidizing fluids can be used. General limits on the acid concentrations are 60 percent hydrofluoric acid, 20 percent nitric acid, 96 percent sulfuric acid.

5.18 How is impregnation of the graphite done? What properties should the chemicals used have?

Graphite has pores, the sizes of which depend upon the microstructure of graphite. Air and other gases may get adsorbed during the manufacturing of graphite. These are first removed by placing the graphite in an autoclave and evacuating, thus sucking out the air and the adsorbed gases (Hillard, 1970). Then the resin is admitted under vacuum, and pressure is increased so that the resin enters the pores. Excess resin is drained out. The temperature of the autoclave is raised to a high value where thermal polymerization of the resin occurs. The process may be repeated to achieve satisfactory impregnation. There are also other methods of impregnating graphite.

Any impregnating chemical should (i) have high thermal and chemical stability similar to that of graphite under conditions of use; (ii) be able to enter the pores easily and polymerize there without setting up stresses, and (iii) not be very costly. Generally, furamic and phenolic resins and PTFE (Teflon)-based chemicals are used as the impregnating agents.

5.19 What factors should be kept in mind when constructing graphite heat exchangers?

These are the following:

1. Graphite is an anisotropic substance. Its thermal conductivity is higher along the planes compared to the value perpendicular to the planes. Therefore, the holes should be drilled in such a manner so that advantage is taken of the high value of its thermal conductivity along the planes.
2. Graphite has high compressive strength and low tensile strength. It is also brittle. Hence, extreme care should be exercised during fabrication, transport, installation, and operation of the heat exchanger.

5.20 What are the types of units formed out of graphite?

The various types are the following:

1. Shell-and-tube type: These have graphite tubesheet and tubes, while the shell is made of steel. The tubes are quite thick (e.g., $\frac{7}{8}$ in or 22.2 mm I.D. and $1\frac{1}{4}$ in or

31.7 mm O.D.; $1\frac{1}{2}$ in or 38.1 mm I.D. and 2 in or 50.8 mm O.D.). These are used in the fertilizer, heavy chemicals, and synthetic fiber industries. The operational limits are 75 psig (620 kPa) with liquids and 50 psig (450 kPa) with steam and 180°C (355°F). Surface area up to 14,000 ft^2 (1300 m^2) can be obtained.

2. Rectangular or cubic block type (Fig. 5.4): A solid block of graphite is perforated with parallel round holes between two pairs of opposite faces (one pair for each liquid). End plates of steel are clamped on each side of the block by means of tie bars to reinforce the graphite block, thus taking advantage of its high compressive strength. Inlet and outlet nozzles are attached to the end plates. These can be arranged to have multi-pass units. The end plates and nozzles must also be compatible with the corrosive nature of the fluids. Some such materials are high alloy steels or carbon steel with a proper lining. The unit so produced is a compact one.
3. Modular block type (Fig. 5.5): These are cylindrical units (modular blocks) with drilled cross holes. The number of modules can be changed to change the heat transfer area as the need arises. The process fluid flows axially, while the service fluid flows sideways through each block, thus giving a cross-flow operation. The holes can be circular or noncircular as the need be.
4. Cascade type: These are long tubes of large diameter (1–4 in or 25–100 mm) arranged in a vertical cascade, connected with U-bends to cool acids and other

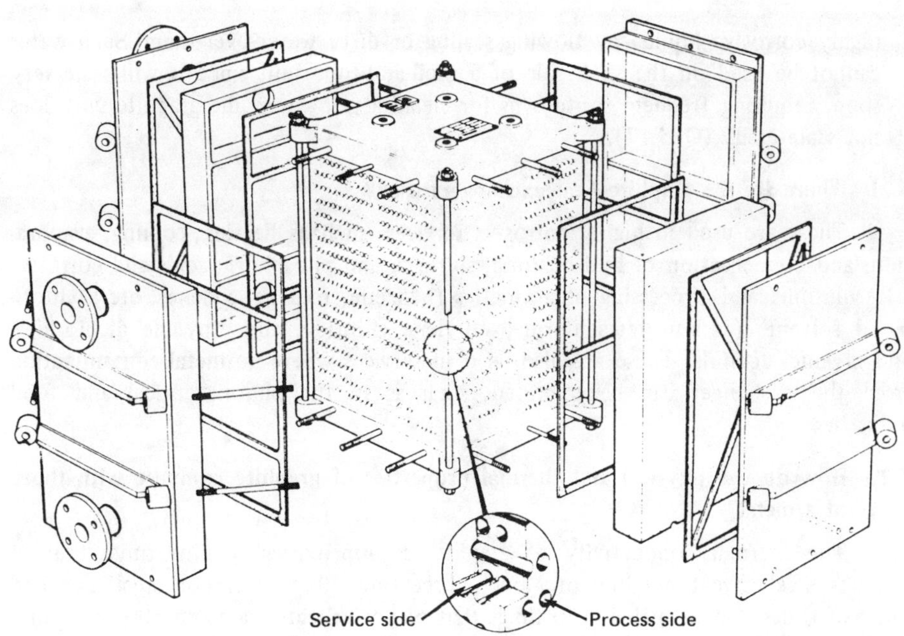

Figure 5.4 Rectangular block graphite heat exchanger. *[Excerpted by special permission from CHEMICAL ENGINEERING (Dec. 23, 1974) Copyright © 1974, by McGraw-Hill, Inc., New York, NY 10020.]*

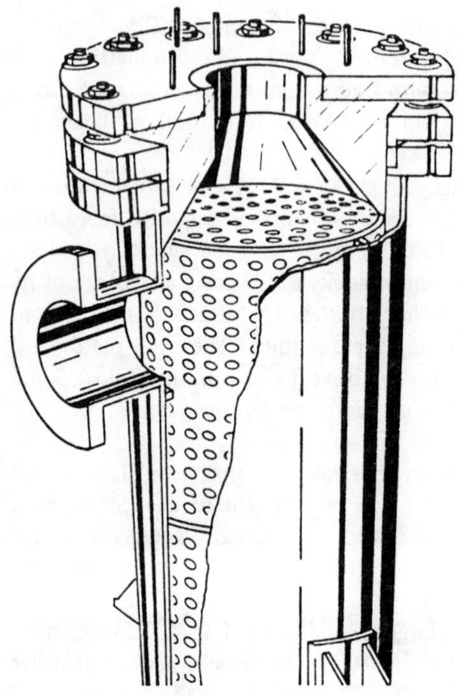

Figure 5.5 Modular block graphite heat exchanger. *[Excerpted by special permission from CHEMICAL ENGINEERING (Jan. 20, 1975) Copyright © 1975, by McGraw-Hill, Inc., New York, NY 10020.]*

highly corrosive liquids by flowing scaling or dirty water over them. Such water cannot be used on the shell side of a shell-and-tube unit since it will scale very soon, requiring frequent shutdowns for cleaning. However, the graphite unit does not scale easily (Q. 5.17).

5.21 Where are the graphite heat exchangers used?

These are used in highly corrosive services, such as heating, cooling, evaporation, and condensation of hydrochloric, sulfuric, and phosphoric acids and corrosive salt solutions and processing of aromatics (benzene, toluene, xylenes, etc.), chlorinated solvents (carbon tetrachloride, ethylene chloride, and ethylene dichloride), and organic alcohols. These are also used in services where no metal contamination from the exchanger can be tolerated, such as in the pharmaceutical and food industries.

5.22 How do the physical and thermal properties of graphite compare with those of a metal?

The thermal conductivity of graphite is approximately nine times that of stainless steel, five times that of Monel, three times that of carbon steel, two and one half times that of nickel, 120 times that of borosilicate glass and 800 times that of PTFE (Teflon), the other highly corrosion-resistant materials. It is immune to thermal shocks within its operating limits of 180°C (355°F) and 75 psig (620 kPa) liquid and 50 psi (450 kPa) steam. It is unaffected by fatigue. Its tensile strength is lower than that of the metals.

Glass Heat Exchangers

5.23 What are glass heat exchangers, and where are they used?

These are low surface area heat exchangers with straight glass tubing and a glass or steel shell. The tubesheets are made of glass-filled TFE (tetrafluoroethylene, a fluorocarbon plastic), glass-filled TFE with an embedded steel plate, or high alumina ceramic. The tubes are sealed with an O-ring in a groove in the tubesheet. Each tube hole in the tubesheet is threaded part way and a threaded TFE fluorocarbon plastic bushing or ferrule is tightened down on the O-ring (Fig. 5.6). In low-pressure usage, a silicone rubber PTFE-covered sealer sheet may be used instead of the O-ring.

These are used in corrosive services such as acid cooling and halogen and organic solvents condensing and cooling.

5.24 What are the principal advantages and disadvantages of glass heat exchangers?

The principal advantages are the following:

1. Resistant to highly corrosive fluids. One can have corrosive fluids on both the shell and the tube side by using a glass shell.
2. Very smooth surface gives almost no fouling. To a large extent, this compensates for the effect of the lower thermal conductivity of glass compared to that of the metals, since the latter scale more.
3. Easy cleaning, maintenance, and assembly: Any scale that might still deposit can be cleaned by using strong chemicals. Furthermore, since O-rings and ferrules are used at the tube-to-tubesheet joints, leaky or bad tubes can be easily replaced.
4. Where even slight metal contamination would be harmful, such as in pharmaceutical, food, and beverage industries, total glass construction (both shell and tubes) can be used.
5. Cost per unit surface area is lower than that of the metallic units (two-thirds the cost of aluminum unit and one-fifth that of a stainless steel one) (Curwen and Newberry Ltd., 1980).
6. Visual insight into the process occurring is possible. Formation of scale or any other type of fouling is easily seen and detected early. Therefore, corrective measures can be taken early.
7. Full vacuum service is possible on both the shell and tube side.
8. In many cases, the glass tubes can be replaced with metal tubes if necessary without any modification of the internals or the tubesheet.

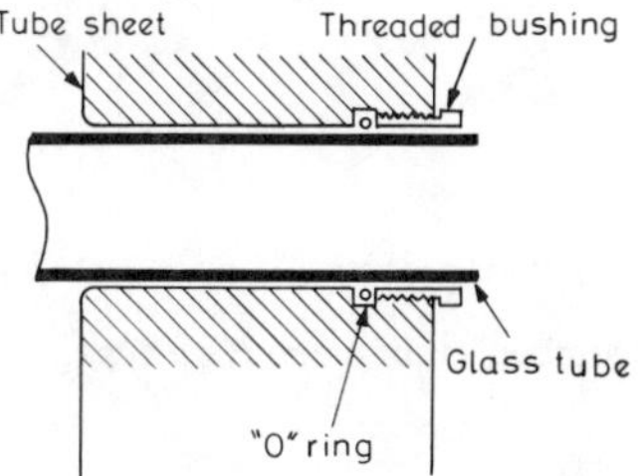

Figure 5.6 Glass tube-to-tubesheet joint in glass heat exchangers. *(Courtesy of Corning Glass Works.)*

The principal disadvantages are the following:

1. Available in small sizes (low surface areas, up to approximately 550 ft^2 or 51 m^2) and only fixed tubesheet type with straight tubes
2. Limited temperature range (–40° to 375°F or –40° to 180°C)
3. Limited pressure range (full vacuum to 75 psig or 620 kPa, decreasing with increasing shell diameter)
4. Brittle and very low impact strength
5. Since their operating pressure range is outside the purview of the ASME Codes, there is no standard way to certify or inspect the mechanical design and fabrication or assembly. One has to rely on the methods adopted by the manufacturer, and these vary with the manufacturer.

Teflon Heat Exchangers

5.25 What are Teflon heat exchangers and where are they used?

These are heat exchangers made from Teflon (PTFE, polytetrafluoroethylene), a highly chemically inert substance. These are used in very corrosive environments, such as in cooling acid plants, metal-treating industries, and heat recovery from corrosive fluids. These can also be used in condensing duties. Thin Teflon tubes [O.D. = 0.1 in (2.5 mm), 0.125 in (3.18 mm), and 0.25 in (6.35 mm); I.D. = 0.08 in (2 mm), 0.1 in (2.5 mm), and 0.2 in (5.08 mm)] are formed into "honeycombed" tubesheets at their ends by fusing the tubes together (Fig. 5.7). This tube bundle is then either used as a coil in cooling tanks (Fig. 5.8), or placed in a steel shell (Fig. 5.9). Bundles are sealed within the shells by means of O-rings. Perforated support plates (made of Teflon) are used to maintain some distance between the tubes. These plates are kept at a fixed distance by stainless steel tie-rods coated with Teflon. Depending upon the situation at hand, corrosive fluid can flow through the tubes or outside the tubes (as in the tanks). The flow outside the tube bundle can be parallel or in cross flow (Smith, 1974).

While PTFE is the most popular material, other fluorochemicals are also being used due to their easy machinability, hence the ease in producing parts and intricate shapes. Some such fluorochemicals are fluorinated ethylene propylene (FEP),

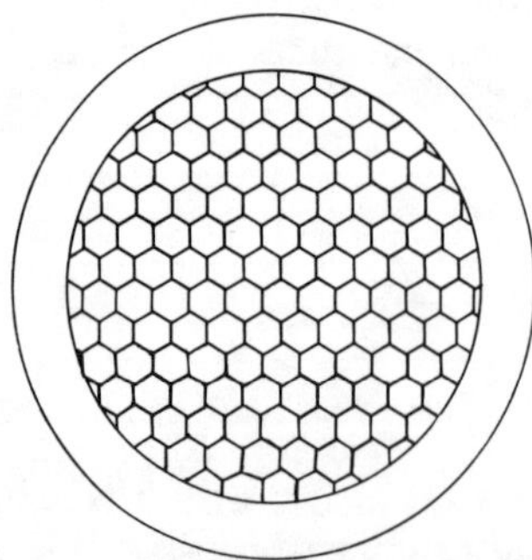

Figure 5.7 Teflon "honeycombed" tubesheet. *(Courtesy of E. I. duPont.)*

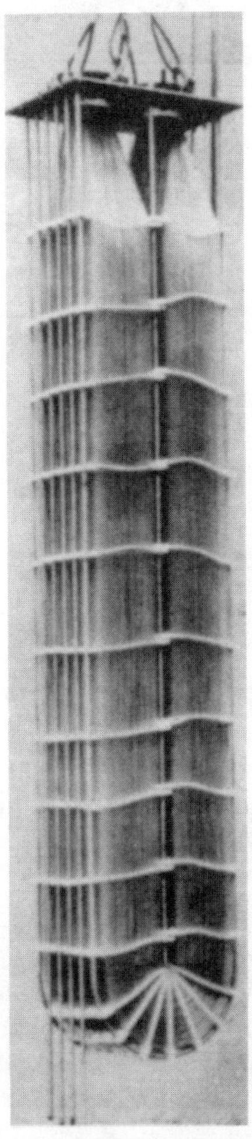

Figure 5.8 Teflon tank coils. *(Courtesy of E. I. duPont.)*

perfluoroacetyl (PFA), ethylene tetrafluoroethylene (ETFE), ethylene-chlorine-tetra-fluoroethylene (E-CTFE), polyvinylidenefluoride (PVDF). Among these, PVDF is getting to be more popular because of its utmost ease in machining and joining to produce complex components. The tube diameter can be much larger than that with PTFE. The various fluorochemicals noted above differ in their melting points, corrosion-resistant properties, ease in moulding and machining, and cost. It must be mentioned that, while a number of these approach PTFE in their corrosion-resistant properties, none exceeds it because, in PTFE, the basic carbon chain is covered with a layer of fluorine molecules that protects the carbon chain from attack. None of the other fluorocarbons have such a total covering of the carbon chain.

5.26 What are the principal advantages and disadvantages of Teflon heat exchangers?

The principal advantages are the following:

Figure 5.9 Teflon shell-and-tube heat exchanger. *(Courtesy of E. I. duPont.)*

1. Resistant to highly corrosive fluids: One can have corrosive fluids on either or both sides of the tubes. When the corrosive fluid is in the shell, the latter may need to be coated with Teflon.
2. Very smooth surface gives almost no fouling. This, combined with the very thin tube walls, compensates to a large extent the effect of lower thermal conductivity of the fluorocarbons.
3. These can be used as coils or as shell-and-tube heat exchangers.
4. Larger heat transfer area than with the glass units is possible.
5. It is significantly cheaper compared to the graphite units when the initial and maintenance costs are taken into account (duPont de Nemours International, 1976).

The principal disadvantages are the following:

1. Available in restricted range of sizes
2. Limited temperature range: 350°F (175°C) maximum
3. Limited pressure range: 135 psig (1030 kPa) internal and 65 psig (550 kPa) external pressure in small diameter tubes at 100°F (38°C). These reduce to 70 psig (585 kPa) and 30 psig (310 kPa), respectively at 200°F (93°C). Pressures are lower for larger diameter tubes (Perry and Chilton, 1973). Vacuum is not used on the tube side.
4. Limited pressure differential between the two fluids, or else the tubes might burst.
5. As is the case with the glass units, one has to rely on the methods adopted by the manufacturers, since these units are outside the purview of the Codes.
6. Cannot handle fluids with suspended particles because of the small tube diameter. Hence, prescreening of the fluid stream (with attendant pressure drop) is required.

7. Difficult to detect a leaky tube. Tube replacement is not possible. It can only be plugged. This might, in due time, affect the heat transfer area significantly if many tubes are plugged and a new tube bundle may have to be installed.
8. Deposits formed can only be cleaned by chemicals or by backflushing. Mechanical means, rods, etc., cannot be used for fear of rupturing the tubes.

Composite Materials Heat Exchangers

5.27 What are composite materials?

These are materials made of a metal and a nonmetal to give it the desired properties for specific applications. Some composites are reinforced concrete (RCC), fiber reinforced plastic (FRP), and mesh-plastic composite (MPC). The last one is claimed to be useful as a heat exchanger material.

5.28 What is a mesh-plastic composite?

It is a composite of woven wire mesh embedded in a plastic manufactured by extruding a plastic sheet with a wire mesh in a die. The plastic surrounds the wire mesh, and there are no air inclusions in the plastic. Figure 5.10 shows a cross section through such a composite.

5.29 What are the advantages of the mesh-plastic composite?

The advantages are the following:

1. Light weight compared to an equivalent strength metal unit
2. Thin wall compared to an equivalent strength pure plastic unit
3. Corrosion resistant since fluids only contact the plastic
4. Low scale formation since the plastic is very smooth, and it also does not corrode
5. Thermal conductivity higher than that of the plastic alone since the plastic thickness is small and the wire mesh provides good thermal conductivity (Fig. 5.11) (Anonymous, 1980)
6. With a very thin plastic covering, the fluid flows over the contours of the wire mesh, thus promoting turbulence and hence increasing the rate of heat transfer (Whitfield, 1981).
7. It can be made into intricate shapes, such as spiral-and-plate types and shell-and-tube units.
8. It can be used at low temperatures and pressures, as a solar heat collector, in evaporation duties in desalination plants, etc.

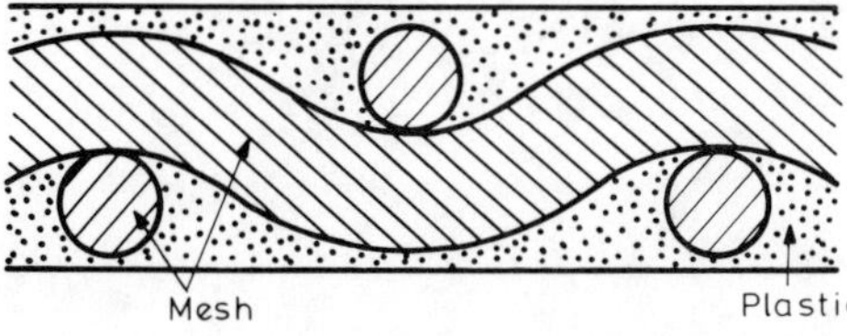

Figure 5.10 Metal–plastic composite (MPC). *(Courtesy of Morgan-Grampian, Ltd.)*

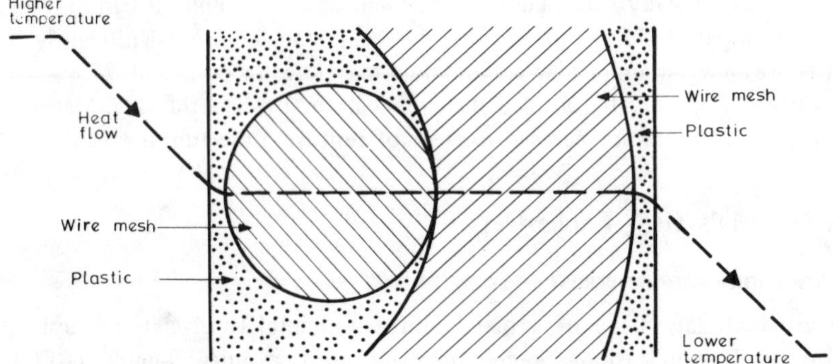

Figure 5.11 Heat flow path in metal–plastic composite. *(Courtesy of Morgan-Grampian, Ltd.)*

9. Depending upon the need, various combinations of wire mesh materials (copper, stainless steel, bronze, etc.) and polymers (Teflon, polypropylene, epoxy, etc.) can be used.
10. Although its thermal conductivity is lower than that of the metal, the overall heat transfer coefficient is not adversely affected since the film resistances are lower due to turbulence at the surface (item 6 above) and the fouling resistances are lower due to smooth surface (item 4 above). Metal–plastic composite is a new development and has not yet seen widespread usage. Hence, the limits on the operating temperature and pressure ranges for different combinations of metal–plastics are still under investigation. It does have a bright future, since it fulfills the industries' needs.

NUCLEAR POWER PLANT HEAT EXCHANGERS

5.30 Where are the heat exchangers used in a nuclear power plant?

It depends on the type of nuclear power plant, viz. boiling water reactor (BWR), pressurized water reactor (PWR), whether these use light water or heavy water as a moderator, and the particular design used. The following is a general list of units used, comprising all types of nuclear power plants. In any given plant, some of these might be absent. Multiple units (and stand-by units) are generally used for many of the services. The usage are the following (Ajeet et al., 1977; Oak Ridge National Laboratory, 1975):

1. Steam generators
2. Primary coolers
3. Moderator heat exchangers
4. Emergency condensers
5. Letdown coolers
6. Regenerative exchangers
7. Residual heat removal
8. Spent fuel cooling

9. Component cooling
10. Sampling coolers
11. Shutdown coolers

The design conditions for these heat exchangers vary a lot depending upon their use. For example, the design pressure varies from 100 to over 2700 psig (690–18,600 kPa) and the design temperature from 125° to 680°F (50°–360°C). Accordingly, the number of shell and tube passes, materials of construction, and types of heat exchanger also vary. Table 5.2 summarizes these for the light-water-moderated reactors that use enriched uranium.

5.31 In what important design aspects do the heat exchangers in the nuclear power plants differ from those in the fossil fuel power plants?

Heat exchangers for use in the nuclear power plants should have structural integrity against damage by the nuclear fission products; material selection, fabrication, and welding procedures are closely examined; inspection and testing are tightly controlled; detailed seismic analysis is required; no chloride ion is permitted in water used for hydrotesting since it may lead to corrosion failure; nuclear regulatory commission (or similar other bodies) have to be kept informed of the progress made, etc. The requirements in fossil fuel power plants are less rigid.

5.32 List the criteria for the classification of an exchanger for the nuclear service.

There are two criteria:

1. Whether the components of the exchanger fall within the scope of Section III Nuclear Power Plant Components of the ASME Boiler and Pressure Vessel Code (ASME, 1980)
2. Whether the exchangers must be designed to meet Class-I seismic requirements

5.33 What are the various classes of construction under Section III of the ASME Boiler and Pressure Vessel Code?

There are four sets of construction rules in Section III. These are designated as Code Classes 1, 2, 3, and MC. These apply to pressure-retaining or pressure-containing components of a nuclear power system and containment system including the heat exchangers. These do not provide guidance in the selection of the code class for a specific system. The selection of the code class is based upon the applicable safety criteria or the requirements of the regulatory authorities.

5.34 What are the materials used most frequently in nuclear heat exchangers? What are their characteristics?

The materials used most frequently are the following:

1. Stainless steels: 300 series (austenitic) since their corrosion rate is rather low
2. Carbon and low-alloy steels: These are principally used to provide strength at lower cost while the corrosion resistance is provided by a stainless steel cladding.
3. Inconel: It is used for tubing since it does not undergo stress corrosion cracking.

Table 5.2 Characteristics of nuclear heat exchangers in power reactor systems

			Tube side				Shell side				
System	Reactor type	Heat exchanger	Design pressure,[a] psig	Design temperature,[b] °F	Number of passes	Material	Design pressure,[a] psig	Design temperature,[b] °F	Number of passes	Material	Arrangement
Reactor coolant	GE BWR	None									
	B&W PWR	Steam generator	2500	650	1	Inconel and SS	1050	600	1	CS	Vertical shell and straight tube with superheater section. Fixed tubesheet and orifice baffles
	CE PWR	Steam generator	2485	650	2	Inconel and SS	1000	550	1	CS	Vertical shell and U-tube with fixed tubesheet, and segmental baffles
	W PWR	Steam generator	2485	650	2	Inconel and SS	1085	600	1	CS	Vertical shell and U-tube, fixed tubesheet, and orifice baffles. Steam swirl vanes and moisture separators
Reactor water cleanup (GE), make-up and purification (B&W), chemical and vol. control (CE), chemical and vol. control (W)	GE BWR	Regenerative	1450	575	Multi	SS	1450	575	3	SS	Horizontal shell and U-tube, fixed tubesheet and segmental baffles
	B&W PWR	None									B&W design does not use a regenerative heat exchanger
	CE PWR	Regenerative	2485	650	4	SS	2735	650	2	SS	Vertical or horizontal shell and U-tube with fixed tubesheet and segmental baffles
	W PWR	Regenerative	2735	650	Multi	SS	2485	650	1	SS	Shell and U-tube

	GE BWR	Nonregenerative	1450	575	Multi	SS	150	370	2	CS	Horizontal shell and U-tube with fixed tubesheet and segmental baffles
	B&W PWR	Letdown coolers	2500	650	1	SS	100	350	1	CS	Heliflow design
		Seal return coolers	100	200	1	SS	100	200	1	CS	Shell and straight tube
	CE PWR	Letdown coolers	650	550	12	SS	150	250	1	CS	Vertical or horizontal shell and U-tube. Segmental baffles and fixed tubesheet
	W PWR	Letdown coolers	600	400	Multi	SS	150	250	1	CS	Shell and U-tube
		Excess letdown coolers	2485	650	Multi	SS	150	250	1	CS	Shell and U-tube
		Seal water	150	250	Multi	SS	150	250	1	CS	Shell and U-tube
Residual heat removal (GE), decay heat removal (B&W), shutdown cooling (CE), residual heat removal (W)	GE BWR	Residual heat removal	450	400	Multi	SS	450	400	1	CS	Vertical shell and U-tube
	B&W PWR	Decay heat removal	150	300	1	SS	300	300	1	CS	Shell and straight tube
	CE PWR	Shutdown cooling	500	400	2	SS	150	200	1	CS	Horizontal shell and U-tube with fixed tubesheet and segmental baffles
	W PWR	Residual heat removal	600	400	Multi	SS	150	200	1	CS	Shell and U-tube
Spent fuel cooling and cleanup	GE BWR	Spent fuel cooling	200	150	2	SS	150	150	1	CS	Horizontal U-tube
	B&W PWR	Spent fuel cooling	125	250	1	SS	150	250	1	CS	Shell and straight tube
	CE PWR	Spent fuel cooling	150	200	2	SS	150	200	1	CS	Horizontal shell and U-tube with fixed tubesheet and segmental baffles

Table 5.2 Characteristics of nuclear heat exchangers in power reactor systems (*Continued*)

			Tube side				Shell side				
System	Reactor type	Heat exchanger	Design pressure,[a] psig	Design temperature,[b] °F	Number of passes	Material	Design pressure,[a] psig	Design temperature,[b] °F	Number of passes	Material	Arrangement
Spent fuel cooling and cleanup (*Cont.*)	W PWR	Spent fuel cooling	150	200	2	SS	150	200	1	CS	Shell and U-tube
Reactor building. Closed cooling (GE) component cooling (B&W), component cooling (CE), component cooling (W)	GE BWR	Reactor building closed cooling	150	200	1	SS	150	150	Multi	CS	Horizontal shell and straight tube with fixed tubesheet and segmental baffles
	B&W PWR	Component cooling	100	125	1	SS	150	150	1	CS	Steel and straight tube with fixed tubesheet and segmental baffles
	CE PWR	Component cooling	150	200	1	SS	150	200	1	CS	Shell and straight tube
	W PWR	Component cooling	150	200	1	SS	150	200	1	CS	Shell and straight tube
Sampling	GE BWR	Main steam	1250	575	1	SS	150	200	1	CS	Shell and coiled tube
	B&W PWR	Pressure sample cooling	2500	670	1	SS	150	150	1	CS	Double pipe
	CE PWR	Reactor coolant sample	2485	650	1	Inconel	150	250	1	SS	Vertical double tube coil. No tubesheet or baffles
	W PWR	Pressure liquid sample	2485	680	1	SS	150	350	1	CS	Shell and coiled tube

Source: Oak Ridge National Laboratory (1975).
[a]Pressure, kPa = psi × 6.895.
[b]Temperature, °C = (°F−32)/1.8.

The important characteristics of interest in a material used for nuclear heat exchangers are the following:

1. Low rate of corrosion: Since the circulating corrosion products can become radioactive, posing danger to plant personnel, it is important that the corrosion rate of the material be very low; also it is essential from the point of view of plant integrity.
2. Weldability: The material should have excellent weldability.

5.35 What coolants are used for heat recovery in a nuclear power plant?

The following coolants may be used:

1. Water
2. Gases (helium, carbon dioxide, etc.)
3. Liquid metals (sodium, potassium, etc.)
4. Organic fluids (diphenyl, terphenyl, etc.)

LIQUID METALS AND MOLTEN SALT HEAT EXCHANGERS

5.36 What are liquid metals and molten salt heat exchangers?

These heat exchangers use liquid metals or molten salt as heat transport media in the high temperature range of 500°-2000°F (260°-1100°C) and also have a high heat transfer coefficient. These are used in nuclear power plants, petroleum refineries, and various chemical plants. Their use is also envisaged in collecting solar energy. The liquid metals employed are sodium, potassium, lithium, lead, bismuth, mercury, sodium-potassium alloys, etc. The molten salts include nitrates, chlorides, fluorides, and hydroxides of the alkali metals.

5.37 What are the principal advantages and disadvantages of liquid metals and molten salt heat exchangers compared to the exchangers using steam or high-temperature gases or heating oils as the heat transfer media?

The principal advantages are the following:

1. The heat capacity of the liquid metals and molten salt being significantly higher than that of the gases and steam, the volume of the former to be circulated is significantly less than the volume of the gases or steam. Hence, the sizes of the pipe and the pump can be reduced significantly.
2. Steam will have to be under very high pressure and hence the pipe, pump casing, and heat exchanger wall thicknesses will be much more than those required for the liquid metals and molten salt systems. If high-temperature gases are used, a large area will be needed since their heat transfer coefficient is low.
3. Heating oils (e.g., Dowtherm) may give coking problems at temperatures above approximately 700°F (370°C). No such problem appears in the liquid metals and molten salt systems, which can be operated at much higher temperatures.

4. Thus, the liquid metals and molten salt systems have the advantages of both water and gas cooling in the sense that these have high heat capacity and are used at low pressures. Table 5.3 compares high-temperature liquid units with the conventional ones. The former are very small in size and weight, which can be a significant advantage when space is at a premium.

Table 5.3 Comparison of high temperature liquid metal heat exchangers with more conventional units

	ART[a] fluoride-to-Nak heat exchanger	ART[a] Nak-to-air radiator	PWR[b]-steam boiler	Calder Hall steam generator
Total cost, $/ft³ ($/m³)	320.00 (11300)	21.50 (759)	26.00 (918)	6.22 (220)
Surface area (external), ft² (m²)	1017 (94.5)	14,900 (1384)	9050 (841)	122,000 (11,334)
Volume of shell, ft³ (m³)				16,300 (461.6)
LMTD, °F (°C)	136 (57.3)	865 (462.8)	35 (1.7)	113 (44.7)
Power density, W/cm³ (Btu/s ft³)	266 (7144)	47 (1262)	9.7 (261)	0.16 (4.3)
W/cm³ °F LMTD (Btu/s ft³ °C LMTD)	1.95 (52.4)	0.054 (1.45)	0.28 (7.52)	0.0014 (0.037)
Cost per unit, $	324,000	320,000	235,585	761,000
Total cost, $/lb heat-transfer matrix ($/kg heat-transfer matrix)	344 (758.4)	107 (235.9)	11[c] (24.25)	2.17[c] (4.78)
Total cost, $/kW heat transferred ($/Btu/s heat transferred)	5.90 (6.22)	5.82 (6.14)	3.06 (3.32)	16.50 (17.4)
Total cost, $/kW for an LMTD = 100°F ($/Btu/s for an LMTD of 37.8°C)	8.02 (8.46)	50.30 (53.04)	1.07 (1.13)	18.70 (19.72)
External dimensions; ft		1.5 × 2.5 × 6	4 × 10 × 28	17.3 × 73
(m)		(0.457 × 0.762 × 1.829)	(1.22 × 3.05 × 8.53)	(5.27 × 22.25)
Volume of heat transfer matrix, ft³ (m³)	7.3 (0.2067)	41.4 (1.1723)	280 (7.9287)	10,000 (283)

(*See footnotes on p. 283.*)

Table 5.3 Comparison of high temperature liquid heat exchangers with more conventional units (*Continued*)

	ART[a] fluoride-to-Nak heat exchanger	ART[a] Nak-to-air radiator	PWR[b]-steam boiler	Calder Hall steam generator
Tube O.D., in (mm)	0.23 (5.84)	0.187 (4.75)	0.75 (19)	1.5 and 2.0 (38.1 and 50.8)
Tube length (mean), ft (m)	5.4 (1.646)	3.0 (0.914)	50.0 (15.24)	150 (45.72)
Number of tubes	3120	5760	921	450
Weight of heat transfer surface, lb (kg)	940 (426)	3000 (1361)	20,700 (9389)	350,000 (158,757)
Weight of casing, lb (kg)			118,100 (53,569)	360,000 (163,243)
Total weight, lb (kg)	2100 (952)	4400 (1996)	138,800 (62,959)	710,000 (322,050)
Energy transferred, MW (10^3 Btu/s)	55 (52)	55 (52)	77 (73)	46 (43.6)
Configuration	Helical spiral tube bundles in spherical shell	Round tubes with plate fin	Axial tubes in cylindrical U-shaped casing	Serpentine finned tubes in cylindrical shell
Shell-side fluid flow	Axial counter-flow	Special two-pass cross-flow	Cross-flow	Cross-flow
Structural material	Inconel	Inconel	Carbon and stainless steel	Carbon steel

Source: Oak Ridge National Laboratory, operated by Union Carbide Corporation for the Department of Energy.

[a]Aircraft reactor test (Oak Ridge National Laboratory: components built and tested, but not reactor).

[b]Pressurized water reactor (near Pittsburgh).

[c]Complete with headers, casing, etc.

The principal disadvantages are the following:

1. They are very corrosive, specially in contact with water and oxygen. Corrosion is by several mechanisms, including direct alloying of the tube material by the liquid metal and/or contaminants in it. Hence, the structural material has to be leakproof. No water vapor or oxygen should be allowed to leak in since they increase the rate of corrosion.
2. Any leakage of liquid metals into the atmosphere may cause fire or toxicity hazards.
3. At shutdown, it should be drained or else the liquid metals will solidify on cooling to the ambient temperature. Similarly, preheating is required before start-up.

Table 5.4 Comparison of heat transfer fluids for high temperatures

	Sodium	NaK (44 wt% K)	HTS[a]	Dowtherm[b]	Mercury	Lead	Pb-Bi eutectic
Melting point, °F (°C)	208 (97.4)	65 (18.3)	290 (143)	55 (12.8)	−37 (−38.3)	622 (328.1)	257 (124.9)
Boiling point, °F (°C)							
14.7 psia (101.3 kPa)	1616 (880.3)	1518 (825.4)	–	485 (251.8)	675 (356.8)	3170 (1749)	3038 (1668)
100 psia (689.5 kPa)	–	–	–	490 (254)	908 (486.4)	–	–
500 psia (3447 kPa)	–	–	–	940 (504)	1194 (645.2)	–	–
Liquid density, g/cc (lb/ft^3) at							
melting point	0.9287 (57.95)	0.91 (56.78)	1.97 (122.93)	–	13.5 (842.4)	10.2 (636.48)	10.0 (624)
60°F (15.6°C)	–	–	–	–	12.3 (767.5)	–	–
1000°F (538°C)	–	0.778 (48.55)	1.69	–	–	–	–
Specific heat, Btu/lb °F (J/kg K)	0.33–0.30 (1382–1256)	0.28 (1172.3)	0.373 (1561.7)	0.63–0.69 (liquid) 500–750 F (2637.7–2888.9) at 260–399 C	0.033 (138.2)	0.034 (142.35)	0.035 (146.5) 290–680 F (143–360 C)
Latent heat of vapor, Btu/lb (J/kg)	1980 (4605480)	– –	– –	140–89 (325640–207014)	127 (295402)	365 (849990)	– –
Thermal conductivity Btu/hr ft^2 °F/ft (J/m s K)	49–38 (84.8–65.8) – –	14.16 (24.5) – –	– – – –	0.105 (0.182) 400°–800°F (204°–427°C) –	5–9 (8.65–15.6) – –	8 (13.85) – –	5.3–6.5 (9.17–11.25) 320–610 F (160–321 C)

Heat transfer	6400 (36341)	35000 (19874)	200–2000	200–300	5700	4100	3700
coefficient	–		(1136–11356)	(1136–1703)	(32366)	(23280)	(21010)
Btu/hr ft² °F			(depending on	(condensing)			
(J/s m² K)	–	–	velocity and	–	–	–	–
at velocity, ft/s	10	10	temperature)	–	10	10	10
(m/s)	(3.048)	(3.048)			(3.048)	(3.048)	(3.048)
Latent heat of fusion, Btu/lb (J/kg)	49 (113974)	–	35 (81410)	–	–	11.3 (26284)	–
Viscosity, centipoise							
(Pa s) at melting point	0.70 (0.0007)	–	19 (0.019)	5 (0.005)	1.9 (0.0019)	2.1 (0.0021)	
600°F (316°C)	0.30 (0.0003)	0.24 (0.00024)	2.9 (0.0029)	0.35 (0.00035)	–	–	1.75 (0.00175)
1000°F (538°C)	–	0.12 (0.00012)	1.1 (0.0011)	–	–	1.7 (0.0017)	1.24 (0.00124)
Probable temperature	250–1500	100–1400	400–1000	360–750	–37 to 1000	650–1700	300–1700
range, °F (°C)	(121–816)	(37.8–760)	(204–538)	(182–399)	(–38.3–538)	(343–927)	(149–927)

Source: From Lyon (1952).
[a]Heat transfer salt composed of $NaNO_2$, $NaNO_3$, and KNO_3.
[b]Eutectic mixture of diphenyl and diphenyl oxide.

5.38 How does the design of a liquid metals/molten salt heat exchanger differ from that of a conventional heat exchanger?

The major differences are the following:

1. In flow through a tube, the axial conduction of heat due to the very high thermal conductivity of liquid metals/molten salt is very important, which is not the case with other fluids. Hence the heat transfer coefficient correlations are different.
2. The thermal resistance of the tube wall is an important parameter, since it is of comparable magnitude to the film resistance offered by the liquid metals/molten salt. This is not the case with the other fluids. Hence, the selection of the tube material is based both on its corrosion and thermal resistance characteristics.
3. Since the film resistance offered by the liquid metals/molten salt is very low, even a thin layer of scale will offer a very high resistance to the overall heat transfer. Hence, the choice of the value of the fouling resistance in a design in this case is more important than is the case with the other fluids.
4. Corrosion and leakage are significantly more detrimental in the case of liquid metals and molten salt compared to that with the other fluids (except the radioactive ones), and hence the choices of materials of construction and the mechanical design and fabrication techniques have to be done very carefully. Pipes and fittings also have to be suitably chosen.

5.39 What factors govern the choice of a specific liquid metal or molten salt system?

The important factors are the following:

1. Melting point
2. Boiling point
3. Liquid properties (density, specific heat, thermal conductivity, and viscosity)
4. Desired temperature of operation
5. Overall heat transfer coefficient

A number of liquid metals, molten salts and Dowtherm have been compared in Table 5.4.

NOMENCLATURE

ART	aircraft reactor test
BWR	boiling water reactor
E-CTFE	ethylene-chlorine-tetrafluoroethylene
ETFE	ethylene tetrafluoroethylene
FEP	fluorinated ethylene propylene
FRP	fiber reinforced plastic
HTS	heat transfer salt
L.D.	Litz Donawitz

LAS	lithium aluminosilicates
MAS	magnesium aluminosilicates
MPC	mesh-plastic composite
NaK	sodium-potassium salt
PFA	perfluoroacetyl
PTFE	polytetrafluoroethylene
PVDF	polyvinylidenefluoride
PWR	pressurized water reactor
RCC	reinforced concrete
TFE	tetrafluoroethylene

REFERENCES

Ajeet, J. B., A. K. Asundi, and P. Jagadishan. Consideration in the Design and Fabrication of Nuclear Heat Exchangers. Paper presented at a seminar on Heat Exchangers and Pressure Vessels, Bombay, April 1977. Ind. Chem. Mfr.: Bombay. 1977.

American Society of Mechanical Engineers. *ASME Boiler and Pressure Vessel Code: Section III.* ASME: New York. 1980.

Anonymous. Metal Plastic Composite Shows Promise in Heat Exchangers. *Des. Eng. (Lond.)* pp. 11 and 15. August 1980.

Coombs, M. G. *Development of a Ceramic Heat Exchanger for a Closed-Cycle Gas Turbine Engine.* Paper No. 79-GT-59. ASME: New York. 1979.

Corning Glass Works. *Model C. Tubular Heat Exchangers,* Bulletin PE-240. Corning Glass Works: Corning, New York. 1975.

Curwen and Newbery Ltd. *EVERTUBE Glass Tube Heat Exchangers,* Catalog. Curwen and Newbery Ltd.: Westbury, Wiltshire, England. 1980.

duPont de Nemours International. *Heat Transfer Products of TEFLON FEP,* Bulletin A-98366. duPont de Nemours International: Geneva, Switzerland. 1976.

E. I. duPont de Nemours Co. *Heat Exchangers: Shell-and-Tube Type,* Bulletin E-09547. E.I. duPont de Nemours & Co. Inc.: Wilmington, Delaware. 1977.

Fraas, A. P., and M. N. Ozisik. *Heat Exchanger Design.* Wiley: New York. 1965.

Hillard, A. *Third Conference on Industrial Carbons and Graphite,* pp. 562–564 (1970). Soc. Chem. Ind.: London. [Reference noted from Chisholm, D. (ed.). *Developments in Heat Exchanger Technology 1.* Applied Science: London. 1980.]

Larson, R., and F. Kreith (eds.). *Workshop on the Research Needs of Heat Transfer Processes and Equipment.* SERI: Golden, Colorado. 1981.

Linde Aktiengesellschaft. *Tubular Heat Exchangers,* Brochure VI.3.7.e/77. Londe AG: Munich. 1977.

Lyon, R. N. (ed.). *Liquid Metal Hand Book.* ONR, AEC and Bureau of Ships, Superintendant of Documents: Washington, D.C. June 1952.

Oak Ridge National Laboratory. *Design Guide for Heat Transfer Equipment in Water-Cooled Nuclear Reactor Systems,* ORNL TM-3578. Oak Ridge National Laboratory: Oak Ridge, Tennessee. 1975.

Perry, R. H., and C. H. Chilton (eds.). *Chemical Engineers' Handbook,* 5th ed. McGraw-Hill: New York. 1973.

Scaccia, C., and G. Theoclitus, Heat Exchangers–Types, Performance and Applications. *Chem. Eng.* 87(20):121–132. 1980.

Smith, F. T. Cross-Flow Heat Exchange through Bundles of Fluorocarbon Resin Tubes. *Heat*

Transfer 1974. Proceedings of the Fifth International Heat Transfer Conference, Tokyo, September 1974. Vol. 5, p. 228. Hemisphere: Washington, D.C. 1974. D.C. 1974.

Tennery, V. J. United States Ceramic Heat Exchanger Technology–Status and Opportunities. In *Heat Exchangers: Theory and Practice.* J. Taborek, G. Hewitt, and N. H. Afgan (eds.). Hemisphere: Washington, D.C. 1983.

Whitfield, M. Composite Alternatives in Heat Exchanger Materials. *Process Eng.* No. 1:57. 1981.

Wolff, F., D. V. Eyre, and M. Grenier. Oxygen Plants: Ten Years of Development and Operation. *Chem. Eng. Prog.* 75(7):83–87. 1979.

CHAPTER

SIX

PLATE AND SPIRAL PLATE HEAT EXCHANGERS

INTRODUCTION

- Shell-and-tube exchanger is the most versatile process equipment used under various conditions. However, there are some situations where it is either not suitable, such as in the dairy industry where low hold-up volume and fast chilling are required, or is rather costly, such as in cooling duties using sea water requiring titanium tubes. In such cases, a plate heat exchanger (PHE) proves quite useful. There are other PHE applications too. While they have been used for a long time in the dairy industry, their use in the process industries is of a relatively recent origin.

Sometimes there are fluid streams containing suspended fibers or particles that have to be cooled. The tubes of a shell-and-tube unit are prone to fouling and plugging due to the suspended material. In such cases, a spiral plate heat exchanger (SPHE) can be used with advantage since it does not easily get plugged. There are other SPHE applications too.

Thus, it is important that both the PHE and SPHE be discussed in detail regarding their construction, uses, limitations, and comparison. This chapter is devoted to them. While we first discuss the PHE, the tables given during the discussion on PHE contain information on the SPHE also, for the sake of completeness and to avoid repetition later.

PLATE HEAT EXCHANGERS

Advantages, Applications, and Limitations

6.1 What is a plate heat exchanger (PHE)?

It is an exchanger consisting of flow passages made of corrugated plates wherein the two fluids flow in countercurrent fashion on either side of each plate. A PHE has a stack of plates clamped together with gaskets between them (Fig. 6.1). The distance between the plates is decided by the geometry of the corrugations. On bolting up, the corrugations on successive plates form narrow flow channels.

6.2 What are its advantages?

These are the following:

1. It is easily assembled and disassembled.
2. It can be easily cleaned both chemically and mechanically.
3. The heat transfer area can be changed as required to satisfy heat duty or process changes by changing the number of plates.
4. It can be used for multiple services at the same time.
5. Any leak is immediately detected, since all the plates are vented to the atmosphere and the leaking fluid falls on to the floor instead of mixing with the other fluid; the bad gasket can be easily changed.
6. The heat transfer coefficient is larger and hence the heat transfer surface area required for a given heat duty is smaller than in a shell-and-tube unit.
7. The floor space required is usually half or even less than that required by a shell-and-tube exchanger for the same heat duty.
8. It has less fouling deposit due to high turbulence, even at low Reynolds number.
9. Very close temperature approach can be obtained if desired.
10. It meets the most stringent sanitary and hygenic conditions.
11. The hold-up volume is low.
12. When the fluids are in countercurrent flow, with equal number of passes for each fluid, the LMTD is almost fully utilized.
13. When construction materials other than the carbon steels are required, the PHE is usually more economical than a shell-and-tube exchanger.

6.3 What are some of the applications of the PHE?

The majority of the PHEs are used because of the high turbulence they create at a low Reynolds number. Some of the successful applications are in the following:

1. Breweries and wineries, for pasteurization of beer and wines
2. Yeast factories, for cooling of fermenting tanks, yeast concentrates, and molasses
3. Food and dairy industries, for pasteurizing, heating, and cooling
4. On ships, to cool lube oil and jacket cooling water, etc.
5. Polymerization industries, for monomer stripping, solids concentrating, and drying of emulsion polymerization products.

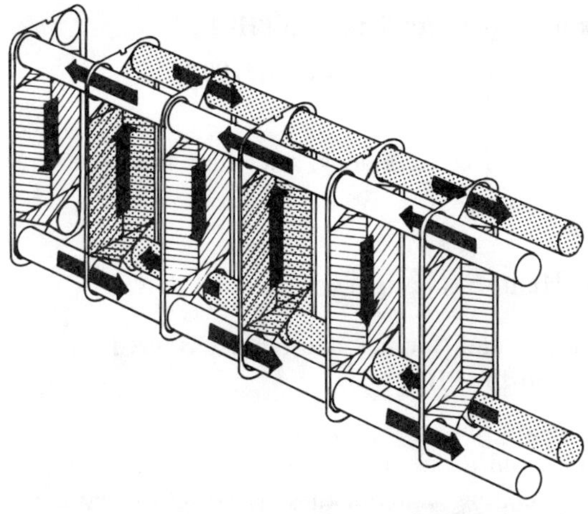

Figure 6.1 Plate heat exchanger (PHE). *(Courtesy APV Company, Inc., B1.)*

6. Synthetic fiber and rubber industries
7. Paper and pulp mills, for assorted duties
8. Aluminum industry, for cooling of rich sodium aluminate solution and heating of strong caustic lye
9. Centralized cooling systems using sea water on off-shore oil-drilling platforms and in industries located near the coast.
10. In secondary circuits of nuclear power plants
11. Many other industries, such as sulfuric acid and steel

6.4 What are the limitations of the PHE?

The PHE is not suitable for the following:

1. Pressure >300 psi (2.07 MPa) due to limits on plates, frames, and gasket sealing.
2. Temperature >175°C (350°F) due to the gasket material. In small units compressed asbestos gaskets may be used to 260°C (500°F).
3. Very viscous fluids and fluids whose viscosity changes radically during passage through the unit, since flow maldistribution may result
4. Very low flow rates (0.1 m/sec or 0.33 ft/sec velocity) which result in flow maldistribution (see also Q. 6.29).
5. Very high flow rates unless the pressure drop in the manifolds or ports is kept to within 20-25 percent. Consider using more than one PHE.
6. Air coolers, since the pressure drop is high.
7. Gas-gas service due to high pressure drops.
8. Water-water duties where carbon-steel shell-and-tube unit might do. Sometimes a PHE may be suitable even in such a case. An economic analysis should be done.
9. Liquids containing large suspended particles or fibers.
10. Condensing duties at low pressures since the pressure drop is high.
11. Evaporator service, although some plate units with limited duty are in use now. (APV, B1)

6.5 What factors influence the maximum pressure limit of a PHE?

These are one or both of the following (Auth and Loiacono, 1978):

1. Plate deformation due to high fluid pressure
2. Gasket retainment in grooves providing effective sealing, since high fluid pressure tries to open the gasketed joint

Comparison with Shell-and-Tube Heat Exchangers

6.6 Can the same corrosion, erosion, and fouling allowances as used in the shell-and-tube exchangers be used in the PHE?

It would actually be disastrous to use the same values.

For corrosion, whereas a rate of 5 mil (0.125 mm)/year may be acceptable in most shell-and-tube exchangers where the tube wall thickness is 50 mil or more, in the PHE with plates generally of only 25 to 30 mil thickness, a maximum corrosion of usually only up to 2 mil/year is acceptable. Hence the plate material should be suitably chosen. An advantage of the PHE is that the whole surface is likely to corrode evenly since the flow is uniform, unlike the shell-and-tube units where corrosion is generally localized.

For erosion damage, while in the tubes only the leading edge and entry region may be susceptible due to the high turbulence and flow contraction, in a PHE these conditions prevail all over the plate. Erosion in a given PHE is likely to result from operation far above the design capacity and with fluids containing abrasives. Hence, the plate material should be chosen to resist both the corrosion and erosion damages. Hastelloy and titanium are among such materials.

For fouling, because of the high turbulence and lack of stagnant regions in a PHE, the fouling factors used should be about one-fifth of those used in the shell-and-tube exchangers. Plate heat exchangers can be cleaned easily both chemically and mechanically.

6.7 Why is fouling lower in the case of a PHE compared to that in a shell-and-tube heat exchanger?

The several factors that influence fouling in a PHE and help keep it low are the following (Raju and Bansal, 1980):

1. Highly turbulent flow keeps boundary layer very thin and particles in suspension. This occurs despite the fact that the average velocities in a PHE are 0.5–0.8 m/sec (1.6–2.6 ft/sec) while those in the tubular units are 1 m/sec (3.3 ft/sec) or higher.
2. Low residence time for crystal formation
3. Good fluid distribution with almost no dead zones or low velocity zones
4. Higher alloy plates that corrode at a very low rate. Thus, the corrosion product is almost absent and this reduces the fouling. In shell-and-tube exchangers, the corrosion product can settle down as particulate deposit, and the corroded sites provide favorable points for the fouling deposit.

Table 6.1 Typical fouling factors for plate heat exchangers

Fluid	Fouling factor, m² °C/W[a]
Water	
Demineralized or distilled	0.00017
Soft	0.00034
Hard	0.00086
Cooling water (treated)	0.00069
Sea (coastal)	0.00086
Sea (ocean)	0.00052
River, canal	0.00086
Engine jacket	0.00103
Oils, lubricating	0.00034 to 0.00086
Oils, vegetable	0.00034 to 0.00103
Solvents, organic	0.00034 to 0.00052
Steam	0.00017
Process fluids, general	0.00017 to 0.00103

Source: Raju and Bansal (1983).
[a]1 m² °C/W = 5.678 ft² hr °F/Btu.

5. The plate surface though corrugated is very smooth and hence deposits take longer time to form.
6. Simplicity in cleaning the deposit by chemical and/or mechanical means allows for lower design fouling factors. Some typical values of the fouling factors for an economic pressure drop of 29.4 kPa (4.27 psi) are given in Table 6.1.

Various design considerations for reducing different kinds of fouling in a PHE have been discussed by Novak (1983).

6.8 What is the size of the particles that can pass through the space between two adjacent plates in a PHE?

Figure 6.2 shows a typical configuration of grooves between two adjacent plates of a PHE. The largest particle size to pass through the space is approximately

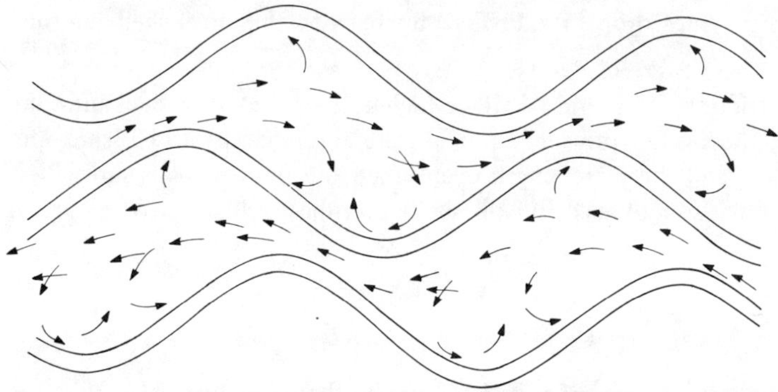

Figure 6.2 Available space for particulate flow in a PHE. *(Courtesy APV Company, Inc., B1.)*

equal to the depth of the corrugations. However, for ease of flow, the particle size should be kept about half the depth of the corrugations. Larger particles should be screened or filtered out; or other types of exchangers, such as the spiral plate ones (SPHE) (Q. 6.46) should be used instead.

6.9 How do the PHEs and shell-and-tube exchangers compare as far as their use under different conditions is concerned?

This is summarized in Table 6.2.

6.10 How do the PHEs and shell-and-tube exchangers compare as far as their cleaning and maintenance are concerned?

It is summarized in Table 6.3 for a general situation. The arguments may vary slightly depending upon the specific shell-and-tube construction.

6.11 How doe the PHEs and shell-and-tube exchangers compare in cost?

The cost depends on the type of construction material, and a comparison can be made only for the situations where a PHE can be used (Q. 6.9). Table 6.4 gives a comparison of the relative costs. The relative values change with the size (Fig. 6.4). This is only an approximate figure since the cost of a shell-and-tube unit also depends upon the type of the front and rear heads, tubesheets, etc.

6.12 Are there some points for comparison between the PHEs and shell-and-tube exchangers not covered in Tables 6.2–6.4?

Yes, and these are given in Table 6.5.

6.13 How does the friction factor in a PHE compare with that in a shell-and-tube exchanger?

The friction factor for the turbulent flow in a PHE is 10 to 400 times more than that for flow inside the tubes of a shell-and-tube exchanger at the same Reynolds number (Marriott, 1971). Relations between the Reynolds number, friction factor, and the heat transfer (Colburn) factor are shown in Fig. 6.5 for a typical PHE.

6.14 How does the pressure drop in a PHE compare with that in a shell-and-tube exchanger?

Although the friction factor in a PHE is higher than that in a shell-and-tube exchanger (Q. 6.13), the overall pressure drop is generally lower in a PHE since the nominal velocity is low and the plate length over which the fluid flows is short.

An overall comparison of a PHE and shell-and-tube exchanger for a typical case is given in Table 6.6.

Parts of a PHE

6.15 What are the important parts of a PHE? Describe their functions also.

These are shown in Fig. 6.6 and described in Table 6.7.

Table 6.2 Comparison of the uses of SPHE, PHE, and shell-and-tube heat exchangers

Item	Spiral plate heat exchangers	Plate heat exchangers	Shell-and-tube exchangers
Pressure limit	1500 kPa (215 psi)	2100 kPa (300 psi)[a]	No limit
Temperature limit	No limit	150°C (300°F), up to 260°C (500°F) for specific duties[b]	No limit
Low flow rates	Induces turbulence at Re >800	Preferable since induces turbulence at low Re (⩾10) because corrugations break down the boundary layer. Hence heat transfer coefficient is high. However, very low flow velocities (⩽0.1 m/sec, 0.033 ft/sec) result in maldistribution and low efficiency	Induces turbulence at high Re (>2000) in the tubes
Gaseous flow	Preferable in cross-flow units where ΔP is within limits	ΔP is high because of limited entry port diameter and small gap between the plates. Hence restricted use	Preferable since ΔP is within limits
Condensing	Preferable since ΔP is within limits	ΔP is high for the same reasons as above, and also due to induced turbulence in the condensate, although for highly corrosive liquids, it may prove more effective	Preferable since ΔP is limited
High viscosity fluids	Preferable. Film coefficient is higher than in shell-and-tube unit	Flow maldistribution is possible specially when being cooled[a]	Preferable
Cooling duties using sea water	Cost can be very high	Preferable since cost is within limits	Cost can be very high since costly materials are used
Very corrosive liquids where no metals can be used	Cannot be used	PHE cannot be constructed. In some cases tantalum may be used	Graphite, glass, or Teflon units are used
Heat recovery	Preferable since close temperature approach is possible	Preferable since close temperature approach of up to 1°C (2°F) is possible and up to 90% heat may be recovered[a,b]	Close temperature approach requires very large surface area
Heat-sensitive liquids and dairy products	Hold-up volume is high	Preferable since the hold-up volume is low	Hold-up volume is high
Air cooling	Not economical	Cannot be done economically	Finned tubes with fans blowing air are used

(*See footnotes on p. 296.*)

Table 6.2 Comparison of the uses of SPHE, PHE, and shell-and-tube heat exchangers (*Continued*)

Item	Spiral plate heat exchangers	Plate heat exchangers	Shell-and-tube exchangers
General duties within the applicable limits of temperature and pressure for PHE	Preferable with slurries and highly fouling liquids	Preferable if any metal or alloy except carbon steel is required	Preferable in most cases where carbon steel can be used
Streams with suspended particles	Most efficient	Can be used only with very small-size particles, or else the space between the plates will get clogged	Can be used although suspended particles should be avoided if possible
Limited pressure drop case	Design can be done to honor the pressure limit	Friction factor is high, although the flow path length is small; hence a very careful design may meet the limited pressure drop requirement	Preferable since the design can be suitably modified
Liquids forming explosive mixture with air or containing toxic/radioactive material	Preferable	Avoid since leakage is into the atmosphere	Preferable (except in air coolers)
Liquids forming explosive mixture with one another	Preferable since there is no interleakage	Preferable since there is no interleakage	Double tubesheet will have to be used. Cost is high
Heat exchange between three or more fluids	Not possible	Easily possible with connector plates (Fig. 6.3)[c]	Not easily possible
Evaporation	Can be used as kettle and thermosyphon reboiler (although not done often)	Can be used for low evaporation duties	Can be used for all evaporation duties
Unequal flow rates of similar fluids	Flow channels can be sized to suit the flow rates and no extra inefficiency is introduced by unequal flow rates	Most efficient for equal flow rates, although it will function without substantial LMTD penalty for unequal flow rates up to 1.4 : 1. Outside this range, unsymmetrical channel arrangement may be used to reduce LMTD penalty.	Flow areas can be sized to suit the flow rates,and no extra inefficiency is introduced by unequal flow rates

[a] Raju and Bansal (1980).
[b] APV (B2).
[c] Raju and Bansal (1983).

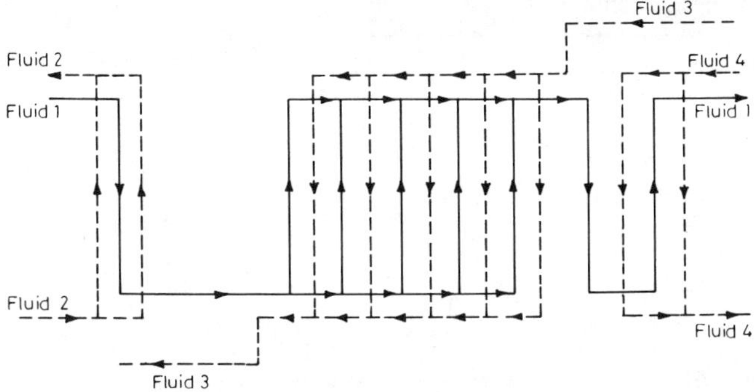

Figure 6.3 Multiple duties.

6.16 What are the typical range of dimensions of a PHE?

These are given in Table 6.8.

6.17 Do codes exist for the manufacture of a PHE?

There are no codes for the most important part of a PHE, vis. the plates. This is due to the fact that the data and the techniques are proprietary of a few companies only. Only the load-bearing components, such as the frame, can be designed as per the existing codes for the pressure vessels (APV, B1).

Table 6.3 Heat exchanger inspection, maintenance, and cleaning

	Spiral[a]	Plate[a]	Shell-and-tube[a]
Fouling inspection			
One side	b	a	b
Both sides	b	a	b or d
Leakage inspection			
One side	b	b	a
Both sides	b	b	a
Corrosion inspection			
One side	b	a	b
Both sides	b	a	b or d
Chemical cleaning			
One side	a	a	b
Both sides	a	a	b or d
Manual cleaning			
One side	b	a	b
Both sides	b	a	b or d
Extension	d	a	d
Repair	c	a	b

Source: Walker (1980) and A. Moleterno, personal communication (1982).

[a]Symbols: a, very good; b, acceptable; c, poor; d, impossible.

Table 6.4 Relative costs of heat exchangers

Plate: stainless steel: rubber gaskets	1
Spiral plate: all carbon steel	1.25
Shell-and-tube: all carbon steel	0.93
Plate: stainless steel: compressed asbestos fiber gaskets	1.3
Shell-and-tube: carbon steel shell, stainless steel tubes	1.5
Spiral plate: all stainless steel	2.5
Shell-and-tube: all stainless steel	2
Plate: titanium plates	2
Plate: Hastelloy plates	8

Source: Walker (1980) and A. Moliterno, personal communication (1982).

Materials Used

6.18 Can all the materials that are used in the shell-and-tube exchanger be used in a PHE?

No. The use in a PHE is restricted to only such materials as can be cold pressed. This rules out carbon steel, glass, graphite, etc. In fact, carbon steel can be pressed but will require a different set of tools than that required for the other materials. Furthermore, a thick plate will have to be used to provide for the corrosion allowance.

6.19 What materials are used in a PHE?

The plates can be made of all the common metals and alloys whose thin sheets can be cold-worked. This includes stainless steel, titanium, nickel, aluminum brass, Incoloy, Hastelloy, etc. In a few cases, tantalum and Monel are also used. However, carbon steel is not used as pointed out earlier. Since no welding is involved, the weldability of the plate material does not matter. Table 6.9 lists some of the materials used for different applications.

The frames are generally made of carbon steel with a chemically resistant paint. Alternately, these can be made of stainless steel if necessary.

The gaskets are made of various elastomers for general services and operating

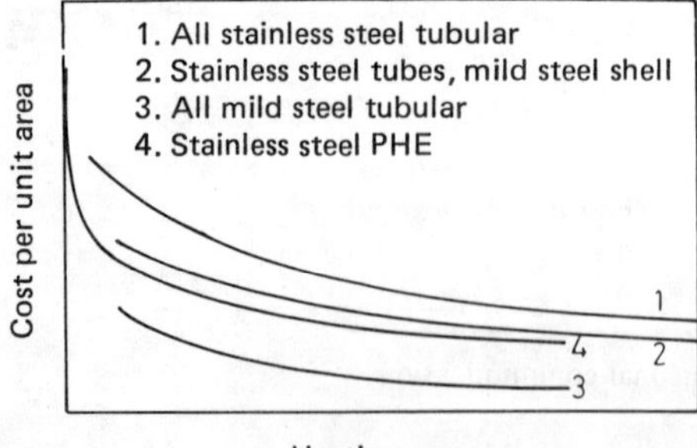

Figure 6.4 Cost comparison.

Table 6.5 General comparison among SPHE, PHE, and shell-and-tube heat exchangers

Item	Spiral plate heat exchanger	Plate heat exchanger	Shell-and-tube exchanger
Space	Occupies less space	Occupies less space	Occupies more space
Foundations	Generally sits over other equipment	Since the unit is generally lighter, no elaborate foundations are needed	Strong and deep foundations are needed
Adaptability to changes in heat duty	It cannot easily be adapted to changes in heat duty	It can be adapted for more or less heat duty with ease	Such adaptability is not easy
Heat losses	Heat losses are not significant since cooler fluid can be placed in the outer channel	Since only the plate edges are exposed, negligible heat is lost to the surroundings. No insulation is needed.	Heat losses are not significant unless temperature on the shell side is very high (or if very low, then heat gain occurs). Insulation may be used then.
Gasket material	Asbestos, spiral wound metal	Elastomers, asbestos	Asbestos, PTFE, metallic, elastomers
Maximum flow rates	0.1 m^3/sec (13,000 ft^3/hr)	0.7 m^3/sec (89,000 ft^3/hr)	Very high
Liquids with large amounts of suspended solids	Excellent	Likely to clog the passage if particle size is large	Likely to clog the tubes
Number of passes possible	Single	Multi-pass	Multi-pass
Maximum surface area (approximate)	250 m^2 (2700 ft^2)	2500 m^2 (27,000 ft^2)	Very high
Additional area easily provided in the same unit	No	Yes	No

Table 6.5 General comparison among SPHE, PHE, and shell-and-tube heat exchangers (*Continued*)

Item	Spiral plate heat exchanger	Plate heat exchanger	Shell-and-tube exchanger
Shipment	Fully assembled and tested unit is shipped	Generally, the completely assembled and pressure tested unit is shipped unless it is too large to be handled by the available transport. In such a case, the plates, frame, and connections are crated in pieceform and shipped. Assembly and testing is done at site.	Generally, the completely assembled and pressure-tested unit is shipped. It is heavy and bulky and requires careful handling.
Cleaning	Complete cleaning on both sides in most units	Complete cleaning on both sides. U-type installations (Fig. 6.10) has all four connections at one end and no piping need be removed for maintenance and cleaning.	In fixed tubesheet units, the shell side cannot be completely cleaned.
Stagnant zones	No stagnant zones since fluid flows over the total surface	No stagnant zones since fluid flows over the total surface	Possibility of stagnant zones exists behind the baffles, thus part of the surface may not be fully effective in heat transfer
Fouling deposit	Very low due to unique flow path	Very low since fluid is in turbulent flow	Relatively higher fouling since the intensity of turbulence is lower
Space for maintenance and repair	No extra space needed except to remove the covers	No extra space needed	Space at least equivalent to the tube length is needed in front of the exchanger
Easy access for cleaning	Yes	Yes	Not in all units

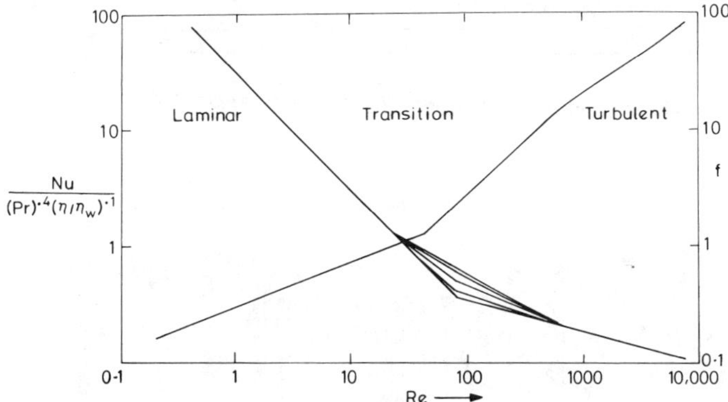

Figure 6.5 Friction factor and heat transfer coefficient for a typical PHE.

pressures up to 300 psi (2.07 MPa). For corrosive liquids and temperature up to 500°F (260°C), compressed asbestos fiber (CAF) gaskets are used (APV, B2). Table 6.10 lists the characteristics of common gasket materials.

6.20 Why is a higher alloy generally needed for the plate heat exchanger compared to shell-and-tube exchanger for the same set of fluids?

It is because the plates are rather thin, and one cannot allow the same corrosion rate as for a shell-and-tube exchanger. Hence, for the plates a higher alloy with lower corrosion rate is used. For example, while type 316 stainless steel might

Table 6.6 Comparison between PHE and shell-and-tube exchanger for a typical case[a]

Item	PHE	Shell-and-tube exchanger
Heat transfer area, ft² (m²)	4820 (448)	13,100 (1217)
Heat transfer coefficient, (clean) Btu/hr ft² °F (W/m² °C)	734 (4168)	386 (2192)
Heat transfer coefficient, (design) Btu/hr ft² °F (W/m² °C)	641 (3640)	271 (1539)
Effective mean temperature difference, °F (°C)	10 (5.5)	8.7 (4.8)
Pressure drop; hot/cold, psi (kPa)	4.6/4.7 (31.7/32.4)	15/15 (103.4/103.4)
Fouling resistance: hot/cold, hr ft² °F/Btu (m² °C/W)	0.0001/0.0001 (0.00057/0.00057)	0.0005/0.0005 (0.0028/0.0028)
Pass arrangement	1/1	4 on tube side, 3 shells in series
Material of construction	Stainless steel	Mild steel
Approximate price (for the particular year)	$66,000	$91,800

Source: APV (B3).

[a]Duty: To heat 1,300,000 lb/hr (163.7 kg/sec) of water from 73° to 97°F (23.8° to 36.1°C) using an equal amount of hot water at 107°F (41.7°C) with available pressure loss of 15 psi (103.4 kPa) for both the streams. Heat recovery = 70.5%, NTU = 2.4.

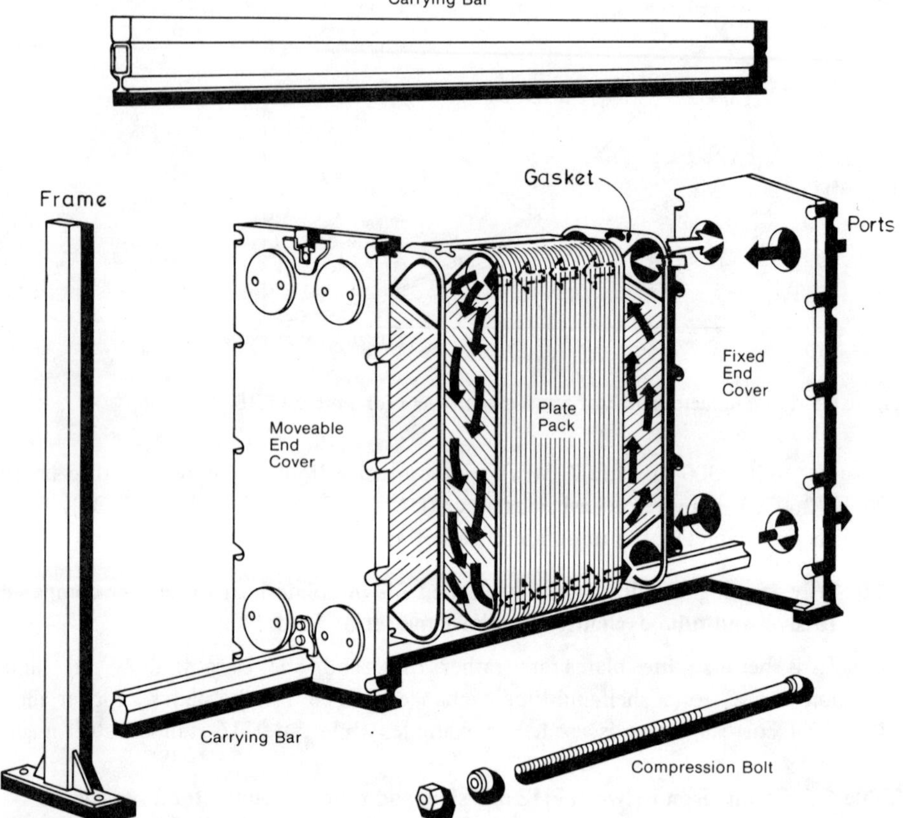

Figure 6.6 Parts of a PHE.

be used for the tubes of a sulfuric acid cooler, one might be required to use an alloy containing 25 Ni, 20 Cr, 4 Mo, 2 Cu alloy or even Incoloy 825 (40 Ni, 25 Cr, 3 Mo, 2 Cu) (Cowen, 1980) for the same duty in a PHE. Also the plates are not stress relieved and are therefore prone to stress corrosion. Hence, a higher alloy is used since it is subject to less stress corrosion.

6.21 Is the reverse also true, that is, a lower alloy might do for a PHE compared to that for the tubes?

There are some such situations also. Due to "concentration" corrosion occurring in the scale deposit on a tube, such as a steam heating coil sitting at the base of a storage vessel, the tubes fail very quickly if they are the type 316 stainless steel. Hence, these have to be made out of Carpenter 20 or Incoloy 825 alloys. However, this whole heater can be replaced by a type 316 stainless steel PHE, since the latter induces turbulence at low Reynolds number and works at low temperature difference, and hence no concentration corrosion occurs (Cowen, 1980).

Manufacture of a PHE

6.22 How is the plate of a PHE made?

The plate is pressed out of a thin, flat sheet 0.5 to 1 mm (20 to 40 mil) thick, depending upon the pressure and corrosion conditions. The pressing gives it the needed corrugations that impart turbulence to the fluid and help the plate to withstand distortion due to high pressures. Deep depressions are formed all around the plate perimeter so that the gaskets can be put in them to seal the plate assembly. There are four circular ports, one in each corner of the plate for the entry and exit of the fluids. The two fluids flow in countercurrent, one on either side of each plate. The plates are made in different sizes and materials with different types of corrugations that are proprietary of the manufacturers. The thermal performance also varies with the type of corrugations.

6.23 What are the important effects of corrugations?

These are the following:

1. Corrugations break down the boundary layer, produce turbulence at low Reynolds number, and significantly increase the heat transfer coefficient.
2. Corrugations between adjacent plates contact at definite points to form flow channels and increase the rigidity of the thin plates to withstand high pressures and to minimize plate deflections.

Table 6.7 Important parts of a PHE and their functions

Part	Function
Plates	Provide flow passages for fluids and surface area for heat transfer
Gaskets	Keep the two fluids from mixing or leaking out, and help in distributing them between alternate plates
Ports	Provide entry and exit areas for the fluids
Frame	Holds the whole unit together
Pressure plates or end plates	Provide pressure to keep the gasketed spaces sealed after the unit has been bolted up. These do not contact the fluids
Tightening bolts or compression bolts	These are tightened with nuts to keep the whole unit intact, leakproof, and in working condition. Larger units may be tightened hydraulically. The plates are compressed to a specified plate pack length. For those types that are not in metal-to-metal contact at all times (as is the majority), maximum and minimum plate pack lengths are given by the manufacturer. When the minimum is reached, the gaskets are replaced.
Carrying bars	There are two carrying bars: an upper and a lower one. The plates slide over these and are kept in upright position. The upper one carries the weight of the plates. The carrying bars are usually longer than the stack of plates so that one can loosen the bolts and slide the plates to inspect and clean them.
Connecting plates	Provide for multiple heat duties in the same PHE (Fig. 6.3). These do not contact the fluids.

Table 6.8 Typical range of dimensions and other useful data[a]

Plate	
Thickness	0.5–1 mm[b] (0.02–0.04 in)
Width	0.05–1.2 mm (0.16–4 ft)
Height	0.3–3.5 m (1–11.5 ft)
Area	0.032–3.63 m^2[c] (0.35–40 ft^2)
Height of corrugations	3–7 mm (0.12–0.28 in)
Contact points	For every 150–2000 mm^2 (0.2–3 in^2)
Gap between plates	1.5–5 mm (0.06–0.2 in)
Total surface area[d]	Up to 2500 m^2 (27,000 ft^2)
Number of plates/unit	Up to 700
Surface area/volume	120–225 m^2/m^3 (37–69 ft^2/ft^3)
Hydraulic mean diameter	3–10 mm (0.12–0.4 in)
Pressure plate thickness	20–100 mm ($\frac{3}{4}$–4 in)
Tightening bolt diameter	12–75 mm ($\frac{1}{2}$–3 in)
Carrying bar	
Length	Up to 3 m (10 ft)
Width	25–150 mm (1–6 in)
Thickness	25–350 mm (1–14 in)
Entry and exit port diameters	Up to 400 mm (16 in)
Gasket	
Thickness	Equal to the corrugation depth
Width	Normally 2 to 3 times the thickness
Flow rates	
Liquid	0.7 m^3/sec (89,000 ft^3/hr), maximum linear port velocity 7 m/sec (23 ft/sec) for water, 3 m/sec (10 ft/sec) for corrosive liquids
Gases at Standard Conditions	3 m^3/sec (380,000 ft^3/hr)
Clearance between the bottom of plates and the ground	0.1–0.2 m (0.3–0.6 ft)
Operation	
Pressure	0.1–1.5 MPa (15–215 psi)
Temperature	–25° to 150°C (–15° to 300°F) with rubber gaskets –40° to 260°C (–40° to 500°F) with compressed asbestos fiber (CAF) gaskets
Performance	
Temperature approach	As low as 1°C (2°F)
Number of transfer units (NTU)	0.3 to 4 per pass
Heat recovery	As high as 80% in a 2-pass PHE. Higher recovery is possible with a larger number of passes but may not be economical
Heat transfer coefficients	3000 to 7000 W/m^2 °C (530 to 1230 Btu/hr ft^2 °F) with water–water units

Source: Raju and Bansal (1980) and Raju and Bansal (1983).

[a]Due to continuous developments, the latest information should be obtained from the supplier when considering these units.

[b]With very close-pitched corrugations and where a low design pressure is called for, 0.3 mm (0.012 in) may be used.

[c]Bell (1981).

[d]It is the developed surface area after pressing and is always more than the flat plate area of the same overall dimensions.

Table 6.9 Some material selections for PHE

Application	Material
Sea or brackish water	Titanium
All other water	Type 316 stainless steel
Dilute acids, up to 70°C (160°F)	Titanium, titanium–palladium alloy
Concentrated H_2SO_4	Hastelloy–C
Chloride solutions	Stainless steel for very dilute (<200 ppm Cl_2), titanium for the rest
Caustic solutions (50 to 70%)	Nickel, Incoloy 825
Wet chlorine, chlorinated brines, hypochlorite solutions	Titanium

Source: Raju and Bansal (1980) and Cowen (1980).

Table 6.10 Characteristics of some gasket materials

Gasket material	Temperature °C (°F)	Service	Remarks
Styrene-butadiene rubber	To 85 (185)	General purpose for aqueous systems.	
Acrylonitril-butadiene rubber	To 140 (285)	Excellent resistance to aqueous systems, fats and aliphatic hydrocarbons	
Ethylene-propylene ter polymer	To 150 (300)	High temperature resistance for a wide range of chemicals	
Resin-cured butyl rubber	To 150 (300)	Excellent resistance to wide range of chemicals including acids, alkalies, ketones, and amines	Poor resistance to fats
Silicone rubber	Low	Limited to sodium hypochlorite and low temperatures generally	
Fluoro-elastomer	Above 150 (300)	Excellent for oils	Use limited by high cost
Compressed asbestos fiber (CAF)	To 260 (500)	Resistance to organic chemical mixtures	Plates and frames must be stronger because of poor elasticity. The use of CAF is discouraged since it is difficult to maintain the seal.

Source: Raju and Bansal (1980). Excerpted by special permission from *Chemical Engineering* August 11, Copyright © 1980, by McGraw-Hill, New York, N.Y. 10020.

6.24 How are the gaskets made and used?

Gaskets are made in one piece to fit into the deep grooves around the perimeter of the plate and around the port holes. Gaskets are a very important part of the heat exchanger; the total performance of the PHE depends on their integrity. They are fitted into the grooves either mechanically or by means of an adhesive. In the case of the compressed asbestos gaskets, a release agent is applied to the face of the gasket in order to prevent its sticking to the opposite plate.

6.25 How is intermixing of streams avoided in a PHE?

Gaskets are arranged such that there is a double seal between the two fluids and vents that open to atmosphere are molded into the gasket (Fig. 6.7). In the event of either of the streams leaking past its gasket seal, the pressurized fluid leaks through the vents into the open atmosphere, thus avoiding intermixing of the fluids and giving a visual indication of the leakage as well.

Fluid Flow and Pressure Drop

6.26 What is the flow arrangement in a PHE?

The flow arrangement on the two sides of any plate is generally countercurrent. The overall flow pattern for the PHE can be changed depending upon the thermal requirements. A few important flow patterns are shown in Fig. 6.8. The basic patterns are the following:

1. Series flow where the total fluid passes over each plate and changes direction over each subsequent plate. This is feasible only with small flow rates, and is rarely used.
2. Looped flow where the fluids are divided into substreams that mix before exiting from the PHE. This is done for high flow rates.
3. Complex flow that is made up of a combination of the series and looped flows. Strong efforts are made to obtain a symmetrical system having the number of channels per pass and the number of passes the same for each fluid. In some cases, a more economical design is obtained by using an unsymmetrical pass system, despite the fact that part of the plate pack will be in co-current operation. Large correction to LMTD is required in such cases. However, it is more than compensated for by an increase in the heat transfer coefficient and proper utilization of the available pressure drop.

6.27 How would you arrange the plates and port holes for two passes for each of the fluid (2/2 PHE), three passes for one and one pass for the other fluid (3/1 PHE), and one fluid doing two heat duties?

These are shown in Fig. 6.9.

6.28, How do the number of passes in a PHE compare with those in a shell-and-tube exchanger?

The number of tube-side passes can be compared with the number of passes for any one fluid in the PHE. The number of tubes in a pass can be compared to

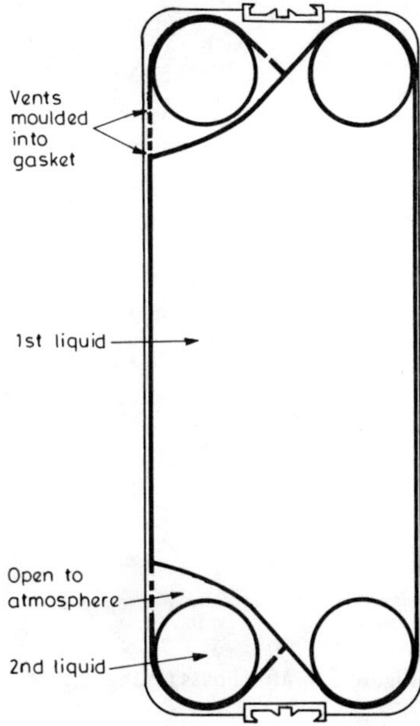

Figure 6.7 Gasket arrangement on a plate. *(Courtesy APV Company, Inc., B2.)*

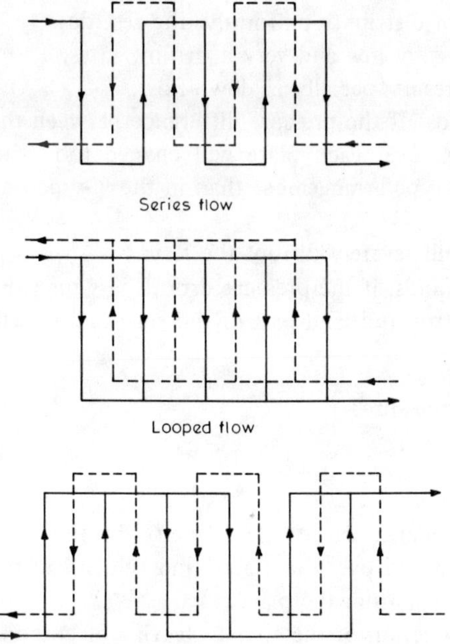

Figure 6.8 General flow patterns.

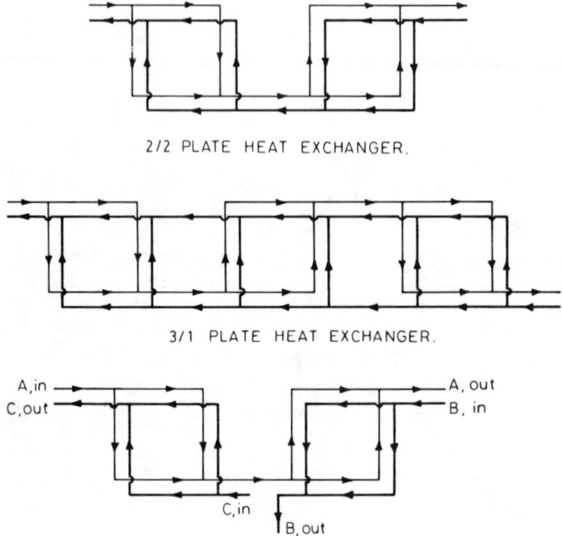

Figure 6.9 Multi-pass arrangements.

the number of flow passages in one pass of a PHE (APV, B3). The number of shell-side passes cannot be directly compared since the fluid flows many times over the tube bundle in cross-flow.

6.29 What factors influence the flow distribution over the plates?

These are the following:

1. Viscosity: Highly viscous liquids may not distribute uniformly, the rest do.
2. Flow rate: Maldistribution is possible in very low and very high flow rates.
3. Long plates: Some maldistribution can result specially in down-flow.
4. Pressure difference between the manifolds: If the pressure difference between the two manifolds changes, the flow rates over each plate will change too. This situation is encountered more in the Z-type arrangement than in the U-type one (Fig. 6.10).
5. Presence of air pockets in the unit: It will severely disrupt the flow.
6. Pressure drop: In downward flow of liquids, if the pressure drop is less than the available static head, the plate will not run full, and part of the surface area will be wasted.

6.30 What factors influence the manifold pressure?

These are the following:

1. Frictional losses in the manifold.
2. Momentum changes due to velocity change since the velocity in the inlet manifold reduces as fluid flows out of it and over the plates, and velocity in the exit manifold increases as the fluid enters it from the plates.
3. Momentum changes due to change in direction of the streams leaving or entering the manifolds (Acrivos et al., 1959).

Thermal Design

6.31 What are the currently available thermal design capabilities for a PHE?

There is no definitive open literature on the topic that can be cross-checked. Hence, the following is based on private communications from very reputed manufacturers. Their claims to design include the following:

1. No phase change systems involving liquids and/or gases.
2. Condensation of vapors with or without inerts, condensation of mixed vapors with miscible or immiscible condensates, condensation of superheated vapors, partial condensation, or any combination of the above.
3. Vaporization of pure liquids and mixtures.
4. In all the above, both manual and computer methods are used. If the physical properties do not change considerably when the fluid is heated or cooled, an overall heat transfer coefficient is generally used. Otherwise the local values of heat transfer coefficient are calculated, and a zone-wise procedure is adopted.

6.32 What factors contribute to the high heat transfer coefficient in a PHE?

1. High turbulence at low Reynolds number due to the breaking of the boundary layer by the corrugations.
2. Lower fouling resistance.
3. Thin plates and hence low resistance to heat transfer.

These factors result in a high heat transfer coefficient. For example, for water-water, similar flow rates, an overall log mean temperature of about 40°C (104°F), similar pressure drops of about 100 kPa (14.5 psi), and an overall heat transfer coefficient in the range of 3500 to 7000 W/m^2 °C (616 to 1232 Btu/hr ft^2 °F), including an allowance for the effects of fouling, can be obtained.

6.33 Under what situation is the PHE most efficient?

It is most efficient when the ratio of volumetric flows on either side of a plate is between 0.7 and 1.4 and both the fluids flow through an equal number of

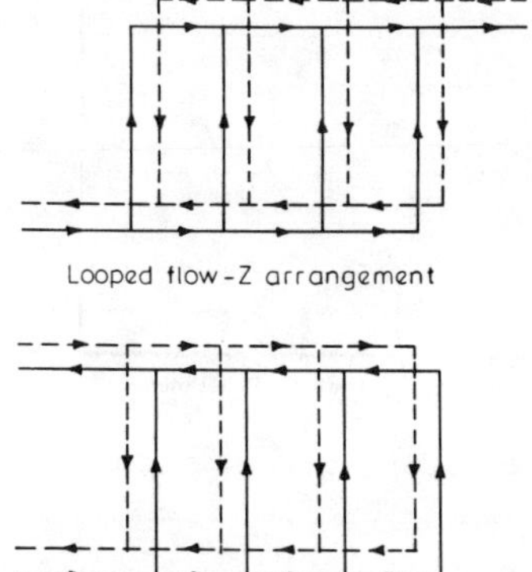

Figure 6.10 Z- and U-type flows.

passes. For other systems, a considerable LMTD correction factor has to be used.

The extent of LMTD correction depends upon several factors, including the number of NTU per pass and the number of channels per pass. The exact methods are proprietary of the manufactures (see Q. 6.38 for NTU).

6.34 What is the information needed to size a PHE?

It is primarily the same as for a shell-and-tube unit. A typical specification sheet is shown in Fig. 6.11. The only important information different from the specification sheet for a shell-and-tube exchanger (Fig. 1.37) concerns the percentage and nature of the undissolved solids.

6.35 How is the PHE designed?

Most of the information concerning the calculations of the heat transfer coefficient and pressure drop are proprietary to the few companies that make the PHE. An overall survey of the available relations in open literature is given by Raju and Bansal (1983). These employ the general type of Nusselt relationships with different coefficients and exponents, depending upon the types of plate used. Relations are given for the laminar and turbulent flows, and also for the non-Newtonian fluids. Correlations for the pressure drop are also given. The design calculations can be done by either the LMTD or the NTU method. However, all these are for the preliminary sizing, and for an accurate design, the manufacturers should be consulted.

Company			Phone
Address			Contact
City	State	Zip	Position

Description and chemical composition		HOT FLUID		COLD FLUID
Rate of flow		GPM		GPM
Inlet temperature		°F		°F
Outlet temperature		°F		°F
Maximum permissible pressure drop		psi		psi
Maximum working pressure		psi		psi
Specific gravity				
Specific heat at mean temperature				
Thermal conductivity at mean temperature				
Viscosity at several temperatures in operating range	cps @ °F cps @ °F	cps @ °F cps @ °F	cps @ °F cps @ °F	cps @ °F cps @ °F
Percent of undissolved solids				
Nature of solids (i.e. fibrous, powder, size)				
Flow fluctuations, if any				

Preferred plate material: ☐ 316 stainless steel ☐ titanium ☐ Monel ☐ Incoloy 825 ☐ Hastelloy C Other:

Remarks:

Figure 6.11 Specification sheet. *(Courtesy APV Company, Inc., B4.)*

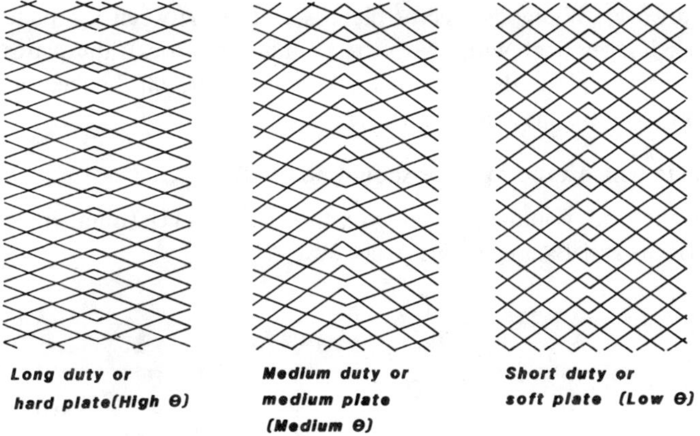

Figure 6.12 Short, long, and medium duty plates. *(Courtesy Alfa-Laval, A2.)*

Plate Characteristic and NTU (Θ)

6.36 Are all the plates similar?

No. The plates are of three types as far as the thermal duty is concerned (see Fig. 6.12).

1. Short duty plates: Also known as soft plates, they give a low heat-transfer coefficient and a low pressure drop and they have high flow rates. They are usually wide and short in length and are used for general duties. The angle of chevrons (embossing) is low.
2. Long duty plates: Also known as hard plates, they give a high heat-transfer coefficient and a high pressure drop and they have low flow rates. They are usually narrow and long and are used when temperature approach is small so as to achieve high heat recovery. The angle of chevrons (embossing) is high.
3. Medium duty plates: These are plates with intermediate properties and angles of chevrons.

Besides the overall dimensions of the plates, the type of corrugations also affects the thermal performance. Generally, a mixed pack of the above types is used to match both the heat transfer and pressure drop requirements. The flow over the plates is not exactly diagonal since it requires separate manufacture of the right-hand and the left-hand plates for the dual channels, while the thermal advantage gained is insignificant (APV, B1). In the vertical flow case, the plates only need to be rotated by 180° to make the dual channels.

6.37 What is a thermal plate?

It is a plate that transfers heat. In a PHE, the end plates do not transfer any heat, they simply contain the fluid and prevent it from touching the pressure plates, which are generally of mild steel and may not be suitable for the particular liquid. Thus, in a PHE with a total of 20 plates, only 18 are thermal plates regardless of the number of passes. However, if one connector plate is used in between the plate pack so that one fluid exchanges heat with two others (multiple duties), the plates on either side of the connector plate also do not transfer any heat, and hence, out of a total of 20 plates, only 16 are thermal plates.

6.38 How does one decide which type of plate should be used?

The most convenient and popularly used concept is the NTU (number of transfer units) given by (Raju and Bansal, 1983; Alfa-Laval, A1)

$$\mathrm{NTU} = \frac{\Delta T}{\Delta T_{\mathrm{lm}}} \tag{6.1}$$

$$= \frac{2A_{\mathrm{p}}U}{MC_{\mathrm{p}}} \tag{6.2}$$

where ΔT = temperature change in one channel
ΔT_{lm} = log mean temperature difference between two adjacent channels
A_{p} = plate surface area
$2A_{\mathrm{p}}$ = surface area of one flow channel (made up of two plates)
U = overall heat transfer coefficient
M = mass flow rate in one channel
C_{p} = heat capacity

NTU for a plate is also known as its thermal length, Θ, or performance factor or temperature ratio, TR. Its value for the same plate varies with the heat duty. For example, if NTU for water is 2.0 that for a viscous oil may be 0.6 only.

For the whole PHE, the NTU is defined as

$$\mathrm{NTU} = \frac{UA_{\mathrm{t}}}{MC_{\mathrm{p}}} = \frac{\Delta T}{\Delta T_{\mathrm{lm}}} \tag{6.3}$$

where A_{t} = total area of all the thermal plates ($=A_{\mathrm{p}}N$)
N = number of thermal plates

In the case of unequal heat capacity flow streams, the NTU value is different for the two streams. The design is then based upon the conditions of the pressure drop limiting side. The soft plate has low NTU (low Θ), while the hard plate has high NTU (high Θ). A medium plate has medium NTU. As discussed in Q. 6.36, a mixed pack of two or more types is used to match the required heat duty and pressure drop. An example illustrates this well (Raju and Bansal, 1983): 400 m^3/hr (14,126 ft^3/hr) of drier scrubber water is to be cooled from 57° to 43°C (124.6° to 109.4°F) using cooling water at 30°C (86°F). The pressure drop allowed is 0.15 MPa (21.75 psi) on both sides.

	High NTU	Medium NTU	Mixed NTU
No. of plates	211	181	147
Pressure loss, MPa	0.15	0.8	0.15
(psi)	(21.75)	(11.6)	(21.75)
Remarks	Too much surface	Too much surface, and pressure drop not fully utilized	Pressure drop fully utilized (70% high NTU, 30% medium NTU plates)

In the high NTU case, too much surface results because the flow rate over these plates is small for the given pressure drop and hence more channels are needed to accommodate the total flow.

In the medium NTU case, the pressure drop is not fully utilized because of the low resistance offered by the plates. The number of plates is indicative of the large area needed to compensate for the lower heat-transfer coefficient in these plates so that the required heat duty is achieved.

In the mixed NTU case, the high NTU plates provide a high heat-transfer coefficient but low flow rate, and the remaining fluid is forced over the medium NTU plates, resulting in the same pressure drop as for the high NTU plates (since the two must match). The heat transfer coefficient is also higher over these medium NTU plates compared to the earlier case due to higher flow rate. Thus, the total number of plates is the least in this case.

Another example of the use of mixed NTU plates is given in Alfa-Laval (A1). It is pointed out that generally a mixture of only two of the three kinds (high, medium, and low NTU plates) suffices for any job, and it is not necessary to mix all the three kinds.

6.39 Is a correction factor applied to LMTD in a PHE?

Yes. A small correction factor is required even when both fluids undergo equal number of passes, contrary to the case of a shell-and-tube unit (1-1 E type or 2-2 F type). This is due to the fact that (i) the flow is not exactly in-line countercurrent; (ii) in the end zones, the heat is transferred only on one side; and (iii) with even number of plates, one fluid has an extra stream (or flow passage) over the other. The correction factor depends on several other factors, such as the number of passes, NTU per pass, and the number of channels per pass. If the number of passes for both fluids is the same and the channel number is reasonable (not too small), then for a rough calculation, the LMTD correction factor can be taken as unity. However, for unequal passes, the correction factor can be significant (Fig. 6.13).

6.40 How is the number of passes decided?

For a given heat duty, one calculates the NTU by Eq. (6.3). The maximum number of NTU for the hard plates is 4 and for the soft plates is 0.3. Hence, if the NTU required is, say 11, and if the hard plates are used one would require three

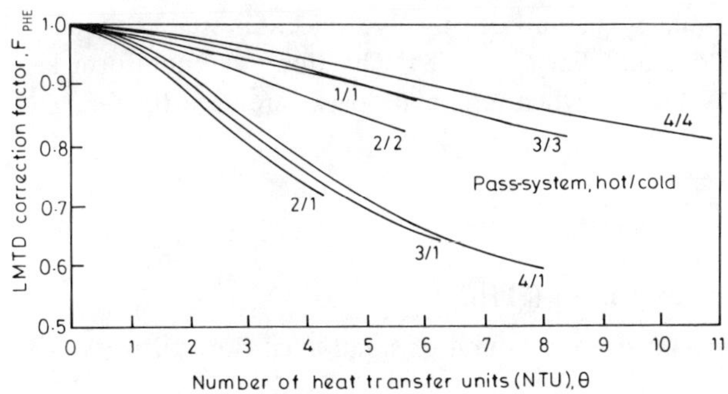

Figure 6.13 Correction factor for multi-pass system.

passes. Of course, these days one uses a mixture of hard and soft or medium plates to utilize the total allowable pressure drop effectively.

6.41 What factors should be kept in mind when sizing up a PHE?

These are the following:

1. LMTD correction as discussed in Q. 6.39.
2. Unequal flow distribution in channels due to significant pressure loss in the manifold and in the feed ports (see also Q. 6.29).

PHE Evaporator

6.42 How does a PHE evaporator work?

A PHE evaporator is essentially either a climbing-and-falling film type of unit or a falling film type with short retention time. The plates are flat and are assembled in units of four. They have large ports to handle the vapor, and the plates are spaced relatively farther apart to minimize the pressure drops both on the boiling and condensing sides. These are used under vacuum or at slightly above atmospheric pressures. Large evaporation rates can be obtained by putting multiple units in parallel. These are primarily used in milk, fruit juice, and distillery units. These do have the usual advantages of a PHE, such as low liquid hold-up, a high heat-transfer coefficient, and easy installation and maintenance.

Major Problems in PHE

6.43 What are the major problems in PHE?

These are the following:

1. Limits on operational pressure and temperature.
2. High pressure drop in the manifold resulting in unequal distribution of the fluid in different flow passages. The contribution of the upper ports to the pressure drop is different from that of the lower ports, and it also varies with the Reynolds number (Price and Fattah, 1978). A good design should keep the port pressure drop to within 20 percent of the total.
3. Nonuniform flow over the plates when the fluid is highly viscous. The flow is not fully developed even more than halfway up the flow channel at low Reynolds number, as determined qualitatively by dye injection studies on single flow channels of plastic (Price and Fattah, 1978). The flow is nonuniform also when the flow rates are very high or when long plate packs are used (Q. 6.29).

SPIRAL PLATE HEAT EXCHANGERS

6.44 What is a spiral plate heat exchanger (SPHE)?

A SPHE is an exchanger consisting of spiral or spiral/axial flow passages (Fig. 6.14).

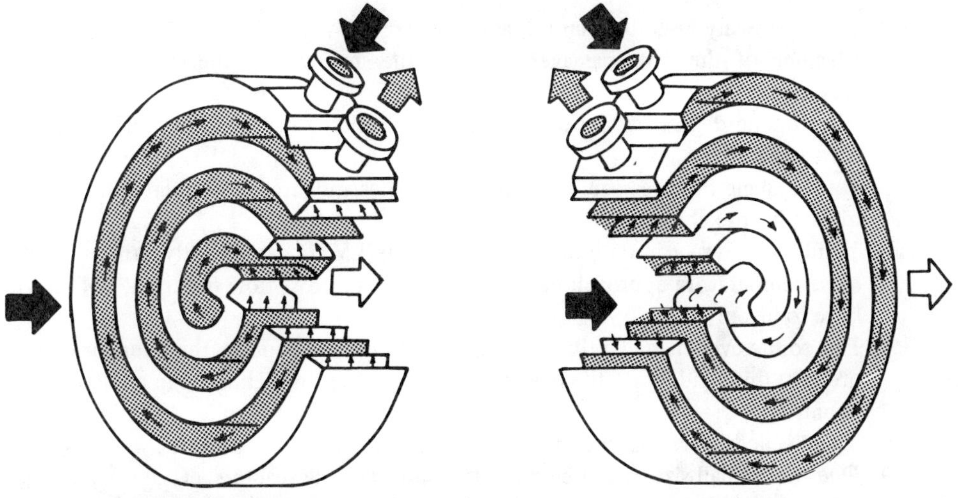

Figure 6.14 Spiral plate heat exchangers (SPHE).

Manufacture of a SPHE

6.45 How is a SPHE manufactured?

A pair of long metal strips are rolled around a retractable split mandrel. The spacing between the two rolled strips is maintained by means of spacer studs welded to the strips. The spiral channels that are formed are closed on their ends in order to contain their respective fluids. One lateral end of each channel is welded shut, while the other is faced to a true plane by grinding, and then sealed with a cover and full-cover gasket. This arrangement is reversed in the second channel to prevent interleakage between the two streams. The covers contain two of the four nozzles, while the other two are on the spirals (Fig. 6.14). The covers are attached to the spiral elements by means of forged hook bolts and adapter.

Uses, Advantages, and Limitations

6.46 Where can a SPHE be used?

It can be used in situations involving slurries up to 50 weight percent, e.g., sewage sludge, metal ore suspensions, and heavy fouling fluids; condensation under vacuum; as gas cooler; and it can be mounted directly over reactors using flanges and avoiding the intermediate piping. It is also used where space is limited (Alfa-Laval, A3).

6.47 What are the advantages of a SPHE?

These are the following:

1. Wide variation in the size between 5 and 2700 ft^2 (0.5 to 250 m^2) and the channel width from $\frac{1}{4}$ to $1\frac{1}{4}$ in (5 to 30 mm).

2. It can be easily assembled and disassembled.
3. Cleaning of the flow passages is easy either by opening the end covers or by flushing.
4. The fouling deposit is low due to the the scrubbing (self-cleaning) effects of the flow streams.
5. Due to single flow passage, it can handle slurries and larger particulates better than a plate heat exchanger can.
6. Fluids are always in counterflow. Hence, the LMTD can be fully utilized, and a close temperature approach can be achieved. In a cross-flow situation, LMTD will have to be corrected.
7. Due to the ever-changing flow direction in the spiral channels, the boundary layers are very thin, and hence the heat transfer coefficient is high, even in a viscous flow situation.
8. There is no differential thermal expansion between the two spirals.
9. Due to its small size, it can be mounted on top of the reactors, etc., with a flanged joint, thus eliminating the intermediate piping. A 100 m^2 (1076 ft^2) effective surface is contained in a unit of 1 m (3.28 ft) in diameter and 1.5 m (4.92 ft) in length.
10. Due to the welding of the flow passages on one side and a full gasket on the other side, there is no interleakage of the fluids.
11. Although it may be equally or more expensive than a shell-and-tube exchanger having the same area, the high heat-transfer coefficient and low fouling-resistance and low maintenance requirements make it more economical overall for the conditions for which it can be used.

6.48 What are its limitations?

These are the following:

1. It is useful only up to a pressure of 1.5 MPa (215 psig) since each turn has to contain the pressure. Temperature limit is as per the stress values prescribed by the applicable code. Units are being used for cooling H_2 gas at 2000°F (1100°C).
2. Its surface area is limited to approximately 250 m^2 (2700 ft^2) (Alfa-Laval, A3) due to the limits on the availability of the metal sheets and on mounting heavy units above the reactors.
3. The heat transfer coefficient in liquid-liquid flow is generally limited to 2900 W/m^2 °C (500 Btu/ft^2 hr °F).
4. The flow capacity is limited to approximately 0.11 m^3/sec (14,000 ft^3/hr).
5. It cannot be used for cyclic temperature services unless metal spiral wound gaskets are used. The expansion and contraction of the cover would destroy other gaskets which rub against it.
6. Unlike a shell-and-tube unit, where only the tubes may see the corrosive liquid, in a SPHE both the plates have to be the same since they each see both the fluids.

Parts and Materials

6.49 What are the important parts of a SPHE?

These are the following:

1. Spiral passages made by rolling the metal sheets for the flow of the fluids.
2. Gaskets to contain the fluids.
3. End covers to hold the unit together.
4. Inlet and outlet ports for the fluids.

6.50 What materials can be used in a SPHE?

These can be any that can be rolled and welded, such as carbon steels, stainless steels, Hastelloy-C, Monel, nickel, nickel and copper alloys, aluminum alloys, and titanium. The choice depends upon the corrosion requirements and service temperatures (Gupta, 1979).

Different Kinds

6.51 What are the different kinds of SPHE?

There are three principal kinds, depending upon whether one or both of the fluids are in spiral flow.

These are the following:

1. Spiral–spiral flow [Fig. 6.15(a)]: Both the fluids are in spiral flow. These are generally used when both the fluids are liquid. The hot fluid enters at the center and flows from the inside outward to the periphery where it exits. The cold fluid enters at the periphery and flows toward the center where it exits.
2. Spiral-cross flow [Fig. 6.15(b)]: One fluid is in spiral flow and the other is in cross-flow parallel to the axis of the spiral element. These are generally used when

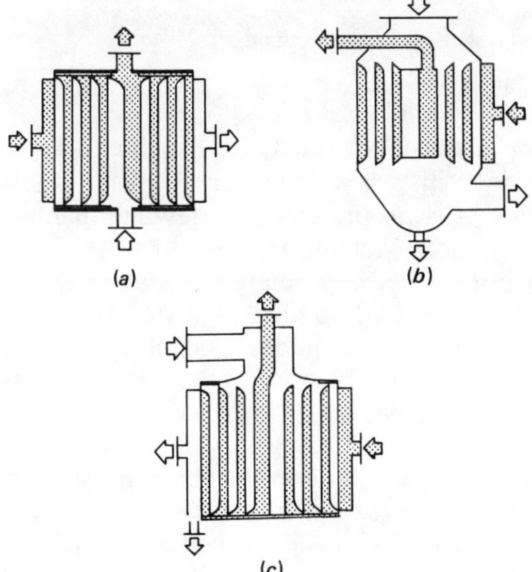

Figure 6.15 Different kinds of SPHEs. (*a*) Spiral–spiral flow. (*b*) Spiral–cross flow. (*c*) Spiral flow–combination of cross-and-spiral flow. *(Courtesy Alfa-Laval, A2.)*

one of the fluids is a vapor or a two-phase mixture that is put in the cross-flow while the second fluid, a liquid, is put in the spiral flow. The spiral flow passage is welded shut on both sides, and hoods are used for the inlet and exit of the vapor stream. Thus, very high flow velocities are obtained in the spiral passages along with a very low pressure drop in the cross-flow stream. An extra outlet is provided for the drainage of the condensate. The spiral passage cannot be cleaned mechanically in this case.

3. Spiral flow–combination cross-and-spiral flow [Fig. 6.15(c)] : This is used when one of the fluids is a mixture of vapor and noncondensables. The coolant enters at the periphery and flows through the spirals toward the center and leaves through a nozzle passing through the top of the vapor inlet dome, which is removable. The broad flange of the dome shields the top of several outermost turns of the vapor passages. The central section under the dome acts as a condenser, and the condensate falls to the bottom and travels countercurrent to the coolant, getting subcooled before exiting. The noncondensables enter the outer spiral passages and get cooled while traveling outward. In the process, any condensate entrained also separates and falls to the bottom, where it joins the remaining condensate and flows out.

6.52 Is the SPHE mounted with its axis horizontal or vertical?

It depends on the service and the type of fluids. If it is a liquid–liquid service and any of the two liquids has suspended materials, the axis should be horizontal, since the flow will then carry any settled solids out of the unit. If, however, the unit is required to be drained completely or one of the fluids is a vapor, then the axis should be vertical. In some cases, the available space may also dictate the choice.

Comparison with the PHE and the Shell-and-Tube Exchanger

6.53 Compare a SPHE with a PHE and a shell-and-tube exchanger.

These comparisons are given in Tables 6.2–6.5.

• The discussion in this chapter has brought out the salient features of the PHE and SPHE. While their utility is unquestioned within their applicable limits, their detailed design is proprietary of a few companies. There are many other types of plate heat exchangers, some of which have been dealt with in Chapter 4. A popular variety emerging is lamella heat exchanger (LHE), which consists of a set of welded plates or flattened tubes welded together and placed inside a cylindrical shell. The flow passages thus formed are called lamellae. Figure 6.16 shows a cross section. One fluid flows inside the lamella, and the other, outside. No baffles are used. Flow is countercurrent and single pass. Temperature and pressure limits are similar to those of a PHE, due to the characteristics of the gasket material. While it has the advantage of high turbulence at low Reynolds number, it cannot be disassembled to be cleaned as a PHE. Its design is also proprietary of a few companies.

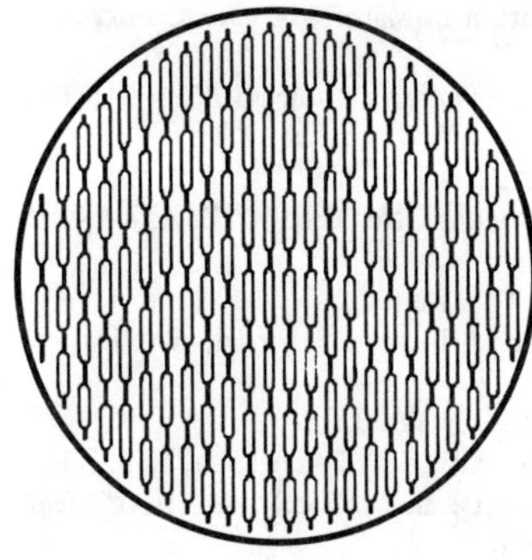

Figure 6.16 Lamella heat exchanger (LHE). (*Courtesy Alfa-Laval,* A2.)

NOMENCLATURE

A_p	plate surface area
A_t	total area of all thermal plates
C_p	heat capacity
CAF	compressed asbestos fiber
f	friction factor
F_{PHE}	LMTD correction factor for plate heat exchanger
LHE	Lamella heat exchanger
LMTD	log mean temperature difference
M	mass flow rate in one channel
N	number of thermal plates
NTU	number of transfer units
ΔP	pressure drop
SPHE	spiral plate heat exchanger
S & T	shell-and-tube
ΔT	temperature change in one channel
ΔT_{lm}	log mean temperature difference between two adjacent channels
TR	temperature ratio
U	overall heat transfer coefficient

Greek Letters

Θ	number of transfer units or thermal length
η	viscosity at bulk temperature
η_ω	viscosity at tube wall temperature

REFERENCES

Note: Company bulletins cited here do not have publication dates; they have been given serial numbers to distinguish them.

Acrivos, A., B. D. Babcock, and R. L. Pigford. Flow Distribution in Manifolds. *Chem. Eng. Sci.* 9:112-124, 1959.

Alfa-Laval. *Alfaflex–The Story of a New Generation of Heat Exchangers,* Bulletin PC 60515 E2. Alfa-Laval: Lund, Sweden. A1.

Alfa-Laval. *Heat Exchanger Guide.* Alfa-Laval: Lund, Sweden. A2.

Alfa-Laval. *Heat Exchanger for Air Conditioning Systems,* Brochure No. IB61003E. Alfa-Laval: Lund, Sweden. A3.

APV. *APV Process Equipment,* Bulletin CP-475, The APV Co.: Tonawanda, N.Y. B1.

APV. *APV. Paraflows, Plate Heat Exchangers for Chemical and Industrial Duties,* Publication A340Y. The APV Company: Sussex, England. B2.

APV. *Heat Transfer Handbook.* APV Equipment Inc.: Tonawanda, N.Y. B3.

APV. *APV Paraflow Plate Heat Exchanger Design Questionnair,* APV Co.: Tonawanda, N.Y. B4.

Auth, W. J., and J. Loiacono. Plate and Spiral Heat Exchangers. In *Practical Aspects of Heat Transfer, A CEP Technical Manual.* AIChE: New York. 1978.

Bell, K. J. Plate Heat Exchangers. In *Heat Exchangers: Thermal-Hydraulic Fundamentals and Design.* S. Kakac, A. E. Bergles, and F. Mayinger (eds.). Hemisphere: Washington, D.C. 1981.

Cooper, A., A. M. Cocks, and A. C. Henton. Improved Water Utilization with Plate Heat Exchangers. *Heat Trans. Eng.* 1(1):31–41. 1979.

Cooper, A., J. W. Suitor, and J. D. Usher. Cooling Water Fouling in Plate Heat Exchangers. *Heat Trans. Eng.* 1(3):50–55. 1980.

Cowen, C. T. Choosing Materials of Construction for Plate Heat Exchangers–I, II. In *Process Heat Exchange.* V. Cavaseno (ed.). McGraw-Hill: New York. 1980.

Gupta, J. P. *Heat Exchanger Design–A Practical Look,* 2d ed. C. S. Enterprises: Delhi, India. 1979.

Marriott, J. Where and How to Use Plate Heat Exchangers. *Chem. Eng.* 78(7):127-134. 1971.

Novak, L. Fouling in Plate Heat Exchangers and Its Reduction by Proper Design. In *Heat Exchangers: Theory and Practice.* J. Taborek, G. Hewitt, and N. H. Afgan (eds.). Hemisphere: Washington, D.C. 1983.

Price, A. F., and A. F. M-A. Fattah. Hydrodynamic Characteristics of a Plate Heat Exchanger Channel. *Trans. I. Chem. E.* 56:217-228. 1978.

Raju, K. S. N., and J. C. Bansal. Consider the Plate Heat Exchanger. *Chem. Eng.* 87(16):133-144. 1980.

Raju, K. S. N., and J. C. Bansal. Plate Heat Exchangers and Their Performance. In *Low Reynolds Number Flow Heat Exchangers.* S. Kakac, R. K. Shah, and A. E. Bergles (eds.). Hemisphere: Washington, D.C. 1983.

Walker, P. Heat Exchanger Selection-Plates or Shell-and-Tube. *Process Eng.* Jan. 61-62. 1980.

CHAPTER

SEVEN

HEAT EXCHANGER DYNAMICS AND OPTIMIZATION

INTRODUCTION

• Process plants are becoming increasingly more complex and bigger in size. Any change, intentional or otherwise, in the operation of a heat exchanger will produce a change downstream that is likely to have important consequences. Hence, it is necessary to know the effects of such changes by a dynamic study of the system in order to modify the design, if necessary.

The large-size units involve a lot of capital and operating costs. Hence, it is essential that they be optimized. In optimization studies, the desirability of a fast response and more stable steady state (as obtained from the dynamic analysis) act as constraints. This points out the connection between dynamics and optimization.

With an increasing interest in keeping costs low, efforts have been made to interconnect all the heat exchangers in a plant in a sort of network and optimize them all together, thus reducing the number of units as well as the cost of utilities. This is known as the heat exchanger network optimization, and it is achieving greater importance in the new plants.

Therefore, in this chapter, we shall discuss the heat exchanger dynamics and optimization. The brief treatment here is meant to serve only as an introduction. For details, the references cited should be consulted.

HEAT EXCHANGER DYNAMICS

7.1 What is heat exchanger dynamics?

During start-up, shutdown, emergency and upset conditions, and deliberate changes in the flow rates and/or the inlet temperatures, the heat exchanger operation becomes transient, which either results in a new steady state or in a return to the original steady state. The time-dependent study of the changes produced in the exchanger due to these transients is known as the heat exchanger dynamics. When it is done theoretically (analytically and/or numerically), it is also called "modeling" or "simulation" of a heat exchanger. Sometimes, a physical model is also used to "simulate" the unit.

7.2 What are the types of changes that can happen in a heat exchanger?

The types of changes possible in a heat exchanger are the following:

1. Impulse change in the inlet flow rate and/or the inlet temperature of either or both the fluid streams. While theoretically an impulse has an infinite magnitude and lasts for zero time, in practice it has a finite magnitude and lasts for a finite, though short time, Δt (Fig. 7.1). Examples of deliberate impulse changes introduced into a heat exchanger would be a sudden change in the flow rate (and hence in the flow velocity) introduced for a short time to shear and dislodge the scale deposit or a sudden change in the tube-side fluid temperature for a short time to crack and dislodge the scale deposit.
2. Step change in the inlet flow rate and/or the inlet temperature of either or both the fluid streams. This can happen due to various reasons, such as changes in the process requirements or one exchanger doing the work of two units while the second one is out of service. When the second unit is put back in service, the first unit again undergoes a step change. Similarly, start-up and shutdown also result in step changes.

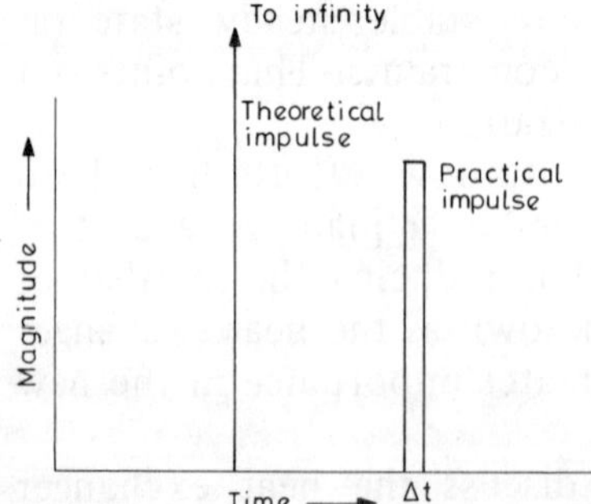

Figure 7.1 Impulse input.

3. Sinusoidal changes in the inlet flow rate and/or the inlet temperature of either or both the fluid streams. This is not usually encountered in industry but is used in simulation studies to determine the frequency response of a unit and thus to predict its transient behavior in general.
4. Gradual fouling up of the heat transfer surface due to scale deposit that reduces the overall heat transfer coefficient with time.
5. Gradual reduction in the heat transfer area due to the plugging of the tubes as and when they start to leak.
6. Gradual blanketing of the heat transfer surface due to the noncondensable gases present in a condensing stream or released from liquids due to change in temperature and if the venting is inadequate.

The study of the dynamics is usually concerned with the first three items, although a good analysis can take into account the rest of the items also, even though their time constant is rather large.

7.3 What are the advantages of studying the dynamics of an exchanger?

Study of the dynamics of an exchanger gives information regarding the changes in the flow rates and/or temperatures in the unit as well as at the exit for a given change in the input conditions. This information is useful in evaluating the following:

1. The effects on the fluid streams (rate of heat transfer, pressure drop, freezing, polymerization fouling, reduction in scale deposit, vibration damage, etc.)
2. The effects on the processes downstream
3. The effects on any auxiliary controls required
4. The new steady state and the time required to reach it
5. Or the time required to return to the old steady state
6. The changes required in the steady state operating conditions so that the operation is more stable or the lag time is minimal. This will affect the size of the equipment for a given heat duty and hence should be evaluated even before the equipment is ordered. For example, to reduce the lag time in the response to a change in temperature, high velocity of the fluid will be required with a smaller unit but at a higher operating cost. Effects on the heat transfer coefficient and pressure drop of changes in several design parameters have been discussed in Chapter 2.
7. Whether the system is statically stable (i.e., after a disturbance from a steady state, the system tends to move toward the original steady state) or is dynamically stable (i.e., after a disturbance from a steady state, the system will ultimately attain the original steady state, the convergence procedure may be periodic or aperiodic in nature). Dynamic stability is more important to determine since, if a system is dynamically stable, it will also be statically stable but not vice versa. These stabilities in a boiling system, such as steam generators, but not vice versa. The stabilities in a boiling system, such as steam generators, have been dealt with at length by Fraas and Ozisik (1965) and Spigt et al.
8. The thermal stresses within the exchanger and loads on the bolts making a

flanged joint, etc. Possible mechanical failures can thus be predicted, and design modifications made to prevent them.

The study of the dynamics is more important in the nuclear power plants where severe transients may lead to accidents. These must be postulated before hand and the response of the exchanger analyzed since such transients impose severe demands on the unit.

7.4 What are the factors on which the dynamic or transient response of a heat exchanger depends?

The factors are the following:

1. The steady state conditions, viz. flow rates, temperatures at the inlet and outlet, heat transfer coefficients, fouling resistance, pressure drop, etc.
2. The nature and magnitude of disturbance [impulse, step, sinusoidal or a combination (Q. 7.2)] and the parameters disturbed, viz. temperatures and/or flow rates
3. The size and construction features of the unit, viz. dimensions and materials of construction of the shell and the tubes, number of the shell and tube passes, number of tubes, tube pitch and layout angle, kind of baffle, baffle cut and baffle spacing, etc. These determine the changes in the flow velocities and hence in the heat transfer coefficients.
4. The thermal conductivity and heat capacity of the tube material.
5. The scheme of fluid flow, fiz. countercurrent, co-current, cross-flow, mixed flow, etc. This determines the effective temperature driving potential.
6. Whether any phase change is occurring

The dynamic analysis of a unit having single phase on both sides (e.g., liquid-liquid or liquid-gas exchangers) is the simplest, a unit with single phase on one side and double or two-phase on the second side (e.g., steam generator, surface condenser) is more complex, while a unit with double phase on both the sides [e.g., steam condensing on one side and Freon evaporating on the other side (Fig. 3.26)] is the most complicated one. The last case has been dealt with by Mayinger and Shult (1981). Different flow regimes exist in the two-phase flow and are difficult to evaluate (Figs. 3.9–3.13, 3.24, and 3.25). In the two-phase flow, instabilities arise due to the very nature of the flow and not necessarily always due to changes at the inlet end. Hence, it is outside the purview of this chapter. Interested readers should consult authoritative references and books, such as Bergles et al. (1981).

7.5 How is the dynamic analysis done?

It is done either experimentally, or by a mathematical model, or both.

1. Experimental methods: Experiments are carried out on a model or prototype (full size) by deliberately inducing disturbances in the inlet fluids and measuring the temperature and flow rates both within and at the outlet of the unit. From this, the results of interest are derived. The experimental set-up, including all the instrumentation and data-logger, etc., become quite elaborate and expensive.

2. Mathematical model: The behavior of the exchanger is expressed in terms of separate mathematical equations for mass, energy, and momentum balances for both the fluids on either side of the tube wall. The thermal resistance of the tube wall is also taken into account. These are then dedimensionalized and solved for different disturbances (boundary conditions) at the inlet. A strategy for mathematical model development is shown in Fig. 7.2. The solution can be achieved by any of the following methods (Gutierrez and Cooper, 1979): (a) direct analytical integration, (b) numerical solution by computer, (c) analog solution on an analog computer, or (d) assimilated solutions based upon regression analysis.

The assumptions made are generally the same as for the steady state analysis of the unit (Chapter 2). The solutions for the impulse and step inputs are done in time domain where the changes in the output are measured as a function of time. The solutions for the sinusoidal inputs are done in frequency domain where a transfer function is obtained for the system, which gives the amplification ratio and the phase angle or phase lag between the values of the perturbed parameter at the inlet and at the outlet. Frequency response is also used when an overall dynamic analysis of several pieces of equipment in a plant is required (Shah, 1981).

The response of temperature at the exit to a change in the inlet conditions is not instantaneous because of the thermal capacitances of the fluids and of the tube wall as well as the resistances to heat transfer. This lag affects the performance of the whole plant and hence its control strategy. On the other hand, for a change in the

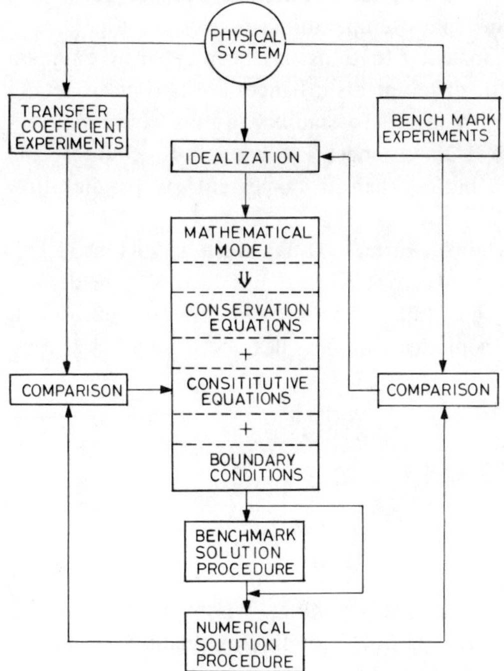

Figure 7.2 Strategy for model development. *(From Banerjee and Hancox, 1979.)*

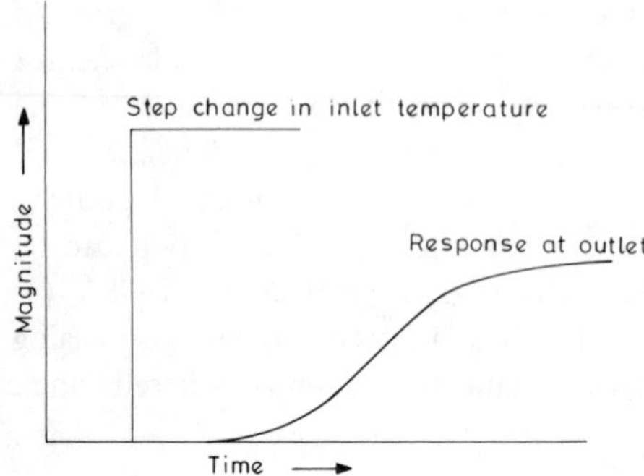

Figure 7.3 Response to a step change in inlet temperature.

flow of an incompressible fluid, the response is instantaneous because of the fixed volume of the unit.

Both lumped parameter and distributed parameter approaches can be used. The latter uses a large number of increments and is more complex, but gives more accurate results and also gives the temperature profiles within the unit, while the former generally uses only one or two sections and gives the response at the outlet and ignores the thermal capacity of the heat exchanger surface.

Since a large number of variables are involved and a variety of disturbances are possible, no general solution can be obtained. Cases of dynamic analysis of specific types of heat exchangers are occasionally reported in the literature, e.g., Kays and London (1964), Myers et al. (1967), Yamashita et al. (1978), Forghieri and Papa (1979), Ravindran and Rao (1978), and Tan and Spinner (1978). A typical plot for a step change in the inlet temperature is given in Fig. 7.3. A numberical solution to some of the transient situations has been given by Patankar and Spalding (1974). Response plots for different disturbances are also given.

Kanoh (1980) points out that with regard to stability, a pure countercurrent exchanger (1-1, 2-2) is better than a 1-2 exchanger involving both counter- and co-current flows, which, in turn, is better than a co-current or parallel flow exchanger.

For the dynamic response of extended surfaces (fins), Kern and Kraus (1972) should be consulted. They consider various types of fins, but do not consider the complete exchangers containing the finned tubes. The experimental approach being more cumbersome, generally a mathematical model has been used by most investigators.

HEAT EXCHANGER OPTIMIZATION

7.6 What is optimization?

Optimization is an attempt to arrive at the best solution of a given problem as judged by a testing criterion or objective function, whose value is either minimized or maximized by using a search or optimum finding technique that also satisfies the constraints on the system. Expressed mathematically

Minimize $f(\mathbf{X})$ (7.1)

Subject to $g_i(\mathbf{X}) \leqslant 0$ for $i = 1, 2, \ldots, m$ (7.2)

and $h_i(\mathbf{X}) = 0$ for $i = m + 1, m + 2, \ldots, m + n$ (7.3)

The vector **X** contains the variables that have to be optimized in order to minimize (or maximize) the objective function $f(\mathbf{X})$ while satisfying the constraints $g_i(\mathbf{X})$ and $h_i(\mathbf{X})$. The objective function and the constraints may be linear or nonlinear functions of the variables.

7.7 What objective functions are used in heat exchanger optimization?

In a given situation, it may be any one of the following:

1. Minimum capital cost
2. Minimum total cost [capital investment plus operating and maintenance costs (Fig. 7.4)]
3. Minimum size for installation in a nuclear power reactor building, space vehicles, automobiles, etc.
4. Minimum weight for use in space vehicles, automobiles, etc.
5. Maximum heat transfer for a specified size of the shell
6. Maximum reliability in nuclear plants, space vehicles, etc. (Redundancy is also used to achieve it.)

In most process industries minimum cost is generally the objective function.

7.8 What are the factors that influence the cost of an exchanger?

Virtually every item connected with the design, fabrication, installation, operation, and maintenance affects the cost of an exchanger. Some of the important items are the following:

1. Size
2. TEMA class (Class R requires more thorough examination and testing, and also the choice of flange type is restricted)
3. Whether vibration and/or seismic analysis required, since it adds to the design cost
4. Materials of construction of the shell, tubes, and other internals

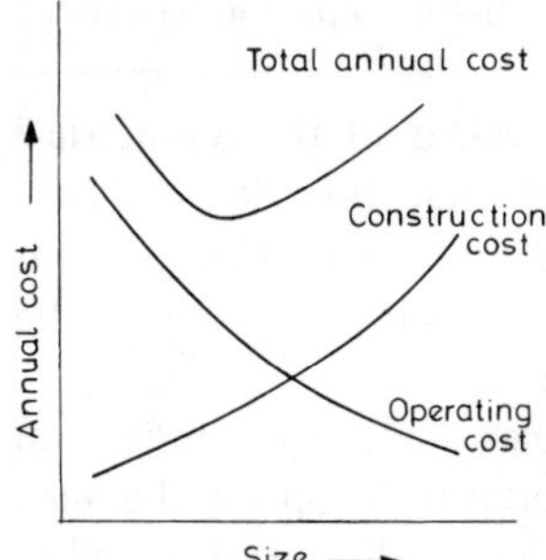

Figure 7.4 Minimum total cost.

5. Transportation cost
6. Installation: vertical or horizontal, with the former being generally costlier
7. Pressure drop on shell and tube sides, pump efficiency, unit cost of power
8. Physical properties of the fluids, since these help determine the heat transfer coefficients and pressure drop
9. Variation in the service fluid inlet temperature over the year and the unit cost of service fluid
10. Maintenance cost
11. Cost of political influence in as far the decisions are affected by nontechnical reasons (LeGoff and Giulietti, 1983; LeGoff, 1980)
12. Interest on capital
13. Rate of depreciation

All of these need to be included in the objective function when the total cost is to be minimized. This is not an easy affair since so many factors are involved, and some are difficult to put in a mathematical form. Besides, these change with time and geographical location, and hence the costs need to be evaluated again each time and at each location.

Articles on process equipment costs appear periodically in various journals (Hall et al., 1982). These account for some of the above factors, such as the size, type, and materials of construction.

7.9 What is meant by constraints on the variables?

Constraints limit the values of the variables. There are several types of constraints:

1. On the variable range: $a \leqslant x \leqslant b$, where x is a variable and a and b are the lower and the upper limits, respectively. For example, the tube pitch (p_t)-to-tube diameter (d_o) ratio generally has a range $1.1 \leqslant p_t/d_o \leqslant 1.5$, baffle spacing (BP) to shell diameter (d_s) ratio generally has a range $0.2 \leqslant \mathrm{BP}/d_s \leqslant 1.0$.
2. Equality constraints: $y = c$, where y is a variable and c is its fixed value. For example, the heat load in an exchanger is specified, and the design must satisfy that requirement; the tube diameter may be fixed because of the plant practices and the sizes used in the exchangers already in service.
3. Inequality constraint: $z \leqslant d$, where z is a variable and d is an upper limit, with no lower limit. For example, the permissible pressure drop has an upper limit; the shell diameter may have an upper limit due to the installation space or transportation restrictions.
4. Explicit constraints: These are the ranges of the fixed values of the geometrical or construction variables. For example, the tube diameter, tube length, tube pitch-to-diameter ratio, layout angle, fin geometry and number of fins per unit length, shell type, shell diameter, number of tube-side and shell-side passes, baffle type, and baffle cut and spacing, are explicit constraints.
5. Implicit constraints: These are the process-related variables that are dependent on the explicit constraints. For example, the heat duty, pressure drop on the shell and tube sides, fast response time to a change in the inlet conditions, desirability

of a stable steady state, and service fluid flow rates (steam, cooling water) are the implicit constraints that are prescribed in the problem specification and are to be satisfied after the explicit constraints have been satisfied. The weight of a unit is also an implicit constraint because it is calculated after the explicit ones have been satisfied.

For a real optimum design, the number of constraints should be kept to a minimum and their ranges made as wide as possible.

7.10 What are the essential requirements of an optimization procedure in general?

For a large number of optimization procedures, these are the following:

1. The independent variables can vary over a wide, limitless range without any constraint on their values.
2. All the variables are continuously differentiable and cannot generally have discrete values, although it is possible to take these into account.
3. The objective function can be expressed as a mathematical equation that is differentiable. The first derivative of the equation equated to zero gives the optimum value with respect to the variable used in the differentiation.

7.11 How is the optimization of a heat exchanger different than a general study of optimization?

As far the above three general requirements of optimization are concerned, the heat exchangers have their own peculiarities as noted below:

1. The different variables and parameters do not have a limitless range. Their range of values is guided by the industrial practices, availability of materials, space limitations, governing Codes and TEMA standards (Tubular Exchanger Manufacturers Association, 1978).
2. Due primarily to the above reasons, the variables do not have continuous values but have discrete, finite values. Hence these cannot be differentiated. Furthermore, some variables are not truly independent. They depend upon one another, such as the baffle cut and baffle spacing.
3. The objective function, which generally is the total cost, depends on the performance of an exchanger. However, the performance of an exchanger cannot be put in an exact mathematical equation form as a function of the different variables. Hence no differentiation of the objective function with respect to a variable is easily possible.

Thus, in all the three important aspects, the optimization of a heat exchanger differs from that of a general system.

7.12 What methods are used in heat exchanger optimization?

As discussed in Q. 7.11, heat exchanger optimization is different than the general optimization situations. Furthermore, with a large number of explicit and implicit constraints to be satisfied (Q. 7.9), the problem becomes nonlinear, and

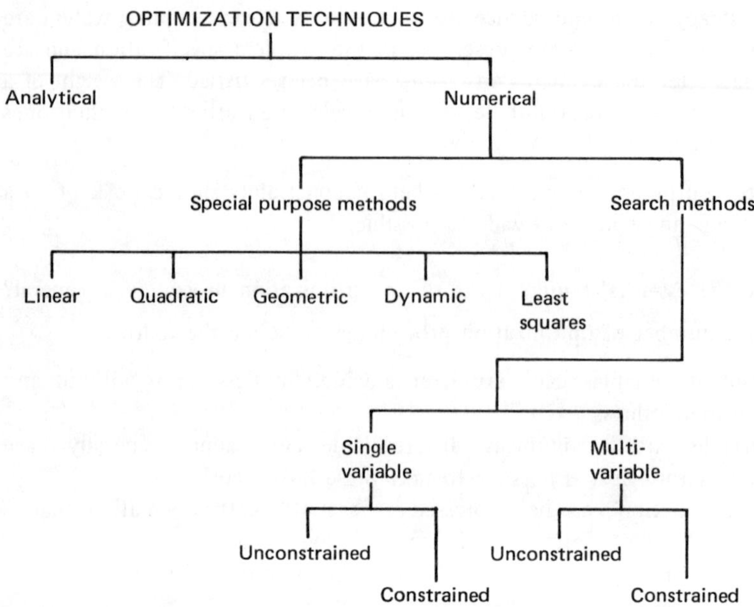

Figure 7.5 Optimization techniques. *(From Kuester and Mize, 1973.)*

there is no single method that is suitable for all the situations. Figure 7.5 summarizes the various techniques of optimization.

The analytical techniques cannot be used since the functional dependence cannot be expressed in the form of an equation (Q. 7.10). Recourse is, therefore, taken to the numerical techniques used on a digital computer. These are divided into two groups (Kuester and Mize, 1973):

1. Special purpose methods: These require specific problem structure and have a rather limited use, if any, in the heat exchanger optimization.
2. Search methods: These are used when the functional dependence is not known. The subdivisions refer to single and multi-variable problems, both unconstrained and constrained. The heat exchanger optimization is a constrained, multi-variable problem, although some of the methods transform it into an unconstrained problem.

There are several algorithms for each of the methods under special purpose and search methods. We will be concerned only with those related to the constrained, multi-variable problems. For the rest, Kuester and Mize (1973), Shah et al. (1979), Fox (1971), Wilde and Beightler (1967), and Himmelblau (1972) should be consulted.

7.13 What are the methods for constrained, multi-variable problem optimization?

The popular optimization algorithms for the constrained, multi-variable problems are the following:

1. Penalty function methods: These convert the constrained problem into a sequence of unconstrained problems by modifying the objective function with a penalty applied to the inequality and equality constraints in a particular manner

$$P = f - r \sum_{i=1}^{m} \ln g_i + \sum_{i=m+1}^{m+n} \frac{h_i^2}{r} \tag{7.4}$$

   where f = original objective function
   P = modified objective function
   r = positive constant
   g_i = inequality constraints in Eq. (7.2)
   h_i = equality constraints in Eq. (7.3)

   Then the unconstrained multi-variable methods are used. Initially a value of r is chosen that may or may not be a feasible solution. If the point obtained is not feasible, a penalty is applied to the modified objective function to force the search back into the feasible region. As the algorithm progresses, r is reevaluated in a monotonically decreasing sequence. As r approaches zero under suitable conditions, P approaches f, and the problem is solved. The advantage of this method is that no feasible starting solution is necessary. The flow sheet for this algorithm is shown in Fig. 7.6.

   Another method where the constrained problem is made unconstrained uses the Fletcher-Powell method to solve the unconstrained problem. It is not a penalty function approach. However, in this all the constraints are of equality type and hence is not a very useful method for heat exchanger optimization.

2. Feasible direction method: It is not a penalty function approach and requires a

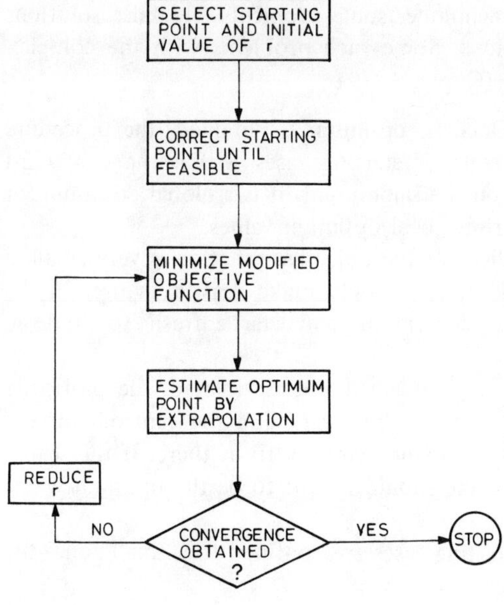

Figure 7.6 Flowsheet for penalty function method algorithm. *(From Kuester and Mize, 1973.)*

starting point satisfying the inequality constraints, i.e., a feasible, nonoptimum, solution. It uses one-dimensional search methods. If a constraint is violated, a point is located on that limiting value of the constraint, and the search is moved to a new direction, moving away from the violated constraint until another constraint is violated or a one-dimensional search minimum is located (Shah et al., 1979).

3. Complex method: This does not convert the constrained problem into an unconstrained one but uses the unconstrained methods with checks to see if any constraint is violated. If this occurs, the point is relocated inside the constrained region in a given manner. Both the inequality and equality constraints can be handled. It is also called a case study method and is used often. The only disadvantage is that it requires $N + 1$ (preferably $2N + 1$) starting feasible cases that satisfy the explicit and the implicit constraints, N being the number of the explicit variables. However, when one feasible case is formulated, the rest can be formulated easily based upon the first case and the user's judgment.

   Once the requisite number of feasible cases is obtained, the objective function is evaluated at each point. The worst point is rejected and is replaced by a point that is αd away from the centroid of the remaining points, where α is a constant (=1.3) and d is the distance between the worst point and the centroid. The new point is located in the direction opposite to that of the worst point (Fig. 7.7). The new point is checked for violation of the explicit and implicit constraints, and suitable corrective measures are taken as warranted. Convergence is assumed if the objective function values at each point are within a small interval for a given number of iterations. A flow sheet for this algorithm is given in Fig. 7.8.

7.14 What are the possible pitfalls in the complex method?

In any numerical optimization technique, some problems and false solutions are likely to come up in a few situations. Some such problems with the complex method and their possible solutions are (Gupta, 1979):

1. Convergence on a local optimum: Once an optimum is obtained, the procedure should be repeated using different sets of starting cases within the constraint limits. If all the solutions converge on the same point, it is a global optimum, or else one has to decide among the various local optimum values.
2. Moving on a flat surface: If the gradient of the objective function is very small, a local optimum will result. To check it, one should make drastic changes in the values of the variables within the constraints and start afresh to find an optimum.
3. Convergence on a boundary: If an explicit constraint is violated, the particular variable should be put midway between the boundary and the centroid instead of putting it at the boundary, which might prematurely fix it there. If the search still fixes the value at the boundary, one should take it to be the optimum value for that variable.
4. Collapse of the search area: If the starting cases are confined to a small zone, the

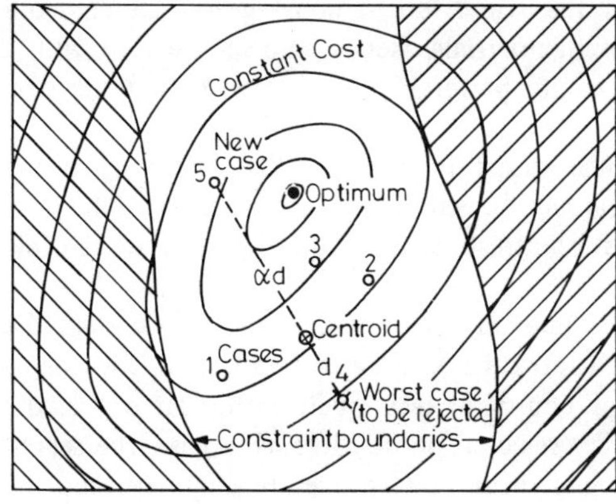

Figure 7.7 Illustration of complex method. *(Courtesy Palen et al., 1974; AIChE Symposium Series.)*

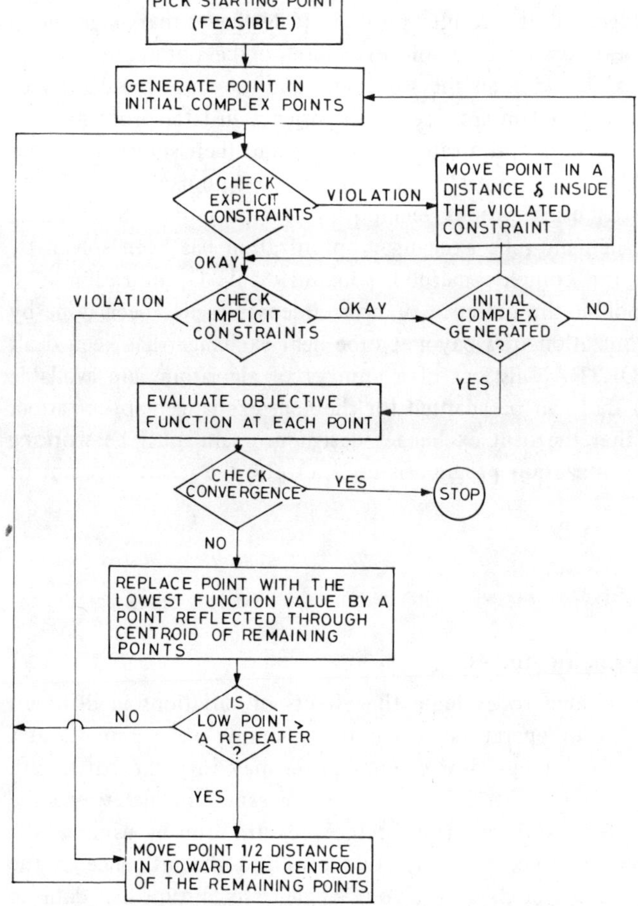

Figure 7.8 Flow sheet for complex method algorithm. *(From Kuester and Mize, 1973.)*

search area will be very small, and the search might converge on the centroid producing little variations in the subsequent trials. Hence, one should start with a large spread of the feasible solutions and preferably use $\alpha = 2$ so that the new cases produce a wider search region.

5. Centroid is worse than the individual cases: In such a situation or when the centroid violates any of the constraints, the choice of a new case would pose problems, and one may have to start afresh using a different set of feasible solutions.
6. Repeated transfer along one straight line: If the new point is worse than the earlier rejected point, the search will move back along the same straight line, a little away from the earlier rejected point but not much better. (If $\alpha = 1$, the search will move exactly to the earlier rejected point.) In such a case, the search should be moved away from the region near that straight line by rejecting the second worst point instead of the worst one.

In spite of the above pitfalls, it is still a good method for the heat exchanger optimization. One should get all the results printed. It is likely that engineering judgment might dictate the adoption of a solution close to the optimum, but not the optimum itself. Besides, knowing all the solutions, one can trace the history of optimization of the particular problem and use the experience in later problems.

Occasionally, heat exchanger optimization problems and their solutions appear in the literature. A problem of compact heat exchanger optimization has been solved by Afimiwala (1976) using various techniques as reported by Shah et al. (1979). One problem of shell-and-tube exchanger optimization has been solved by Palen et al. (1974) using the complex method [Box (1965)]. Optimization of a condenser for a single vapor in the presence of noncondensables has been done by Reddy et al. (1977). Optimization of a bayonet tube heat exchanger has been dealt with by Baum (1979). FORTRAN listings of a number of algorithms are available (Kuester and Mize, 1973) that can be adapted for the heat exchanger optimization. It should be emphasized that the heat exchanger design program must be working satisfactorily before the optimization program is used, or else the result may not be the optimum.

PARAMETRIC STUDIES

7.15 What is meant by parametric studies?

These are the studies done to evaluate the effects of variations in different parameters on the design and/or operation of a unit. Generally, one parameter at a time is varied through its total range. For example, one may vary the baffle cut, while keeping the rest of the geometrical parameters the same and determine the influence on the heat transfer coefficient and on the pressure drop by using a heat exchanger design computer program. Similarly, one can find the influence of the baffle spacing, seal strips, tube pitch and layout angle, etc. From the data so obtained, one can determine the most optimum values of the different parameters. Webb and Scott (1980) have done parametric studies on internally finned tubes.

7.16 What are the optimum conditions for a single-segmental baffled exchanger?

The single-segmental baffled exchanger continues to be the most used because of a lot of experience with its design and operation. The optimum conditions for a single-segmental baffled design based on varied experience of a large number of designers and users are the following:

1. Baffle cut: 25 percent
2. Baffle spacing: It should be such that the velocity of the fluid through the baffle window is of the same order of magnitude as the fluid velocity in cross-flow over the tube bundle. Usually the baffle spacing is more in the laminar flow than in the turbulent flow in order to reduce the shell-to-baffle clearance leakage stream in the former case.
3. Seal strips: One pair of seal strips should be used for every four to seven tube rows in all floating-head-type designs.
4. Tube length and diameter: Within the permissible pressure drop, one should use the smallest tube diameter and maximum tube length. It results in the maximum area in a smaller diameter shell. It is more economical also to fabricate a smaller diameter and longer shell compared to a larger diameter and shorter shell. Of course, the fouling and cleaning requirements should be kept in mind.
5. Tube layout: Generally 30°, however it is also influenced by the cleaning and maintenance requirements.

HEAT EXCHANGER NETWORK OPTIMIZATION

7.17 What is heat exchanger network optimization?

In any large plant, there are a number of process streams that either need to be heated or cooled as per the process requirements. In general, one would use steam or other heat transfer media to heat the cold streams, and cooling water or air to cool the hot streams. However, if the streams that need to be cooled down could transfer their energy to the streams that need to be heated up, a large amount of savings would result from the reduced requirements of the service fluids (steam, water) and the number of heat exchangers, heaters, and coolers would also be greatly reduced, thus saving both the capital and the operating costs. Such a combination of units, which, working in unison, keeps the plant going, is called a heat exchanger network. Optimization of such a system is called the heat exchanger network optimization and yields a minimum number of heat exchangers and the lowest possible amount of service utilities requirements.

7.18 What are the advantages and disadvantages of heat exchanger network optimization?

The advantages are the following:

1. Minimum capital and operating costs since the number of units is reduced. It also cuts down on the foundations, maintenance, etc.

2. Minimum external energy consumption in terms of steam
3. Minimum loss of energy to cooling water or air

Items (2) and (3) are very important due to the increasing concern about the direct energy costs.

The disadvantages are the following:

1. In a large plant, some of the streams may be located at a considerable distance away. Bringing them together to exchange heat may require modifying the plant layout, laying long pipelines, and high pumping power, which may not be economical.
2. Intermixing of two fluids due to leakage of one into the other may cause a hazardous situation in some cases.
3. Interconnection of various streams affects the start-up and shutdown operations, response to emergency and upset conditions, and reliability and control of the plant. Generally, it is more feasible to do network optimization in a new plant than in an existing plant.
4. The assumptions regarding the overall heat transfer coefficients, type of units, etc., place a severe constraint on the reliability of the final result obtained by the heat exchanger network optimization. Each unit in the network needs to be looked at separately (see Q. 7.20).

7.19 How is heat exchanger network optimization done?

There are several methods available in the literature, such as Nishida et al. (1977) and Linnhoff and Flower (1978), and more are being evolved. Basically, the idea is to match the capability of the hot streams to give energy with the requirements of the cold streams needing energy such that the highest possible temperature changes, satisfying the requirements, are obtained. If the final required temperatures are not achieved, then steam and cooling water are used, as the case may be, to achieve them. Srinivasan (1979) has compared the algorithmic evolutionary synthesis method given by Nishida et al. (1977) and the temperature interval evolutionary method given by Linnhoff and Flower (1978) and found the final networks and their costs calculated by the two methods to be the same. The two methods are quite involved (commensurate with the problem they attempt to solve). and they cannot easily be explained without a full discussion and examples. Therefore, the original references should be consulted.

7.20 Is each heat exchanger in the resulting network also an optimum?

Not necessarily. Among the important assumptions made in the network optimization methods currently available are the following:

1. The overall heat transfer coefficients are the same for all the units of any given type, viz. all the heat exchangers have one value of the overall heat transfer coefficient, all the coolers have one value, and all the heaters have one value, although the value for the coolers may be different than that for the heaters or the heat exchangers. It is, however, well known that the heat transfer coeffi-

cients vary significantly depending upon the properties, temperatures, and the flow rates and will not be the same for all the coolers and all heaters, etc.

2. The cost is taken as an exponent of the area of the units. This may be a reasonable first approximation if all the units had similar construction features and materials of construction. However, the operating conditions, differential thermal expansion, heat load, etc., determine the type of construction to be used in each case, and corrosion and fouling tendencies, besides many other factors, determine the material of construction (Chapter 10). Thus, all the units doing the same kind of service may not be similar, and hence their costs would vary significantly. Therefore, it is necessary to look at each unit individually after the network optimization has been completed and design it in an optimum fashion using one of the methods stated earlier, but keeping the entry and exit temperatures the same as obtained by the network optimization for each unit. It may be that the final cost then comes out to be more than the cost predicted by the network optimization. During individual unit optimization, the points mentioned under "disadvantages" in Q. 7.18 should be kept in mind in order to obtain viable units and to ensure a smooth running of the plant.

● This has been a brief introduction to the vast topic of dynamics and optimization of heat exchangers. As the sizes increase and costs escalate, more and more emphasis will need to be placed on such studies. Attempts to modify the existing techniques occasionally appear in the literature, e.g., Prasad and Saibabu (1982). Newer techniques too are being evolved.

NOMENCLATURE

a lower limit on a variable
b upper limit on a variable
BP baffle spacing or baffle pitch
c fixed value of a variable
d upper limit on a variable; distance between worst point and centroid
d_0 tube outer diameter
d_s shell inside diameter
f function to be minimized for optimum design
g_i inequality constraints on optimization
h_i equality constraints on optimization
i index
m index
n index
N number of explicit constraints
p_t pitch
P modified objective function
r positive constant

Δt time interval
X vector containing the variables to be optimized
y variable
z variable

Greek Letters

α a constant
δ distance

REFERENCES

Afimiwala, K. A. Ph.D. thesis, Mechanical Engineering Department, State University of New York at Buffalo, New York. 1976.

Banerjee, S., and W. T. Hancox. Transient Thermohydraulics Analysis for Nuclear Reactors, Keynote Lecture. *Heat Transfer 1978. Proceedings of the Sixth International Heat Transfer Conference, Toronto, August 1978.* 6:311–337. Hemisphere: Washington, D.C. 1979.

Baum, A. J. Optimum Design of a Bayonnet Tube Heat Exchanger. *Heat Transfer 1978. Proceedings of the Sixth International Heat Transfer Conference, Toronto, August 1978.* 4:285–289. Hemisphere: Washington, D.C. 1979.

Bergles, A. E., J. G. Collier, J. M. Delhaye, G. F. Hewitt, and F. Mayinger.*Two-Phase Flow and Heat Transfer in the Power and Process Industries.* Hemisphere: Washington, D.C. 1981.

Box, M. J. A New Method of Constrained Optimization and Comparison with Other Methods. *Computer J.* 8:42–52. 1965.

Forghieri, R., and G. Papa. Dynamic Response of Counter-Current Heat Exchangers to Temperature Disturbances and Step Flow Variations. *Heat Transfer 1978. Proceedings of the Sixth International Heat Transfer Conference, Toronto, August 1978.* 4:291–295. Hemisphere: Washington, D.C. 1979.

Fox, R. L. *Optimization Methods for Engineering Design.* Addison-Wesley: Reading, Massachusetts. 1971.

Fraas, A. P., and M. N. Ozisik. *Heat Exchanger Design.* Wiley: New York. 1965.

Gupta, J. P. *Heat Exchanger Design–A Practical Look,* 2d ed. C. S. Enterprises: Delhi. 1979.

Gutierrez, H. J., and A. R. Cooper. Heat Exchanger Process Dynamics Review. *Chem. Eng. J.* 17:13–18. 1979.

Hall, R. S., J. Matley, and K. J. McNaughton. Current Costs of Process Equipment. *Chem. Eng.* 87(7):80–116. 1982.

Himmelblau, D. M. *Applied Nonlinear Programming.* McGraw-Hill: New York. 1972.

Kanoh, H. Some Aspects of Dynamics and Control of Heat Exchangers. In *Simulation of Distributed-Parameter and Large Scale Systems,* S. G. Tzafestas (ed.). North-Holland: Amsterdam. 1980.

Kays, W. M., and A. L. London. *Compact Heat Exchangers,* 2d ed. McGraw-Hill: New York. 1964.

Kern, D. Q., and A. D. Kraus. *Extended Surface Heat Transfer.* McGraw-Hill: New York. 1972.

Kuester, J. L., and J. H. Mize. *Optimization Techniques with Fortran.* McGraw-Hill: New York. 1973.

LeGoff, P. The Energetic and/or Economic Optimisation of Heterogeneous Reactors. *Chem. Eng. Sci.* 35:2029–2063. 1980.

LeGoff, P., and M. Giulietti. A New Energy-Cost Characteristic Diagram for Optimising Heat Exchangers. In *Heat Exchangers: Theory and Practice.* J. Taborek, G. Hewitt, and N. H. Afgan (eds.). Hemisphere: Washington, D.C. 1983.

Linnhoff, B., and J. R. Flower. Synthesis of Heat Exchanger Networks–I and II. *AIChE J.* 24(4): 633–642 and 642–654. 1978.

Mayinger, F., and M. Schult. Dynamic Behaviour of Double-Phase-Change Heat Exchangers. In *Heat Exchangers: Thermal-Hydraulic Fundamentals and Design.* S. Kakac, A. E. Bergles, and F. Mayiner (eds.). Hemisphere: Washington, D.C. 1981.

Myers, G. W., J. W. Mitchell, and R. F. Norman. The Transient Response of Crossflow Heat Exchangers, Evaporators and Condensers. *Trans. ASME, J. Heat Trans.* 89:75–80. 1967.

Nishida, N., Y. A. Liu, and L. Lapidus. Studies in Chemical Process Design and Synthesis–III: A Simple and Practical Approach to the Optimal Synthesis of Heat Exchanger Networks. *AIChE J.* 23(1):77–93. 1977.

Palen, J. W., T. P. Cham, and J. Taborek. Optimization of Shell-and-Tube Heat Exchangers by Case Study Method. *AIChE Symp. Ser. No. 138.* 70:205–214. 1974.

Patankar, S. V., and D. B. Spalding. A Calculation Procedure for the Transient and Steady-State Behavior of Shell-and-Tube Heat Exchangers. In *Heat Exchangers: Design and Theory Sourcebook.* N. Afgan and E. U. Schlünder (eds.). McGraw-Hill: New York. 1974.

Prasad, K. B. S., and K. B. S. Saibabu. A Probabilistic Approach to Modify the Complex Method. *Computers Chem. Eng.* 6:97–99. 1982.

Ravindran, P. V., and C. D. Rao. Heat Exchanger Dynamics and Parameter Dependence on Flow Rates. *Indian J. Tech.* 16(1):1–4. 1978.

Reddy, R. K., A. Husain, and Y. B. G. Varma. Optimal Design of Condensers. *J. Sci. Indust. Res.* 36(8):370–378. 1977.

Shah, R. K. The Transient Response of Heat Exchangers. In *Heat Exchangers: Thermal-Hydraulic Fundamentals and Design.* S. Kakac, A. E. Bergles, and F. Mayinger (eds.). Hemisphere: Washington, D.C. 1981.

Shah, R. K., K. A. Afimiwala, and R. W. Mayne. Heat Exchanger Optimization. *Heat Transfer 1978. Proceedings of the Sixth International Heat Transfer Conference, Toronto, August 1978.* 4:185–191. Hemisphere: Washington, D.C. 1979.

Spigt, C. L., F. van der Walle, and H. I. Bardoux. Prediction of Performance Characteristics of Heat Exchanging Equipment. In *Heat Exchangers: Design and Theory Sourcebook.* N. Afgan and E. U. Schlünder (eds.). McGraw-Hill: New York. 1974.

Srinivasan, S. Optimization of A Heat Exchanger Network, M. Tech. Thesis, Chemical Engineering Dept. I.I.T. Kanpur, India. 1979.

Tan, K. S., and I. H. Spinner. Dynamics of a Shell-and-Tube Heat Exchanger with Finite Tube-Wall Heat Capacity and Finite Shell-Side Resistance. *I & EC Fundamentals* 17:353–358. 1978.

Tubular Exchanger Manufacturers' Association. *Standards of the Tubular Exchanger Manufacturers' Association,* 6th ed., TEMA: Tarrytown, New York. 1978.

Webb, R. L., and M. J. Scott. A Parametric Analysis of the Performance of Internally Finned Tubes for Heat Exchanger Application. *Trans. ASME, J Heat Trans.* 102(1):38–43. 1980.

Wilde, D. J., and C. S. Beightler. *Foundations of Optimization.* Prentice-Hall: Englewood Cliffs, New Jersey. 1967.

Yamashita, H., R. Izumi, and S. Yamaguchi. Analysis of the Dynamic Characteristics of Cross-Flow Heat Exchangers with Both Fluids Unmixed. *Bull. JSME* 21(153):479–485. 1978.

CHAPTER

EIGHT

CODES AND STANDARDS

INTRODUCTION

• A pressure vessel, if it ruptures, can cause harm to life and property by its pieces moving great distances at high speeds and/or by contaminating the environment with toxic or radioactive elements. All through the nineteenth century and the early parts of the twentieth century, badly designed and fabricated boilers exploded, killing a number of people. In 1911, the American Society of Mechanical Engineers (ASME) appointed a committee "to formulate standard specifications for the construction of steam boilers and other pressure vessels and for their care in service" (Chuse, 1977). The preliminary report of the committee prepared in 1913 became, after many discussion and meetings, the first "ASME Rules for Construction of Stationary Boilers and for Allowable Working Pressures," 1914 edition, as adopted in the spring of 1915. Since then it has been expanded and issued at regular intervals. Various other countries have also issued codes for the same purpose (Evans, 1978).

A heat exchanger is also a pressure vessel. Hence, its construction as a pressure envelope is guided by the codes. In addition, it has many internal parts (Chapter 1) necessary for its working as a heat exchanger. There are no codes yet governing their complete designs and manufacturing tolerances, etc. A number of very experienced heat exchanger manufacturers got together to formulate general rules under the title *Standards of the Tubular Exchanger Manufacturers' Association* (abbreviated as TEMA Standards). The first edition was issued in 1941 and the sixth in 1978. The TEMA Standards are not codes but they must be followed if so specified by the buyer. Most buyers world-wide do specify them. A brief and informative history of the development of the TEMA Standards has been provided by Ringo (1980).

While the TEMA Standards are put together solely by the manufacturers' representatives, the American Petroleum Institutes' Standards (API Standard 660) (American Petroleum Institute, 1976) are developed by the representatives of both the manufacturers and the users. Hence, these standards are wider ranging and more general in nature. They point out many areas where the customer must provide specific information so that the fabricated unit is more efficient and relatively trouble-free.

Thus, the codes and the standards given guidance for the mechanical design and structural integrity of the unit. Their objective is best described by the following paragraph (ASME, 1950):

> The objective of Code rules and Standard (apart from fixing dimensional values) is to achieve minimum requirements for safe construction, in other words, to provide public protection by defining those materials, design, fabrication and inspection requirements whose omission may radically increase operating hazards. . . . Experience with Code rules has demonstrated that the probability of disastrous failure can be reduced to the extremely low level necessary to protect life and property by suitable minimum requirements and safety factors. Obviously, it is impossible for general rules to anticipate other than conventional service, ... Suitable precautions are therefore entirely the responsibility of the design engineer guided by the needs and specifications of the user.

In this chapter, we shall discuss the salient features of the codes, and TEMA and API Standards. Brief mention of the HEI (Heat Exchange Institute) Standards will also be made. Sections II, III, V, IX, and XI of the ASME Codes are discussed in Chapters 5, 9, 10, 11, and 12, whereas large number of points related to the TEMA Standards are discussed in Chapters 1, 2, 9, 10, 11, 13, and 14. Therefore, only such topics as are not discussed elsewhere are discussed in this chapter.

CODES

8.1 What is a pressure vessel?

Based on the general usage of the term, a pressure vessel is a closed container of limited length (in contrast to very large length of piping), the walls of which are subjected to a net differential pressure, whether internal or external.

8.2 Which vessels are subject to internal pressures?

A vast majority of vessels are subject to internal pressure. Some examples are the storage tanks, gas cylinders, heat exchangers, reboilers, etc.

8.3 Which vessels are subject to external pressures?

Jacketed vessels and vessels under vacuum, such as surface condensers (Chapter 3) are subject to external pressures. Also, the vessels used in deep sea surveying are under external pressure. Some vessels may accidently be subject to external pressure if an internal vacuum is created due to fast drainage without adequate venting or the lowering of temperature that might condense the vapor inside. Vessels designed to withstand several atmospheres of internal pressure may not withstand 1 atm of external pressure.

8.4 What is a pressure vessel code?

It is a set of rules and recommendations for material selection, design, fabrication, testing, and inspection that outlines a discipline within which the pressure vessel has to be manufactured. This set of rules is popularly known as "code," and in formulating these rules, the endeavor is to lay down the minimum standards to ensure safe use of the vessel within the design conditions.

8.5 Is code a legal document?

Code becomes a legal document in a city/state/country if the government concerned passes an appropriate legislation making it a legal requirement.

8.6 What are the more popular codes?

The most popular code is the ASME Code (American), which is followed in the Unites States, most of Canada, and in many other countries. There are other codes followed in some countries. Such codes have been issued by Britain, France, Germany, Holland, India, Italy, Japan, Spain, Sweden, and the International Standards Organization.

8.7 What are the salient differences in the pressure vessel codes issued by different countries?

It is a very vast topic, and reference should be made to authoritative articles for detailed information (Evans, 1978; Mahajan, 1979). (See also *Hydrocarbon Processing,* volume 50, number 6, June 1971 for several articles on this subject). Basically, the codes differ in their legal status in their own countries; range of applicability with regards to pressure, diameter, volume, temperature, material of construction, etc.; whether applicable for external pressures and stationary and/or mobile units; whether design stresses are based upon the tensile stress or yield stress; the safety factor considered in the design; requirements if a vessel is fabricated in one country and imported into another with different codes; etc. Some codes also give details about the procedures to be adopted in welding, inspection and testing, stamping symbols, details in the stamping form on the vessel, etc.

• The remaining discussion below on codes is primarily confined to the ASME Codes because of their wide spread usage.

8.8 What are the various sections of the ASME Codes?

There are eleven sections, some of which have subsections also as given below:

Section I: Power Boilers
Section II: Material Specifications
Part A: Ferrous Materials
Part B: Nonferrous Materials
Part C: Welding Rods, Electrodes, and Filler Metals
Section III: Division 1 and Division 2
Subsection NCA: General Requirements of All Subsections
Division 1: It has the following subsections
Subsection NB: Class 1 Components
Subsection NC: Class 2 Components
Subsection ND: Class 3 Components
Subsection NE: Class MC Components
Subsection NF: Component Supports
Subsection NG: Core Support Structures
Appendices for Subsections NB to NG
Division 2: Code for Concrete Reactor Vessels and Containments
Section IV: Heating Boilers
Section V: Nondestructive Examination
Section VI: Recommended Rules for Care and Operation of Heating Boilers
Section VII: Recommended Rules for Care of Power Boilers
Section VIII: Pressure Vessels
Division 1
Division 2 (Alternative Rules)
Section IX: Welding and Brazing Qualifications
Section X: Fiberglass-Reinforced Plastic Pressure Vessels
Section XI: Rules for Inservice Inspection of Nuclear Power Plant Components–Division 1

8.9 Which section of the ASME Code is used most often?

ASME Code Section VIII Division 1 (ASME, 1980) is used most often since pressure vessels are encountered in almost all the industries.

8.10 How often is the ASME Code issued?

Because of the technological developments and new design situations that often arise, the new editions of the ASME Codes are presently issued every 3 years. In order to keep the users informed, addenda are issued every 6 months to update the codes between two consecutive editions.

8.11 What are the main differences between ASME Code Section VIII Division 1 and Division 2?

The differences refer to the applicability with regards to the pressure and temperature, permissible stresses and stress analysis reports, materials permitted, and their testing and fabrication details (Mahajan, 1979; S. N. Sengupta, 1982, personal communication), etc. These are given in Table 8.1.

Table 8.1 Main differences between ASME Code Section VIII, Division 1 and Division 2

	Section VIII, Division 1	Section VIII, Division 2[a]
Permissible stress	$\frac{1}{4}$ of tensile strength (safety factor of 4)	$\frac{1}{3}$ of tensile strength (safety factor of 3, therefore thinner wall than in Division 1)[b]
Range of pressure	Up to 3000 psi (20684 kPa)	Pressure can exceed 3000 psi (20684 kPa). For very high pressure some deviations from these rules may be necessary
Range of temperature	At elevated temperatures also	Below creep point
Structural quality plate AS-283	Permitted	Not permitted
Cryogenic service	Impact test of material not required	Impact test of material required
Secondary stresses, fatigue, cyclic conditions, etc.	Ignored	Taken into account
Full radiography examination	Not always required	Almost always required
Complete design report and stress analysis	Not required	Required
Alternate service and location for the vessel	Allowed	Not allowed
Cost of material	More	Less
Cost of engineering design, analysis, etc.	Less	More
Material testing	As per material specification requirements	More stringent in selection of test pieces, ultrasonic testing, impact testing, etc.
Mobile vessels	Permitted	Not permitted
Fabrication requirements	Less strict than Division 2	More strict than Division 1. Some requirements are (i) generally, nozzle to shell attachment welds should be full penetration welds; (ii) minimum requirements laid down for rounding the edges of nozzles/vessels at the opening; (iii) more thorough NDE of welds required

[a]A summary of situations when Division 2 should be used and when it should not be used is given by Smolen and Mase (1982).

[b]The difference in the allowable stress for Division 1 and Division 2 decreases as the design temperature increases.

8.12 What is a safety factor? Why is it required?*

It is the ratio between the material strength at the design temperature to the maximum stress level permitted in the material at that temperature. The material strength may be the ultimate tensile strength or the yield strength (proof stress). At elevated temperatures, creep strength has also to be considered (see Chapter 9 for definitions).

The safety factor is required to allow for uncertainties in the loading conditions, load distributions, stress concentration factors, and variations in the material properties. Hence, the safety factor is also called a factor of uncertainty (Fraas and Ozisik, 1965). In the ASME Code Section VIII, Division 1 it used to be five but was change to four in 1930's. In the ASME Code Section VIII, Division 2, it is three.

8.13 How do the safety factors used in some of the popular Codes differ from one another?

The safety factors may be applied to the "ultimate tensile strength" and/or the "yield strength or proof strength." Their values in eight different national codes are given in Table 8.2.

*Q. 8.12 and Q. 8.13 have been contributed by S. N. Sengupta (1982, personal communication).

Table 8.2 Safety factors in different national codes

	Safety factor applied to	
Code	Ultimate Tensile Strength	Yield Strength or proof stress
ASME Section VIII, Division 1	4	1.5[a]
ASME Section VIII, Division 2	3	1.5[a]
AD Merkblatt (German Code)	3.5[b]	1.5
BS 5500 (British Code)	2.35	1.5
ANCC (Italian Code)	2.0[c]	1.5
SNCT (French Code)	3 or 2.4[d]	1.6 or 1.5 or 1.9[d]
IS 2825 (Indian Code)	3	1.5
Swedish Pressure Vessels Code	–	1.5

[a]Though the safety factors on the yield strength is the same in these two codes, the tabulated "maximum allowable stress" values are higher in the case of Division 2 compared to Division 1 in the range where yield strength governs.

[b]This safety factor on the ultimate tensile strength is applied only in cases where the material has no guaranteed yield strength.

[c]This safety factor on the ultimate tensile strength is applied only in cases where the ratio of yield strength to ultimate tensile strength exceeds 0.75.

[d]Depending upon whether or not the gas/steam pressure vessel is subjected to French regulation and also whether manufactured with or without inspection.

Note: The various Codes also differ in classification of vessels in different classes as far as their joint efficiency, extent of radiography, design temperature and pressure are concerned and also the corrosion allowance required. A good summary is given by Bhattacharya (1977).

8.14 Since a safety factor of four is used in Division 1 compared to a safety factor of three in Division 2, will a vessel fabricated under Division 1 meet the requirements of Division 2?

Not necessarily so since the restricted choice of materials, the closer inspection, and detailed stress analysis requirements for Division 2, as noted in Q. 8.11, may not be met by a vessel fabricated under Division 1.

8.15 What conditions are not covered in the ASME Codes?

The ASME Codes provide adequate safety under most circumstances. However, these cannot take into account all the conditions that a pressure vessel might encounter under abnormal situations not envisaged at the design stage. Some such abnormal situations (Perry and Chilton, 1973; Faupel, 1979; Ford et al., 1981) are unusually low temperatures, unusual thermal cycling and stresses, very high pressures, local hot spots, explosions, runaway reactions, unusual vibrations of the vessel due to wind, attack by the material contained in the vessel, vessels of noncircular cross sections, etc. The user and the designer must be careful about such situations and should attempt to account for them at the design stage itself.

Furthermore, the ASME Codes cover only the new construction exclusively. For maintenance, inspection, and repair, NBIC and the American Petroleum Institute have proposed codes to be followed by the process industry (Buhrow, 1980).

8.16 What vessels are omitted from construction under the rules of the ASME Code Section VIII Division 1?

These are the following (ASME, 1980):

1. Vessels covered in other sections
2. Fired process tubular heaters
3. Parts of mechanical devices in which the stresses and design considerations derive from the mechanical functions such as pumps, compressors, and engines.
4. Piping systems
5. Piping components
6. Water vessels of 120 gal (0.45 m^3) capacity or less
7. Hot water supply storage tanks limited to 200,000 Btu/hr (58,800 W) heat input, a maximum water temperature of 200°F (93°C) and 120 gal (0.45 m^3) maximum nominal capacity.
8. Vessels of less than 15 psi (103.4 kPa) internal or external pressure with no limitation on size
9. Vessels of less than 6 in (152.4 mm) inside diameter with no limitation on pressure

8.17 What are the various Subsections, Parts, and Appendices of the ASME Code Section VIII Division 1?

There are three Subsections (A, B, and C) and thirteen parts followed by eight tables, thirteen mandatory and eighteen nonmandatory appendices as detailed below:

Subsection A (Part UG) gives general requirements

Subsection B covers methods of fabrication (Part UW: Welding, UF: Forging, UB: Brazing)

Subsection C covers the classes of materials (Part UCS: Carbon and Low-Alloy Steels, UNF: Nonferrous Materials, UHA: High-Alloy Steels, UCI: Cast Iron, UCL: Clad Plates, UCD: Cast Ductile Iron, UHT: Ferritic Steels, ULW: Layered Construction, ULT: Low Temperature Construction)

Mandatory appendices (1–13) give details of the procedures frequently referred to in the Code

Nonmandatory appendices (A–Z with I, J, N, O, Q, U, V, and X omitted) give information and suggested procedures that are helpful, although not essential

8.18 What are the different Parts and Appendices of Section VIII Division 2?

It has eight parts followed by ten mandatory and nine nonmandatory appendices—Part AG: General Requirements, AM: Material Requirements, AD: Design Requirements, AF: Fabrication Requirements, AR: Pressure Relief Devices, AI: Inspection and Radiography, AT: Testing, AS: Marking, Stamping, Reports, and Records.

Mandatory Appendices (1-6, 8-10, and 18) give details of the procedures frequently referred to in the code.

Nonmandatory Appendices (A–I) give details of procedures which are helpful though not essential and also give examples of application of code formulas, SI units, etc. The number of appendices may vary with the newer editions of the code.

8.19 What are design pressure, maximum allowable working pressure, calculated test pressure, and operating pressure for a pressure vessel?

The definitions, as per the ASME Code Section VIII Division 1 are the following (ASME, 1980):

1. Design pressure: This is the pressure used in the design to determine the minimum thickness of the vessel exclusive of the corrosion allowance.
2. Maximum allowable working pressure: This is the maximum gauge pressure at the top of the vessel under designed operating conditions. It is the least of the values found for the maximum allowable working pressure for any of the essential parts of the vessel.
3. Calculated test pressure: This it the highest permissible internal pressure as determined by the design formulae using nominal thicknesses with corrosion allowances included and using the stress values for the temperature of the test.
4. Operating pressure: This is the pressure at the top of the vessel under normal operation. It does not exceed the maximum allowable working pressure.

8.20 What are required thickness, design thickness, and nominal thickness for a pressure vessel?

The definitions are the following (ASME, 1980):

1. Required thickness: Minimum thickness as computed by the application of the code formulae before the addition of the corrosion allowance.

2. Design thickness: Required thickness plus corrosion allowance
3. Nominal thickness: Commercially available minimum thickness $\geqslant$ design thickness

8.21 What are design temperature and operating temperature for a pressure vessel?

The definitions are the following (ASME, 1980):

1. Design temperature: This is the temperature of the part as used in the design. It is the highest probable mean metal temperature for temperatures above −20°F (−28.9°C) and the lowest probable mean metal temperature for temperatures below −20°F (−28.9°C).
2. Operating temperature: This is the actual temperature under normal operation.

8.22 What are the various classes of construction under Section III of the ASME Boiler and Pressure Vessel Code?

There are four sets of construction rules in Section III. These are designated as Code Class 1, 2, 3, and MC (Q. 8.8). These apply to pressure-retaining or pressure-containing components of a nuclear power system and containment system. These do not provide guidance in the selection of the code class for a specific system. The selection of the code class is based upon the applicable safety criteria or the requirements of the regulatory authorities (Oak Ridge National Laboratory, 1975).

8.23 Into what class of construction do the heat exchangers in the nuclear service fall?

The heat exchangers fall into Code Class 1, or Code Class 2 of Section III of the ASME Code (Oak Ridge National Laboratory, 1975). Code Class 1 applies to the vessels designed to contain the nuclear fuel and reactor coolant. These are constructed as per Subsection NB of Section III. Code Class 2 vessels are constructed as per Subsection NC of Section III, although they could also be made as per Subsection NB of Section III (see also Chapters 5 and 9).

8.24 In what important aspects does Section III of the ASME Code differ from Section VIII?

Section III provides rules for the minimum requirements for the design, fabrication, inspection, and certification of vessels, parts of vessels, materials used, etc., for nuclear services while Section VIII is for nonnuclear services. Section III requires the most extensive quality control, nondestructive examination, detailed stress analysis, and documentation of all aspects of testing, quality control, analyses, etc., and the availability of the records for the life of the vessels. It is the most restrictive of all the codes (Yokell, 1978).

8.25 What are the various ASME stamping symbols?

Stamping symbols are marked on the vessels fabricated as per the code requirements. A list of ASME stamping symbols and what they signify is given below (Perry and Chilton, 1973):

U: Built in accordance with the code by an authorized fabricator and inspected by an authorized inspector
UM: As above but exempted from inspection. This applies to certain small vessels.
N: Nuclear vessel
H: Heating boiler
S: Power boiler
M: Miniature boiler
A: Boiler assembly
UV: Relief valve
NV: Nuclear safety valve
RP: Reinforced plastic vessel

8.26 What information does the ASME Stamping Form on the pressure vessel contain?

A typical Stamping Form is shown in Fig. 8.1. The details are described below. Numbers 1 to 6 refer to locations on the figure.

1. Stamp *National Board* and the National Board number at top of form if National Board inspection is required.
2. Stamp *Part* under code symbol on those parts of a vessel for which partial data sheets are required.
3. Stamp type of construction and radiographic and postweld heat treatment status of main longitudinal and circumferential joints under code symbol by means of the following letters:

   W: arc or gas welded
   S: seamless
   RES: resistance welded
   F: forged welded
   B: brazed
   RT1: completely radiographed
   RT2: partially radiographed

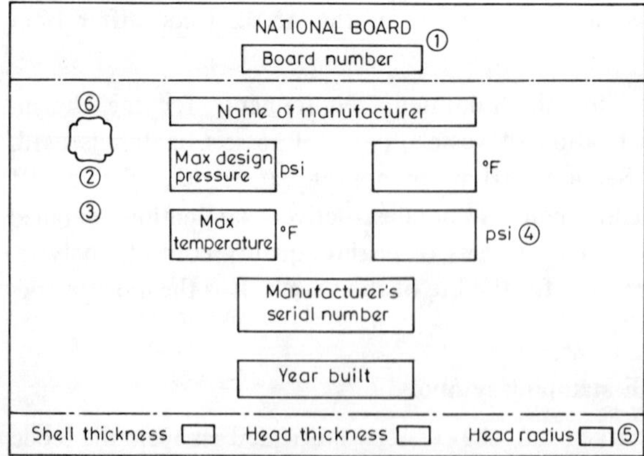

Figure 8.1 Stamping form. *(From* Pressure Vessels: The ASME Code Simplified, *5th ed. by R. Chuse. Copyright © 1977. Used with permission of McGraw-Hill Book Company.)*

RT3: spot radiographed
RT4: part of vessel radiographed where none of RT1, 2, or 3 apply
HT: completely postweld heat-treated
PHT: partially postweld heat-treated

Vessels intended for special service must also be appropriately stamped, separated by a hyphen after lettering as above.

L-lethal service
UB-unfired boiler
D-direct firing

4. Maximum temperature seldom required if less than 650°F (343°C) (Code Paragraph UG-116, footnote 1). For vessels operating below –20°F (–28.9°C), stamping must also give minimum allowable temperature and the minimum and maximum temperatures supplied. If the material has passed the required impact tests, the letters *IT* should be placed after the figures for minimum temperature. When in accordance with Code Paragraph UCS-66 or UHA-51 no impact test is made, stamp *Maximum Allowable Pressure* for coincident temperature of –20°F (–28.9°C) and above. For the operating pressure at coincident temperature below –20°F (–28.9°C), such marking is followed by stamping UCS-66 or UHA-51.
5. Pennsylvania requires shell and head thickness and head radius.
6. Stamp *User* above Code symbol if vessel is inspected by the user's inspection.

8.27 How does the ASME grant and renew its stamp to a fabricator?

A team of inspectors authorized by the ASME inspects very thoroughly the manufacturing facilities, qualifications of the personnel, and, most importantly, the quality control program of the fabricator. Only on their full satisfaction is the stamp(s) granted. For the renewal of the stamp, a triannual inspection is carried out. The UM stamp is renewed annually.

TEMA, API, AND HEI STANDARDS

8.28 What is the scope of application of the TEMA Standards?

TEMA Standards are applicable to shell-and-tube heat exchangers up to the following (Fig. 8.2):

1. An inside diameter of 60 in (1524 mm)
2. A maximum product of nominal diameter (inches) and design pressure (psi) of 60,000 (10500 mm × MPa) or
3. A maximum design pressure of 3000 psi (20,684 kPa). These limit the shell wall thickness to approximately 2 in (76.2 mm). These hold for all three classes of construction R, C, and B. For larger sizes, the Recommended Good Practice (RGP) Section 12 (TEMA, 1978) should be consulted.

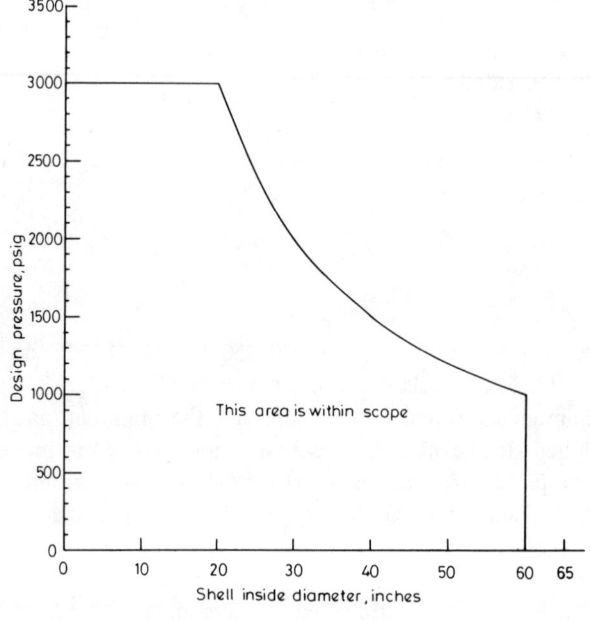

Figure 8.2 Scope of TEMA Standards. *(From Rubin and Gainsboro, 1978.)*

8.29 What are the broad topics covered in the TEMA Standards?

The following is a list of the broad topics covered in the TEMA Standards (1978), and of the chapters in this book where parts of such topics are discussed:

1. Nomenclature (Chapter 1)
2. Fabrication tolerances (Chapter 10)
3. General fabrication and performance information (Chapters 11 and 14)
4. Installation, operation, and maintenance (Chapter 14)
5. Mechanical standards: R, C, and B classes (Chapters 1 and 9)
6. Material specifications (Chapter 10)
7. Thermal standards (Chapters 2 and 13)
8. Physical properties of fluids
9. General information (Chapter 1)
10. Recommended good practice (Chapter 8)

8.30 What are the R, C, and B classes of construction of heat exchangers as specified in the TEMA Standards?

TEMA Class R Standards specify design, fabrication, and materials of unfired shell-and-tube heat exchangers for the generally severe requirements of petroleum and related processing applications. This class is also used in nuclear power plants for the exchangers that do not see any radioactivity.

TEMA Class C Standards specify design, fabrication, and materials of unfired shell-and-tube heat exchangers for the chemical process service. Some examples are surface condensers in thermal power plants, units used in pharmaceutical industry, general applications, etc.

TEMA Class B Standards specify design, fabrication, and materials of unfired shell-and-tube heat exchangers for the generally moderate requirements of commercial and general process applications. Some examples are polymer and textile industry and chemical industry.

8.31 What are the differences between R, C, and B classes of the TEMA Standards?

The differences in R, C, and B classes of the TEMA Standards are given in Table 8.3.

8.32 What is the Recommended Good Practice (RGP) section of TEMA Standards, 6th edition?

It provides the designer with additional information related to the design of shell-and-tube heat exchangers not covered by or beyond the range of the mandatory TEMA Standards. The information in the RGP section is to be taken as a guide only and is not a mandatory requirement of the TEMA Standards.

8.33 Why do the TEMA Standards specify a minimum shell thickness that, for low pressures, may be excessive compared to the value obtained by the Code design formulae plus the corrosion allowance?

The minimum thickness specified by TEMA is adequate for at least 150 psig (1135 kPa) design pressure when constructed from carbon-steel or equivalent materials. This results in excessive thickness for low-pressure design (Ringo, 1980). However, it is desirable because, with very thin shells, many fabrication problems are encountered in bending thin plates to a round shape within the permissible tolerance, and distortions might result from handling, supporting, and stacking. Furthermore, the excess metal in a thicker shell provides reinforcement for the shell openings and protection from external loads.

8.34 When does the TEMA Standards supplement or override the ASME Code specifications?

The ASME Code provides for the thickness of the pressure boundary only, such as the shell, front and rear heads, flanges, and covers (K. P. Singh, 1981, personal communication). The rest of the information comes from the TEMA Standards. This includes the thickness of the pass partition plates and baffles, baffle spacing, tube-to-baffle hole clearance, shell-to-baffle clearance, drill drift, dimensional tolerances, tubesheet design, tube-to-tubesheet joint, nozzle location, nozzle extension into the shell and angularity (Yokell, 1978), gasketed joint details, etc.

In situations where the specifications are provided both by the codes and the TEMA Standards, the latter generally overrides the former. Some such situations are the out-of-roundness in a rolled plate, channel cover thickness, etc.

The interrelationship between codes, standards, and customer specifications were brought out in a seminar during the ASME Winter Annual Meeting (1979). A number of papers were presented by contractors, designers, fabricators, and users (De Haan, 1979; Smith, 1979). These point out that besides the codes and TEMA

Table 8.3 A comparison of R, C, and B classes of TEMA Standards[a,b]

Paragraph	Topic	R	C	B
1.12	Definition	For the generally severe requirements of petroleum and related processing applications	For the generally moderate requirements of commercial and general process applications	For general process service
1.51	Corrosion allowance on carbon steel	$\frac{1}{8}$ in	$\frac{1}{16}$ in	$\frac{1}{16}$ in
2.2	Tube diameters	$\frac{3}{4}$, 1, $1\frac{1}{4}$, $1\frac{1}{2}$, and 2 in O.D.	R+ $\frac{1}{4}$, $\frac{3}{8}$, $\frac{1}{2}$, and $\frac{5}{8}$[c]	R+ $\frac{5}{8}$
2.5	Tube pitch and minimum cleaning lane	1.25 × tube O.D., $\frac{1}{4}$ in minimum lane on a square layout	R+ $\frac{5}{8}$ in tubes may be located 1.2 × tube O.D.	R+ pitch may be $\frac{13}{16}$ in in 12 in and smaller shells for $\frac{5}{8}$ and $\frac{15}{16}$ in for $\frac{3}{4}$ tubes
3.13	Minimum shell diameter	8 in	6 in	6 in
4.42	Longitudinal baffle thickness	$\frac{1}{4}$ in minimum	$\frac{1}{8}$ in alloy, $\frac{1}{4}$ in carbon steel	Same as TEMA Class C
4.71	Minimum tie-rod diameter	$\frac{3}{8}$ in	$\frac{1}{4}$ in in 6–15 in shells	$\frac{1}{4}$ in in 6–15 in shells
5.11	Floating-head cover cross-over area	1.3 times tube flow area of one tube pass	Same as tube flow area	Same as tube flow
5.31	Lantern ring construction	375° F maximum. 300 psi up to 24 in diameter shell, 150 psi for 25–42 in shells, 75 psi for 43–60 in shells	600 psi maximum	Same as TEMA Class R
6.2	Gasket materials	Metal jacketed or solid metals for (a) internal floating-head cover (b) 300 psi and up (c) all hydrocarbons	Metal jacketed or solid metals for (a) internal floating-head cover (b) 300 psi and up Asbestos permitted for 300 psi and lower pressures	Same as TEMA Class C
6.32	Peripheral gasket contact surface	Flatness tolerance ± $\frac{1}{32}$ in maximum	No tolerance specified	No tolerance specified
7.131	Minimum tube-sheet thickness with expanded tube joints	Outside diameter of the tube	0.75 × tube O.D. for 1 in and samller, $\frac{7}{8}$ in for $1\frac{1}{4}$ O.D., 1 in for $1\frac{1}{2}$ O.D., 1.25 in for 2 O.D.	Same as TEMA Class C
7.44	Tube hole grooving	Two grooves	Above 300 psi design pressure and above 350° F design temperature: two grooves	Same as TEMA Class R

(See footnotes on p. 355.)

Table 8.3 A comparison of R, C, and B classes of TEMA Standards[a, b] (*Continued*)

Paragraph	Topic	R	C	B
7.51	Length of expansion	Smaller of 2 in or tubesheet thickness minus $\frac{1}{8}$ in	Smaller of 2 in or tubesheet thickness minus $\frac{1}{8}$ in	Same as TEMA Class R
7.7	Tubesheet pass partition grooves	$\frac{3}{16}$ in deep grooves required	Over 300 psi $\frac{1}{16}$ in deep grooves required or other suitable means for retaining gaskets in place	Same as TEMA Class C
9.3	Pipe tap connections	6000 psi coupling with bar stock plug	3000 psi coupling	3000 psi coupling with bar stock plug
9.32	Pressure gauge connections	Required in nozzles 2 in and up	Shall be specified by purchaser	Same as TEMA Class R
9.33	Thermometer connections	Required in nozzles 4 in and up	Shall be specified by purchaser	Same as TEMA Class R
10.1	Minimum bolt size	$\frac{3}{4}$ in	$\frac{1}{2}$ in recommended, smaller bolting may be used	$\frac{5}{8}$ in

Source: Rubin, F. L. (1980); with permission of Gulf Publishing Co., Houston, Texas.
[a]TEMA (1978).
[b]See specific paragraphs in the TEMA Standards for further details and exceptions.
[c]It implies that all the tube diameters allowed in Class R are permitted in Class C, further $\frac{1}{4}$, $\frac{3}{8}$, $\frac{1}{2}$, and $\frac{5}{8}$ in diameter tubes are also permitted in Class C.

Standards, API Standards 660 (American Petroleum Institute, 1976) should also be considered since they are more practically oriented.

8.35 What are the broad topics covered in the API Standard 660?

The following is a list of the broad topics covered in the API Standard 660 (American Petroleum Institute, 1976):

1. General
2. Proposals
3. Drawings and other data required
4. Guarantee
5. Design
6. Materials
7. Fabrication
8. Inspection and test
9. Preparation for shipment
10. Supplemental requirements
11. Appendices

8.36 What are the general differences between the TEMA Standards and the API Standard 660?

As stated in the beginning, the TEMA Standards are put together only by the manufacturers' representatives, while the API Standard 660 has the benefit of inputs both by the manufacturers' and the users' representatives. Hence, while the TEMA Standards may attempt to safeguard the interests of the manufacturers only, the API Standards distribute the responsibility. Furthermore, the API Standards are a little more rigid. For example, they prohibit packed floating-head construction and make welding neck flanges mandatory for pressures greater than 150 psi (1135 kPa).

8.37 What are the relevant HEI Standards?

The Heat Exchanger Institute has standards for feedwater heaters and surface condensers, among others. They concern themselves with the thermal and hydraulic factors instead of the mechanical design details. The Heat Exchanger Institute has also recently issued standards for power heat exchangers (Heat Exchange Institute, 1980) (see Q. 1.2).

8.38 What are the important items related to heat exchanger design that are not covered by the ASME Code, TEMA, API, or HEI standards?

The ASME Code and the standards do not specify how to analyze the external piping loads on the exchanger nozzles so that the design could be modified to ensure that the exchanger is not overstressed (De Haan, 1979). These do not also consider the effects of other external loads, such as earthquakes, wind, ladder, or platform supports. The unsteady state situations are not dealt with. Design of the split-ring exchangers is not specified by any code or standards, while those of the floating head is covered only by the ASME Code (Smith, 1979).

8.39 What are the various organizations whose codes and standards play a role in the design and/or fabrication of heat exchangers and pressure vessels?

A number of countries have issued codes. Some of the more popular ones, along with the addresses, are the following:

ASME Code: American Society of Mechanical Engineers
345 E. 47th Street
New York, New York 10017

British Code: British Standards Institution
2 Park Street
London WIA 2 BS, England

Dutch Code: Ministry of Social Affairs
Christoffel Plantijnstraal
The Hague, The Netherlands

German Code: Carl Heymanns Verlag KG,
Gereonstrasse 18-32, D-5000 5 Koln,
Federal Republic of Germany

Indian Code: Indian Standards Institution
Bahadurshah Zafar Marg
New Delhi-110002, India

Italian Code: Casa Editrice Luigi Di G.
Pirola, via Comelica (P.O. Box 3680)
Milano, Italy 24688

Japanese Codes: There are three different ones:

1. Japanese Pressure Vessel Code: Japan Boiler Association, 5-35-4 Shiba, Minato-Ku, Tokyo, Japan
2. Japanese Industrial Standard Construction of Pressure Vessels: Japan Standards Association 4-1-24 Akasaka, Minato-ku, Tokyo, Japan
3. Japanese High Pressure Gas Control Law: The Institution for Safety of High Pressure Gas Engineering, 32-Nishikubotomoe-Cho, Minato-ku, Tokyo, Japan

Swedish Code: Tryckkarlskommissionen
IVA, P. O. Box 5073, S-10242
Stockholm, Sweden

Standards related to various aspects of design and fabrication have been issued by many organizations. Some of these are the following:

Tubular Exchanger Manufacturers' Association. (TEMA), 25 North Broadway, Tarrytown, New York 10591

Expansion Joint Manufacturers' Association (EJMA), 25 North Broadway, Tarrytown, New York 10591

American Society for Testing and Materials (ASTM), 1916 Race Street, Philadelphia, Pennsylvania 19103

American National Standards Institute (ANSI), 1435 G Street N.W., Washington, D.C. 20005

American Welding Society (AWS), 550 N.W. 42nd Avenue, Miami, Florida 33127

American Petroleum Institute (API), 2101 L. Street N.W., Washington, D.C. 20037

Heat Exchanger Institute (HEI), 1230 Keith Building, Cleveland, Ohio 44115

American Society for Non-destructive Testing (ASNT), 4153 Arlinggate Plaza, Columbus, Ohio 43228

REFERENCES

American Petroleum Institute. *API Standard 660,* 3d ed. API: Washington, D.C. 1976.

American Society of Mechanical Engineers. *Pressure Vessel and Piping Design: Collected Papers,* p. 689. ASME: New York. 1950.

American Society of Mechanical Engineers. *ASME Boiler and Pressure Vessel Code Section VIII Division 1.* ASME: New York. 1980.

Bhattacharya, N. Manufacturers' Problems in Following Codes for Pressure Vessels and Heat

Exchangers. In *Seminar on Pressure Vessels and Heat Exchangers*. Indian Chemical Manufacturer: Bombay, India. 1977.

Buhrow, R. P. New API Vessel Inspection Code. *Hydrocarbon Processing*, 59(5):135–136. 1980. It refers to National Board Inspection Code, A Manual for Boiler and Pressure Vessel Inspectors; API RP-510, Maintenance Inspection, Rating and Repair of Pressure Vessels.

Chuse, R. *Pressure Vessels: The ASME Code Simplified*, 5th ed. McGraw-Hill: New York. 1977.

De Haan, J. A. A Contractor's Viewpoint of the Interrelationship between Codes, Standards and Customer Specifications for Process Exchangers. In *The Interrelationships between Codes, Standards, and Customer Specifications for Process Heat Transfer*. ASME: New York. 1979.

Evans, F. L. A Special Report: Worldwide Pressure Vessel Codes. *Hydrocarbon Processing* 57(12):89–98. 1978.

Faupel, J. H. Pressure Vessels of Noncircular Cross Section (Commentary on New Rules for ASME Code), *Trans. ASME, J. Press. Vessel Tech.* 101(3):255–267. 1979.

Ford, Sir H., E. H. Watson, and B. Crossland. Thoughts on A Code of Practice for Forged High-Pressure Vessels of Monobloc Design. *Trans. ASME, J. Press. Vessel Tech.* 103(1):2–9. 1981.

Fraas, A. P., and M. N. Ozisik. *Heat Exchanger Design*. Wiley: New York. 1965.

Heat Exchanger Institute. *Standards for Power Plant Heat Exchangers*, 1st ed. HEI: Cleveland, Ohio. 1980.

Mahajan, K. K. *Design of Process Equipment: Selected Topics*. Pressure Vessel Handbook Publ.: Tulsa, Oklahoma. 1979.

Oak Ridge National Laboratory. *Design Guide for Heat Transfer Equipment in Water-Cooled Nuclear Reactor Systems*, ORNL TM-3578. ORNL: Oak Ridge, Tennessee. 1975.

Perry, R. H., and C. H. Chilton, *Chemical Engineers' Handbook*, 5th ed. McGraw-Hill. 1973.

Ringo, D. Heat in History: The Tubular Exchanger Manufacturers' Association, Inc. *Heat Trans. Eng.* 2(1):69–73, 1980.

Rubin, F. L., and N. R. Gainsboro, *Practical Heat Transfer: Shell-and-Tube Heat Exchangers*, AIChE Today Series. AIChE: New York. 1978.

Rubin, F. L. What's the Difference between TEMA Exchanger Classes? *Hydrocarbon Processing* 59(6):92. 1980.

Smith, H. E. The Interrelationship of Codes, Standards and Customer Specifications for Process Heat Exchangers. In *The Interrelationships between Codes, Standards, and Customer Specifications for Process Heat Transfer*, ASME: New York. 1979.

Smolen, A. M., and J. R. Mase, ASME Pressure-Vessel Code: Which Division to Choose? *Chem. Eng.* 89(1):133–136. 1982.

Tubular Exchanger Manufacturers' Association. *Standards of Tubular Exchanger Manufacturers' Association*, 6th ed. TEMA: Tarrytown, New York. 1978. Abbreviated as TEMA Standards. Addenda and Errata issued 1982.

Yokell, S. Design and Fabrication of Heat Exchangers. In *Practical Aspects of Heat Transfer*. AIChE: New York. 1978.

CHAPTER

NINE

MECHANICAL DESIGN; STRESS, SEISMIC, AND FAILURE ANALYSIS

INTRODUCTION

• The structural integrity of a pressure vessel and heat exchanger depends on proper mechanical design arrived at after detailed stress analysis keeping in view all the static and dynamic, and steady and transient loads. However, should a unit still fail, the failure analysis must be done thoroughly to learn its causes and to avoid its recurrence. In this chapter, therefore, the questions will focus on these topics.

MECHANICAL DESIGN

General

9.1 What is meant by mechanical design of equipment?

It is the design of equipment to withstand the expected loads, stresses, and corrosion due to fluids and environment so that the equipment will perform satisfactorily all through its expected service life at a minimum total cost (capital cost plus maintenance/repair cost).

9.2 What is the minimum information necessary for the mechanical design of heat exchangers?

The minimum information necessary is the following:

1. Thermal (process) design details in the form of a specification sheet (Fig. 1.37)
2. Type of TEMA shell, channel, and bonnet (Fig. 1.1)
3. Number of shell and tube passes
4. Number, type, size, and layout of tubes
5. TEMA Class (Q. 8.31)
6. Diameter and length of shell, channel, and bonnet
7. Corrosion properties of the fluids and the environment in which the unit will be installed and the expected service life
8. Materials of construction (Q. 10.3)
9. Flow rates to size the nozzles and to determine if damaging vibrations would occur
10. Design temperatures and pressures
11. External pressure if the equipment is under external pressure or is under internal vacuum
12. Worst case coincident conditions of temperature and pressure
13. Nozzle, wind, and seismic loads
14. Impact loads including rapidly fluctuating pressures
15. Superimposed loads, such as insulation, piping, stacked units, etc.
16. Reactions of supports
17. Thermal stresses
18. Special requirements of restrictions by the purchaser on available space, piping layout, location of supports, type of material, servicing conditions, etc.
19. Codes to be followed
20. Vertical or horizontal installation, i.e., the operating position
21. Cost of various materials to choose among alternatives

Besides the above, other information such as radiographic examination, gasket type, flange facing, and so forth, also influence the mechanical design.

9.3 How is the mechanical design done?

It is done as per the procedure given in the applicable codes and standards. For special designs, for which no guidance is provided by the codes and standards, the procedure may be arrived at by mutual discussions between the buyer and the vendor. Approval of the same may then be obtained from the appropriate code committees.

9.4 Which are the paragraphs in the codes and TEMA Standards that govern the design of the pressure vessels and shell-and-tube heat exchanger?

Paragraphs pertaining to the design of a pressure vessel according to the ASME Code Section VIII, Division 1 (ASME, 1980a), the most popular code, are given in Fig. 9.1. Equivalent paragraphs from the codes of other countries are available in the literature (Rubins, 1978).

Paragraphs pertaining to the design of a shell-and-tube heat exchanger according to the TEMA Standards (1978) are given in parentheses in Fig. 9.2. Some parts are not shown in this figure. Refer to Fig. 1.2 for them. For some parts, the mechanical design has to be done using the applicable codes. The paragraphs listed in such cases give general guidelines. R, C, B refer to the three TEMA Classifications (1978). G refers to the general paragraphs in the beginning of the TEMA Standards.

Now we shall discuss the salient aspects of the design of important parts of heat exchangers and pressure vessels.

Shell, Channel, and Bonnet

9.5 What factors decide the thickness of shell, channel, and bonnet?

Various types of shells, channels, and bonnets are shown in Fig. 1.1. Their thickness depends upon the design pressure, allowable stress for the material of construction at the design temperature, radius of the unit, efficiency of the welded joint, pass partition groove depth, and the corrosion allowance required. The actual thickness may be greater than the calculated one in some cases since the material is available only in standard thicknesses. Depending upon the pressures on the shell side and the tube side, the shell and the channel may be of different thicknesses.

9.6 How is the thickness determined for a unit under external pressure?

There is no straightforward formula as in the case for the design under internal pressure because the buckling has to be considered too. The procedure to be followed is given in the ASME Code Section VIII, Division 1 paragraph UG-28 (American Society of Mechanical Engineers, 1980a). This requires the use of various charts in Appendix 5 of that code.

Heads

9.7 What are the various kind of heads that can be used as end closures on a shell as per the ASME Code?

These are the following (Fig. 9.3):

1. Flat cover
2. Flanged and dished (torispherical)
3. Elliptical dished (ellipsoidal)
4. Hemispherical
5. Flanged and conical dished (toriconical)

9.8 What are the general uses of different heads?

The general uses of the different heads are the following (Yasuda et al., 1976):

1. Flat cover: These are easy to fabricate in any thickness from plates or forgings and hence are extensively used from low- to high-pressure services. Their thickness is significantly greater than that of the cylindrical section to which these are attached as a closure.

Figure 9.1 ASME Code Section VIII, Division 1 (1980a): Important paragraphs. *(Reprinted by permission of Missouri Boiler & Tank Co., St. Louis, MO.)*

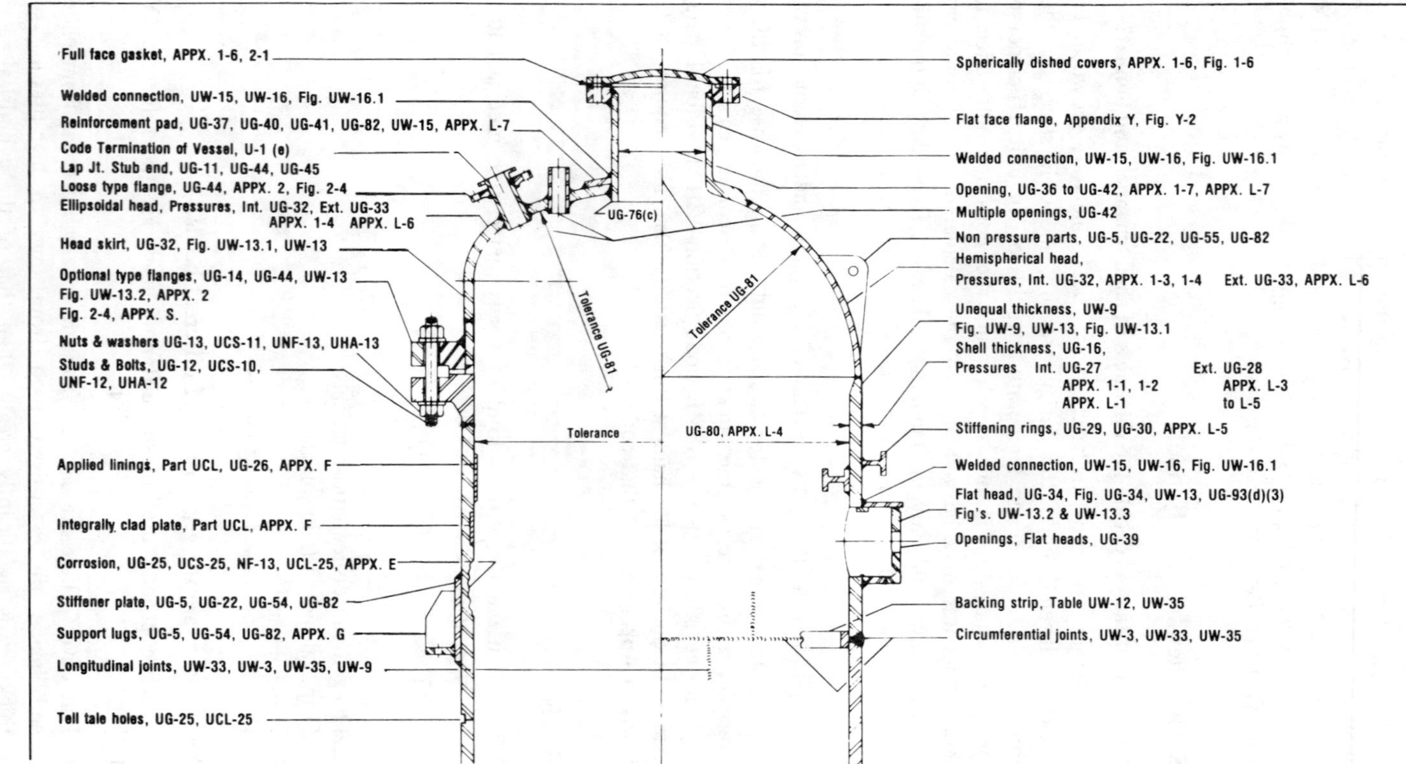

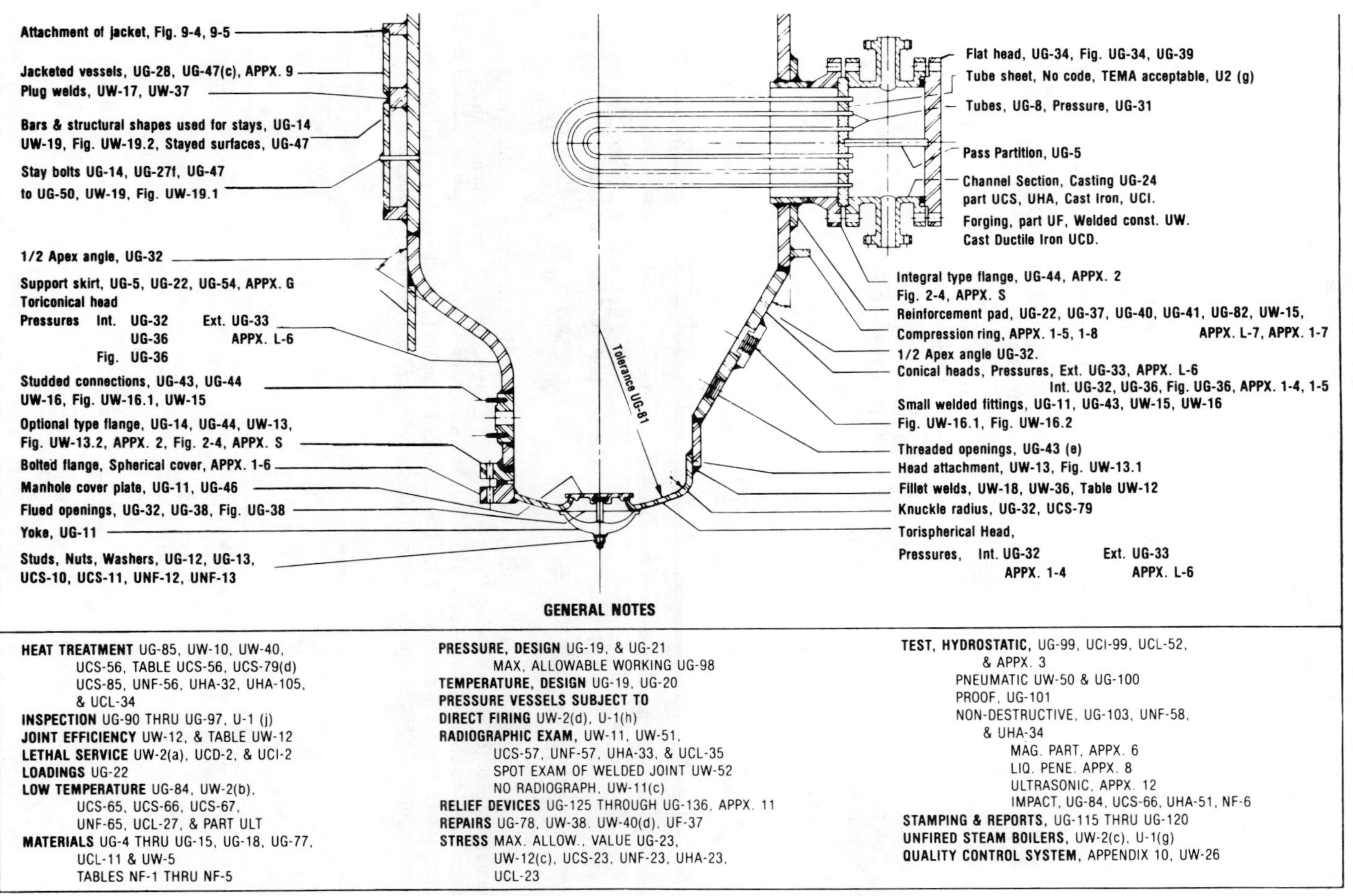

GENERAL NOTES

HEAT TREATMENT UG-85, UW-10, UW-40, UCS-56, TABLE UCS-56, UCS-79(d) UCS-85, UNF-56, UHA-32, UHA-105, & UCL-34
INSPECTION UG-90 THRU UG-97, U-1 (j)
JOINT EFFICIENCY UW-12, & TABLE UW-12
LETHAL SERVICE UW-2(a), UCD-2, & UCI-2
LOADINGS UG-22
LOW TEMPERATURE UG-84, UW-2(b), UCS-65, UCS-66, UCS-67, UNF-65, UCL-27, & PART ULT
MATERIALS UG-4 THRU UG-15, UG-18, UG-77, UCL-11 & UW-5 TABLES NF-1 THRU NF-5

PRESSURE, DESIGN UG-19, & UG-21 MAX, ALLOWABLE WORKING UG-98
TEMPERATURE, DESIGN UG-19, UG-20
PRESSURE VESSELS SUBJECT TO DIRECT FIRING UW-2(d), U-1(h)
RADIOGRAPHIC EXAM, UW-11, UW-51, UCS-57, UNF-57, UHA-33, & UCL-35 SPOT EXAM OF WELDED JOINT UW-52 NO RADIOGRAPH, UW-11(c)
RELIEF DEVICES UG-125 THROUGH UG-136, APPX. 11
REPAIRS UG-78, UW-38, UW-40(d), UF-37
STRESS MAX. ALLOW., VALUE UG-23, UW-12(c), UCS-23, UNF-23, UHA-23, UCL-23

TEST, HYDROSTATIC, UG-99, UCI-99, UCL-52, & APPX. 3
PNEUMATIC UW-50 & UG-100
PROOF, UG-101
NON-DESTRUCTIVE, UG-103, UNF-58, & UHA-34
MAG. PART, APPX. 6
LIQ. PENE. APPX. 8
ULTRASONIC, APPX. 12
IMPACT, UG-84, UCS-66, UHA-51, NF-6
STAMPING & REPORTS, UG-115 THRU UG-120
UNFIRED STEAM BOILERS, UW-2(c), U-1(g)
QUALITY CONTROL SYSTEM, APPENDIX 10, UW-26

1. Stationary Head—Channel (RCB-8.1)
2. Stationary Head—Bonnet (RCB-8.1)
3. Stationary Head Flange—Channel or Bonnet (RCB-10)
4. Channel Cover (RCB-8.2)
5. Stationary Head Nozzle (RCB-9)
6. Stationary Tubesheet (RCB-7)
7. Tubes (RCB-2)
8. Shell (RCB-3)
9. Shell Cover (RCB-3.2)
10. Shell Flange—Stationary Head End (RCB-10)
11. Shell Flange—Rear Head End (RCB-10)
12. Shell Nozzle (RCB-9)
13. Shell Cover Flange (RCB-10)
14. Expansion Joint (RCB-7.19)
15. Floating Tubesheet (RCB-7)
16. Floating Head Cover (RCB-5.11)
17. Floating Head Flange (RCB-10)
18. Floating Head Backing Device (RCB-5.13)
19. Split Shear Ring (RCB-9.5)
20. Slip-on Backing Flange (RCB-10)
21. Floating Head Cover—External (RCB-3.2)
22. Floating Tubesheet Skirt (RCB-5.24)
23. Packing Box Flange (RCB-10)
24. Packing (RCB-5.23)
25. Packing Follower Ring
26. Lantern Ring (RCB-5.31)
27. Tie Rods and Spacers (RCB-4.7)
28. Transverse Baffles or Support Plates (RCB-4)
29. Impingement Baffle (RCB-4.6)
30. Longitudinal Baffle (RCB-4.42)
31. Pass Partition (RCB-8.13)
32. Vent Connection (E2.33, RCB-7.182, RCB-9.31)
33. Drain Connection (E2.34, RCB-7.182, RCB-9.31)
34. Instrument Connection
35. Support Saddle (G-7.1)
36. Lifting Lug (G-7.2)
37. Support Bracket (G-7.1)
38. Weir
39. Liquid Level Connection

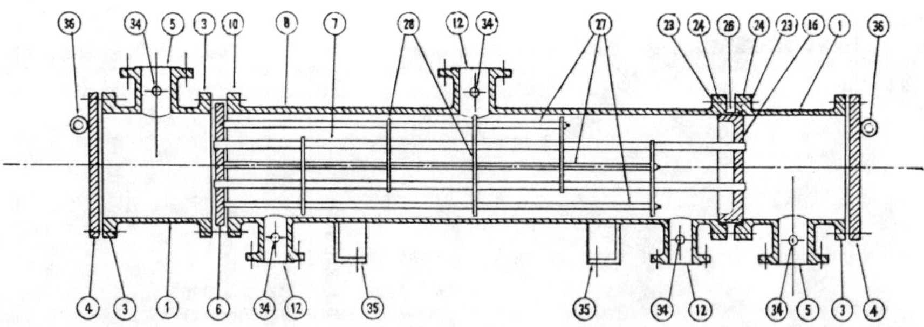

Figure 9.2 TEMA Standards (1978): Important paragraphs are given in parentheses. Some parts are not shown in the above drawing. Refer to Fig. 1.2 for them. For some parts, the mechanical design has to be done using the applicable codes. The paragraphs listed in such cases give general guidelines. RCB refers to the three TEMA Classifications. G refers to the General paragraphs in the beginning of the TEMA Standards. *(From Tubular Exchanger Manufacturers' Association, 1978.)*

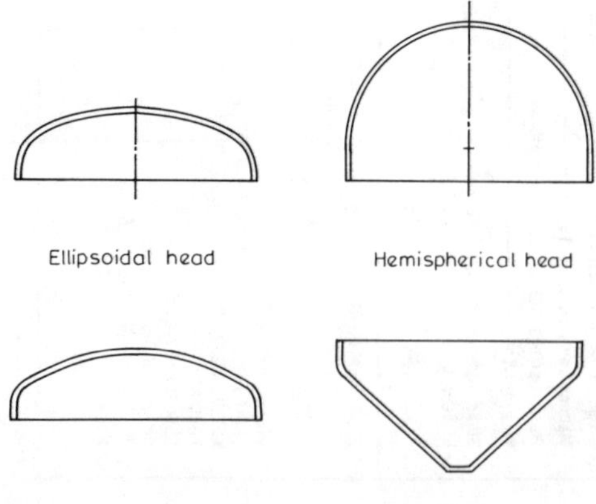

Figure 9.3 Different kinds of head.

2. Flanged and dished (torispherical): These are widely used for the low-pressure service. Their thickness is greater than that of a cylindrical shell to which they are attached.
3. Elliptical dished (ellipsoidal): These are widely used for the low- to intermediate-pressure services. When the minor-to-major axes ratio is 0.5 (most common), the head thickness is almost the same as that of the cylindrical shell. This simplifies the joining of the two and minimizes the discontinuity stresses at the joint.
4. Hemispherical: These are used for the high-pressure service since their thickness is about half that of a cylindrical shell.
5. Flanged and conical dished (toriconical): These are used for low- and intermediate-pressure service with the half-apex angle generally no more than 30°. A knuckle portion is provided to minimize the discontinuity stresses where it joins the shell.

9.9 How does one decide among the different heads?

The head that is the lowest in cost and meets the code requirements should be used. The total cost of a head depends upon the pressure, diameter, thickness, material of construction, and ease of fabrication. The ellipsoidal and hemispherical heads have the least weight per unit volume and better stress distribution. The torispherical head is easier to fabricate and is less deep. Most often the ellipsoidal and torispherical heads are used.

Flanges

9.10 Why are flanges used?

These are used to connect two sections by bolting them together. These permit easy assembly and disassembly of the sections. Both the sections must have a flange attached to their joining end.

9.11 Where are flanged joints used in heat exchangers and pressure vessels?

Flanged joints are used to connect together the following:

1. Channel and channel cover
2. Shell and channel
3. Shell and rear head
4. Inlet and outlet nozzles for the shell-side and tube-side fluids to the pipes carrying the fluids
5. Manholes and peepholes and their covers
6. Drain, vent, and safety relief connections on the shell and channel to the connecting pipes in cases where the fluids cannot be discharged into the atmosphere

9.12 What is the minimum size of an opening that requires a flanged joint?

Flanged joints are generally used for sizes greater than $1\frac{1}{2}$ in (38 mm) nominal pipe diameter (Brownell and Young, 1959).

9.13 What are the different kinds of flanges?

See Q. 1.96.

9.14 What factors influence the choice of a flange in a specific case?

If 100 percent radiography is required (see Chapter 11), a welding neck type (integral type) flange is used. These are very costly. When 100 percent radiography is not required, other types may be used since they are cheaper. The welding neck flanges must be properly aligned on the shell and the channel. Since these are the strongest among all the flanges, these are used where leakage or failure might prove disastrous. The loose-type flanges are used for moderate services where the pressure and temperature fluctuations, vibrations, and shock are not expected to be severe (Bhattacharyya, 1976). If, because of the fluid corrosive conditions, exotic materials such as titanium are required, one may use them to make a lap while carbon-steel may be used to make a slip-on flange (Fig. 9.4).

9.15 What factors enter into the design of the flanges?

These are design pressure and temperature, shell or pipe diameter, allowable stresses for flange and bolt materials, gasket (material, type and dimensions), whether the gaskets are within the inner bolts or straddling the bolts, diameter of location of gasket load reaction, pass rib area, operating load, load to seat the gasket, actual bolt area, bolt load, moments on flange in operating conditions and in gasket seating conditions, and shape constants.

9.16 What are the different stresses that should be considered in flange design?

These are longitudinal hub stress, radial flange stress, tangential flange stress, allowable stresses for the flange material at the atmospheric and design temperatures, and allowable stresses for the bolt material at the atmospheric and design temperatures (Mahajan, 1979). The first three are calculated both for the operating and gasket seating conditions. Depending upon the type of flange, some of these may be zero.

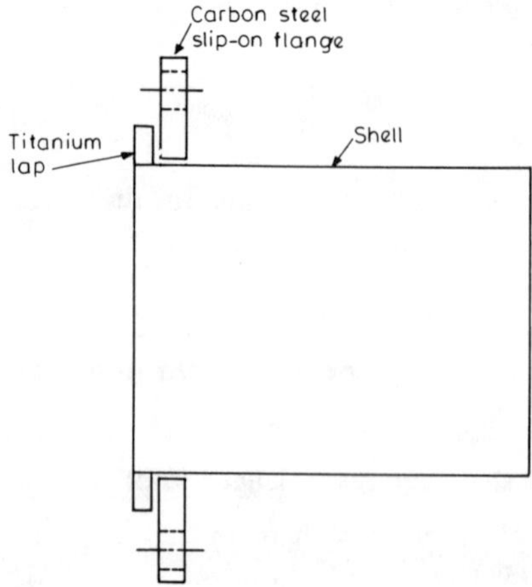

Figure 9.4 Slip-on flange.

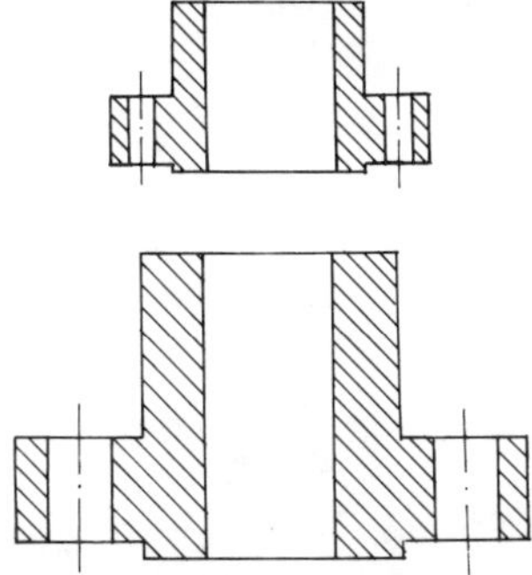

Figure 9.5 Flanges with different ratings.

9.17 How does a flanged joint affect the shell?

It sets up bending stresses in the shell, and hence the shell must be thickened in the vicinity of the flange. In thin vessels, it causes no serious problem. However, in thick vessels with basic wall thickness of up to 8 in (203.2 mm) as in the pressurized water reactors (PWR), the flanged joint costs almost one-third of the total cost of the vessel (Fraas and Ozisik, 1965).

9.18 What is meant by rating of a flange?

It is the pressure that a flange can withstand at a given temperature. Thus, both the temperature and pressure should be known before choosing a flange. For example, according to the American National Standards ANSI B16.5-1977, a nominal 150 psi (1034 kPa) carbon-steel flange can take 225 psi (1551 kPa) pressure at 250°F (121°C) and only 100 psi (689.5 kPa) pressure at 750°F (399°C). Similarly, a nominal 600 psi (4137 kPa) carbon-steel flange can take 1440 psi (9653 kPa) pressure at 100°F (37.8°C) and only 445 psi (3068 kPa) at 900°F (482°C). Tables of pressure-temperature relationships for different flanges are available in standard references, such as Baumeister (1978).

9.19 In what respects do the flanges with different ratings (for a given size pipe or shell) differ from one another?

Flanges with different ratings are shown in Fig. 9.5 and differ from one another in the following:

1. Thickness
2. Number of bolts
3. Bolt circle diameter (circle passing through the center-line of the bolts)
4. Bolt diameter
5. Flange diameter

9.20 What are the different ratings for steel and cast iron flanges?

The ratings are governed by the Standards of the American National Standards Institution (ANSI).

For the steel flanges in sizes up to and including 24 in (610 mm) ANSI B 16.5-1977 designates 150, 300, 400, 600, 900, 1500, and 2500 psi (respectively,

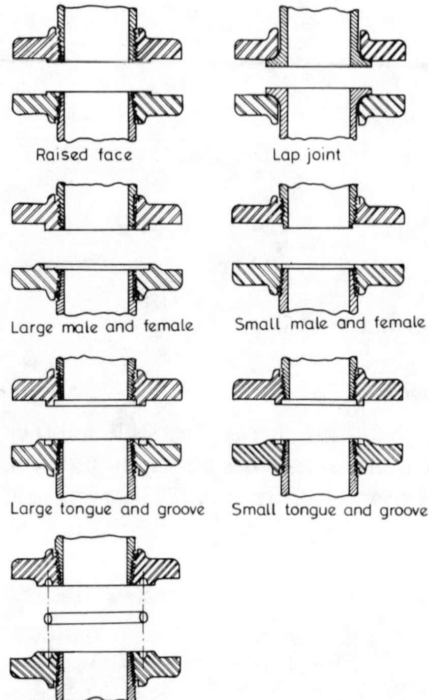

Figure 9.6 Flange facings. *(From Brownell and Young, 1959.)*

1034, 2068, 2758, 4137, 6205, 10,342, and 17,237 kPa) ratings, while API 605 (3d edition, February 1980) can be used in sizes 26 through 60 in, inclusive (660–1524 mm, inclusive) for 75, 150, 300, 400, 600, and 900 psi (respectively, 517, 1034, 2068, 2758, 4137, and 6205 kPa).

For the cast iron flanges, ANSI B 16.1-1975 designates 25, 125, 250, and 800 psi (respectively, 172, 862, 1724, and 5516 kPa) ratings. These are used in the low-pressure services.

The ratings correspond to the service pressures at specified service temperatures. The pressure rating decreases as the temperature increases since the allowable stress for the flange material decreases with an increasing temperature (Brownell and Young, 1959).

9.21 What are the different types of flange facings?

The following facings are in general use (Fig. 9.6):

1. Plain straight face: General services
2. Plain face, serrated, or V-grooved: Concentric grooves are cut on the plain face. These are used when thick and loosely woven fibrous or soft metallic gaskets are prescribed for some services. The serrations help keep the gasket in place.
3. Male-and-female: It has a recess in one flange and a corresponding raised face on the other. These are used on the hydraulic lines and sometimes also on the high-pressure steam lines.

4. Tongue-and-groove: The tongue or the raised face and the groove or the recess are narrow rings located between the bolt holes and the port. Uses are similar to the male-and-female type. Besides, these can also be used for the vacuum service and for the external pressure (Bhattacharyya, 1976).
5. Raised face: There is a high, narrow, raised ring on each of the mating flanges. It is specially useful for the high-pressure lines since no mating of the flanges is required, and the gasket is automatically centered.

9.22 Are the mating flanges on the shell and the channel in a heat exchanger similar?

While the bolt circle, bolt hole, bolt material, and the gasket characteristics are the same for both the flanges, the other dimensions of the flanges and their materials may be different depending upon the temperature, pressure, and corrosion characteristics of the two fluids.

9.23 Can the flanges on the two sides of a shell be of different sizes?

Yes, e.g., in the case of an outside-packed floating head (Fig. 1.1).

9.24 What material is used for flanges and bolts?

Generally, carbon-steels of suitable grades are used for the two, e.g., SA-105 for the flanges and SA-193 grade B7 for the bolts.

9.25 Are the flanges on the shell of the same material as the shell?

Yes, since both the shell and the flange face the same fluid. However, a slip-on flange may be of a different material as long as the lap is of the same material as the shell (Q. 9.14).

Bolts and Studs

9.26 What is the difference between a bolt and a stud?

A bolt has a head on one end and threads on the other end that are used for tightening a nut.

A stud has no head, but has threads on both ends that are used for tightening a nut on each end. Studs are used for high-temperature services and for better tightening in large units (Fig. 9.7). The bearing surface of the nut that contacts the flange should have a good machine finish.

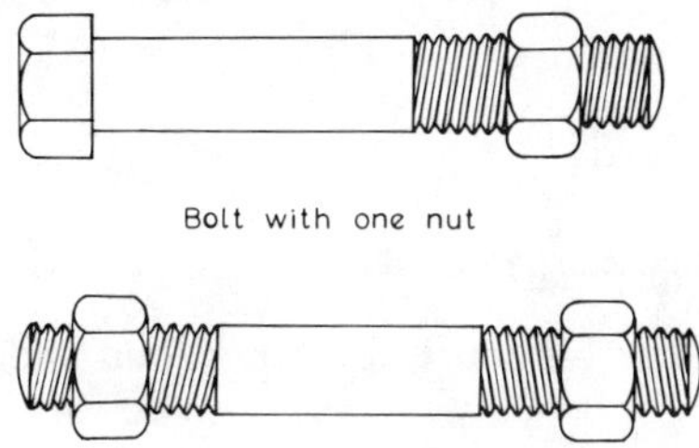

Figure 9.7 Bolt and stud with nuts.

9.27 What is the clearance between a bolt and a bolt hole?

Generally $\frac{1}{8}$ in (3.18 mm) diametral or $\frac{1}{16}$ in (1.59 mm) radial clearance.

9.28 Why are bolts used in multiples of four in a flange?

So that the fittings on the unit can be rotated, if required, to face in any quarter.

9.29 Why is it important to have a minimum spacing between two consecutive bolts?

It is required to accommodate the nuts on each of the bolts and to put a wrench on any one of them to tighten or loosen it. The spacing also gives the necessary strength to the unpierced section (ligament) between two consecutive holes. One can use more than the minimum spacing if necessary.

9.30 When are washers used?

These are used to distribute load if the nut on tightening would dig into the body of the flange. Washers are made of softer material to be effective. However, the situation does not arise too often for their use.

9.31 How are the bolts designed?

Bolt design is included in the design of the flanges. The tensile stress of the bolt material is the chief item involved in the design of the bolts.

Gasketed Joints

9.32 Where are the gaskets used in heat exchangers and pressure vessels?

Gaskets are used in all the bolted joints between two metal surfaces to make the joints leak-tight. Such joints are the shell-to-channel, channel-to-channel cover, nozzle-to-piping, etc.

9.33 What is the difference between a gasket and a packing?

Both seal the space between two metal pieces to make them leak-tight. However, the packings allow some movement of one metal piece relative to the other, while the gaskets do not. Some examples of the use of the packings are in the floating rear-hear constructions (Fig. 1.1), and in chemical reactor vessels and fermentors with a mechanical agitator.

9.34 What materials are used as gaskets?

These are given in Table 9.1.

9.35 What factors decide the choice of a gasket material?

These are the pressure, temperature, and corrosive nature of the fluid and the length of service desired.

Table 9.1 Gasket materials[a]

Gasket material	Remarks
Self-energizing types	
O-rings, metallic, elastomer, other gasket types considered as self-sealing	
Elastomers without fabric or a high percentage of asbestos fiber	
Below 75 Shore Durometer	
75 or higher Shore Durometer	
Asbestos with a suitable binder for the operating conditions	$\frac{1}{8}$ in (3.2 mm) thick
	$\frac{1}{16}$ in (1.6 mm) thick
	$\frac{1}{32}$ in (0.8 mm) thick
Elastomers with cotton fabric insertion	
Elastomers with asbestos fabric insertion, with or without wire reinforcement	3-ply
	2-ply
	1-ply
Vegetable fiber	
Spiral-wound metal, asbestos filled	Carbon
	Stainless or Monel
Corrugated metal, asbestos inserted or corrugated metal, jacketed asbestos filled	Soft aluminum
	Soft copper or brass
	Iron or soft steel
	Monel or 4–5% chrome
	Stainless steels
Corrugated metal	Soft aluminum
	Soft copper or brass
	Iron or soft steel
	Monel or 4–6% chrome
	Stainless steels
Flat metal, jacketed asbestos filled	Soft aluminum
	Soft copper or brass
	Iron or soft steel
	Monel
	4–6% chrome
	Stainless steels
Grooved metal	Soft aluminum
	Soft copper or brass
	Iron or soft steel
	Monel or 4–6% chrome
	Stainless steels
Solid flat metal	Soft aluminum
	Soft copper or brass
	Iron or soft steel
	Monel or 4–6% chrome
	Stainless steels
Ring joint	Iron or soft steel
	Monel or 4–6% chrome
	Stainless steel

Source: American Society of Mechanical Engineers (1980a).

[a]Temperature limits (Brownell and Young, 1959): Paper, cloth, rubber gaskets up to 250°F (121°C); asbestos-composition gaskets up to 650°F (343°C); metallic gaskets for higher temperatures.

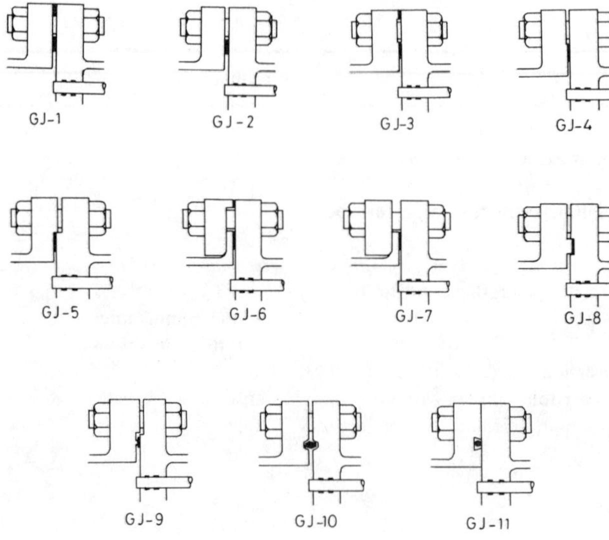

Figure 9.8 Gasketed joints for integral tubesheet designs. *(Patterson-Kelley Co., Division of HARSCO Corp.)*

9.36 What are the various kinds of gasketed joints?

These are shown in Fig. 9.8. The contact surfaces may be flat or grooved. In a removable tubesheet exchanger, the joints gasketed with the shell and with the channel may be similar or different (Fig. 9.9).

9.37 Describe the applications of the various gasketed joints?

These are given in Table 9.2.

9.38 How are the gaskets kept in place?

The gaskets are kept in place by the compressive loading through bolting of the gasket and the mating surfaces.

9.39 What factors should be considered in the proper design of a gasketed joint?

These are the following (Baumeister, 1978):

1. Flanges should be rigid enough so as not to bow under bolting and fluid pressure.

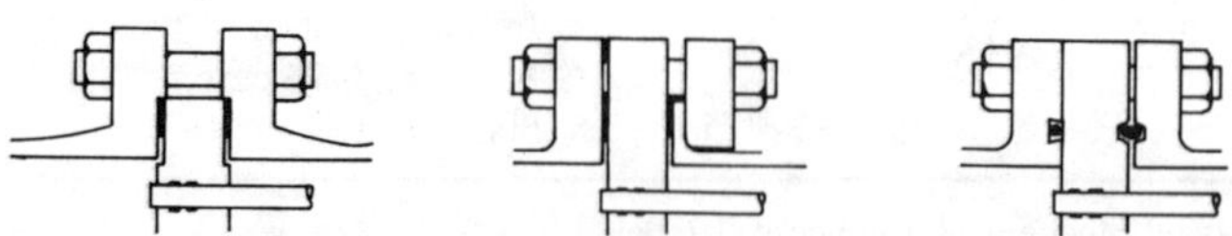

Figure 9.9 Gasketed joints combinations for removable tubesheet designs. *(Patterson-Kelley Co., Division of HARSCO Corp.)*

Table 9.2 Application of gasketed joints

Temperature class	Pressure class
T-1 Up to 250°F (121°C)	P-1 Up to 150 psig (1.136 MPa)
T-2 Up to 375°F (190°C)	P-2 Up to 350 psig (2.515 MPa)
T-3 Above 375°F (190°C)	P-3 Above 350 psig (2.515 MPa)

Gasket joint number	Design	Description	Recommended gasket material	Usual application	Remarks	Gasket jointing
GJ-1	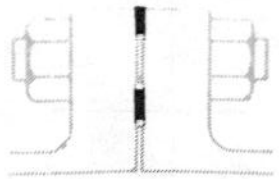	Flat face with full face gasket	Compressed asbestos	T-1 liquids, T-2 vapors, P-1. Good for material with low ductility	Least expensive, gaskets are located on bolts which ensure alignment of pass ribs	GJ-1
GJ-2	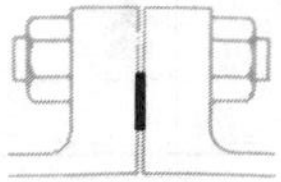	Unconfined flat ring gasket contained entirely within bolt holes	Compressed asbestos	T-2 liquids and vapors. All pressure classes	Most generally used (except for GJ-1). Gasket materials that are hard to seat require thicker flanges and heavier bolts	GJ-2
GJ-3	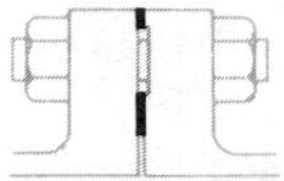	Relieved flat face with full face gasket	Compressed asbestos	T-1 liquids, T-2 vapors, P-1	Relieving the flange permits minimum flange thickness. Gaskets are located on bolts which ensure alignment of pass ribs	GJ-3
GJ-4	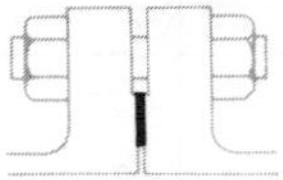	Unconfined flat ring gasket contained entirely within bolt holes	Compressed asbestos	T-2 liquids and vapors. All pressure classes	Most generally used (except for GJ-1). Gasket materials that are hard to seat require thicker flanges and heavier bolts	GJ-4
GJ-5	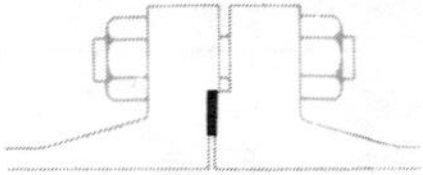	Semi-confined flat ring gasket	Compressed asbestos	All temperature and pressure classes	Confinement of gasket provides additional safety against gasket blowing out. Confinement also provides axial location of gasket	GJ-5
GJ-6	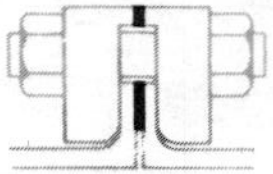	Van Stone with outside bearing edge and full flat face gasket	Compressed asbestos	T-1 liquids, T-2 vapors, P-1	Outside bearing edge permits minimum flange thickness. Gaskets are located on bolts which ensure alignment of pass ribs, and, in addition, this design permits economy of alloy material by use of a carbon-steel backing flange	GJ-6

Table 9.2 Application of gasketed joints (*Continued*)

Temperature class	Pressure class
T-1 Up to 250°F (121°C)	P-1 Up to 150 psig (1.136 MPa)
T-2 Up to 375°F (190°C)	P-2 Up to 350 psig (2.515 MPa)
T-3 Above 375°F (190°C)	P-3 Above 350 psig (2.515 MPa)

Gasket joint number	Design	Description	Recommended gasket material	Usual application	Remarks	Gasket jointing
GJ-7	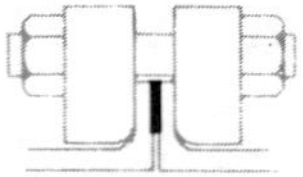	Van Stone (or lap joint) with unconfined flat ring gasket	Compressed asbestos	T-1 liquids, T-2 vapors, P-1	Permits economy of alloy material by use of a carbon-steel backing flange. Backing flange thickness will be heavier than GJ-6	GJ-7
GJ-8	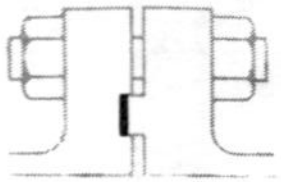	Tongue and groove joint with fully confined flat ring gasket	Any gasket material	All temperature and pressure classes	Relatively expensive, especially on multi-pass units where grooves are required to accommodate pass partitions. Grooves completely confine and protect gasket. Grooves provide positive alignment of gasket and pass partitions	GJ-8
GJ-9	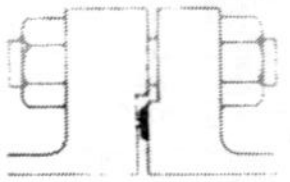	Semi-confined metal or metal jacketed flat ring gasket	Any ductile metal is used for solid gasket and for jacket with compressed asbestos filler	All temperature and pressure classes	Metal for solid gasket or jacket must be compatible with process fluid. Gasket is relatively expensive and is not reuseable	GJ-9
GJ-10	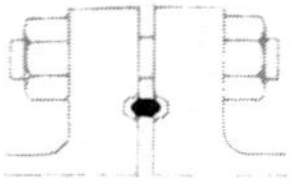	Ring joint gasket	Any ductile metal	All temperature and pressure classes	Most expensive. Requires precise matching of groove to accommodate ring. Cannot be used for multi-pass without special pass chamber construction	GJ-10
GJ-11	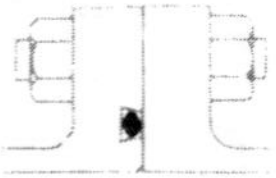	O-Ring gasket joint	Materials presently available to 500°F	All pressure classes. Temperature is limited by O-ring material. Good for materials with low ductility	Requires precise machining of O-ring groove and good surface finish. No gasket seating load on bolts. Can be used for multi-pass if slight leakage across pass partition is permissible	GJ-11

Source: Patterson-Kelley Co., Division of HARSCO Corp.

2. Flange surface finish should be correctly chosen for the given gasket material and fluid pressure (Q. 9.21).
3. Bolts should be pretightened to seat the gasket, causing it to flow to fill surface irregularities but not be so tight as to crush the gasket. The bolt loading should be enough to seal the joint during the operation too when the fluid pressure tries to separate the joint.

9.40 Are gaskets reusable?

Generally, they should be replaced each time a gasketed joint is opened. In cases such as plate heat exchangers, the same gaskets can be used many times since they only compress but do not flow under the bolt loading used.

9.41 What are design seating stress and gasket factor?

The design seating stress or load, y, is the load required to seat the gasket properly. The gasket factor, m, refers to the operating conditions when the fluid pressure tries to separate the joint, and hence the gasket must have residual pressure (or stress) in excess of the fluid pressure to still seal the joint. It is the ratio of the gasket stress when the vessel is under pressure to the internal pressure in the vessel.

9.42 What are self-energizing gaskets?

These seal better as the pressure in the system increases because they get further compressed (Yasuda et al., 1976).

9.43 How are the gaskets manufactured?

The asbestos gaskets are manufactured out of a full sheet by cutting out an inner circle so that no joint is made. If it is a multi-pass unit, then the appropriate segments are removed to get a one-piece gasket with ribs.

The spiral metallic gaskets do have a joint since these are made out of metal strips preformed into a definite shape. In the case of multi-pass units, appropriate ribs are welded to the circular gasket.

Tubesheet

9.44 What is the importance of the tubesheet?

See Q. 1.77.

9.45 Whar are the uses of tubesheets in different kinds of exchangers?

See Q. 1.78.

9.46 Are tubesheets always round?

See Q. 1.79 and 9.65.

9.47 What are the different kinds of tubesheets?

These are the following (Fig. 1.1):

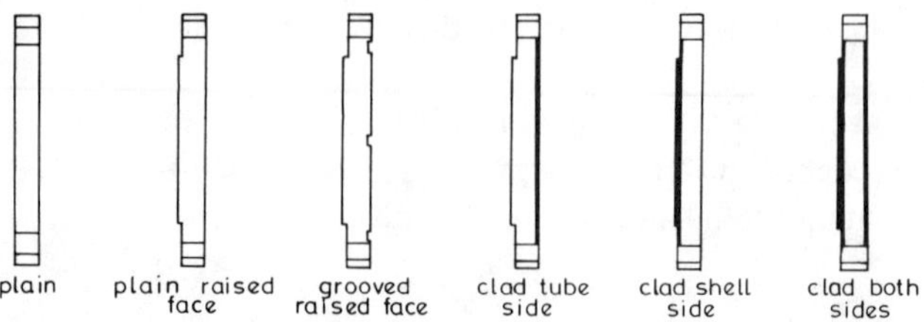

Figure 9.10 Tubesheet facings. *(Patterson-Kelley Co., Division of HARSCO Corp.)*

1. Fixed tubesheets: These are welded to the shell.
2. Removable tubesheets: These are used with U-tubes. The tube bundles can be taken out.
3. Floating tubesheets (floating rear heads): These are used with straight tubes where differential expansion is significant. The front tubesheet has to be removable type. The tube bundle can be taken out.

9.48 What are the various kinds of tubesheet facings?

These depend upon the kinds of fluids handled, any cladding required, pass partition plates, kind of gasketed joint used, etc. Figure 9.10 illustrates the various tubesheet facings.

9.49 What are the advantages of using a removable tubesheet of the same diameter as the flange?

It allows the shell and the tube side to be tested individually in less time (Fig. 9.11). Furthermore, since enough solid rim is available, the lugs to move the tubesheet and the tube bundle can be put on it. The above advantages do not exist when the tubesheet is shorter than the inner edges of the flange bolt holes (Fig. 9.12), which is cheaper.

9.50 How are the tubesheets designed for a shell-and-tube exchanger?

These are generally designed as per the TEMA Standards (1978). This is probably the largest single contribution of the TEMA Standards to the manufacture of the shell-and-tube heat exchangers. A book containing sample problems including

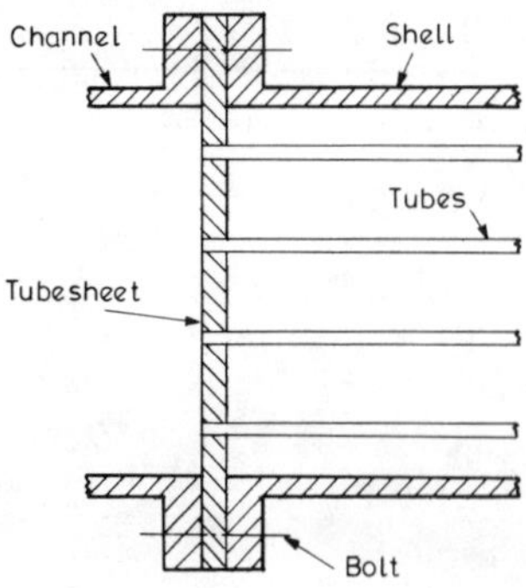

Figure 9.11 Removable tubesheet of the same diameter as the flange.

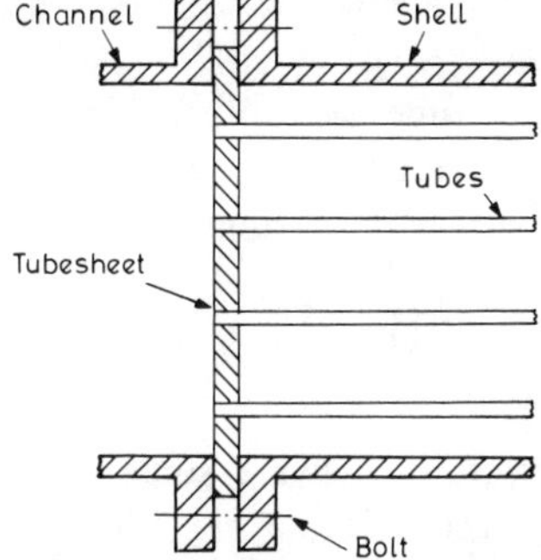

Figure 9.12 Removable tubesheet shorter than the inner edges of the flange bolt holes.

those of the tubesheet designs under various conditions has been released (TEMA, 1980). Computer programs to do all those designs are also available (Gupta and Sivasankaran, 1981). The British and the German codes give tubesheet design methods that differ from the TEMA method.

9.51 What stresses act on a tubesheet?

These are the following:

1. Bending stresses due to bolting
2. Shear stresses at the contact point with the flange surface
3. Shell-side and tube-side pressures
4. Bending moment and shear force due to the tubes
5. Thermal stresses

9.52 How are the tubesheets designed for the surface condensers?

These are generally designed as per the HEI Standards (Heat Exchange Institute, 1980).

9.53 What is the minimum spacing or ligament required between two adjacent tube holes in the tubesheet?

It should be enough so that the two holes will not merge into one another due to drill drift or due to grooving the tubes into the tubesheet.

9.54 Discuss the joints used for joining the tubes to the tubesheets.

These are (Fig. 9.13) the following:

1. Rolled joints: The tubes are rolled into the holes to seal against any leakage and intermixing of the tube-side and shell-side fluids. These are used for less critical

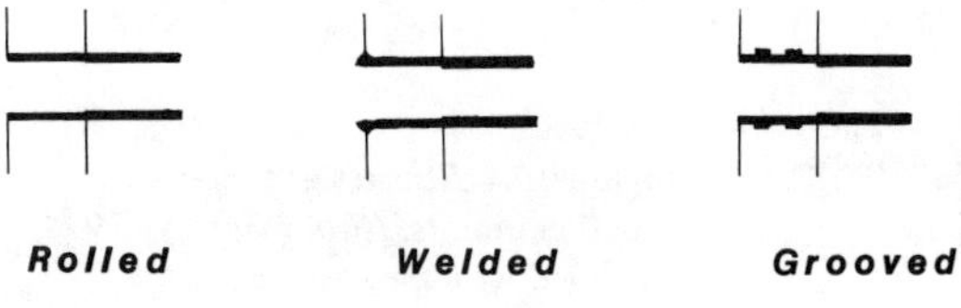

Figure 9.13 Three kinds of tube-to-tubesheet joints. *(Patterson-Kelley Co., Division of HARSCO Corp.)*

and general applications where the thermal transients are not encountered, or else these joints will get loosened.

2. Welded joints: The rolled tube ends are welded to the tubesheet to provide an ultimate seal against leakage at the joint. These are also used when the thermal transients are present or the ligament available is less than that required for an effective rolling.
3. Grooved or serrated joints: The rolled tubes are grooved into the tubesheet to provide a strong mechanical seal.

In some nuclear and other very critical applications, buyers sometimes specify the joint to be both grooved and welded. The extra safety, if any, due to this is of doubtful value.

The tubes in the above joints can be projecting from the tube hole, flush, or flared.

9.55 What stresses act on the tube-to-tubesheet joints?

These are the following:

1. Residual stresses in the tube wall and the ligament due to the expansion of the tube by mechanical rollers that deform the tube plastically to fill the gap between the tube and the tube hole. The stress is compressive in the tube wall and in the ligament close to it, and it is tensile in the rest of the ligament (Fig. 9.14).
2. Stresses due to the pressure of the shell-side and tube-side fluids (Paragraphs R, C, B: 7.25) (TEMA, 1978).
3. Thermal stresses in the fixed tubesheet units (Paragraphs R, C, B: 7.23) (TEMA, 1978).

9.56 How do the tube holes influence the strength of the tubesheet?

Removal of the material due to drilling causes a loss of strength of the tubesheet. However, the tubes do stiffen the tubesheet to a certain extent. The calculation procedure for the overall effects are available in American Society of Mechanical Engineers (1977) for circular holes on equilateral triangular pitch.

9.57 What materials are used for tubesheets?

Any material compatible with the corrosion requirements of the tube-side and shell-side fluids can be used. The stronger the material, the less is its required thickness for the particular design pressure and temperature conditions.

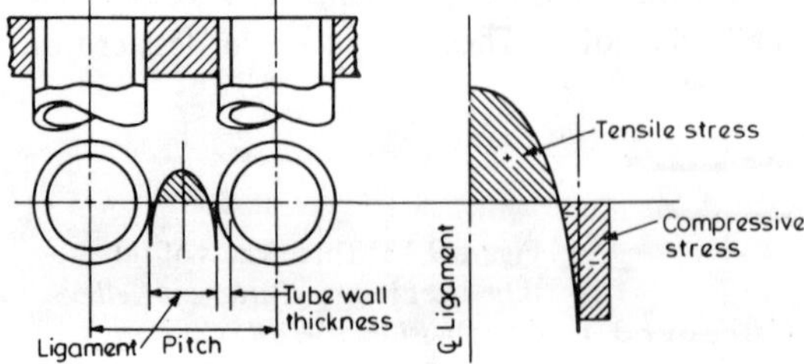

Figure 9.14 Stresses at tube-to-tubesheet joints. *(From Harvey, 1974, modified with permission.)*

9.58 What is a clad tubesheet?

It is generally a carbon-steel tubesheet with a cladding of another material to withstand the corrosion. Cladding may be on one or both sides of the tubesheet, depending upon the corrosion protection requirements. The cladding is done by explosion bonding of the clad material to the base plate. Another way to achieve protection against corrosion is to deposit by welding the required thickness of the material by using consumable electrodes. This is called weld overlay.

9.59 Does cladding influence the mechanical strength of a tubesheet?

Generally, the influence is ignored in order to be on the conservative side. However, if a shear test is performed between the base plate and the cladding according to the Code (American Society of Mechanical Engineers, 1980a), then one can take credit for the cladding in the mechanical strength of the tubesheet. However, no credit is taken for the weld overlay.

9.60 How do the tubesheets for the floating rear heads differ from those at the front end?

In most cases, the floating tubesheet is smaller in diameter than the shell inner diameter in order to facilitate its movement. The interleakage of the two fluids is prevented by a suitable packing in the rear-head types P and W and by the gasketed flange joints in types S and T (Fig. 1.1). On the other hand, the front end tubesheets are greater than the shell inner diameter.

Double Tubesheet

9.61 What are double tubesheets?

See Q. 1.80.

9.62 Where is the double tubesheet construction used?

See Q. 1.81.

9.63 What factors should be kept in mind when designing a double tubesheet?

The material of the outer tubesheet should be compatible with the tube-side fluid, while that of the inner tubesheet should be compatible with the shell-side fluid. However, the most important consideration is that since the two tubesheets will expand differently (in the radial direction), they will stress the tubes. These stresses, as well as the normal compressive/tensile stresses on the inner tubesheets, should be considered in the design.

9.64 Are the two tubesheets of the same material?

Generally they are of different materials since, while the inner one sees only the shell-side fluid, the outer one sees only the tube-side fluid. Hence, the two have to satisfy separate temperature, pressure, and corrosion requirements.

9.65 What are curved tubesheets? What are their advantages?

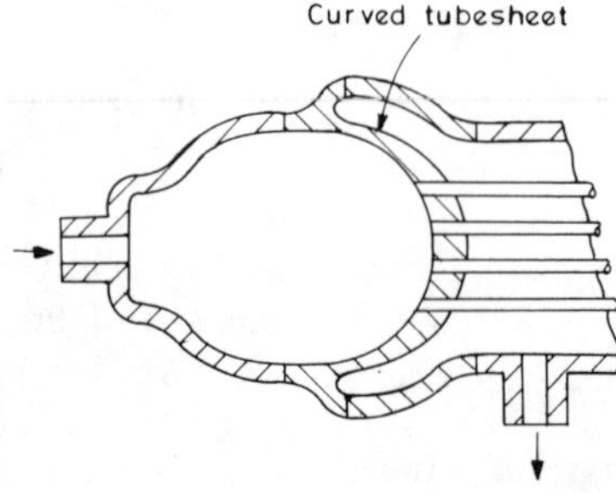

Figure 9.15 Curved tubesheet. *(Courtesy Joseph Oat Corporation, 1980.)*

The surface of such tubesheets is not plane but is curvilinear in three dimensions. Figure 9.15 shows one typical case. These hold the thermal stresses to a minimum during sudden and intense temperature transients where a conventional tubesheet might otherwise develop cracks.

Nozzles

9.66 What are the functions of nozzles?

See Q. 1.90.

9.67 What are the different kinds of nozzles?

See Q. 1.91 and Q. 1.92–1.94.

9.68 What criteria are used to size the nozzles?

Uses of nozzles have been enumerated in Q. 1.90. They are sized differently for different purposes:

For the process fluid inlet and outlet, the nozzle sizes are based upon the permissible pressure drop or fluid velocity, taking into account any erosion and vibration damage that might be caused to the tubes directly underneath the nozzles.

For the safety and relief valves, their sizes are based upon the overpressure to be released, external fire considerations, etc.

For the drain and vent connections, their sizes are based upon the number of such connections in between the baffles, in channels, etc. and the time allowed to drain/vent the whole unit.

For the instrument ports, their sizes are based upon the instrument connection sizes.

9.69 What information is required for designing a nozzle?

The information required is the following:

1. Number and orientations of the nozzles
2. Size of pipe to be attached to the nozzle
3. Self-reinforcing-type nozzle or one with a reinforcing pad
4. Forces and displacements to which the nozzle will be subjected

5. Space available to install the unit and piping, if it is a restricted space or if the units are stacked
6. Velocity of the fluid through the nozzle so that the tube bundle underneath does not vibrate and there is no erosion damage

9.70 What are the characteristics of a good nozzle design?

A well-designed nozzle should have a very low pressure drop (only a few percent of the total shell-side pressure drop for liquids and up to 25 percent for gases, since this pressure drop does not aid in heat transfer), should give a good flow distribution over the tube bundle, should not have excessive discontinuity stresses (Q. 9.134), should be able to take up the loads and moments due to piping and earthquakes, should be sufficiently strengthened by proper reinforcement to take care of the local stresses, and should provide easy accessibility to connect or disconnect the piping.

9.71 How are the nozzles attached to a vessel?

These are attached by arc or gas welding (Chapter 10). Small diameter nozzles may be threaded into the vessel wall (Para UG-43); minimum requirements for the attachment welds are given in Paragraphs UW-16 (American Society of Mechanical Engineers, 1980a).

9.72 What loads act on a nozzle?

Loads on a nozzle are generally considered only in the nuclear heat exchangers and not in the process industry heat exchangers. These loads are considered only when supplied by the buyer. These generally refer to the forces and moment in all the three directions due to the connected piping.

9.73 Is the actual shape of the hole for the nozzle attachment in the shell or head always round?

This refers to the shape of the cut-out portion if spread out on a flat surface. The shape depends on the curvature of the surface to which the nozzle is attached. The shape can be circular, elliptical, or obround (see Fig. 9.16).

The circular opening has the minimum stress concentration, while the obround one has the maximum of the three.

9.74 What are the various possible nozzle orientations?

These are given in Fig. 9.17.

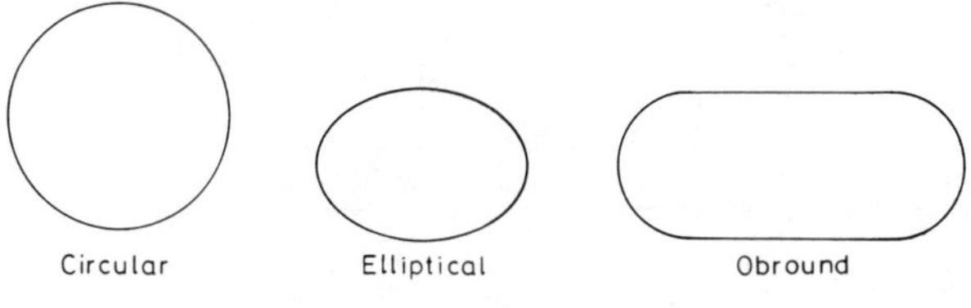

Figure 9.16 Shapes of nozzle holes.

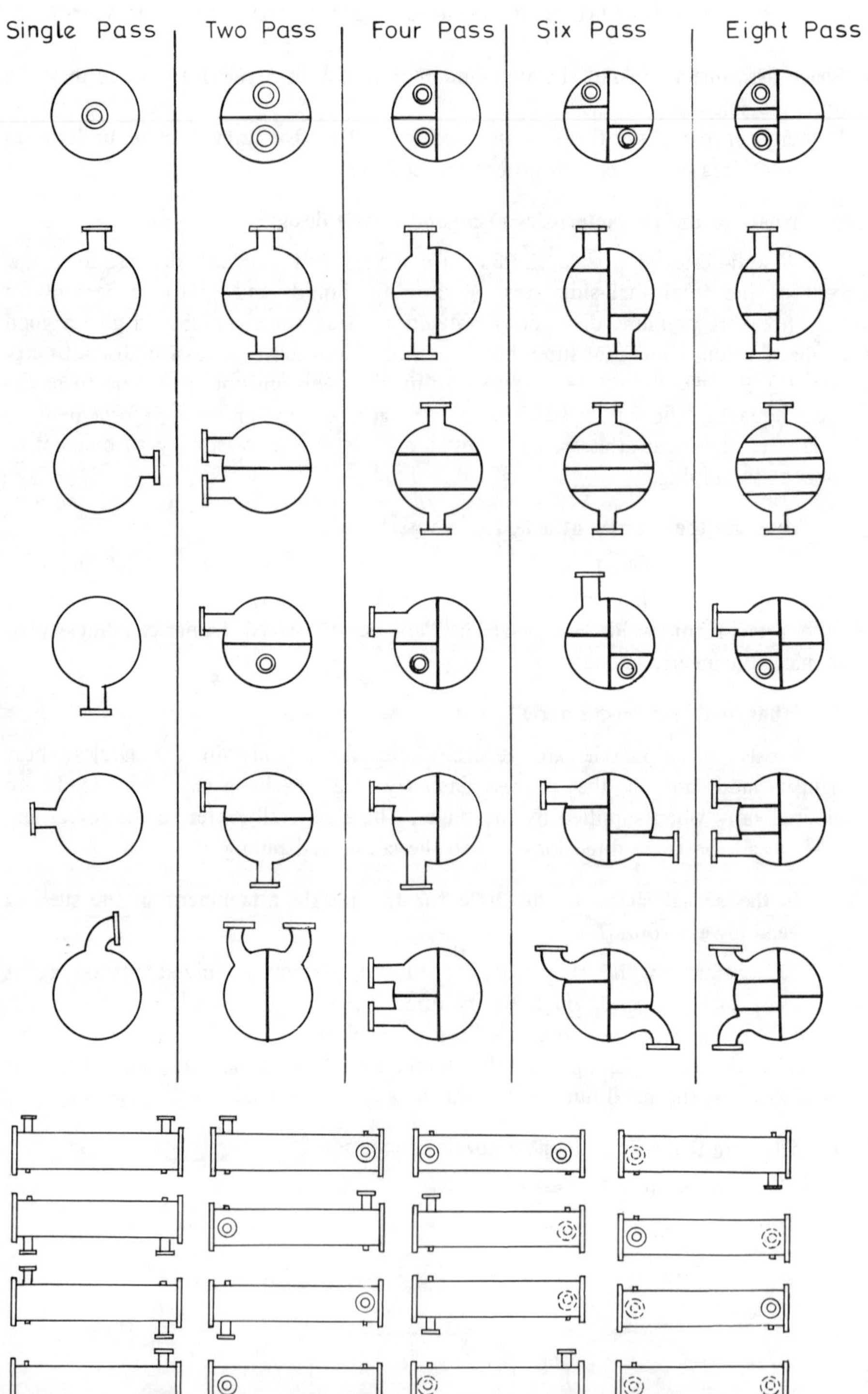

Figure 9.17 Nozzle orientations. *(Patterson-Kelley Co., Division of HARSCO Corp.)*

9.75 What factors influence the choice of the nozzle orientation?

These are the following:

1. Number of tube-side and shell-side passes
2. Location of the pass partition plates
3. Location of the rear end of the U-tube bundle
4. Piping connections
5. Clearance available with respect to other equipment and buildings
6. Nozzle loadings
7. Vertical or horizontal exchanger mounting
8. Ease of access to the tubes, etc.

9.76 How does the cutting of a hole in the shell affect the strength of the shell?

It produces stress concentration in the shell wall near the hole, which weakens the shell (Fig. 9.18). The stress is maximum at the edge of the hole and decreases to slightly less than half the value at a distance equal to the hole radius. The stress values vary with the type of hole (circular, elliptical, or obround) and the type of shell (cylindrical or spherical).

9.77 How can these stresses be reduced?

The stresses can be reduced by the following (Sivasankaran and Gupta, 1983b):

1. Using a thick shell: This is a rather costly way.
2. Using a thick nozzle (self-reinforcing nozzle)
3. Using a reinforcing pad in the vicinity of the nozzle to thicken the shell locally

Choice between (2) and (3) depends on the customer's preferences, application of the unit, whether radiographic examination is required or not, etc. In both, the material cut-out is replaced by an equal amount and placed within the limits of the reinforcements. Figure 9.19 shows three types of nozzle mountings out of which the butt weld type is the costliest and can be examined 100 percent radiographically. It is used in the nuclear power plants, hydrocracker reactors, and other places where stringent quality control is a must.

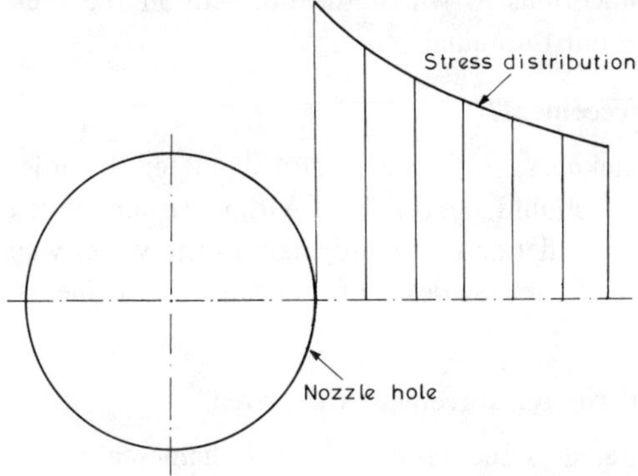

Figure 9.18 Stress concentration near the nozzle.

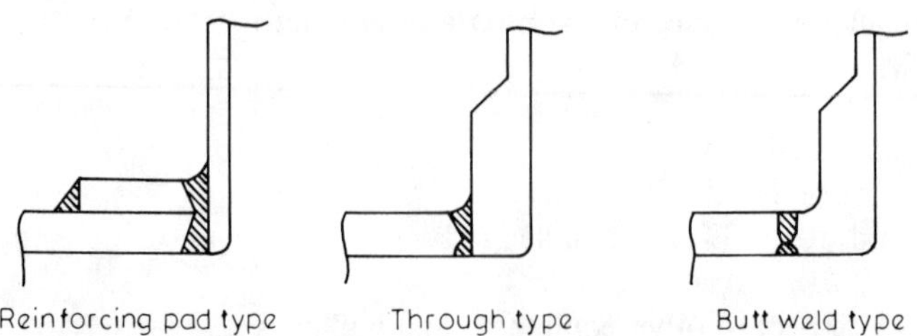

Figure 9.19 Three ways to connect nozzle to the shell. *(From Yasuda et al., 1976.)*

The above is based upon the area equalization method used in the ASME Code. The British Code uses controlled maximum stress method where the reinforcement is provided such that the maximum stress is kept within limits, and the German Code uses the experimental yield method for this purpose (Nichols, 1971). For details, the specific codes should be consulted.

9.78 What factors must be kept in mind in reinforcing a nozzle opening?

These are the following (Harvey, 1974):

1. The reinforcement should be sufficient to compensate for the weakening effects of the opening, yet should not change the general dilation and strain pattern of the shell.
2. The reinforcing material should be placed immediately adjacent to the opening, and its profile and contour should be such as not to introduce an overriding stress concentration.

9.79 What sized holes are exempt from reinforcement?

According to Paragraph UG-36 (American Society of Mechanical Engines, 1980a), the following are exempt from reinforcement:

1. 3 in (76.2 mm) nominal pipe size in vessel shells or heads $\frac{3}{8}$ in (10 mm) thick or less
2. 2 in (50.8 mm) nominal pipe size in vessel shells or heads over $\frac{3}{8}$ in (10 mm) thick
3. Threaded, studded, or expanded connections in which the hole cut in the shell or head is not greater than 2 in (50.8 mm) nominal pipe size

9.80 What is meant by limits on reinforcement?

A reinforcing pad has certain thickness, surface area and diameter. In order for the reinforcement to be effective, it should be confined within certain limits. These limits refer to the distances both parallel and perpendicular to the vessel wall and are given in Paragraph UG-40 (American Society of Mechanical Engineers, 1980a).

9.81 How is the cross-sectional area of the reinforcement calculated?

For vessels with internal pressure, it is the product of the diameter of the opening (in corroded condition) and the shell thickness required to withstand the internal pressure. Further details are given in Paragraph UG-37 (American Society of Mechanical Engineers, 1980a).

For vessels under external pressure, the reinforcement is 50 percent of the above when the thickness of the shell is the one to support the external pressure.

For vessels alternately under internal and external pressures, the reinforcement should satisfy both the criteria.

9.82 Does excessive reinforcement cause any harm?

It creates hard spot on the vessel and does not allow its natural growth under pressure (Harvey, 1974). Local overstressing results, and the vessel may get damaged when pressurized. Thus, the end results are the same as in under-reinforcement.

The limits on nozzle diameter to which the above rules apply are given in Appendix I Paragraph (1-7) (American Society of Mechanical Engineers, 1980a).

9.83 Should reinforcement be done on the outside or inside of the shell?

It can actually be done on either side. However, it is more convenient to do on the outside. Furthermore, the reinforcing pad material on the outside does not have to meet the corrosion requirements of the contained fluid. Hence, carbon-steel is used in most instances as long as it satisfies the welding requirements to the nozzle and the vessel.

9.84 Is the reinforcement required in a head different than that in a shell?

For most heads, except flat heads, the reinforcement is calculated as for a shell. However, for a flat head, the area of reinforcement is calculated according to Paragarph UG-39 (American Society of Mechanical Engineers, 1980a). If the diameter of the opening exceeds 50 percent of the head diameter, Appendix 2 (American Society of Mechanical Engineers, 1980a) should be followed.

9.85 How is the reinforcing pad attached to the shell?

It is formed into the same contour as the shell with a hole to match the nozzle outer diameter and is welded to the nozzle and the shell. A tell-tale hole is provided in the pad in order to release the gases entrapped during welding and to check for the proper welding of the pad to the vessel.

9.86 If two nozzles are close together, how is the reinforcement done?

If the distance (ligament, L) between the two nozzles (Fig. 9.20) is at least equal to the nozzle opening diameter, d_i, then each nozzle acts independently. That is, when

$$\frac{L}{d_i} \geqslant 1 \tag{9.1}$$

the average membrane stress in the vessel wall is not increased by the presence of the reinforced nozzles, and the effect is only of the local stress concentration. When the reinforced nozzles are closer than this, the stress in the ligament is greater than the basic membrane stress in the vessel. Although the shape of the stress

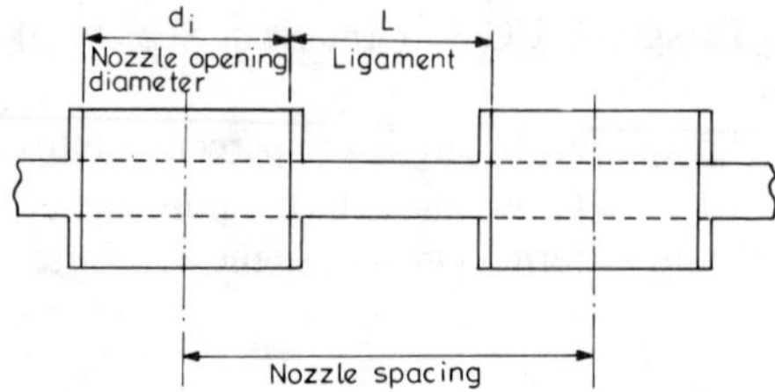

Figure 9.20 Multiple nozzles. *(From Harvey, 1974, modified with permission.)*

distribution remains essentially unaltered, its magnitude is increased. To reduce it, the vessel wall should be increased in thickness by $1.0/(L/d_i)$. Paragraph UG-42 (American Society of Mechanical Engineers, 1980a) gives the code requirements for such a situation.

9.87 What minimum distance should be kept between a nozzle and a nearby flanged joint?

The distance should be sufficient so that the whole length of the flange bolts can be taken out from one side without touching the nozzle. The heat of welding of the flange may influence the welded joint of the nozzle if the two are very close together in thick shells, and a minimum distance is necessary to avoid this.

9.88 How do the nozzles influence the performance of an exchanger?

The pressure drop in the nozzles does not help in heat transfer and hence should be minimized by a proper design (Fig. 1.22). Furthermore, the tubes underneath the nozzles face the fury of the entering flow and are liable to damaging vibrations and erosion. Ways to reduce the damaging vibrations are discussed in Chapter 15.

Expansion Joints

9.89 What are expansion joints?

See Q. 1.97.

9.90 What are the principal kinds of expansion joints?

See Q. 1.98.

9.91 What factors affect the choice of an expansion joint for a specific application?

Flanged and flued type are used when the movement expected is small ($\frac{1}{8}$–$\frac{3}{16}$ in or 3.175–4.76 mm) (Kopp and Sayre, 1952); while, the bellows types are used when the movement expected is large, as in the pipelines.

9.92 Do expansion joints make a weak joint?

See Q. 1.99.

9.93 What standards guide the design of the expansion joints?

EJMA Standards (Expansion Joint Manufacturers' Assocation, 1980) are used to design the bellows expansion joints, while the method of Kopp and Sayre (1952) is used to design the flanged and flued expansion joints. A detailed stepwise method (Sivasankaran, 1981) and computer program (Sivasankaran and Gupta, 1983a) for the design are also available.

9.94 What information is needed for designing an expansion joint?

The information necessary is fluid properties, corrosion tendency, environmental conditions, testing and operating pressures and temperatures, movement or maximum differential expansion, whether insulation is needed, and cyclic temperature changes (magnitude and number of cycles). Furthermore, the acoustic vibrations in the system should be avoided (Chapter 15). A specification sheet is given in Fig. 9.21 (see also Ratnasamy, 1981).

9.95 What are the characteristics of a good expansion joint design?

An expansion joint should be thin and flexible. It should withstand both corrosion and stresses due to fatigue since, if it fails, the exchanger has to be taken out of service for repair or replacement. The expansion joint is generally used for shell-side pressure not exceeding 100 psig since, at higher pressure, the joint becomes thick and its flexibility is reduced.

9.96 What is a universal expansion joint?

It has two sets of bellows. It can take lateral deflection, axial movement, angular rotation, or any combination. It is primarily used in piping layouts (Fig. 9.22).

9.97 What is a convolution? How is its number in an expansion joint decided?

It is the smallest flexible unit of a bellows. The number of convolutions is decided based upon the extent of the thermal stresses. If these stresses are high, multiple convolutions are used to withstand the stresses and absorb the movements.

9.98 How are the expansion joints attached to the shell?

These are attached by full penetration welds.

Support Systems

9.99 How are the heat exchangers and pressure vessels supported?

See Q. 1.111.

9.100 How do the supports for the horizontal and vertical installations differ from one another?

In a horizontal installation, two supports equidistant from either end are used (Fig. 1.35). One is rigidly held on the ground, while the other is placed on rollers to allow for the thermal expansion of the shell. On the other hand, in a vertical installation, only one support is used (Fig. 1.36). It could be either near the bottom or near the top. While the wind load is more on a vertical unit than on a horizontal one, both have to be rigorously analyzed for the seismic load.

9.101 What factors should be considered in the design of the support system for a heat exchanger or pressure vessel?

STANDARD EXPANSION JOINT SPECIFICATION SHEET

Customer ______________________ Date ______________

______________________ Proposal No. ______________

Project ______________________ Inquiry/Job No. ______________

______________________ Sheet ______ of ______

ITEM NO.			
QUANTITY			
SIZE			
TYPE OF EXPANSION JOINT			
FLOWING MEDIUM			
FLOW VELOCITY			
FLOW DIRECTION			
DESIGN PRESSURE, PSIG			
INTERNAL TEST PRESSURE, PSIG			
MAXIMUM TEMPERATURE, °F			
MINIMUM TEMPERATURE, °F			
INSTALLATION TEMPERATURE, °F			
AXIAL COMPRESSION, IN			
AXIAL EXTENSION, IN			
LATERAL DEFLECTION, IN			
ANGULAR ROTATION, DEG.			
BELLOWS MATERIAL			
PIPE SPECIFICATION/MATERIAL			
WELD END PREPARATION			
FLANGE SPECIFICATION/MATERIAL			
INTERNAL SLEEVES/MATERIAL			
EXTERNAL COVERS/MATERIAL			
ANCHOR BASE (MAIN/INTERMEDIATE)			
TIE RODS			
MAXIMUM OUTSIDE DIAMETER			
OVERALL LENGTH			
INSTALLATION POSITION (HORIZ/VERT)			

For Additional Data, Use "Supplemental Specification Sheet."

Figure 9.21 Specification sheets for expansion joint. *(© 1980 by Expansion Joint Manufacturers' Assoc., with permission.)*

SUPPLEMENTAL SPECIFICATION SHEET
To be used with Standard Expansion Joint Specification Sheet

Customer______________________________ Date______________

______________________________ Proposal No. ______________

Project ______________________________ Inquiry/Job No. ______________

______________________________ Sheet ________ of ______________

	ITEM NO.			
	PURGE, INSTRUMENTATION CONNECTION			
	VIBRATION AMPLITUDE			
	VIBRATION FREQUENCY			
	MINIMUM I.D.			
	MAX. AXIAL SPRING FORCE (LBS)			
	MAX. LATERAL SPRING FORCE (LBS)			
	MAX. SPRING FORCE END MOMENT (IN-LBS)			
	CYCLIC DESIGN LIFE			
	ASME CODE PARTIAL DATA FORMS REQ'D			
	APPLICABLE CODES AND SPECIFICATIONS			
SPECIAL FLANGE DESIGN	FACING			
	O.D.			
	I.D.			
	THICKNESS			
	B.C. DIAMETER			
	NO. HOLES			
	SIZE HOLES			
	HOLE ORIENTATION			

REMARKS:

Figure 9.21 Specification sheets for expansion joint. *(© 1980 by Expansion Joint Manufacturers' Assoc., with permission.)* ***(Continued)***

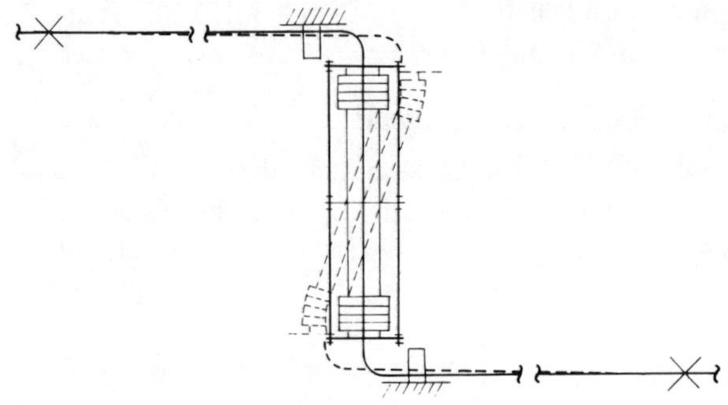

Figure 9.22 Universal expansion joint. *(© 1980 by Expansion Joint Manufacturers' Assoc., with permission.)*

The important factors to be considered are the following:

1. Size of the unit (diameter, length), wall thickness
2. Weight of the exchanger (empty weight, weight full of liquid, insulation if any)
3. Operating conditions (pressure or vacuum, temperature)
4. Properties of the support material (allowable stresses, etc.)
5. Characteristics of the ground where the exchanger is to be installed (load bearing capability, etc.)
6. Stresses due to wind load
7. Stresses due to seismic load
8. Nozzle load and moments, piping arrangement
9. Whether horizontal or vertical installation required
10. Whether more than one unit is to be stacked
11. Floor space available
12. Elevation of the vessel with respect to ground or floor
13. Location of the supports

9.102 What are the characteristics of a good support design?

It should prevent excessive localized stresses in the vessel. It should permit expansion and contraction of the vessel, or else the latter will weaken. Appendix G (American Society of Mechanical Engineers, 1980a) lists a few more general points of interest in this regard.

9.103 How are the support saddles for the horizontal vessels designed?

Horizontal vessels are supported on saddles. An analysis of the stresses introduced in the shell by the saddle supports has been given by Zick (1951). The design requires the calculation of the following:

1. Maximum longitudinal stress
2. Tangential shear stress
3. Circumferential stress at the horn of the saddle
4. Additional stress in the head used as a stiffner
5. Ring compression in the shell over the saddle

Zick has given a nomograph to aid in the design. Detailed derivations and a solved example are given in Brownell and Young (1959).

9.104 How is the location of the saddle supports decided?

Figure 9.23 (Brownell and Young, 1959) shows a general situation. Dimension B in kept less than R in order to take advantage of the stiffening effect of the head. B is often taken to be $0.4R$. Besides, B should never exceed $0.2L$, otherwise the stresses due to the cantilever action will be excessive (Brownell and Young, 1959).

9.105 What are ring stiffeners?

In thin-walled vessels and when the saddles are located away from the head ($B/R > \frac{1}{2}$) (Fig. 9.23), the shell alone is unable to resist the circumferential bending

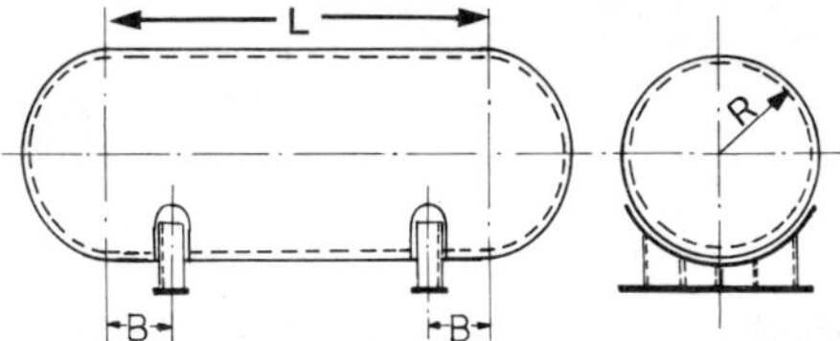

Figure 9.23 Location of saddle supports. *(From Brownell and Young, 1959.)*

moment. Ring stiffeners are then attached to the shell to reduce the load on the shell. These stiffeners can be either internal or external in the case of a heat exchanger (Fig. 9.24). Their design has been discussed by Brownell and Young (1959). Ring stiffeners for vessels under external pressure are discussed in Paragraph UG-29 (American Society of Mechanical Engineers, 1980a).

9.106 In a two-saddle support system, if one support settles more than the other, will the load shared by the two saddles still be equal or not?

It will still be equal. However, if more than two saddles are used and one settles more than the rest, the load sharing between all will not be equal. This is a big disadvantage of using more than two supports for a horizontal unit.

9.107 How are the supports for the vertical units designed?

The various types of supports used for the vertical units have already been discussed in Q. 1.111. The design of these supports is quite involved, and reference should be made to the standard texts such as Brownell and Young (1959) and Harvey (1974), or journal articles such as Singh (1976).

9.108 What are bearing plates? How are they designed?

Bearing plates are used to anchor securely the bottom of the vessel support to the concrete foundation by means of anchor bolts. The design of the bearing plates involves the calculation of the tensile and compressive stresses and the neutral axis. Brownell and Young (1959) and Truong and Truong (1981) should be consulted for details.

9.109 What kind of support is used for units on ships?

Ships have a number of pressure vessels. Nuclear-powered ships have many more, including heat exchangers. These are subject to high horizontal and vertical shock loadings, as well as roll and pitch. For these, skirt supports are preferred (Harvey, 1974).

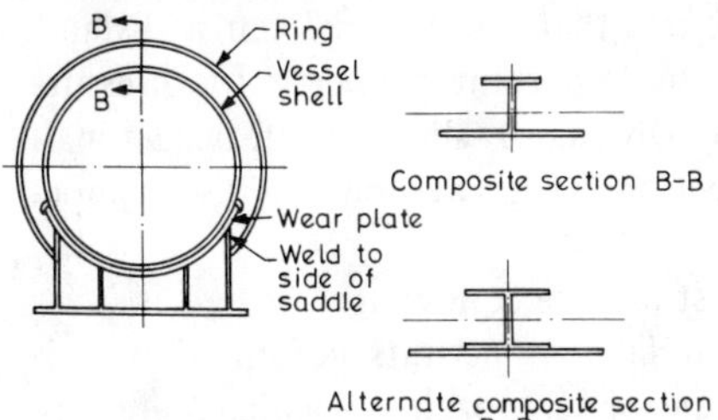

Figure 9.24 External stiffening ring. *(From Brownell and Young, 1959.)*

STRESS ANALYSIS

Definitions

9.110 Define stress and strain.

Stress: External forces applied to a solid body produce internal resistance in the body. The intensity of this internal resistance at a point is called stress. It is expressed as the force per unit area.

Strain: This is defined as the ratio of the change in length to the initial length between two points in a body in the limit as the initial length approaches zero. The change in length may have been produced by external or internal forces or by a change in temperature.

9.111 What is a stress-strain curve, and what are the important points/zones on it?

This curve depends on the kind of material. Figure 9.25 shows an idealized, uniaxial, normal stress-strain curve for a typical carbon-steel. The various important points/zones on it are the following:

Elastic range: Most of the engineering materials have an initial stress-strain relation that may be considered linear, i.e., the stress, σ, is directly proportional to the strain, ϵ:

$$\sigma = E\epsilon \tag{9.2}$$

where E, the proportionality constant, is called the modulus of elasticity and the relation $E = \sigma/\epsilon$ is called Hooke's law. This linear range is called the elastic range, and its end point is called the elastic limit. The higher the value of E, the more stiff is the material and the less deformation it will undergo as compared to a material of low E value. If the stress in the elastic range (Hooke's law region) is relieved, the material returns to its original, unstressed condition without any permanent deformation. The strain in this range is called the elastic strain.

Yield point: This is the stress at which a marked increase in the strain occurs without an increase in the stress. This is also known as the yield stress, and its value is close to the elastic limit. A permanent deformation sets in, and transition occurs from the elastic to the plastic range.

Plastic range: Very little increase in the stress causes a big increase in the strain. The material deforms and absorbs energy resisting rupture. If the applied stress is removed, there is residual strain left in the material. The strain in this range is called the plastic strain.

Ultimate stress: This is the maximum stress that the material can withstand. It is also called the ultimate strength and is an important property for pressure vessel materials. The ASME Code Section VIII, Division 1 allows a safety factor of 4 in determining the allowable stress while Section VIII, Division 2 with its more rigorous procedures, allows a safety factor of 3.

Some other material properties of interest are the following:

Ductility: This is the property of a material that permits deformation. It is important in relieving the high local stresses at joints, nozzles, openings, etc., by

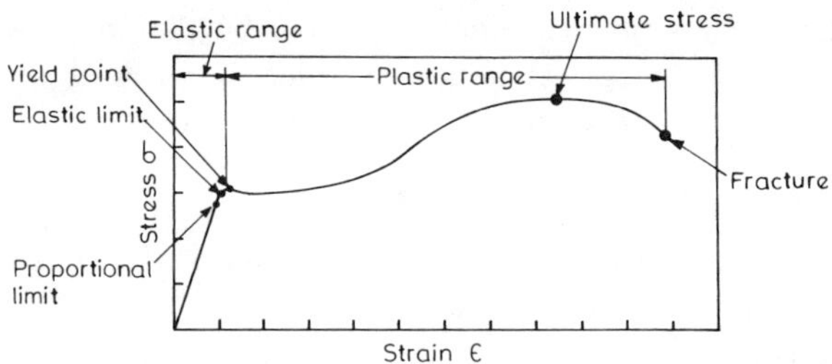

Figure 9.25 Stress–strain curve.

plastic flow of the material. In rolling, forging, extruding, drawing, etc., the ductility allows the material to be worked upon without rupture. Ductility can be increased by heating the material.

Toughness: This is the ability of a material to absorb energy. It is measured by the energy absorbed per unit volume in stressing to rupture. It is the total area under the stress-strain curve (Fig. 9.25). Tough materials have high ductility. Toughness is a very desirable property for vessels or parts subjected to a mechanical or a thermal shock.

Brittleness: Materials with low ductility are brittle. They cannot withstand mechanical or thermal shock, since they cannot deform to absorb energy. There is no appreciable plastic range. Most ductile materials, especially steels, become brittle at temperatures below a critical temperature, called the transition temperature. They lose impact strength and ductility. Hence, for cryogenic applications, the materials should be suitably chosen, and, in operation of such vessels, the thermal and mechanical shocks should be avoided. Heating of the material above the transition temperature helps it recover its ductility and impact strength. The transition temperature (actually a zone) depends upon the grade of steel and plate thickness. It is higher for thicker plates. Below the transition temperature zone, the code allows only 40 percent of the design strength allowed above the transition temperature zone for nonimpact tested carbon steel.

9.112 What are the different stress categories?

In pressure vessel technology, the stresses are divided into various categories based upon the magnitude of threat they pose to the integrity of the structure. This categorization helps a designer/analyst to determine the factor of safety available in various parts of the pressure vessel.

The stresses can be broadly classified into five categories:

1. Primary membrane stresses
2. Primary bending stresses
3. Local membrane stresses
4. Secondary stresses
5. Peak stresses

Detailed discussion on these can be obtained in several specialized references such as Harvey (1974), Singh (1979).

9.113 What is allowable stress?

It is the value of the stress at a given temperature allowed by the code in designing the pressure vessels. The basis to establish this value is outlined in Appendix P in the Code (American Society of Mechanical Engineers, 1980a).

9.114 How does the allowable stress vary with the temperature?

It decreases as the temperature increases. For example, for carbon-steel (SA-516 Gr 70) the allowable tensile stress is 17,500 psi from –20° to 650°F; 12,000 psi at 800°F; 2500 psi at 1000°F (American Society of Mechanical Engineers, 1980a) (respectively, 120.6 MPa from –28.9° to 343°C, 82.7 MPa at 427°C, 17.2 MPa at 538°C). Thus, the shell thickness at 1000°F (538°C) will be much higher than that at 650°F (343°C) if the rest of the terms remain unchanged. Values given in the applicable code at the design temperature have to be followed in an actual design.

9.115 Why is a safety factor used in the mechanical design of the pressure vessels?

The safety factor relates the allowable stress to the elastic limit or the ultimate strength. Usually, the allowable stress is half of the elastic limit or one-fourth of the ultimate strength. It allows for the variation in the material properties, uncertainties in the loading conditions, load distribution, and stress concentration factors (Fraas and Ozisik, 1965).

9.116 What is meant by stress analysis of a unit?

It is finding the stresses at important locations in the unit and checking if they are within the allowable limits. It is usually done in critical applications, such as nuclear or hazardous chemicals applications, and at high pressures and temperatures (Fraas and Ozisik, 1965).

Information Required

9.117 What information is required for the stress analysis of a heat exchanger?

The following information is required:

1. Function of the exchanger
2. Its classification (nuclear, nonnuclear)
3. Environmental conditions
4. Design specifications
5. Mechanical and operational loadings; including vibrations and shocks
6. Transient conditions in terms of pressure and temperature, duration of such transients, number of such occurrences to be considered, and the category of each occurrence (normal, upset, emergency, or faulted)
7. Nozzle reactions (forces and moments)
8. Deadweight loadings (empty, operating, and flooded)
9. Seismic loadings
10. Allowable stress limits and fatigue life requirements for the materials used under various conditions as per the applicable code.

Loads

9.118 What are the various loads that act on an exchanger and should be considered in its design?

The following loads should be considered in the design (Oak Ridge National Laboratory, 1975):

1. Support reactions: These arise due to the supports for the exchanger. These act differently on the horizontal and the vertical exchangers.
2. Impact loads: These arise due to the rapidly fluctuating pressures, thermal shocks or water hammer.
3. Superimposed loads: These arise due to the stacking of units, insulation, piping, ladder, etc.
4. Nozzle loads: The piping connected to the nozzles causes axial, bending, and torsional forces on the nozzles.
5. Wind loads: These are more important for the vertical units and for the horizontal units mounted at the top of columns.
6. Seismic loads: Also called the earthquake loads, these are very important in the high earthquake-prone zones.
7. Thermal loads: These are caused by the differential thermal expansion of the shell and the tube bundle, and the thermal expansion of the tubesheet.

9.119 Where do the various loads act in an exchanger?

These are summarized in Table 9.3.

Areas Requiring Stress Analysis

9.120 What are the typical areas of heat exchanger that require stress analysis?

These are the following (Yasuda et al., 1976; Oak Ridge National Laboratory, 1975):

Table 9.3 Locations of loads

	Load				
Area to be analyzed	Pressure	Dead load	Piping reaction	Earthquake and wind load	Thermal load
Skirt to head junction	0	0		0	0
Nozzle	0		0		0
Internal and external supporting attachments	0	0			0
Head to shell junction	0				0
Shell flange	0				0

Source: Yasuda et al., 1976.

1. Basic pressure-retaining parts: These include the front and the rear ends, shell, tubesheet, etc. Stresses due to steady state, transient, and seismic loads should be calculated in each of these parts.
2. Tubes: Stresses due to internal and external pressures, tube vibrations, and transient and seismic loads should be analyzed.
3. Penetrations: These include the nozzles, manholes, and handholes. Stress analysis due to steady and transient loads is required. Furthermore, analysis is required to check whether adequate reinforcement has been provided or not. The closures for the manholes and handholes should be analyzed for the pressure loads.
4. Pass partition plates or divider plates: Due to different pressures on the two sides, there are stresses developed in these plates that should be analyzed.
5. Internals: The heat exchanger internals should be checked for the stresses due to the seismic loads, pressure, dead weight, and the flow-induced loads.
6. Gross structural discontinuities: Due to different thicknesses and radii of curvature, the head-to-shell joints, nozzle-to-shell joints, and nozzle-to-head joints should be analyzed for the stresses.

9.121 What stresses are analyzed for in a nuclear heat exchanger?

The stress analysis in a nuclear heat exchanger is carried out for the following (Yasuda et al., 1976):

1. Primary stresses due to bending in the main shell
2. Stresses in the bolted joints to determine if leakage might occur
3. Stresses due to nozzle loads at the nozzle-to-shell and nozzle-to-channel junction
4. Stresses due to pressure loads and thermal transients in the tubesheet
5. Stresses due to thermal transients in all areas of the heat exchanger that will see rapid fluctuations in the temperature
6. Stresses near the support-to-shell junction
7. Stresses in the supports due to the seismic load
8. Stresses in the foundation bolts of the heat exchanger

9.122 What are the steps involved in the stress analysis?

The various steps involved according to the Code (American Society of Mechanical Engineers, 1977), are listed in Fig. 9.26.

9.123 What are some special situations that require detailed stress analysis because codes do not consider such cases?

Among numerous such cases, two important ones are:

1. When the tubesheet is welded to both the shell and the channel to avoid leakage at the gasketed bolted joints due to a very high-pressure difference on the two sides of the tubesheet where such leakage may be disasterous, as in the hydrogen recirculation system in a Platformer or Hydrotreater unit
2. When the distance between the nozzle and flange junctions with the shell is reduced to a minimum in certain cases to make heat exchanger inlet and outlet regions more effective

These cases have been dealt with by Mitchell (1973).

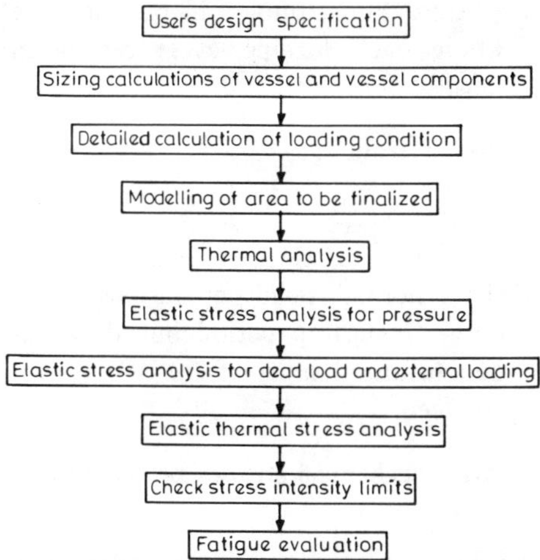

Figure 9.26 Steps in stress analysis. *(From Yasuda et al., 1976.)*

Thermal Stresses

9.124 When are severe thermal stresses likely to be introduced in an exchanger?

This happens when the temperature differential in the unit is above 100°F (37.8°C), resulting in severe differential expansion. Such situations can arise (1) between the shell and the tubes; (2) between the tubes of different tube passes (due to temperature changes and/or flow maldistribution), which puts severe stresses on the tubes and the two halves of the tubesheet; and (3) between two tubesheets of a double tubesheet construction. Design techniques exist to take care of the above provided the designer is aware of the problem. For example, problem (1) could be accommodated by using an expansion joint or a floating head; problem (2) could be accommodated by allowing some flexibility such as simple bends in boiler tubes or serpentive bends in a superheater; problem (3) could be accommodated by a judicious choice of the gap between the two tubesheets and the material of construction of the tubes and the tubesheets.

9.125 How are the transient thermal stresses introduced in a pressure vessel?

These are produced by a nonuniform temperature (T) distribution in the material and/or by different coefficients of linear thermal expansion (α) for the different materials used. Essentially, these prevent the natural growth or contraction of a material due to a temperature change. In cylindrical vessels, it is the αT gradient that governs the thermal stresses. Such gradients can be along the circumference, length, or thickness (radius) of the vessel. These gradients depend upon the temperature, flow rates and the properties of the fluids, inlet and outlet nozzle locations, etc., and vary with time, thus introducing the transient thermal stresses.

9.126 Why are the thermal stresses more troublesome at high temperatures?

They cause creep in the material at high temperatures. The elastic limit or the yield stress of the material is very low at high temperatures. Hence, the material gets into the plastic range to relieve the stresses. Furthermore, if the pressure fluctuates, thermal racheting takes place where each loading cycle results in incremental strain, which is a highly unacceptable situation.

9.127 How can the transient thermal stresses be controlled?

There are two methods:

1. Special design features: Use thermal barriers, sleeves, shrouds, etc. This is a costly method since all these are not needed for normal, steady operations.
2. Special start-up procedure: It results in a longer start-up period but does not require special design features. It is the one followed most often. It does not allow the transient thermal stresses to be produced.

9.128 What is the difference in the thermal and mechanical stresses produced in a material?

When a material is subjected to thermal stresses, it yields to relax such stresses. These are a self-limiting type; they are also called secondary stresses. When such stresses are applied for the first time on a ductile material, it will not rupture. However, it could fail due to repeated thermal stresses that cause fatigue. Furthermore, large distortions due to thermal stresses in certain structures (e.g., turbines) could render them inoperative.

On the other hand, if a material is subjected to a stress beyond its yield strength due to mechanical loading, it will continue to yield until it breaks, since the external load remains constant unless strain hardening or redistribution of stresses limits the deflection. These stresses are not self-limiting.

Transient Analysis

9.129 What are the different types of transient analyses required?

All class I heat exchangers in nuclear service must be analyzed to assure a satisfactory fatigue life. For this, the following analyses are done (Oak Ridge National Laboratory, 1975):

1. Mechanical and thermal transient stresses: The worst temperature distribution for all the significant areas of discontinuity in the structure must be determined for each transient. Stresses resulting from this temperature distribution as well as due to pressure, nozzle loads, and support loads must then be calculated.
2. Stress concentration factors: The effects of discontinuities and stress concentrations at each critical point must be considered in the evaluation of the transient stresses and fatigue life. Such critical points are the junction of a shell to a flange, smooth transition from a cylinder to a cone, and junctions of the nozzles to a shell (see also Q. 9.131 and 9.132).
3. Fatigue evaluation: After the above calculations, fatigue is evaluated for the stresses.

9.130 What is the difference between stress analysis and fatigue analysis?

Stress analysis gives the distribution of stresses in the critical regions of a heat exchanger or pressure vessel, particularly at the discontinuities, to evaluate its safe operation and to modify the design accordingly, if necessary.

Fatigue analysis gives the effects of pressure and temperature cycling on the unit and is very important in order to ascertain the safe operation during the whole life of the pressure vessel.

Both the stress and the fatigue analyses can be done theoretically if enough information exists on the system of interest. For unusual configurations, recourse may be taken to experimental analysis combined with theoretical development.

9.131 What factors should be kept in mind at the design stage to achieve long life of a vessel?

The most important consideration is to have the lowest feasible values of the stress concentration. This requires smoothing out sharp corners by removing some metal and/or changing the radii as is sometimes done in nozzle-to-shell connections (Harvey, 1974).

9.132 What is a stress concentration factor?

Maximum stress in a structure occurs at a notch or at an abrupt change in the cross section. It is greater than the value calculated by the elementary formulae. The ratio of this maximum stress to the nominal stress calculated by the elementary formulae is called the stress concentration factor. It is a constant for the particular shape and is independent of the material, provided the material is isotropic (Baumeister, 1978).

9.133 What harm does the stress concentration cause?

Stress concentration is known as peak stress in the code (American Society of Mechanical Engineers, 1980a), which allows the local peak stresses to be 1.5 times the working stress. However, if the peak stress exceeds this value, the metal is likely to crack.

9.134 What is a discontinuity stress?

For a given thickness, a cylindrical shape under internal pressure undergoes different radial expansion than does a spherical shape or any other shape under the same internal pressure. Thus, when a cylindrical shell with a hemispherical head is pressurized, the diameters of the shell and the head will change by different amounts. However, since the two are welded together, the edges of the two must deform in such a manner as to have identical displacement and slope at their juncture. The stress produced because of this kinematic discontinuity is called the discontinuity stress.

The discontinuity stresses also arise when two bodies of identical curvature but different wall thicknesses, such as two cylindrical shells of the same inner diameter but different thicknesses, are welded together and subjected to the same internal pressure. Here, the thicker shell will expand less than the thinner one.

However, the discontinuity stresses develop at the juncture between the thick and the thin shell for the same reasons as discussed in the previous paragraph.

The code allows the discontinuity stress to be three times the allowable stress, almost equal to the ultimate stress.

9.135 What kind of stress is set up by a flanged joint?

A flanged joint sets up bending stress in the shell. This requires the shell to be thickened in the vicinity of the joint.

Tubesheet

9.136 Discuss the thermal behavior of the tubesheet.

A tubesheet has the shell-side fluid on one side and the tube-side fluid on the other. Thus, it is subjected to two different temperatures on either side. These temperatures cause the tubesheet to expand both radially and axially. Both these movements are constrained by the welding or bolting of the tubesheet to the shell. Thus, stresses are set up in the tubesheet that are also transmitted to the tubes. The situation is worse in a multi-pass unit where the temperature on one or both sides of the tubesheet may have more than one value because of the different fluid temperatures in each pass.

9.137 What factors affect the tubesheet temperature? Why is it important to calculate it?

The shell-side and tube-side fluid temperatures, the number of shell-side and tube-side passes, partial phase change, etc., affect the tubesheet temperature.

It is important to calculate the tubesheet temperature since it determines the shear stresses on the bolts holding the channel, tubesheet, and the shell together. In a double tubesheet heat exchanger, it determines the gap between the two tubesheets. One should keep in mind the probable failure of the cold fluid supply while establishing the design temperature of the tubesheet.

9.138 How are the compressive and tensile loads on the tubes taken care of in the tube-to-tubesheet joints?

For normal operations, simple rolling of the tubes makes the tube-to-tubesheet joints leak-tight. For compressive loads, grooving is necessary, which is done by a grooving tool. For tensile loads, tubes are welded to the tubesheet. Welding gives a smooth contour at the entrance to the tube-side fluid, which reduces the entrance pressure drop and the erosion losses.

9.139 Are there stresses in the removable tube bundle units?

Removable tube bundle units are used to minimize thermal stresses and for ease in cleaning of the outside tube surface. However, these are not free from stresses, although the magnitude is less than that in a fixed tubesheet unit.

There are two kinds of removable bundles:

1. U-tube bundles: Stresses occur in the U-tube bundle when, due to different temperatures of the tube-side fluid in the two passes, the two straight legs of the U-tube expand differently. The baffle holes somewhat constrain the free movement of the U-bend resulting in stresses (Singh and Holtz, 1979).
2. Straight tubes with floating rear heads have similar stresses developed in a multi-pass unit. Besides, these may also distort the rear tubesheet.

9.140 What is the importance of the thermal stress analysis in a fixed tubesheet exchanger?

The thermal stresses in a fixed tubesheet exchanger are transmitted to the tube bundle, which acts as an elastic foundation. These stresses should be superposed on the hydrodynamic loading stresses for a complete analysis. The thermal stress analysis should be made for all the phases of operation, such as the start-up, shutdown, normal operation, upset conditions, and abnormal or emergency conditions.

9.141 What information is required for the thermal stress analysis in a fixed tubesheet exchanger?

The following information is required:

1. Average temperature of the shell and the tubes
2. Linear coefficients of expansion for the shell and tube materials
3. Modulus of elasticity for the shell and tube materials
4. Dimensions of the shell and the tubes

If the calculated expansion of the tubes is greater than that of the shell, the tubes are under compressive stress; conversely, they are under tensile stress.

Supports

9.142 What factors cause stresses in a skirt support for a vertical vessel?

These are the following:

1. Weight of the vessel and contents transmitted to the skirt by the shell
2. Maximum load intensity due to externally applied moments and forces when these are a factor, e.g., wind, earthquake, and piping loads

9.143 What is the accuracy with which the mechanical performance of an exchanger can be predicted?

The mechanical performance relates to the ability of the exchanger to withstand the various loads. This ability can at best be predicted with the same accuracy as the thermal performance, i.e., ±20 percent. The error increases with the complexity of the situation. Therefore, larger safety factors are used in more complex situations.

Experimental Stress Analysis

9.144 What is meant by experimental stress analysis and when is it done?

This means doing the experiments to determine the stresses in a system or in its specific parts. It is done when one is dealing with a unique system and/or operating conditions for which it is difficult to calculate the stresses accurately by the existing theories of elasticity, plasticity, creep, etc. Sometimes, it is also done to test the theoretical calculations and/or the fabrication procedure to the satisfaction of the certifying authorities.

9.145 How is the experimental stress analysis done?

There are essentially three methods (Harvey, 1974):

1. Strain gauges: These are put on the vessel or its scale model to measure directly the strain produced. Both primary direct stresses and secondary bending stresses can be evaluated. There are both mechanical and electrical strain gauges available.
2. Photoelastic: Isotropic and transparent material models of the vessel are built, and stresses in them are measured by passing polarized light. Another method is the pasting of thin sheets of photoelastic plastic on the actual metal structure, and the surface strains transmitted to the plastic by the structure can be measured by reflected light. The measurements show whether the assumptions in the various stress analysis formulae hold or not. Complicated shapes can be analyzed in this manner. More importantly, they show local stress concentrations in regions of abrupt changes in the geometry or in the concentrated loads.
3. Moire's fringe technique: Grids with periodic rulings are marked on the undeformed test piece and on a transparent screen. After the test piece is stressed, the grids on it and the transparent screen are superimposed on one another, and, by the fringes formed due to the displacement of the grids on the test piece, one can determine the strains and calculate the stresses. This is the most adaptable method for evaluating both the pressure and thermal stresses, particularly the high-temperature ones, and can be applied directly to the actual structure.

9.146 What is a stress analysis report? When is it required?

This is a report containing all the calculations and relevant drawings of the components. It should show that the stress intensity and fatigue life are in compliance with the code. A major purpose of such a report is to facilitate an independent review of the stress analysis for its adequacy (Oak Ridge National Laboratory, 1975). It is required for the heat exchangers and pressure vessels used in nuclear and other critical services.

Stress Relieving

9.147 When is it necessary to relieve from equipment or part the residual stresses induced during fabrication?

Stresses are relieved when they would either cause stress corrosion cracking or otherwise result in the mechanical failure of the equipment or of the component. If neither of these is likely to happen, there is no need to relieve the stresses.

9.148 How are the stresses relieved?

For stress relieving, the equipment or the part is heated to a predetermined temperature, held there for a prescribed time, and then cooled gradually in the furnace or in the air or quenched with water for a quick cooling. The exact treatment and schedule of temperature vs. time varies with the equipment and the type of material. This is called post-weld heat treatment (PWHT) and should be carried out by a properly trained person. It affects the mechanical properties of the equipment. It is required for parts and equipment made by hot forming and by electro-slag welding and sometimes even for those made by cold forming. As an example, the TEMA Standards (1978) require heads made of carbon-steel to be stress relieved by PWHT.

It is not done for austenitic stainless tubes because the temperature may reach the sensitizing zone with a resultant possible redistribution of carbon. In such cases, the ultrasonic vibration method may be used for stress relieving.

Good control of the temperature and the cooling rate is required during PWHT to prevent distortion of the unit and damage to its mechanical properties (Yamada and Minamimoto, 1976).

Sometimes pre-weld heat treatment (preheating) of the material is done to circumvent PWHT (Q. 9.150).

9.149 What is low temperature post-weld heat treatment? When is it used?

This is a post-weld heat treatment at a temperature of 200°-300°C (390°-570°F) instead of the usual temperatures of 800°C (1470°F) or higher.

It is recommended for use in the prevention of transverse cracks in heavy section butt weldments. It helps in reducing the residual stresses and diffusible hydrogen in the weldments.

9.150 What is preheating? How does it differ from PWHT?

Preheating refers to heating a metal to above the ambient temperature, usually from 125° to 500°F (50° to 260°C) prior to welding so that the thermal strains due to welding are minimized. The rate of cooling of the metal is reduced so that the metal can achieve a more favorable grain structure, minimizing hard zones and cracks.

The need for preheating depends on the chemical analysis of the materials being welded, their thicknesses, etc. Sometimes, preheating allows a thicker plate to be welded without the need of PWHT. It also removes any moisture present, which reduces the probability of hydrogen getting into the weld and producing embrittlement, cracking, and porosity. The rate of cooling of a welded joint (which had been preheated before welding) is slowed down allowing more time for metal transformation to occur to a favorable structure so that the cracks in weldments are minimized. The minimum requirements for preheating are given in American Society of Mechanical Engineers (1980b).

SEISMIC ANALYSIS

9.151 What is seismic analysis?

It is the evaluation of the stresses in the vital areas of heat exchangers and pressure vessels due to the inertia loads produced by the seismic (earthquake) acceleration of the ground. Some such areas are the bolted flange joints, tube-to-tubesheet joints, support structures, etc.

9.152 In what kinds of installations is the seismic analysis done?

It is done in the installations whose safety under seismic conditions is required. Examples are components in the nuclear plants that handle radioactive material, vessels containing toxic chemicals, and tall buildings in high earthquake-prone zones.

9.153 What are the different seismic classifications for a nuclear power plant?

All structures and equipment in a nuclear power plant are classified as Class I, Class II, or Class III on a seismic basis.

Class I structures and components are those whose failure could cause a significant release of radioactivity or are vital to a safe shutdown of the plant. They require dynamic analysis to ensure their safe working under the expected seismic conditions. Furthermore, they should permit safe shutdown under twice the basic seismic load applicable to the site [see also American Society of Mechanical Engineers (1980c)].

Class II structures and components are those that are necessary for the operation of the plant but not for a safe shutdown.

Class III structures and components are those that are neither necessary for the operation nor for the safe shutdown of the plant. Their failure will not result in the release of radioactivity.

9.154 How does an earthquake affect a heat exchanger or pressure vessel?

An earthquake, in essence, produces inertia loads on the unit. It may cause a bolted joint to leak or some other joints to break. The typical frequency of ground motion during an earthquake is approximately 25 cycles/sec. If the fundamental frequency of vibrations of an exchanger (considered in flexural mode as a rigid beam) is less than or equal to 25 cycles/sec, the unit will amplify the ground motion and undergo severe stresses. However, if the fundamental frequency of the unit is high, it will not undergo severe stresses.

9.155 Are the seismic effects different on the horizontal and vertical units?

Yes, since a horizontal unit has two supports that will accelerate at different times depending upon the direction of the earthquake. Hence, differential stresses are likely to be severe. A vertical unit, though has only one support, yet it undergoes severe oscillations at the unsupported end because of its large distance from the supported end. In either case, joints might leak unless properly designed.

9.156 How is the seismic analysis done?

Stresses produced due to the seismic acceleration are analyzed in the shell, flanges, supports, etc. One can do a static analysis where the ground acceleration is taken in three directions (x, y, z). It is a conservative method and increases the costs of material and fabrication significantly. One can, instead, do a dynamic analysis wherein the response spectra of the earthquake is used. The analysis is quite involved and is less conservative and more realistic. The cost of doing the analysis is greater while that of the material and fabrication is less compared to the static analysis case.

9.157 Describe the seismic analysis of a heat exchanger.

Earthquakes cause stresses and deformation throughout the exchanger. Hence, the unit should be designed to withstand the expected earthquakes. In an approximate seismic analysis, the exchanger is assumed to be rigid. The dynamic forces acting on it are equal to those associated with the acceleration (both horizontal and vertical) of the base. In a more detailed analysis, the equipment is assumed to be flexible with large differential motion between the equipment and the base.

9.158 Generally what kinds of recommendations are made after a seismic analysis?

If the unit as designed will not be able to withstand the stresses caused by the seismic loads, the recommendations generally are to (1) increase the wall thickness and (2) decrease the unsupported span (overhang). These are intended to increase the fundamental natural frequency of the equipment (see Q. 9.154 also).

9.159 What are the characteristics of an earthquake?

Earthquakes are either tectonic or volcanic in nature. These originate at a point called its epicenter or focus and travel radially, producing horizontal and vertical ground accelerations. The travel is described in the form of "P" and "S" waves. Vibrations thus produced are random in nature and are recorded on instruments called seismographs. The intensity of the earthquake is measured on Richter scale or modified Mercalli intensity scale. The scale is logarithmic, meaning thereby that an earthquake of intensity of 7 is 10 times more powerful than an earthquake of intensity 6. Those of intensity 6 or above are regarded as major earthquakes.

9.160 What are the different seismic zones?

Each country is divided into approximately five zones depending upon the intensity of the earthquakes originating in that zone. The higher the intensity, the higher the zone number. As more data are collected, the zone boundaries are periodically modified. A dynamic seismic analysis has to be done for building to be constructed or equipment to be installed in the high-intensity zones.

FAILURE ANALYSIS

9.161 What is failure analysis?

Pressure vessels and heat exchangers do fail mechanically sometimes, although with more detailed analysis at the design stage and more stringent quality control, the probability of such failures is very low. However, when failures do occur, their analysis with a view toward future guidance is called the failure analysis. Failure here refers to excessive deformation or fracture occurring in the equipment.

In intricate designs, it is better to do the failure analysis (or conversely, safety analysis) at the design stage itself by considering all the causes that might result in the failure of a unit and the consequences of such a failure with regard to human safety and loss of production.

Articles on failure analysis appear in various journals and are presented in conferences. A number of them, including case histories, have been compiled by the American Society for Metals (1974).

9.162 What factors generally cause the mechanical failure of a heat exchanger?

Among the several factors that can cause a heat exchanger to fail mechanically are the following:

1. Corrosion of parts in excess of the corrosion allowances
2. Excessive thermal or mechanical stresses acting singly or in combination
3. Unexpected seismic load
4. Blocked safety relief valves due to their not being opened for a long time
5. Vibration failure of the tubes
6. Inferior workmanship and poor inspection
7. Defective or unsuitable materials of fabrication and welding
8. Poor maintenance
9. Collapse of a poorly designed saddle support
10. Undue loads on nozzles due to poor alignment of the piping and an improper bolt tightening procedure

9.163 What are the different types of fractures that can occur in a unit?

The following types of fractures can occur in a unit:

Brittle fracture: The material cannot withstand any mechanical or thermal shocks. There is no measurable deformation (no ductility). The tear or crack is long and pieces may fly off. While some materials are brittle at the room temperature (e.g., cast iron), others become brittle below a transition temperature [e.g., mild steel becomes brittle at $-20°$ to $-100°F$ ($-28.9°$ to $-73.3°C$), while it is ductile above $-20°F$ ($-28.9°C$)]. In situations where it is important, the transition temperature too should be specified along with the chemical composition of the steel. Full load should not be applied on a vessel until it is above its transition temperature.

Ductile or plastic fracture: The material deforms beyond its plastic limit and fails. It usually tears the vessel in certain area. This happens above the transition temperature.

Fatigue fracture: This is caused by cyclic conditions of temperature or mechanical load. Repeated start-ups can build in unanticipated stresses, strains, and excessive thermal gradients leading to failure.

Corrosion failure: This is due to the loss of the material or material strength caused by environmental factors (see Chapter 13 for discussion on corrosion).

9.164 What factors contribute to the brittle fracture?

The following are the contributory factors:

1. Low temperature, below the transition temperature of the metal
2. Composition and heat treatment of the material as received
3. Fabrication process of the pressure vessel/heat exchanger
4. Changes in the brittle fracture characteristics of the material after the vessel has been installed. The transition temperature gets increased by cold deformation or straining of the material. This happens at local areas of high stress during the normal operations. Such areas are at the support, welded baffles, attachment welds, etc. Hence, if a vessel is subjected to overpressure during one start-up, it may not withstand the same overpressure in future start-ups because the transition temperature may have gone above the normal operating temperature at certain localized points.

9.165 How can brittle fracture be differentiated from ductile one by looking at the fractured material?

In the case of a ductile fracture, there is reduction in the cross section and elongation in the length, whereas in a brittle fracture there is no such change in the dimensions and it is a clear break (Fig. 9.27).

9.166 What is Charpy impact strength?

This is a measure of the ductility and toughness of the material. It actually measures the energy required to break a square cross-section specimen with a specified notch. The details regarding the test are available in the American Society of Mechanical Engineers (1980d) Code Section II and many related books.

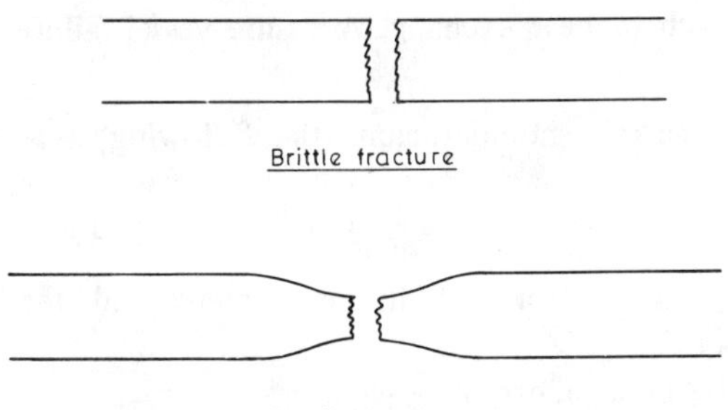

Figure 9.27 Brittle and ductile fractures.

9.167 What is creep deformation? What factors contribute to it?

When materials are subjected to an external stress, they undergo instantaneous deformation that may be completely elastic or partly elastic and partly plastic (permanent). The materials would continue to deform with time if the applied stress is maintained. Such a time-dependent deformation is called creep. The nature of the creep deformation is permanent. The rate at which creep deformation occurs is primarily a function of the temperature, stress, and the nature of the material. It is significant at temperatures $>0.3T_m$, where T_m is the melting temperature of the material. Creep, therefore, becomes a significant factor in elevated temperature applications.

9.168 Define elastic and plastic deformations. What factors affect the nature of deformation?

On the application of external forces or loads, solids can undergo deformation (strain) that may be elastic or plastic in nature:

Elastic deformation: If the material recovers its original shape and size after the removal of the external forces, then the deformation under their action is called elastic deformation.

Plastic deformation: If the material does not recover its original shape and size after the removal of the external forces but, instead, a permanent change is produced, it is called plastic deformation.

The factors that affect the nature of the deformation are the following:

1. Nature of the external forces
2. Rate of loading
3. Temperature
4. Micro-structure of the material

9.169 What is meant by thermal cycling? How does it affect a unit?

Whenever a unit is subjected to a heating and cooling cycle, its structural components are subjected to thermal cycling. Stresses are developed in the unit if the expansion and contraction accompanying the thermal cycling are constrained. The stresses may be so high that they may lead to a distortion of the components and/or cracking. Repeated thermal cycling may lead to thermal fatigue failure due to cumulative damage.

9.170 What should be the basic approach to heat exchanger/pressure vessel failure analysis?

The basic approach to failure analysis should include the following steps (Elayaperumal, 1978):

1. Do not have any preconceived notion about the causes of failure.
2. Study carefully and systematically the history of the equipment and the operating conditions prior to the failure.
3. Go deep into the analysis covering the broad range of possibilities.
4. Choose a proper sequence of tests (see Q. 9.172).

5. Examine the results carefully and systematically.
6. Give specific recommendations for remedial measures.
7. Investigate similar potential failures in the plant.

9.171 What information is required for failure analysis?

Information is required both about the failed component and the failure (Elayaperumal, 1978):

1. About the failed component
   Manufacturer
   Fabrication technique and heat treatment
   Stress distribution
   Quality control and inspection
   User
   Location and function of the component
   Operating conditions (normal, transients)
   Expected life of the component
   Service life at the time of failure
2. About the failure
   Date and time of failure
   Temperature and environmental conditions at the time of failure
   Record of the failed and adjacent areas
   Extent of damage
   Deviations from the normal operating conditions at the time of failure
   Opinions of operating personal regarding failure
   History of earlier complaints/observations

9.172 What tests and examinations should be done to facilitate failure analysis?

Some or all of the following tests may be performed for the failure analysis after the required information on the failed component and the failure, as detailed earlier, has been obtained (Elayaperumal, 1978):

1. Visual examination
2. Macroscopic examination for homogeneity, integrity, and quality
3. Nondestructive examinations (Chapter 11)
4. Fractrographic examination
5. Chemical analysis of the component, corrosion products, coating, etc.
6. Mechanical tests to compare with the design specifications
7. Microscopic examination
8. Microprobe analysis
9. X-ray examination
10. Simulation tests

The results of the above tests and examinations should be analyzed and appropriate recommendations made to prevent recurrences.

Sometimes, reports and articles appear in the journals, and pamphlets are issued by various users and insurance agencies on equipment failure, e.g., Siegele and Sommer (1981), Hanford (1978), and Factory Mutual Engineering (1978). These are very informative.

9.173 What is reliability analysis?

It means analyzing the design for its statistical probability of success to do the intended job during its expected lifetime under the specific load, environment, and the operating conditions (Gangadharan et al., 1975). In essence:

$$\text{Reliability} = 1 - \text{probability of failure}$$

A unit may need proper maintenance for the reliability analysis to hold.

- The discussions in this chapter should serve as an introduction to the topics mentioned in the chapter title. For more details, the references given should be consulted.

NOMENCLATURE

B	distance between the support and nearest tangential point for horizontal vessels (Fig. 9.23)
d_i	nozzle inside diameter (Fig. 9.20)
E	modulus of elasticity
L	ligament between two nozzles (Fig. 9.20)
m	gasket factor
P	outside packed floating head (Fig. 1.1)
PWHT	post-weld heat treatment
PWR	pressurized water reactor
R	vessel radius (Fig. 9.23)
S	floating head with backing device (Fig. 1.1)
T	pull-through floating head (Fig. 1.1)
T_m	melting temperature
W	externally sealed floating tubesheet (Fig. 1.1)
x	direction coordinate
y	design seating stress or load for a gasket; direction coordinate
z	Direction coordinate

Greek Letters

α	coefficient of linear thermal expansion
ϵ	strain
σ	stress

REFERENCES

American Petroleum Institute. *API Standard 605,* 3d ed. API: Washington, D.C. 1980.

American Society of Mechanical Engineers. *ASME Boiler and Pressure Vessel Code, Section VIII, Division 2.* ASME: New York. 1977.

American Society of Mechanical Engineers. *ASME Boiler and Pressure Vessel Code, Section VIII, Division 1.* ASME: New York. 1980a.

American Society of Mechanical Engineers. *ASME Boiler and Pressure Vessel Code, Section IX.* ASME: New York. 1980b.

American Society of Mechanical Engineers. *ASME Boiler and Pressure Vessel Code, Section III.* ASME: New York. 1980c.

American Society of Mechanical Engineers. *ASME Boiler and Pressure Vessel Code, Section II.* ASME: New York. 1980d.

American Society for Metals. *Source Book in Failure Analysis.* ASM: Metals Park, Ohio. 1974.

Baumeister, T. (ed.). *Marks' Standard Handbook for Mechanical Engineers,* 8th ed. McGraw-Hill: New York. 1978.

Bhattacharyya, B. C. *Introduction to Chemical Equipment Design: Mechanical Aspects.* Chemical Engineering Education Development Centre, I.I.T.: Madras, India. 1976.

Brownell, L. E., and E. H. Young. *Process Equipment Design.* John Wiley: New York. 1959.

Elayaperumal, K. Corrosion Failures–Principles of Corrosion and Failure Analyses with Reference to Heat Exchangers and Condensers. In *Heat Exchanger Design–A Practical Look,* 1st ed. J. P. Gupta (ed.). I.I.T.: Kanpur, India. 1978.

Expansion Joint Manufacturers' Association. *Standards of the Expansion Joint Manufacturers Association, Inc.,* 5th ed. EJMA, Inc.: Tarrytown, New York. 1980.

Factory Mutual Engineering Corp. *Preventive Maintenance in Boilers and Machinery* and *A Boiler and Machinery Conservation Reader.* Factory Mutual Engineering Corp.: Norwood, Massachusetts. 1978.

Fraas, A., and M. N. Ozisik. *Heat Exchanger Design.* John Wiley: New York. 1965.

Gangadharan, A. C., et al. Reliability Evaluation of a Sodium Heated Steam Generator. In *Reliability Engineering in Pressure Vessel and Piping.* ASME: New York. 1975.

Gupta, J. P., and S. Sivasankaran. Computer Program for Tubesheet Design. *Hydrocarbon Process.* 60(11):273–276. 1981.

Hanford, R. Analysis of Boiler Drum Failure during Hydrostatic Testing. *J. Product Liability* 2:59–76. 1978.

Heat Exchange Institute. *HEI Standard for Power Plant Heat Exchangers,* 1st ed. HEI: Cleveland, Ohio. 1980.

Harvey, J. F. *Theory and Design of Modern Pressure Vessels,* 2d ed. Van Nostrand Reinhold: New York. 1974.

Joseph Oat Corp. U.S. Patent No. 6207944, June 17, 1980.

Kopp, S., and M. F. Sayre, Expansion Joints for Heat Exchangers, paper presented at the Annual Meeting of ASME, New York, November 1950 (revised May 1952).

Mahajan, K. K. *Design of Process Equipment: Selected Topics.* Pressure Vessel Handbook Publ.: Tulsa, Oklahoma. 1979.

Missouri Boiler and Tank Co. *Quick Reference Guide: ASME Pressure Vessel Code (Section VIII, Division 1).* Missouri Boiler and Tank Co.: St. Louis, Missouri. 1980.

Mitchell, J. G. A Theoretical Examination of the Stress System at a Continuous Cylinder/Tubesheet Junction and a Nozzle/Cylinder Junction. In *Advances in Thermal and Mechanical Design of Shell-and-Tube Heat Exchangers,* NEL Report No. 590. National Engineering Laboratory: Glasgow, Scotland. 1973.

Nichols, R. W. (ed.). *Pressure Vessel Engineering Technology.* Applied Science Publ.: London. 1971.

Oak Ridge National Laboratory. *Design Guide for Heat Transfer Equipment in Water-Cooled Nuclear Reactor Systems,* ORNL TM-3578. ORNL: Oak Ridge, Tennessee. 37830, U.S.A. 1975.

The Patterson-Kelley Co. *Heat Exchangers, Manual No. 700-A.* The Patterson Kelley Co. Inc.: East Stroudsburg, Pennsylvania.

Ratnasamy, F. Better Service from Heat Exchanger Expansion Joints. *Hydrocarbon Process.* 60(1):125–126. 1981.

Rubins, F. L. A Special Report on Worldwide Pressure Vessel Codes. *Hydrocarbon Process.* 57(12):89–98. 1978.

Siegele, D., and E. Sommer. Theoretical Numerical Investigation of Bursting Behaviour of Pressure Vessels. *Ger. Chem. Eng.* 4:348–355. 1981.

Singh, K. P. Design of Skirt-mounted Supports. *Hydrocarbon Process.* 55(4):199–203. 1976.

Singh, K. P. Stress Categories. In *Heat Exchanger Design-II.* J. P. Gupta (ed.). C.S. Enterprises: Delhi, India. 1979.

Singh, K. P., and M. Holtz. On Thermal Expansion Induced Stresses in U-Bend of Shell-and-Tube Heat Exchangers. *Trans. A.S.M.E., J. Eng. Power.* 101(4):634–639. 1979.

Sivasankaran, S. Mechanical Design of Heat Exchangers. M. Tech. Thesis, Chemical Engineering Department, I.I.T.: Kanpur, India. 1981.

Sivasankaran, S., and J. P. Gupta. Computer Program for Expansion Joint Design. *Hydrocarbon Process.* 62(12):93–96. 1983a.

Sivasankaran, S., and J. P. Gupta. Computer Program for Nozzle Design. *Hydrocarbon Process.* 62(5):99–103. 1983b.

Truong, K. T., and T. T. Truong. Bearing Plate Design Aid. *Hydrocarbon Process.* 60(5):165-168. 1981.

Tubular Exchanger Manufacturers' Association. *Standards of Tubular Exchanger Manufacturers' Association,* 6th ed. TEMA Inc.: Tarrytown, New York. 1978. Known as TEMA Standards.

Tubular Exchanger Manufacturers' Association. *Standards of Tubular Exchanger Manufacturers' Association Sample Problem Booklet for Sixth Edition 1978.* TEMA Inc.: Tarrytown, New York, 1980.

Yamada, K., and Y. Minamimoto, *Fabrication and Quality Assurance of Heavy Wall Pressure Vessels,* Hitachi Zosen Technical Report HZTR PVP-6. Hitachi Zosen: Osaka, Japan. 1976.

Yasuda, R., K. Wakaizumi, S. Yasui, and K. Tanaka. *Pressure Vessel Design.* Hitachi Zosen Technical Report HZTR PVP-5. Hitachi Zosen: Osaka, Japan. 1976.

Zick, L. P. Stresses in Large Horizontal Cylindrical Pressure Vessels on Two Saddle Supports. *Welding J. (New York), Res. Supple.* 30:435-S. 1951.

CHAPTER

TEN

MATERIAL SELECTION, FABRICATION, AND WELDING

INTRODUCTION

- Selection of the materials of construction is a very important aspect in the design of a heat exchanger or a pressure vessel. The material must be able to withstand the design and operating conditions and the outside environment throughout the expected life, at a minimum initial cost and with a low maintenance and repair bill. Thus, it is evident that material selection is a specialized field.

The selected materials have to be brought into the desired shape to perform the required job. This requires a high degree of knowledge of the fabrication procedures and associated details. Welding, being the most important activity in the fabrication of the equipment, has become a highly specialized field in its own right, and is continuously being upgraded to meet the increasingly more stringent demands.

Hence, in this chapter, we shall deal with the salient features of material selection, fabrication, and welding.

MATERIAL SELECTION

10.1 What materials of construction are used in heat exchangers and pressure vessels?

Any component or the entire unit can be made of carbon-steels, alloy steels, copper and its alloys, nickel and its alloys, titanium, aluminum, tungsten, or other special alloys. Sometimes, nonmetallic materials, such as glass, Teflon, graphite, or composite materials, are also used (see Chapter 5).

10.2 Discuss the use of some of the important materials.

Stainless steels, carbon- and low-alloy steels, nickel-chromium alloys, nickel-copper alloys, and admirality brass are among the important materials used in the heat exchanger and pressure vessel fabrication.

Stainless steels: Austenitic (300-series) stainless steels are the most frequently used materials in nuclear heat exchangers, and are widely used in other heat exchangers for chemical process industries as well. Their corrosion rate is very low, and these can be produced in controlled composition. When carbon or low-alloy steels are also used along with austenitic stainless steel in the same equipment, the stress-relieving temperature of the welds of the former lies in the sensitization zone for the latter, resulting in the redistribution of carbon and deterioration of the properties of the stainless steel (Q. 9.148). In such a case, one should use extra low carbon austenitic stainless steel Type 304L or stabilized grade such as Type 321. As it turns out, most exchangers of stainless steel do have carbon-steel or low-alloy steel components. Hence, generally Type 304L is used instead of Type 304.

Carbon and low-alloy steels: These are very widely used in the shells, heads, tubesheets, and channels as well as in other parts, such as the baffles and tie-rods. In cases where, due to corrosion, stainless steel or some other expensive material has to be used, cost can be reduced significantly by using carbon- or low-alloy steels clad with stainless steel or other materials (see Q. 10.47 for cladding). In such a case, the strength is provided by the carbon- and low-alloy steels, and the resistance against corrosion is provided by the cladding. Thus, one does not need to purchase a thick sheet or plate of the costlier material. This practice is adopted both in the nuclear and nonnuclear usage of the heat exchangers.

Nickel-chromium-ion alloys (Inconel): These have a very high percentage of nickel (76 Ni, 16 Cr, 8 Fe) and are used where the austenitic stainless steel tubes fail due to stress corrosion cracking or creep. This phenomena, due to the combined action of stress and corrosion, affects the stainless steel in caustic and chloride environments.

Copper-nickel alloys: These are very frequently used for the heat exchanger tubes in seawater service due to their low corrosion rate and overall good performance.

There are some other materials, such as titanium, aluminum, and admirality brass, that are used as the heat exchanger tubing depending upon the situation at hand.

10.3 What factors should be considered in selecting materials for heat exchangers and pressure vessels?

Amongst a host of factors are the following:

1. Compatibility of the materials with the tube-side and shell-side fluids at the design conditions of temperature, pressure, and fluid velocity.
2. Compatibility of the shell, channel, and bonnet materials with the outside environment
3. Compatibility of the shell-side and tube-side materials with one another.
4. Lowest cost for the desired duty over the expected life of the unit
5. Seamless or welded tubing
6. Clad or alloy tubesheet
7. Corrosion allowance
8. A higher alloy might enable the use of thinner wall tubes, thus reducing the exchanger weight and also the pressure drop through the tubes, since the inside cross-sectional area of such tubes is larger.
9. If material is not approved by the code, a time-consuming special case ruling may be required from the code committee.
10. Creep due to long-term, high-temperature operation
11. Tensile strength, shear strength, bending strength, fatigue strength, elastic limit, yield point, wear resistance, properties at high temperature, and reliability
12. Thermal conductivity, heat capacity, density, and coefficient of thermal expansion
13. Capacity for being rolled, castability, forgeability, machinability, weldability, suitability for brazing, etc., depending upon the use
14. Limitations of the design and effects of fabrication methods and heat treatment on the material properties
15. Installation and maintenance requirements
16. Delivery date of equipment, if affected by material availability

10.4 What are the characteristics of the materials used most frequently in the nuclear heat exchangers?

The important characteristics of interest in such materials are the following:

1. Low rate of corrosion: Since the circulating corrosion products can become radioactive posing danger to the plant personnel, it is important that the corrosion rate of the material be very low. This is also required from the point of view of plant integrity.
2. Excellent weldability.

The materials used most frequently in the nuclear heat exchangers are: (a) 300 series (austenitic) stainless steels since their corrosion rate is rather low; (b) carbon- and low-alloy steels, which are principally used to provide strength at lower cost while the corrosion resistance is provided by a stainless steel cladding; (c) Inconel, which is used for the tubing since it does not undergo stress corrosion cracking in most environments.

10.5 Can dissimilar materials be used in an exchanger?

Yes. In fact, except for the wholly carbon-steel exchanger, in most others dissimilar materials are more of a rule rather than an exception. Their choice depends upon the corrosive nature and the process conditions of the shell-side and the tube-side fluids. Usually, the shells are of carbon-steel since it is strong and cheap. Thick tube-sheets and cover plates are also of carbon-steel with an appropriate cladding to combat the corrosion of the tube-side fluid (Q. 10.2). More corrosive fluid is usually put on the tube side since then only the tubes need be made of the corrosion-resistant materials, while the shell can be of carbon-steel.

10.6 What role do the ASTM Standards play in specifying the materials?

ASTM has various committees that evaluate the materials based upon the wide experiences the committee members have. They also give a specific number to the material. The actual choice of a material for a specific application is done by the user based upon the various criteria listed in Q. 10.3.

10.7 How do the specification numbers for different materials and alloys given by the ASME differ from those given by the ASTM?

ASME prefixes an S, e.g., SA202. It generally indicates that the material A202 specified by the ASTM has been adopted by the ASME Code, although it may not be true in all the cases. Not all ASTM specified materials have been adopted by the ASME Code.

10.8 What are the important physical properties of materials?

The following are the important physical properties of materials:

Elasticity is the ability to return to a normal shape after the applied force has been removed. No permanent deformation is set in the material.

Malleability is the property that allows the material to be given different shapes by bending, rolling, or hammering without cracking or breaking. In cases involving repeated rolling, intermediate heat treatment (annealing) may be necessary to prevent cracking.

Hardness is the ability to withstand surface wear or abrasion by another material due to rubbing or impact action.

Tensile strength is the ability to withstand large pulling force without breaking. It is also called tenacity.

Resilience is the ability to store energy when stressed and to release it when the stress is removed.

Homogeneity refers to the uniform grain or fiber of the material. If one cuts a homogeneous material, it will look the same regardless of the direction of the cut.

Creep is the time-dependent inelastic deformation of materials. It occurs at high temperatures. The metal will continue to deform with no increase in stress. The creep strength of many metals and alloys goes down severely as the temperature increases to above 800°F (425°C).

10.9 The codes specify a variety of materials. There are scores of carbon-steels and low-alloy steels too. How do these differ from one another in broad terms?

These may differ from one another in the proportions and kinds of the alloying components used, the type of heat treatment given, strength, and the shape or product form of the material (plate, tube, bar stock, etc.). Some values are given in Table 10.1.

10.10 What is the difference between iron and steel?

Iron, in industrial terminology, generally refers to cast iron, which is an alloy of elemental iron and carbon. Steel is a refined form of iron with carbon in solution as Fe_3C, while cast iron has carbon in the form of graphite besides that in the solution, which gives it a nonuniform structure and makes it brittle.

10.11 What is the difference between low-carbon steel and high-carbon steel?

Low-carbon steel has $\leqslant$0.2 percent carbon, can be easily welded, but cannot be hardened. On the other hand, high-carbon steel has from 0.7 to 2 percent carbon, can be hardened, but cannot be easily welded. Welding results in a change in the structure of the material and redistribution of the carbon and thus affects the strength of steel adversely.

10.12 What is an alloy steel?

It is a steel that contains, besides iron and carbon, one or more of the following alloying materials: aluminum, boron, chromium, cobalt, manganese, molybdenum, nickel, niobium, silicon, titanium, tungsten, vanadium, zirconium, etc. Each of these alloying elements affects the properties of steel. The amounts of the alloying elements added depend on the end use of the steel.

10.13 What important properties must steel have for high-temperature usage?

The four important properties are the following (Greene, 1980):

1. Strength: At high temperature, the strength is time dependent due to creep phenomenon. The creep rate increases with increasing temperature and leads to failure of the material.
2. Thermal expansion: This causes thermal stresses in the system that might lead to failure.
3. Oxidation resistance: Steel gets oxidized and loses strength and thermal effectiveness at high temperature in air, steam, and other atmospheres. Increasing the amount of chromium enhances the oxidation temperature of steel.
4. Thermal conductivity: It is important for effective heat transfer. Alloying elements decrease the thermal conductivity at low temperatures, while their effect is minimal at high temperatures.

These are in addition to the properties of steel considered for operation at or around the ambient temperature.

10.14 What are the different types of steels used for high-temperature service?

These are basically four different types (Mack, 1976):

Table 10.1 **Specifications and properties of some materials**

Material and nominal composition	P number	Product form	Specification number	Grade	Specified minimum yield strength 1000 psi (MPa)	Specified minimum tensile strength 1000 psi (MPa)	Maximum allowable stress, 1000 psi (MPa)	
							−20° to 650°F (−28.9° to 343°C)	800°F (427°C)
Carbon-steel								
C-Si	1	Plate	SA-516	60	32 (220.6)	60 (413.7)	15 (103.4)	10.8 (74.5)
C-Si	1	Plate	SA-516	70	38 (262)	70 (482.6)	17.5 (120.6)	12 (82.7)
C-Si	1	Forgings	SA-181	I	30 (206.8)	60 (413.7)	15 (103.4)	10.8 (74.5)
C-Mn-Si	1	Seamless pipe	SA-106	A	30 (206.8)	48 (330.9)	12 (82.7)	9.3 (64.1)
Low-alloy steel								
C-$\frac{1}{2}$ Mo	3	Plate	SA-204	A	37 (255.1)	65 (448.1)	16.2 (111.7)	15.6 (107.6)
C-$\frac{1}{2}$ Mo	3	Plate	SA-204	B	40 (275.8)	70 (482.6)	17.5 (120.5)	16.9 (106.5)
Cr-Mn-Si	4	Plate	SA-202	B	47 (324)	85 (586)	21.2 (146.2)	12.8 (88.3)
Cr-Cu-Ni-Al	4	Seamless pipe	SA-333	4	35 (241.3)	60 (413.7)	15 (103.4)	
1 Cr-$\frac{1}{5}$ Mo	–	Bolt	SA-193	B7	105 (723.9)	125 (861.8)	25 (172)	21 (144.8)

Source: American Society of Mechanical Engineers (1980a).

1. Carbon-steels: For most usage up to 750°F (400°C) these are the best since they are the cheapest, very strong, and have the highest thermal conductivity. Above this temperature, their creep rate increases and, above 950°F (510°C) they get oxidized too.
2. Chromium-molybdenum steels: These resist oxidation at temperatures up to 1200°F (650°C), but their thermal conductivity is somewhat lower. These are costlier than the carbon steels.
3. Ferritic stainless steels: These can be used up to 1500°F (820°C). These have higher thermal conductivity, lower thermal expansion coefficient, and lower cost than that of other materials for the same service. They can resist oxidation as well as attack from sulfur- and carbon-containing flue gases. Their coefficient of thermal expansion is lower than that of the austenitic stainless steels. These lose strength after very long usage at high temperatures. This should be kept in mind when determining the life expectancy of the equipment.
4. Austenitic stainless steels: These do not lose strength after very long usage at high temperatures and are the strongest among steels for use at high temperatures. These resist oxidation at high temperature and are used in boilers, superheaters, refinery services, etc., but not where chloride is present.

10.15 What alloying element could be used to reduce fatigue in heat transfer equipment operating at high temperatures?

Vanadium is added to steel to increase the endurance limit and to reduce the fatigue.

10.16 What harm does hydrogen cause to carbon steel? What materials should be used if hydrogen is present in the vessel?

Carbon in steel combines with hydrogen to form methane and other products under high pressure and temperature (above 200°C or 390°F). This first results in decarburization followed by localized fissuring of the vessel due to high gaseous pressure, leading to the failure of the vessel.

Chromium- and molybdenum-containing steels are best suited for use with hydrogen. They resist decarburization, fissuring, and the associated loss of strength (Hagel and Miska, 1980).

10.17 What is the creep range of an alloy steel?

It is generally between 400° to 800°C (750° to 1500°F) depending largely on its grade and heat treatment.

10.18 Which steels are preferred in the creep range for the pressure vessels?

They are chromium-molybdenum grades, specially 1 Cr-$\frac{1}{2}$ Mo and 2$\frac{1}{4}$Cr-1 Mo (Hagel and Miska, 1980).

10.19 What is the basis of design in the creep range?

Stress rupture and creep strength guide the design in the creep range. Design stress is usually based upon the average stress for rupture in 1,000,000 hr, with suitable factors of safety (Hagel and Miska, 1980).

Table 10.2 Materials for feed water heaters tubes

Tube material	Maximum metal temperature, °F (°C)
Arsenical copper	400 (204)
Admirality metal B, C, D	450 (232)
Copper–nickel	
90–10	600 (316)
80–20	700 (371)
70–30 annealed	700 (371)
70–30 stress relieved	800 (427)
Nickel–copper	
70–30 annealed (Monel)	900 (482)
70–30 stress relieved	800 (427)
Carbon–steel	800 (427)
Stainless steels	800 (427)

Source: Heat Exchange Institute (1980).

10.20 What materials are used more frequently for the shell and tubes of the feed-water heaters?

The shell is invariably made of carbon steel as are also the heads and the tube-sheets. The tubes are made of different materials depending primarily upon the quality of water and also the service temperature (and hence pressure) as given in Table 10.2.

10.21 What materials are used for steam generator tubes?

The different materials with their temperature limits are given in Table 10.3.

10.22 What materials are used for surface condenser tubes?

The different materials, in order of increasing corrosion resistance, are listed in Table 10.4.

Table 10.3 Materials for steam generator tubes

Ferrous	Temperature limit
Mild and low carbon-steels	<350°C (662°F)
Cr-Mo-V steels	250° to 580°C (482° to 1076°F)
Austenitic stainless steels	580° to 700°C (1076° to 1292°F)
Nonferrous	**Use**
Monel (70 Ni/30 Cu)	For nuclear power plant steam generators
Inconel (70 Ni, 15 Cr, 15 Fe)	
Incoloy (30 to 35 Ni, 19 to 23 Cr, rest Fe)	
Aluminum	
Titanium	

Table 10.4 Materials for surface condenser tubes

Material	Application
Commercial pure aluminum (3 S grade)	Freshwater service
Brass (70 Cu, 30 Zn)	Freshwater service
Admirality metal (71 Cu, 28 Zn, 1 Sn, 0.04 As)	Freshwater service
Cupro-nickel (90 Cu, 10 Ni, or 70 Cu, 30 Ni)	Sea water, 2.5 to 2.7 m/s velocity (8.2 to 8.9 ft/sec)
Modified cupro-nickels	Polluted sea water, 2.8 to 3.5 m/s (9.2 to 11.5 ft/sec)
Titanium	Highly corrosive conditions

10.23 What materials are used for nuclear heat exchangers?

Stainless steels, carbon- and low-alloy steels and Inconel are the ones most used (see Q. 5.34 for more details).

10.24 What important factors should be kept in mind when selecting materials for cryogenic services?

Cryogenic conditions are encountered in air liquefaction, natural gas storage and transport, etc. Most materials become brittle at these temperatures, although their tensile strength and yield strength generally increase. Any notch left over in finish during manufacture becomes the source of initiation of a crack at cryogenic temperatures. Corrosion is usually not a problem with most low-temperature fluids. Therefore, the following factors should be kept in mind when selecting the materials for the cryogenic service (Agarwal, 1977):

1. Thermal expansion and contractions: If the coefficient of thermal expansion is large, significant dimensional changes will occur in the equipment resulting in unwanted stresses. Hence, materials of low coefficient of thermal expansion should be used.
2. Thermal conductivity: Supports, piping, etc., should be of low thermal conductivity so that the heat gain is low. Furthermore, insulation should be used on the total equipment.
3. Strength of material: High strength material, such as 9 percent nickel steel, results in thin sections and is widely used.

10.25 What materials are used at cryogenic temperatures?

The materials used are listed in Table 10.5.

10.26 What decides the minimum thickness of plates used in heat exchangers and pressure vessels?

This is governed by the applicable codes and TEMA Standards and depends upon the material, diameter of the shell, operating pressure and temperature, corrosion allowance, service life, dead loads and other external loadings, testing requirements, upset and emergency conditions, etc.

Table 10.5 Materials for cryogenic usage

Temperature	Materials
>–45°C (–49°F)	Killed carbon-steel Normalized and tempered (N and T) Low alloy steels
>–60°C (–76°F)	Quenched and tempered (Q and T) ASTM A537
>–100°C (–148°F)	3.5 Percent nickel steel
>–100°C (–148°F)	9 Percent nickel steel, austenitic chromium-nickel steel, aluminum, or copper

Source: Hagel and Miska (1980). Reprinted by special permission from CHEMICAL ENGINEERING, 87(15):89–91. 1980. Copyright © 1980 by McGraw-Hill, Inc., New York, N.Y. 10020.

10.27 What are the possible advantages of using lighter sections of thin-walled materials?

The advantages are the following (Hagel and Miska, 1980):

1. Weight of the vessel is reduced
2. Foundation and transportation costs are lowered
3. Costs of forming, welding, heat treatment and stress relieving, erection, and non-destructive testing, are reduced
4. Resistance to brittle fracture is generally improved

10.28 What are the ASME/ASTM specifications for seamless cold-drawn low carbon-steel heat exchanger and condenser tubes?

These refer to the ASME Specifications A179-75. The steel for these tubes is made by the open-hearth, basic oxygen, or electric furnace. The tubes are made by the seamless process and are cold-drawn. They are then heat treated at 1200°F (650°C) or higher. The chemical composition of steel is (in percent): carbon (0.06 to 0.18), manganese (0.27 to 0.63), phosphorus (0.048 maximum), and sulfur (0.058 maximum), the rest being iron. Heat analysis and product analysis are then performed. The mechanical tests required are the flattening, flaring, flange, hardness, and the hydrostatic. These tests are done to ensure that the tube will withstand the operations performed while installing it (bending to make a U-bend, flaring at the tubesheet end, etc.) and also the loads and fluid pressure when it is used.

10.29 What are the different kinds of welded tubings?

Among the several kinds, the more popular ones are the following:

1. As-welded
2. Welded and annealed
3. Welded, bead-conditioned and annealed
4. Welded, drawn, and annealed

10.30 What caution should be exercised in selecting the tube material, thickness, and length?

The tubes should be able to withstand the following:

1. Pressure on the inside and the outside
2. Temperature on both the sides
3. Thermal stresses due to the differential expansion of the shell and the tubes
4. Corrosive nature of both the shell-side and the tube-side fluids

The thermal resistance of the tube plays an important role in equipment involving phase change (condensers, reboilers). Hence, a tube material with high thermal conductivity and compatible with the above four points should be selected. The thickness should be uniform within permissible tolerance and, for the U-bends, minimum or average thickness should be specified. This is to take care of the thinning at the U-bend. Standard lengths should be chosen, otherwise a penalty in price may have to be paid for. The length and the number of tubes should be selected to give the required heat transfer area (based on the tube outer diameter) clear off the tube-sheet, since the tube length in the tubesheet does not contribute to the heat transfer, and it can be significant in high-pressure exchangers. The surface area per unit length of the tubes is tabulated in TEMA Standards and other texts. For the U-tube bundles, the surface area is calculated based upon the mean tube length.

10.31 What effects does the tube material have on the performance of an exchanger?

The tube material affects the following:

1. Resistance to corrosion, erosion, and vibration damage (natural frequency of the tube)
2. Thermal stresses induced due to differential thermal expansion of the tube bundle and the shell
3. Extent of fouling and microbial deposit
4. Resistance to heat transfer
5. Ease of joining tubes to the tubesheet
6. Cost and weight of the unit, etc.

10.32 What material is used for tubes in saltwater coolers?

Salt water can be brackish, brine, or sea water, the last being used a lot in coolers and condensers in plants on the sea coast. Titanium often turns out to be the best. Aluminum bronzes are also used widely if there is no pollution. Compatibility of other materials against various forms of possible attacks has been tabulated by Foxall and Gilbert (1976).

10.33 What tests are conducted for checking the suitability of a material under the desired operating conditions?

The important tests are the following:

1. Chemical analysis to determine its composition
2. Mechanical tests to determine its strength, etc.
3. Checking in detail the reports on the heat treatment done by the manufacturer of the material
4. Visual examination for any nicks, scratches, and notches
5. Nondestructive examinations: (a) radiographic examination, (b) ultrasonic examination, (c) liquid penetrant examination, (d) magnetic particle examination, and (e) eddy current examination. These are discussed at length in Chapter 11.

FABRICATION

10.34 What are the various steps in the fabrication of a heat exchanger?

The following is a list of the steps. Each step may involve several substeps. More than one step may be carried out simultaneously.

1. Preparation of detailed drawings for fabrication signed by a responsible engineer
2. Discussion between the engineer, quality assurance chief, and the workshop incharge about scheduling the job and other salient features
3. Procurement of the raw material, welding electrodes, gaskets, tubes, rear and front heads, nozzles, and flanges, if not made in the workshop
4. Keeping the buyer and/or inspector informed at various steps during fabrication as per the contract
5. Welders' qualifications and welding procedures to be established and approved by the inspector
6. Selection of proper plates for the shell and channel
7. Cutting the plates to size by gas or plasma arc cutting machines
8. Rolling the plates into shell sections (Fig. 10.1) and tack-welding the longitudinal seam; thick plates are hot rolled (Greene, 1980)
9. Checking out-of-roundness (Fig. 10.2) and correcting it, if excessive
10. Welding the shell sections (longitudinal welds)

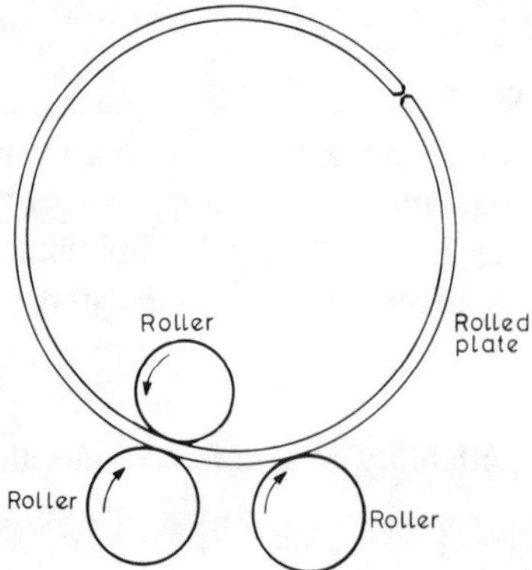

Figure 10.1 Plate rolling.

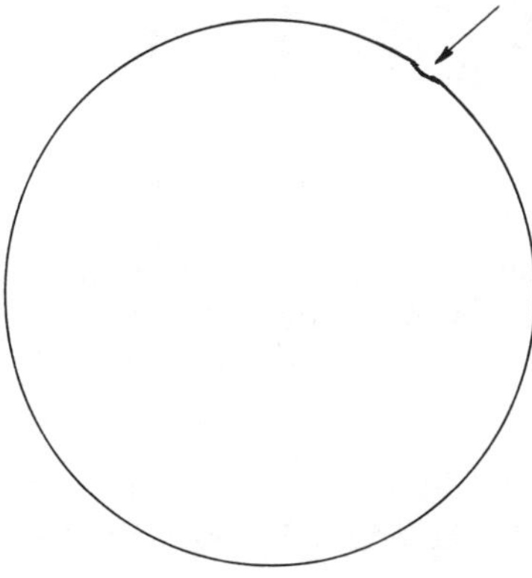

Figure 10.2 Out-of-roundness of shells.

11. Welding the sections together (circumferential welds), keeping the longitudinal seams staggered (Fig. 10.3)
12. Making the heads by cold spinning (if thin) or hot spinning (if thick) or by a forming press (if very thick)
13. Attaching the nozzles and flanges to the shell and heads; nozzle welding is a very important operation.
14. Attaching longitudinal pass partition plate, if any, in the shell/channels
15. Attaching any support brackets, etc., if welded to the shell/channels
16. Stress relieving of shell/channels
17. Preparing the tubesheet and baffles by cutting the plate material to size and drilling, reaming, and chamfering the tube holes. Sometimes, the solid tubesheet may be procured from another vendor.
18. Machining of gasket surfaces in the tubesheet and flanges
19. Preparing a cage for the tube bundle by using one tubesheet, all the baffles and the tie-rods and spacers
20. Inserting the tubes into the cage and making a tube bundle
21. Roller expanding, grooving, or welding the tubes to the tubesheet
22. Sliding the tube bundle into the shell
23. In a fixed tubesheet exchanger, the rear tubesheet is then fixed and tube-to-tubesheet joints completed
24. Attaching the pass partition plates, if any, in the heads

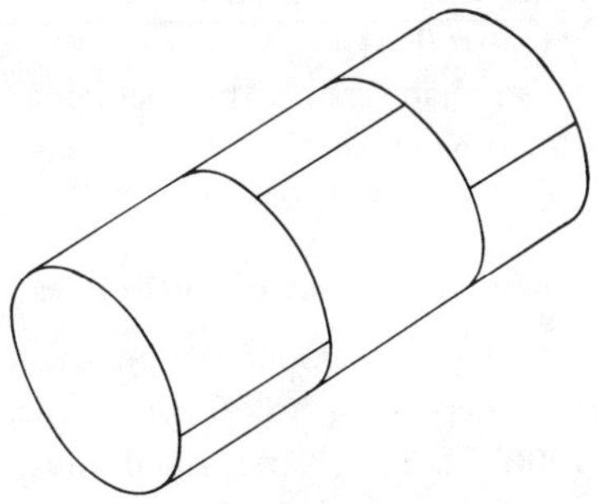

Figure 10.3 Welding of shell sections.

25. Preparing the requisite number of bolts and nuts
26. Putting the unit together and testing for dimensional consistency
27. Soap suds test to check if any tube is left unexpended
28. Hydrostatic testing of the unit and checking for the leaks, if any (Chapter 11)
29. Sometimes pneumatic testing may also be required (Chapter 11).
30. All through, a record of the materials and electrodes used, any deviations from the original item, persons who have done the job, comments of the inspectors, etc., should be maintained.
31. At each step, the fabrication has to be inspected and cleared for further work by the inspector, as agreed to in the contract.

10.35 What are the critical processes performed in the fabrication of a heat exchanger?

These are the following:

1. Welding
2. Tubesheet marking, drilling, reaming
3. Tube-to-tubesheet joints
4. Bolted joints with gaskets

10.36 In what operations are the automatic machines used for quality and/or speedy manufacture of heat exchangers?

Automatic machines are used in a number of operations related to heat exchanger manufacture. Some such automated operations are the following:

1. Plate cutting
2. Edge preparation
3. Plate rolling
4. Welding
5. Tubesheet marking, drilling, and reaming
6. U-tube bending
7. Tube-end rolling

10.37 What machines are used during the fabrication?

Some of these are the following:

1. Automatic plate cutting machine with oxyacetylene and/or plasma arc torches
2. Multiple spindle drilling machine, N.C. (numerically controlled)
3. Automatic deburring machine for drilled holes. It also chamfers on the shell side, bell-mouths on the inlet side, and provides grooves for tube expanding.
4. Submerged arc welding and other welding operations.

10.38 What is the difference between hot forming and cold forming of steel plates?

In cold forming, continuous operation is carried out at room temperature to obtain the required shape (cylinders, end closures) whereas, in hot forming, the operation is done at temperatures around 500°C (900°F). Parts so formed must be fully heat treated to relieve the stresses produced by hot forming. However, in hot forming, roughly double the plate thickness compared to that in cold forming can be bent in a given machine.

10.39 What are manufacturing tolerances?

In the fabrication of any heat exchanger or pressure vessel (any process equipment), it is very difficult to obtain the exact dimensions because of the various reasons such as the following:

1. Rolling of flat plate and its welding not producing a geometrically perfect circular cylindrical shell
2. Vibrations of the drill bit not producing a perfectly straight hole, specially in thick materials
3. During the production of the flat plates and tubes, their thickness is not the same all through because of the particular operations involved
4. The lengths and diameters of various sections in the unit may not turn out to be the exact desired values

The permissible out-of-roundness in item (1), the depature from perfect straightness of the holes in item (2), the variations in the designated thickness of plates and tube wall in item (3), and the deviations in the lengths and diameters in item (4) are called the manufacturing tolerances.

These are governed by the Code (American Society of Mechanical Engineers, 1980a) and TEMA Standards (1978).

10.40 What are the standard tolerances for the external dimensions of a heat exchanger?

The standard tolerances for heat exchanger fabrication are shown in Fig. 10.4. The important point to note is that the tolerances are not cumulative. Besides these, almost each step in the fabrication of an exchanger has its own tolerance limits, e.g.:

1. Ninety-six percent holes in the tubesheet must be perfectly straight (i.e., up to 4 percent holes can have some departure from straightness).
2. An average tube wall thickness has approximately ± 10 percent allowance
3. Out-of-roundness in bending a plate into a cylindrical form is <1 percent, etc.

The designers, fabricators, inspectors, and users must be familiar with these tolerance limits. These are tabulated by TEMA (1978) and by Chuse (1977).

10.41 How are the front and rear heads made?

Depending upon the design conditions, these may be especially spun or formed to get the required shape and special features. In some cases, these may be cast in standard castings. Open-end heads (channels) are made from cylindrical sections or plates. These are closed with easily removable cover plates (Fig. 1.1).

10.42 Are the tube pass partitions always straight?

These can be straight or wavy rib design (Fig. 10.5). The latter can withstand higher pressures using thinner sections.

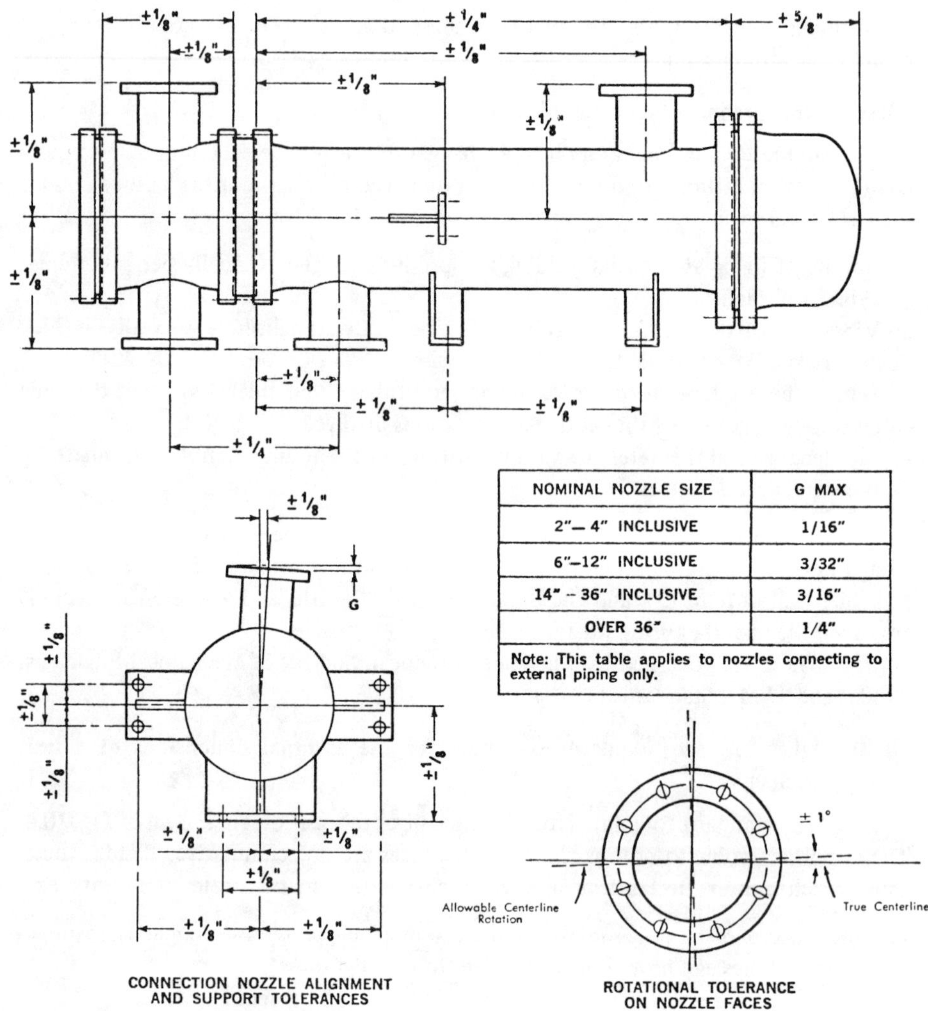

NOMINAL NOZZLE SIZE	G MAX
2"— 4" INCLUSIVE	1/16"
6"–12" INCLUSIVE	3/32"
14" – 36" INCLUSIVE	3/16"
OVER 36"	1/4"

Note: This table applies to nozzles connecting to external piping only.

Figure 10.4 Standard tolerances. *(© by Tubular Exchanger Manufacturers' Association. All rights reserved.)*

10.43 How are the longitudinal baffles placed in the shell?

Among the techniques used to install the longitudinal baffles are the following (Fig. 10.6):

1. Welded seal: Most desirable but often precluded by the specification of a removable tube bundle
2. Multi-flex leaf arrangement
3. Adjustable packing
4. Tongue-and-groove labyrinth or slide-in arrangement

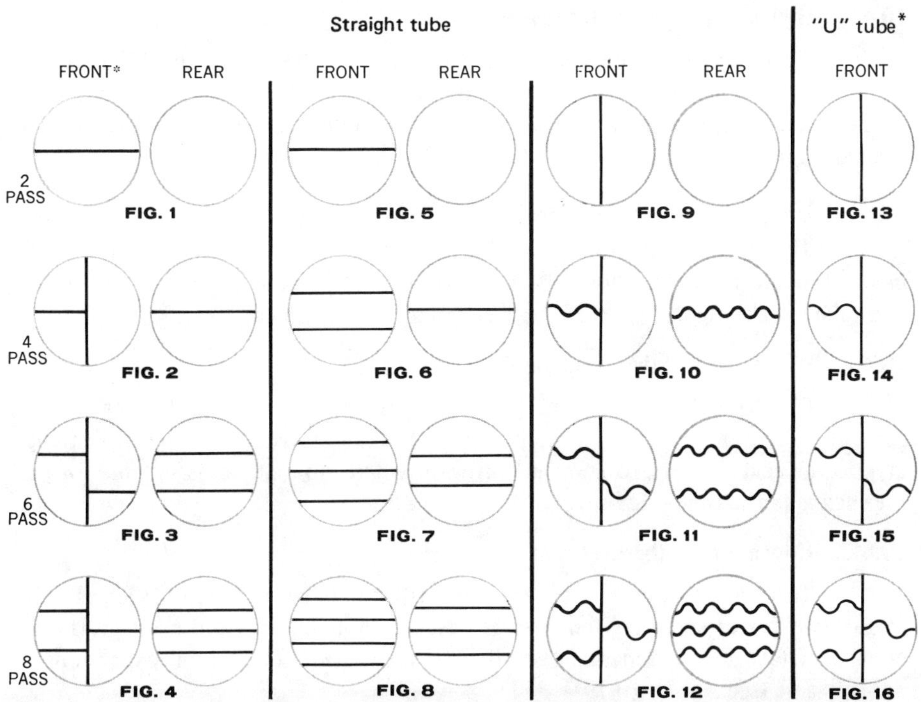

Figure 10.5 Pass partition rib designs. *(Courtesy Patterson-Kelley Co., Division of HARSCO Corp.)*

The latter three arrangements result in liquid leakage from one shell pass to the other whenever a significant pressure differential exists between the two shell passes. This can adversely affect the exchanger performance. Heat transfer by conduction through the longitudinal baffle also has adverse effects on the exchanger performance. These points should be kept in mind.

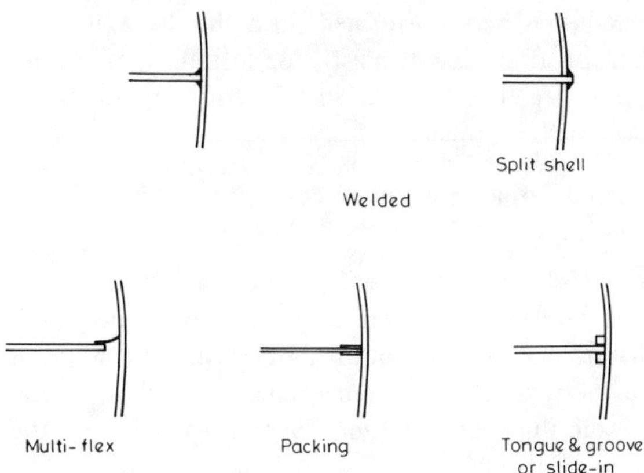

Figure 10.6 Installation of longitudinal baffles.

10.44 How are the nozzles fabricated?

These may be made out of standard pipes or elbows if up to 24 in (610 mm) diameter and used with reinforcing pads if required (Fig. 1.30). Larger diameters need to be made out of plates. Self-reinforcing (or integrally reinforced) nozzles are generally forged and then machined to size.

10.45 What precautions are taken while drilling holes in the baffles?

The hole alignment in all the baffles should be exactly in line with the holes in the tubesheets. There should be no burrs due to drilling. An overlap of at least two tube rows should be provided between two consecutive baffles.

10.46 Why are clad tubesheets used?

Clad tubesheets are used (1) to reduce costs where the cladding provides corrosion resistance while the carbon-steel backing provides the needed strength at a lower cost and (2) to combat corrosion where no material is suitable for both the shell- and the tube-side fluids.

10.47 How are the tubesheet clad?

These are clad using great precautions by bonding techniques, such as rolling of the two plates and heat treatment, explosion bonding, or weld overlay. The two bonded metals do not separate even if their coefficients of thermal expansions are not the same (see also Q. 10.79).

10.48 What factors should be kept in mind when using a double tubesheet?

Since both the tubesheets are at different temperatures, they undergo different radial expansions that cause deflection and shearing action on the tubes, the maximum effect being in the outer tube rows. Differential expansion of the tubes relative to the shell causes compressive stresses. Hence, the rolling of the tubes in the inner tubesheet should be done very carefully.

10.49 How are the tubes bent into a U-shape?

The most common method is roll forming where the tubes are passed between rollers repeatedly until the required bend is formed. In a thin tube, internal support against buckling is provided by filling sand, plastic, or soft alloy of led-tin-bismuth. In a thick-walled tube with a large bend radius, such internal support is not needed. Usually a number of tubes can be bent together.

10.50 Why is the U-bend restricted to a minimum center line radius of two diameters?

When a tube is bent into a U-shape, the outside of the bend is in tension while the inside is in compression. In order to avoid serious damage to the bend, the plastic deformation in the bend has to be kept to within 25 percent. This requires that the minimum center line radius be two diameters. This is also influenced by the tube material, ratio of the tube wall thickness to tube diameter, temper of the material, and the bending tools and die used.

10.51 What tools are used in preparing a tube bundle?

The following tools are used:

1. Tube guide to thread tubes through misaligned tubesheet and baffles
2. Facing tool
3. Flaring tool
4. Beading tool
5. Tools to make tube-to-tubesheet joints

10.52 What sequence of tube insertion in the tubesheet is adopted in making tube bundles for straight tubes and for U-tubes?

In a straight tube bundle, almost any sequence could be adopted, but it is preferable to start from the center of the tubesheet and gradually reach out toward the periphery so that one can keep examining the tubes in the whole length of the bundle.

In a U-tube bundle, the innermost tubes are placed first in the tubesheet since otherwise one would not be able to insert them later if the outer tubes are first put in place. Basically, the U-tubes with the minimum bend radius are placed first, and those with the maximum bend radius the last.

10.53 What are the various ways of joining the tubes to the tubesheets?

The most popular ways are the following:

1. Roller expanding
2. Welding
3. Roller expanding and welding
4. Grooving with one or two grooves
5. Grooving and welding

See Q. 9.54 for description of these joints.

10.54 What damage is caused if the tube-to-tubesheet joint is not properly rolled?

A heat exchanger is only as good as its tube-to-tubesheet joints. Rolled joints are widely used, with or without grooving. Rolling should be very precise with automatic torque controlled motors, otherwise one may have under-rolled or over-rolled joints. Under-rolled joints result in leaks necessitating rerolling. Over-rolled joints work-harden the metal and increase the possibility of stress corrosion cracking; sometimes the ligament between the adjoining tubes breaks; the tubesheet might bow and the diameter of the tubesheet might increase such that it would not fit in the shell (John, 1959). Furthermore, over-rolling in one tube hole can release the contact pressure in the already rolled tubes in the vicinity. For best results, rolling should be simulated on test pieces and the torque required to thin the tube wall 3–4 percent determined and set on the tube rolling tool. Actual amount of thinning required depends upon the tube and tubesheet materials. A detailed discussion of tube-to-tubesheet joints is available in Yokell (1982). See also Paragraph R, C, B-7.5 of the TEMA Standards (TEMA, 1978).

10.55 How are the axial and tensile loads on the tubes taken care of in the tube-to-tubesheet joints?

For normal operations, simple rolling of the tubes makes the tube-to-tubesheet joint leaktight. For axial loads, grooving is necessary and is done by a grooving tool. For tensile loads, tubes are welded to the tubesheet. The welding gives a smooth contour for the entrance of the tube-side fluid, which reduces the entrance pressure drop and erosion losses.

10.56 When is welding of tubes-to-tubesheet done?

A simple rolled tube-to-tubesheet joint may spring a leak due to fatigue or stress relaxation. Hence, for more leak-tightness and for high-temperature operations, welding is done.

10.57 When is brazing of tubes-to-tubesheet done?

Brazing is done when the stress concentration and crevice corrosion cannot be tolerated but neither the temperature is very high nor is the possible leakage of one fluid into the other hazardous. It is generally preferred only in the cases where welding cannot be done, as in joining dissimilar materials such as aluminum and steel. Since the brazing alloys (copper, copper alloys, aluminum alloys, etc.) are brittle, no welding can be done close to a brazed joint because the heat and/or the thermal stresses due to welding may crack the brazed joint.

10.58 What precautions are taken when a bimetallic tube is rolled into a tubesheet?

In a bimettalic tube, the inner material is compatible with the tube-side fluid, while the outer one is compatible with the shell-side fluid. If it is rolled into a tube-sheet as is, the tube-side fluid will contact the outer material at the edges. Hence, the outer material is stripped off for a portion of the tubesheet thickness and a ferrule of the same material as the inner tube is used as a filler so that the tube-side fluid does not contact the outer material of the bimetallic tube (Fig. 10.7).

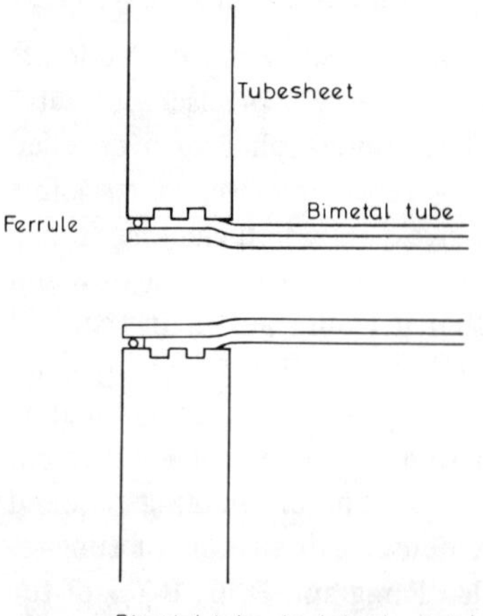

Figure 10.7 Bimetallic tube-to-tubesheet joint with ferrule. *(Courtesy Patterson-Kelley Co., Division of HARSCO Corp.)*

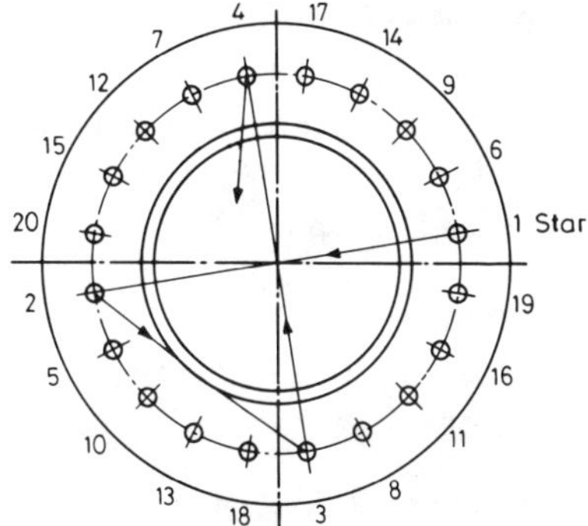

Figure 10.8 Bolt tightening sequence.

10.59 For high-pressure usage, how should the tubes be joined to the tubesheet?

Two popular methods are welding and roller expanding. A welded joint can withstand the maximum temperature. However, a low carbon stainless or Inconel overlay is required before welding tubes to tubesheet to reduce carbon pickup by the welds that will otherwise become brittle and will crack. Tube replacement becomes difficult in a welded joint.

In a roller expanded joint, tube replacement is easy. However, the joint temperature is critical, and, furthermore, an expanded joint cannot be used above 3000–4000 psi (20–28 MPa) because of the increased thickness of the tube that becomes diffficult to expand.

10.60 How should the bolts on an exchanger be tightened?

The bolts should be tightened to produce the minimum seating stress on the gasket surface. The flanged joint should provide for a gasket seating detail such that the gasket is not crushed. A common problem in seating the gaskets with ribs is that the rib gets hung on the pass partition plate. The design of the rib gasket seating surface and the installation procedures should preclude the possibility of ribs getting hung on the pass partition surface. A schematic of the suggested bolt tightening sequence is shown in Fig. 10.8. Furthermore, the same schedule should be used to retighten the bolts after the exchanger has achieved its steady operating temperature, since thermal expansion during the start-up might create a leaky joint.

10.61 How can the pressure vessel for very high temperatures be fabricated?

At very high temperatures, materials weaken and creep rates increase rapidly. Most stress tables go up to 1200°F (650°C) for carbon and low-alloy steels and up to 1500°F (815°C) for stainless steels. Beyond these temperatures, special permission may be required from the regulatory authorities, which is granted only rarely, if ever. The life of the apparatus at very high temperatures is short.

One can use a metal inner lining surrounded by insulation that, in turn, is contained in a pressure vessel. Perforations through the insulation make the differential pressure across the lining nil. Ceramic linings are also used for high-temperature work.

10.62 How are the very large-size, high-pressure vessels made?

These are made of prestressed concrete and are fabricated at site. In very large-diameter vessels, e.g., nuclear reactor buildings, the concrete vessel will have a steel inner lining to contain the pressure.

Multilayer construction is also used. It is discussed in Chapter 12.

10.63 How can complicated shapes be fabricated?

Often thinner curved shapes can replace a thick flat plate (Schroeder, 1979). One way to fabricate such a shape is by suitably sizing and placing explosive charges.

10.64 What guarantees does the fabricator have to give to the buyer?

It can be any depending upon the mutually agreed terms and conditions. TEMA (1978) specifies the following guarantees:

1. Thermal performance: The manufacturer shall guarantee the thermal performance if the thermal rating was done by the manufacturer. Excessive fouling due to deposition of coke, silt, scale, etc., will be the responsibility of the user. If the thermal performance is not satisfactory, a test will be carried out under conditions and procedures mutually agreed to between the buyer and the manufacturer. The above guarantee shall be for a period of 12 months from the date of shipment of the exchanger. Sometimes, it is up to 18 months from the date of shipment or 12 months in actual use, whichever is earlier.
2. Mechanical design: The manufacturer shall guarantee the mechanical design for satisfactory performance at the design conditions as specified on the data sheet. Parts, including tubes, obtained from other suppliers are guaranteed only to the extent of the warranty of the original manufacturer to the heat exchanger fabricator. Damage due to corrosion and vibration, regardless of when it occurs, will not be the responsibility of the manufacturer, except to the extent that any damaged parts within the guarantee period will be replaced FOB. The manufacturer is not responsible for any consequential or indirect damages. Parts fabricated to the buyer's drawing will meet the specified dimensions within the allowable tolerance limits.

10.65 What documents should the fabricator furnish to prove and certify adherance to quality control?

The following documents should be furnished if required by the buyer (Oak Ridge National Laboratory, 1975):

1. Mill test reports of the materials of construction listing such items as the chemical analysis, tensile tests, and Charpy impact tests at the desired temperature
2. Nondestructive examination reports listing the type of examination conducted, the names of the components tested, the data noted or recorded, conclusion drawn, etc.
3. Defect and correction reports listing the defects noted in various parts, procedures used in identifying the defects, corrective actions taken, inspections done to assure the corrections, and formal acceptance of the corrections by the buyer

4. Proof testing reports listing the tests done (e.g., hydrostatic and pneumatic tests, leak tests, and final flow and thermal tests, if required), procedures adopted, data taken, and their interpretation
5. Cleaning reports containing details of the cleaning procedure, cleaning solution, water analysis, and decisions reached
6. Welding procedures, welders' qualifications, and mock-up results
7. Samples of heat transfer tubing, tubesheet, and shell plate materials may also be provided for future reference
8. ASME Code Stamp forms and name plate rubbing

WELDING

10.66 What are the ways by which a metal product can be shaped?

There are four ways (Houldcroft, 1975):

1. Mold casting to the desired shape
2. Hot or cold forging or mechanically forming to the desired shape from solid metal or powder
3. Machining from a solid block
4. Fabricating by joining pieces

10.67 What are the different ways of fabricating a product from pieces?

There are four ways (Houldcroft, 1975):

1. Mechanical jointing (using rivets, bolts and nuts, screws, or interlocking joints)
2. Welding (metallurgical bond by applying heat or pressure)
3. Brazing and soldering (metallurgical bond by inserting a metal having a lower melting point than the parts)
4. Adhesive bonding (using adhesives or resins)

10.68 Why is welding generally preferred over the other ways of fabrication?

Welding is faster than mechanical jointing and gives a stronger product. It also permits greater flexibility in the design.

Welding gives a significantly stronger joint than does brazing, soldering, or adhesive bonding. A welded joint can be used at extremes of temperature, both low and high, while a brazed, soldered, or adhesive bonded joint can only be used at the ambient or near-ambient temperatures.

10.69 How is heat provided in welding?

There are four sources (Fig. 10.9):

1. Gas flame
2. Electric arc
3. Electric resistance
4. Exothermic chemical reaction (thermit welding)

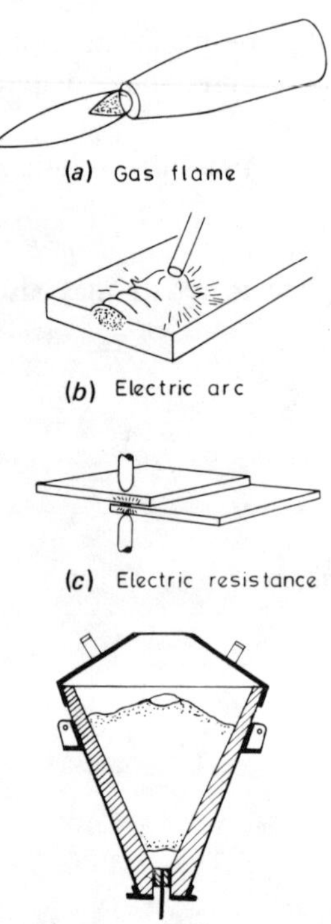

Figure 10.9 Heat sources for welding. *(Reprinted with permission; from Houldcroft, 1975.)*

10.70 What are the various welded joint categories?

Welded joint category indicates the location of a joint in a vessel and not its type, which is discussed in the next question. The different categories are shown in Fig. 10.10, and are described below:

Category A: Longitudinal joints, circumferential joints connecting hemispherical head or flat cover

Category B: Circumferential joints, circumferential joints of torispherical and ellipsoidal heads, and angle joints of not greater than 30°

Category C: Flange joints, Van Stone laps, tubesheets, flat head to main shell

Category D: Seams joining communicating chambers, such as nozzles-to-main shell, nozzles to heads, and so on

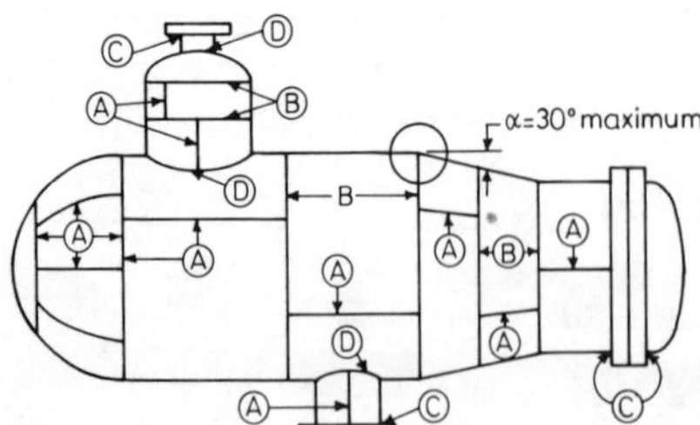

Figure 10.10 Welded joint categories. *(Reprinted with permission; from Chuse, 1977.)*

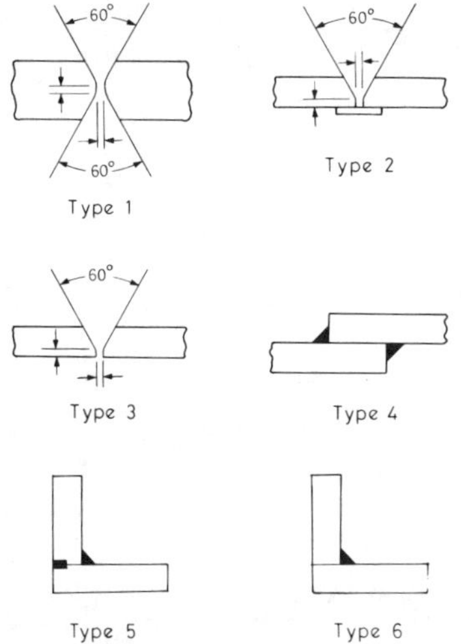

Figure 10.11 Welded joint types.

10.71 What are the various welded joint types?

These are shown in Fig. 10.11 and are described below (Singh, 1979):

Type No. 1: Double welded butt joint, only removable backing strips allowed
Type No. 2: Single welded butt joint, permanent backing strips allowed
Type No. 3: Single welded butt joint, without the use of a backing strip
Type No. 4: Double full fillet lap joints
Type No. 5: Single full fillet lap joints with plug welds
Type No. 6: Single full fillet lap joints without plug welds

See Q. 10.114 for backing strips.

10.72 What are the Standard Welding Symbols?

These have been introduced by the American Welding Society and are shown in Fig. 10.12. These provide a means of explicitly communicating about the type of welding desired on the fabrication drawing itself, ruling out any probable ambiguities.

10.73 List the various important welding processes.

These are the following:

1. Forge welding
2. Fusion welding
3. Thermit welding
4. Oxyacetylene welding
5. Friction welding
6. Explosive welding
7. Electric welding

These are described below. Some of these are not used in heat exchangers and pressure vessels but have been included for the sake of completeness.

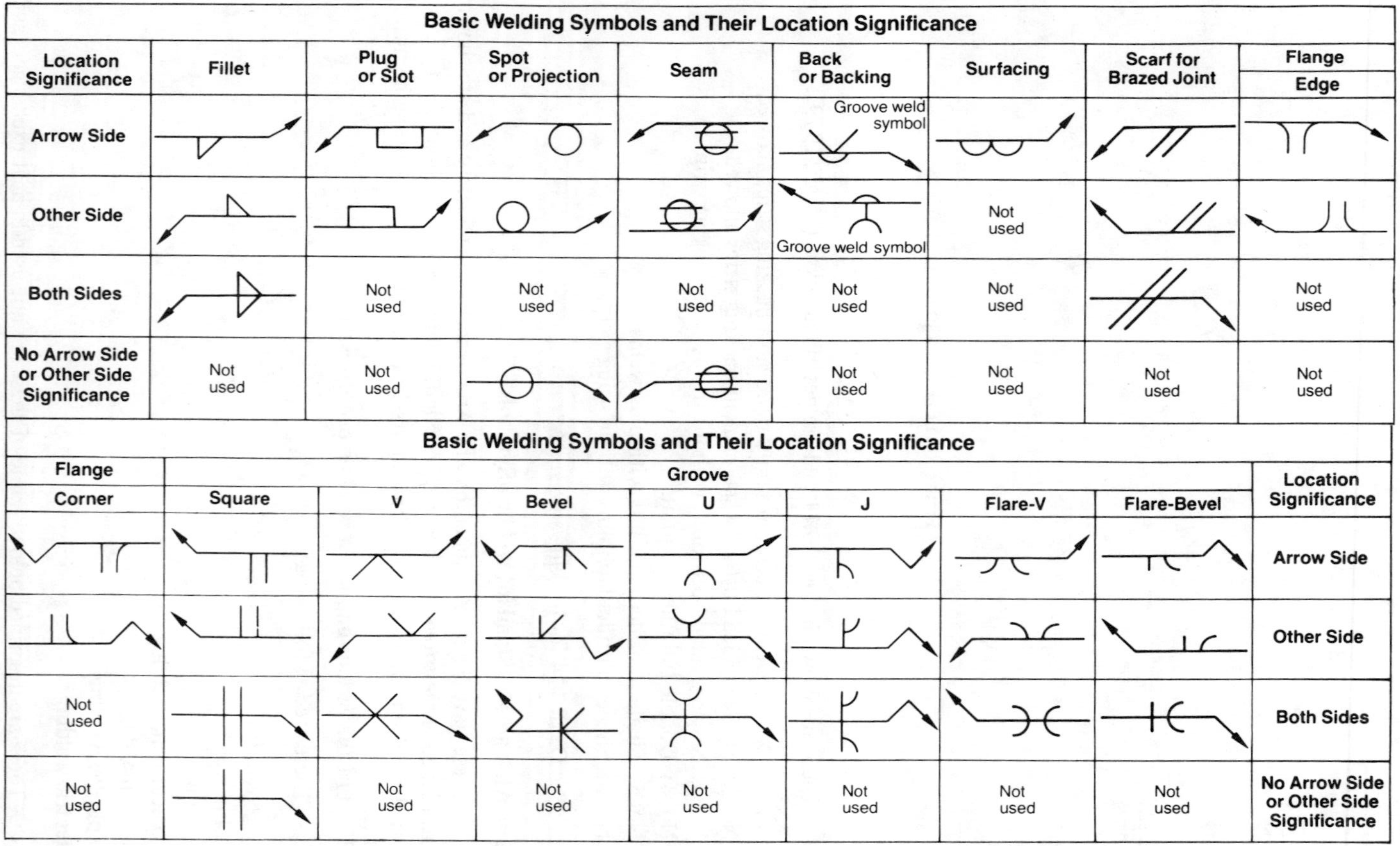

Basic Welding Symbols and Their Location Significance

Location Significance	Fillet	Plug or Slot	Spot or Projection	Seam	Back or Backing	Surfacing	Scarf for Brazed Joint	Flange Edge
Arrow Side					Groove weld symbol			
Other Side					Groove weld symbol	Not used		
Both Sides		Not used	Not used	Not used	Not used	Not used		Not used
No Arrow Side or Other Side Significance	Not used	Not used			Not used	Not used	Not used	Not used

Basic Welding Symbols and Their Location Significance

Flange	Groove							Location Significance
Corner	Square	V	Bevel	U	J	Flare-V	Flare-Bevel	
								Arrow Side
								Other Side
Not used								Both Sides
Not used		Not used	Not used	Not used	Not used	Not used	Not used	No Arrow Side or Other Side Significance

Figure 10.12 Standard welding symbols. *(Copyright © 1980 by American Welding Society, Miami, Florida; used with permission.)*

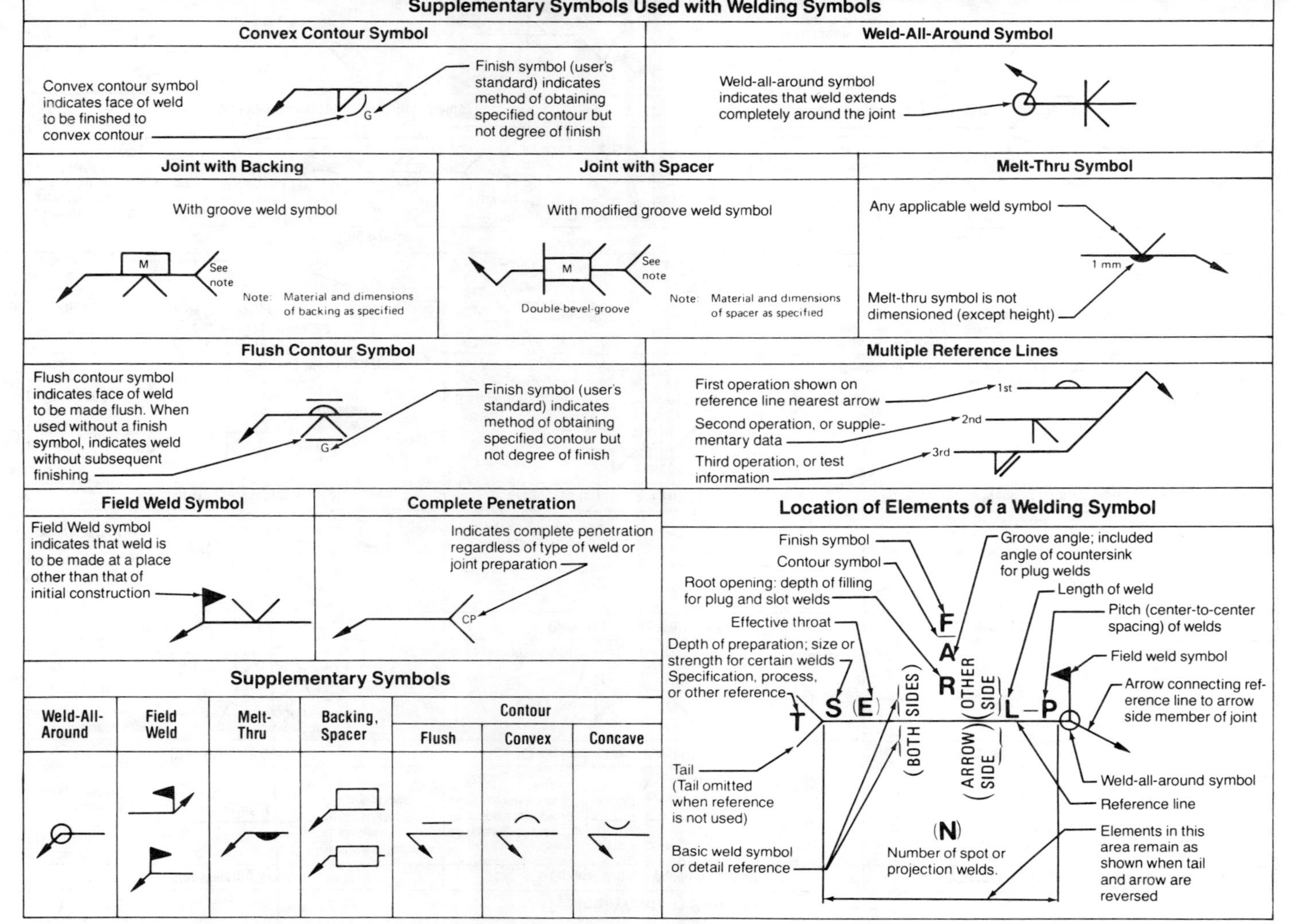
Supplementary Symbols Used with Welding Symbols
Convex Contour Symbol
Convex contour symbol indicates face of weld to be finished to convex contour
G
Finish symbol (user's standard) indicates method of obtaining specified contour but not degree of finish
Weld-All-Around Symbol
Weld-all-around symbol indicates that weld extends completely around the joint
Joint with Backing
With groove weld symbol
M
See note
Note: Material and dimensions of backing as specified
Joint with Spacer
With modified groove weld symbol
M
See note
Double-bevel-groove
Note: Material and dimensions of spacer as specified
Melt-Thru Symbol
Any applicable weld symbol
1 mm
Melt-thru symbol is not dimensioned (except height)
Flush Contour Symbol
Flush contour symbol indicates face of weld to be made flush. When used without a finish symbol, indicates weld without subsequent finishing
G
Finish symbol (user's standard) indicates method of obtaining specified contour but not degree of finish
Multiple Reference Lines
First operation shown on reference line nearest arrow
1st
Second operation, or supplementary data
2nd
Third operation, or test information
3rd
Field Weld Symbol
Field Weld symbol indicates that weld is to be made at a place other than that of initial construction
Complete Penetration
Indicates complete penetration regardless of type of weld or joint preparation
CP
Supplementary Symbols
Weld-All-Around
Field Weld
Melt-Thru
Backing, Spacer
Contour
Flush
Convex
Concave
Location of Elements of a Welding Symbol
Finish symbol
Contour symbol
Root opening; depth of filling for plug and slot welds
Effective throat
Depth of preparation; size or strength for certain welds
Specification, process, or other reference
Tail
(Tail omitted when reference is not used)
Basic weld symbol or detail reference
Groove angle; included angle of countersink for plug welds
Length of weld
Pitch (center-to-center spacing) of welds
Field weld symbol
Arrow connecting reference line to arrow side member of joint
Weld-all-around symbol
Reference line
Elements in this area remain as shown when tail and arrow are reversed
T
S (E)
F
A
R
L–P
(BOTH SIDES)
(OTHER SIDE)
(ARROW SIDE)
(N)
Number of spot or projection welds.

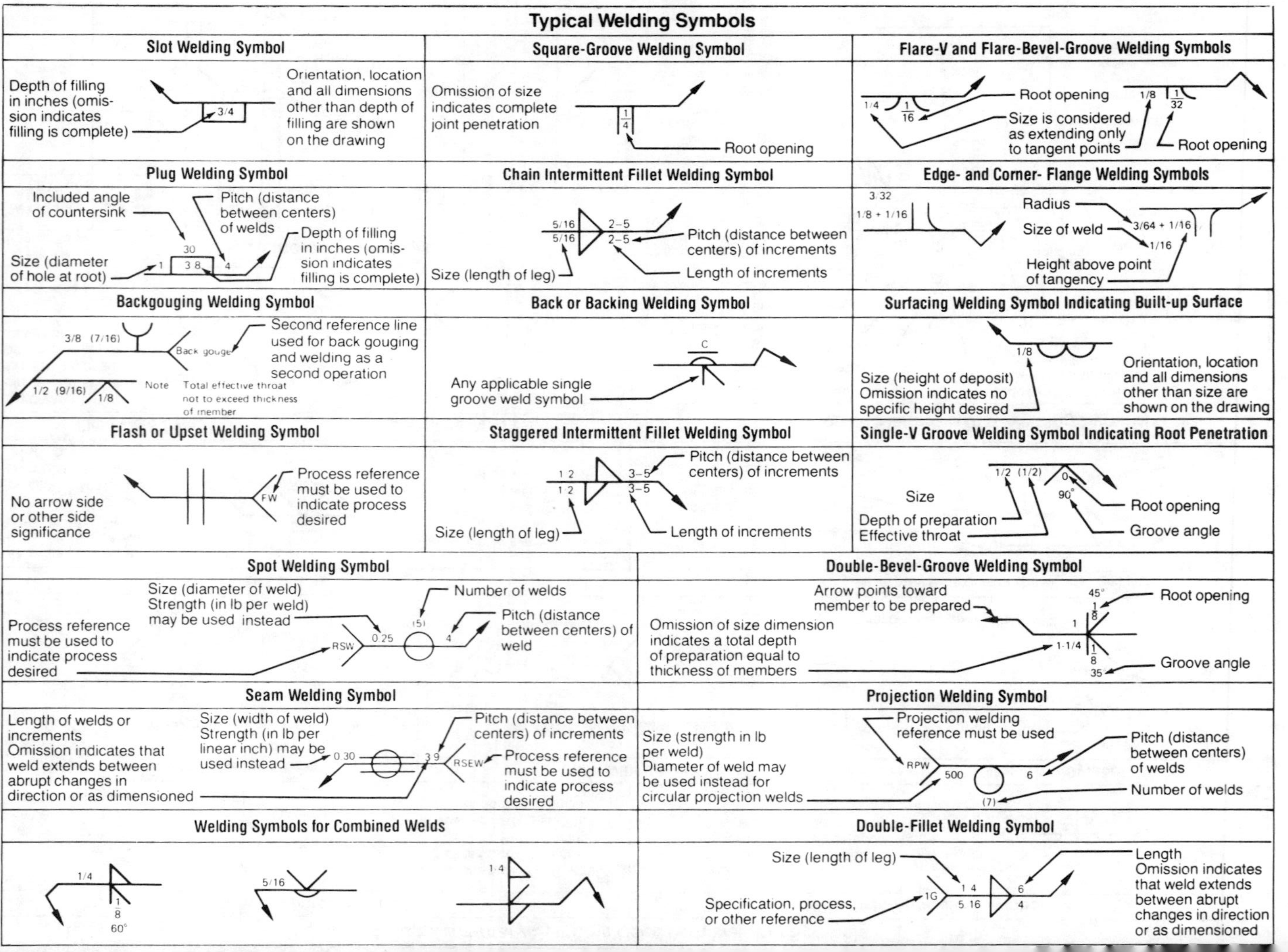
Typical Welding Symbols
Slot Welding Symbol
Depth of filling in inches (omission indicates filling is complete)
3/4
Orientation, location and all dimensions other than depth of filling are shown on the drawing
Square-Groove Welding Symbol
Omission of size indicates complete joint penetration
1/4
Root opening
Flare-V and Flare-Bevel-Groove Welding Symbols
1/4
1/16
Root opening
Size is considered as extending only to tangent points
1/8
1/32
Root opening
Plug Welding Symbol
Included angle of countersink
Pitch (distance between centers) of welds
30
1
3/8
4
Size (diameter of hole at root)
Depth of filling in inches (omission indicates filling is complete)
Chain Intermittent Fillet Welding Symbol
5/16
5/16
2–5
2–5
Pitch (distance between centers) of increments
Length of increments
Size (length of leg)
Edge- and Corner- Flange Welding Symbols
3/32
1/8 + 1/16
Radius
Size of weld
3/64 + 1/16
1/16
Height above point of tangency
Backgouging Welding Symbol
3/8 (7/16)
1/2 (9/16)
1/8
Back gouge
Second reference line used for back gouging and welding as a second operation
Note Total effective throat not to exceed thickness of member
Back or Backing Welding Symbol
C
Any applicable single groove weld symbol
Surfacing Welding Symbol Indicating Built-up Surface
1/8
Size (height of deposit) Omission indicates no specific height desired
Orientation, location and all dimensions other than size are shown on the drawing
Flash or Upset Welding Symbol
No arrow side or other side significance
FW
Process reference must be used to indicate process desired
Staggered Intermittent Fillet Welding Symbol
1/2
1/2
3–5
3–5
Pitch (distance between centers) of increments
Size (length of leg)
Length of increments
Single-V Groove Welding Symbol Indicating Root Penetration
1/2 (1/2)
0
90°
Size
Depth of preparation
Effective throat
Root opening
Groove angle
Spot Welding Symbol
Size (diameter of weld) Strength (in lb per weld) may be used instead
Number of welds
Pitch (distance between centers) of weld
Process reference must be used to indicate process desired
RSW
0.25
(5)
4
Double-Bevel-Groove Welding Symbol
Arrow points toward member to be prepared
45°
1/8
1
1-1/4
1/8
35°
Root opening
Groove angle
Omission of size dimension indicates a total depth of preparation equal to thickness of members
Seam Welding Symbol
Length of welds or increments Omission indicates that weld extends between abrupt changes in direction or as dimensioned
Size (width of weld) Strength (in lb per linear inch) may be used instead
0.30
3-9
RSEW
Pitch (distance between centers) of increments
Process reference must be used to indicate process desired
Projection Welding Symbol
Projection welding reference must be used
Size (strength in lb per weld) Diameter of weld may be used instead for circular projection welds
RPW
500
6
(7)
Pitch (distance between centers) of welds
Number of welds
Welding Symbols for Combined Welds
1/4
1/8
60°
5/16
1/4
Double-Fillet Welding Symbol
Size (length of leg)
Specification, process, or other reference
1G
1/4
5/16
6
4
Length Omission indicates that weld extends between abrupt changes in direction or as dimensioned

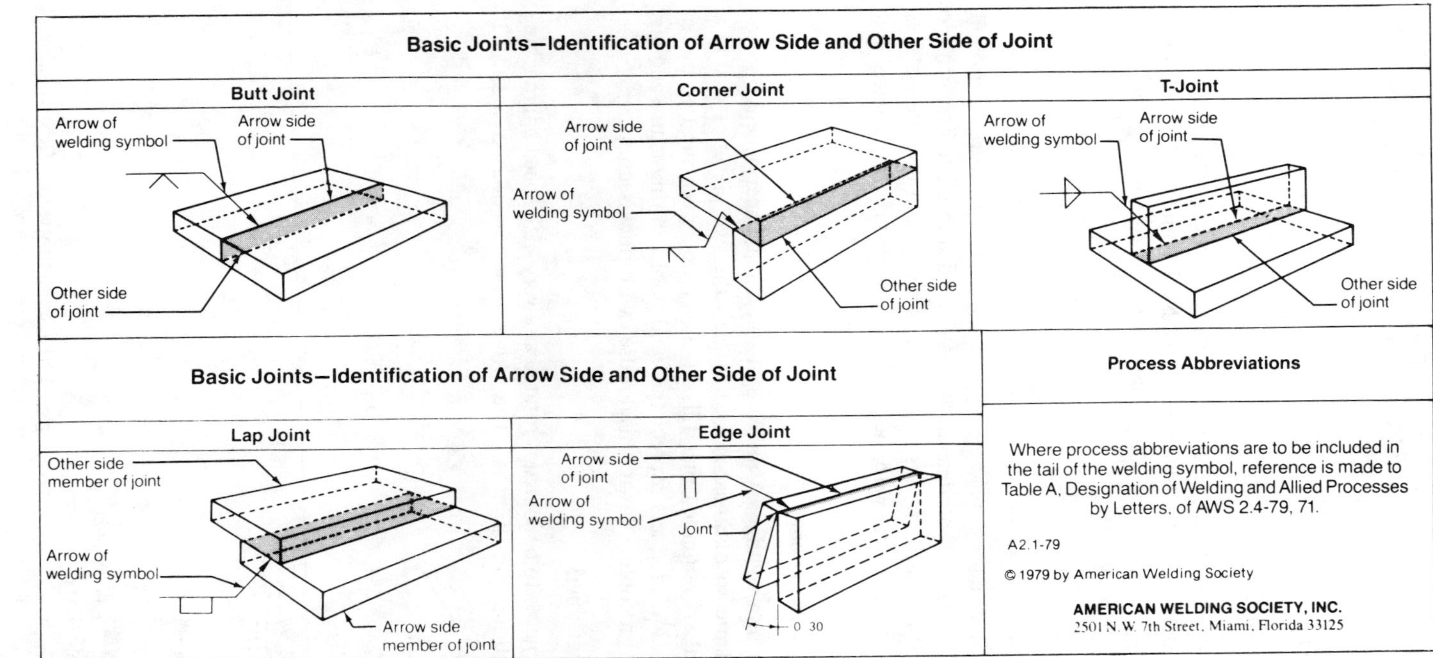

Figure 10.12 Standard welding symbols. *(Copyright © 1980 by American Welding Society, Miami, Florida; used with permission.)* (*Continued*)

10.74 What is forge welding?

It is the joining of two metal pieces by heating them to their forging temperature (above recrystallization temperature), then hammering or pressing them together while still hot. A good forge weld has the same strength as the solid metal (Oak Ridge National Laboratory, 1975).

10.75 What is fusion welding?

It is the joining of two metal pieces by heating their joining ends to a molten stage and then allowing them to cool, forming a joint. No hammering or pressure is needed. A filler metal may or may not be used. This is the most common method used in heat exchangers and pressure vessels.

10.76 What is thermit welding?

A mixture of iron oxide and aluminum powder, called thermit, melts on ignition due to a highly exothermic reaction, giving a high temperature of around 5000°F (2760°C) (Fig. 10.13). The molten thermit is poured into a gap between the two pieces to be joined. It melts the pieces and solidifies with them on cooling, thus making a joint. It is generally used with rails, and not with heat exchangers and pressure vessels.

10.77 What is oxyacetylene welding?

It is a type of fusion welding. A mixture of oxygen and acetylene is burned, and the high intensity flame so obtained is used to melt the two pieces as well as a filler rod of similar metal (if required) that fills in the gap between the two pieces (Fig. 10.14). The combustion products protect the weld pool from moisture and air. A strong joint is obtained on cooling down. It is not used with thick sections.

10.78 What is friction welding?

One of the two pieces to be joined is rotated at a very high speed approaching several thousand revolutions per minute. The other piece is pressed against it

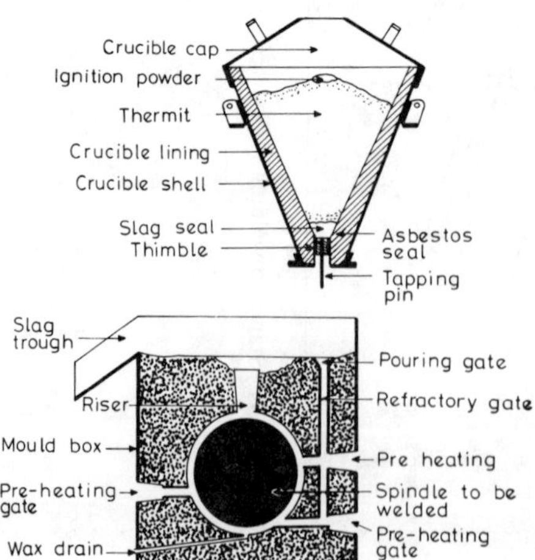

Figure 10.13 Thermit welding. *(Reprinted with permission; from Houldcroft, 1975.)*

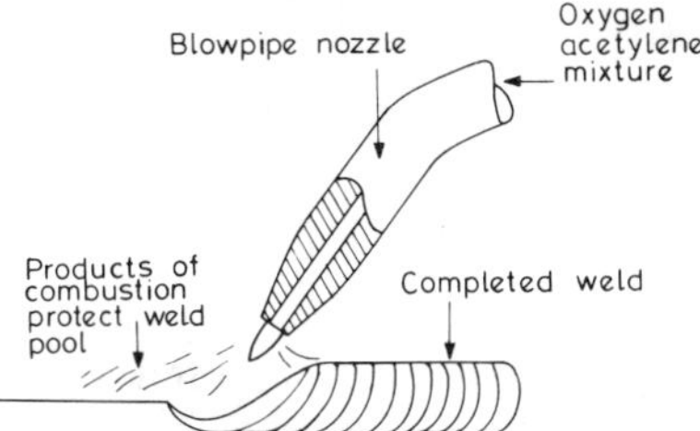

Figure 10.14 Oxyacetylene welding. *(Reprinted with permission; from Houldcroft, 1975.)*

(Fig. 10.15). The friction between the two results in a very high temperature, and the two pieces fuse together. This type of welding can be done in only round shafts with polished ends. A slight variation of the process is called inertia welding. These are not used with heat exchangers and pressure vessels.

10.79 What is explosive welding? What are its effects?

Explosive welding is used to bond two metal pieces together by detonating an explosive charge. This method is used in the bonding of two metal plates and the joining of tubes to tubesheet. In the first case, the charge is placed on top of the plates (Fig. 10.16). When ignited at one end it causes the progressive collapse of the bonding plate on to the lower plate. In the second case, a small charge is placed in the tube near the tube end (Fig. 10.17). When ignited, it causes the collapse of the tube wall around the tube hole in the tubesheet, thus forming a bond.

Tests done with carbon steel base plates on which aluminum plates were explosively bonded showed that the mechanical properties of the base plate were considerably affected. Its tensile strength and ductility were decreased, and the impact ductile/brittle transition temperature was increased. The amount of explosive used has a strong role to play, and it should be kept to a minimum (Loosemore and Peapell, 1979).

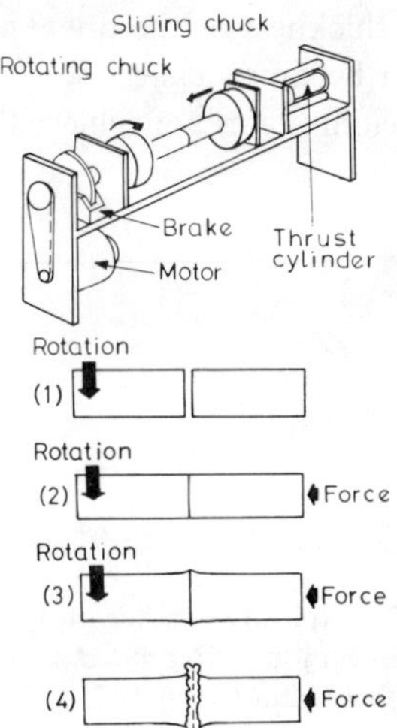

Figure 10.15 Friction welding. *(Reprinted with permission; from Houldcroft, 1975.)*

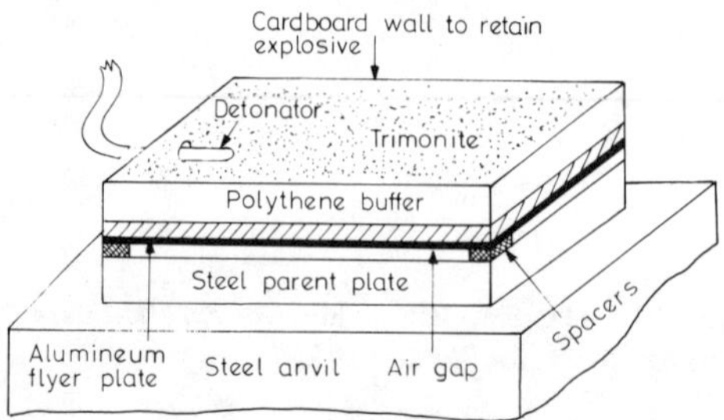

Figure 10.16 Explosive bonding of two plates. *(Reprinted with permission; from Loosemore and Peapell, 1979, p. 225.)*

10.80 What is electric welding?

It uses electric current to weld two pieces. This can be done in three ways:

1. Resistance welding: The parts are raised to fusion temperature by placing them in contact with one another and passing heavy electric current, as shown in Fig. 10.9(*c*). Spot welding of thermocouples and light metal parts is a form of resistance welding. It is very quickly done, and no electrode or flux or third material are involved.
2. Carbon-arc welding: An electric arc is initiated between the metal pieces and a carbon rod, which melts the two pieces as well as a filler rod that fills up the joint. This is not generally used with heat exchangers and pressure vessels.
3. Metallic-arc welding: An electric arc is initiated between the metal pieces and a metallic electrode, which in most cases itself melts and fills up the joint. This is the one used most often in heat exchangers and pressure vessel shells and heads and is described in detail in the next question.

10.81 Describe briefly the various metallic-arc welding processes used in the fabrication of an exchanger.

These are the following (Oak Ridge National Laboratory, 1975; Baumeister, 1978):

1. Shielded metal-arc process: This is the process used most often. The joint surfaces of the base metals are melted by an electric arc established between the base metal and a coated metallic electrode. There is no limit on the thickness or the orientation of the materials that can be joined in this manner. It becomes less economic as the thickness of the material increases since lower welding rates are achieved.

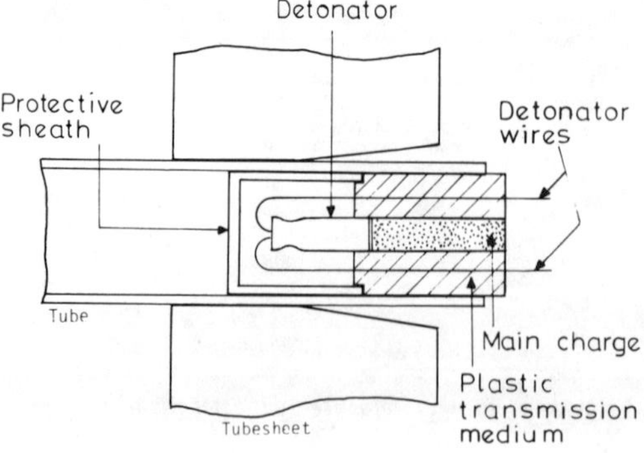

Figure 10.17 Explosive bonding of tube-to-tubesheet joint. *(Reprinted with permission; from Houldcroft, 1975.)*

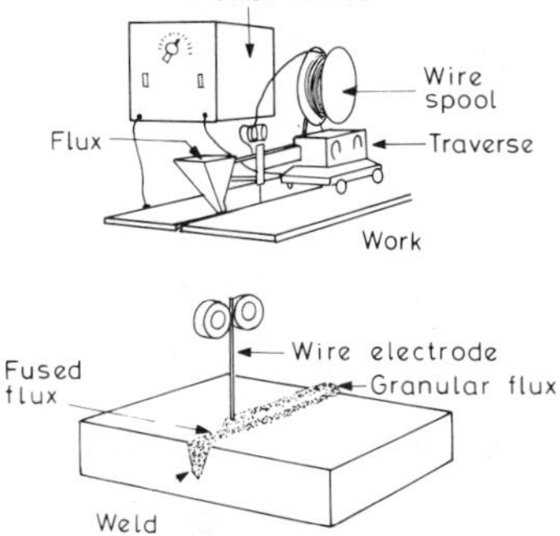

Figure 10.18 Submerged arc welding. *(Reprinted with permission; from Houldcroft, 1975.)*

2. Submerged-arc welding process: In this, the arc is between the bare metal electrode and the base metal. Both the arc and the welding zone are covered with a blanket of granular fusible material called flux in which the arc is submerged (Fig. 10.18). Thick materials can be joined but they must be horizontal. High degree of uniformity can be obtained by using this process in a fully automatic manner that gives a good control over the amperage, voltage, travel speed, etc.
3. Gas metal-arc welding process: Metal-arc inert gas process (MIG) requires an inert gas to shield the arc and the welding zone. The arc is between a continuously fed filler metal wire and the base metal (Fig. 10.19). The inert gases used are argon, carbon dioxide, helium, nitrogen, or their mixtures. Usually, the electrodes in MIG are such that they will produce a weld of the same composition as the base metal with equal or greater strength. Welding can be done in any position. Process is usually semiautomatic or fully automatic.
4. Gas tungsten-arc welding process: It is also called tungsten inert gas (TIG) process. The heat for welding is produced by an arc between the base metal and a non-consumable tungsten or tungsten alloy electrode. An inert gas, such as argon, helium, nitrogen, or their mixture, is used to shield the arc (Fig. 10.20). It is

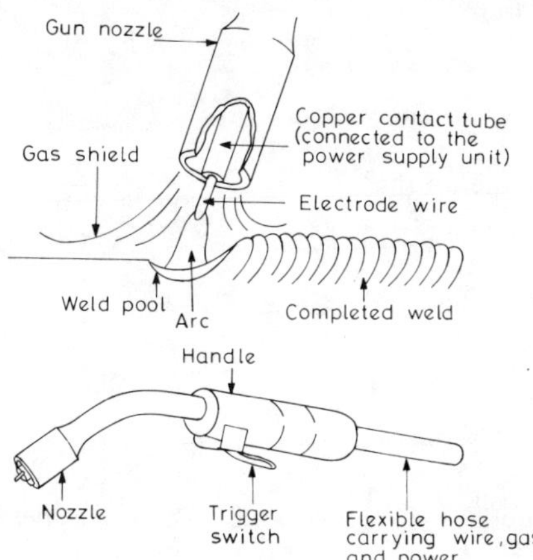

Figure 10.19 Gas metal-arc welding (MIG). *(Reprinted with permission; from Houldcroft, 1975.)*

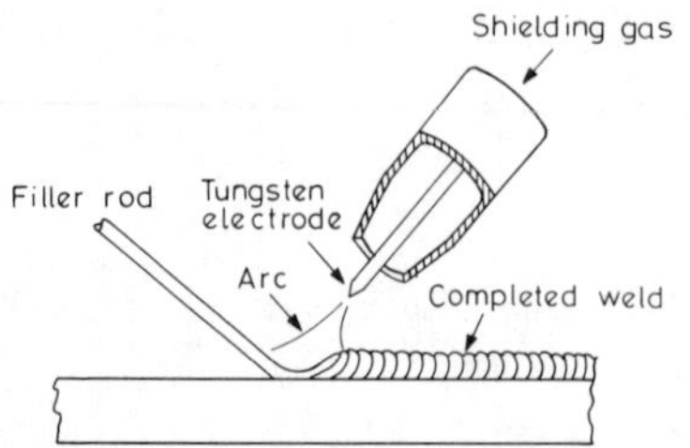

Figure 10.20 Gas tungsten-arc welding (TIG). *(Reprinted with permission; from Houldcroft, 1975.)*

a precision welding process that may be done manually or automatically. In some cases, no filler material is used (autogenous welding).

10.82 What is electroslag welding?

In this a molten slag is used to melt the filler material and the pieces to be welded. The slag is kept molten due to its resistance to the electric current passing between the electrode and the metal pieces. The weld pool is shielded from air by this slag across the full cross section (Fig. 10.21). The electroslag welding is only possible in vertical situations. It is used to weld very thick pieces. The welding is achieved in one pass.

10.83 What factors affect the choice of the welding process? When should such a choice be made?

The choice of the welding process should be made at the design stage itself so that proper access is provided for welding and suitable fabricating material and electrodes are chosen. Among the important factors affecting the choice of the welding process are the following:

1. Type and specification of material
2. Thickness of material
3. Shape of material (rod, plate, shell, etc.)
4. Access to welding zone (corners, restricted spaces, etc.)
5. Welding position (vertical, horizontal, overhead, etc.)
6. Expected efficiency or quality of the welded joint
7. Service temperature of the welded material

More than one type of welding process may be used in one vessel.

10.84 What are the different types of welding operations?

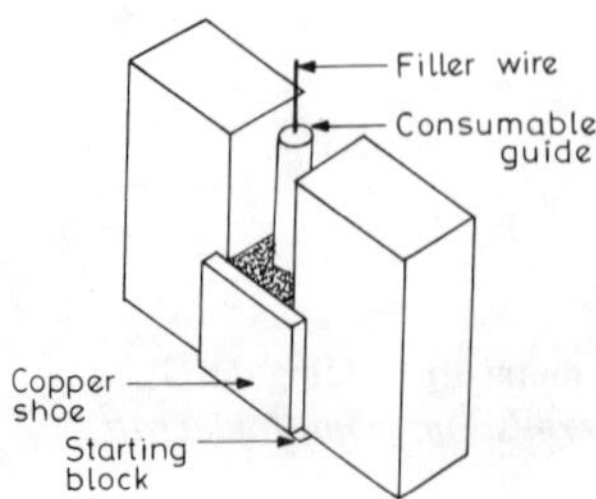

Figure 10.21 Electroslag welding. *(Reprinted with permission; from Houldcroft, 1975.)*

These may broadly be classified into three categories: (1) manual, (2) semi-automatic, and (3) automatic. While the manual operation can reach into corners and restricted spaces, the automatic one can give better quality control and faster rate of welding.

10.85 What are the advantages and disadvantages of automatic welding operation compared to the manual one?

The advantages are the following (Houldcroft, 1975):

1. Continuous working. The electrode is fed continuously. There are no stops and starts, so the weld pool does not solidify during the welding
2. Higher welding current may be used giving deep penetration or faster and constant speed of welding or both.
3. Reduced edge preparation
4. Less distortion of the welded piece, greater uniformity in welding, and fewer defects
5. Reliability and increased productivity

The disadvantages are the following (Houldcroft, 1975):

1. Restricted geometry. It cannot weld complicated shapes or in restricted spaces.
2. Reduced tolerance to variations in the prior fit of the pieces, and thus requires more time
3. Welding variables must be predetermined and set on the machine. These variables are the current or voltage, speed of electrode wire feeding, speed of welding, etc.

10.86 What factors affect the design of a joint for welding?

These are the following (Koopman, 1960):

1. Corrosion and/or erosion: If the joint is to be subjected to corrosion or erosion, the welded joint should present few irregularities or crevices. Full penetration welds perform better.
2. Stresses: Whether tension, compression, shear, bending, or torsion stresses will be applied during service should be known. Furthermore, whether the loads are static or dynamic should be known.
3. Inspection and testing: The welded joint should be so located that it can be inspected and subjected to nondestructive testing for quality control and maintenance requirements.

10.87 What information is required for making a welded joint?

The following information is required:

1. Type, thickness, and shapes of materials to be welded
2. Type of joint, with or without a backing strip
3. Groove design, number of passes of welding, interpass cleaning if required, type of weld (string or weaving bead)
4. Spacing of the pieces, maximum gap at fit up
5. Welding position (horizontal or vertical), up or down welding

6. Composition and size of the welding electrode or wire, single or multiple electrodes, consumable or nonconsumable electrodes, filler wires, fluxes, shielding gas
7. Voltage and current to be supplied, ac or dc current
8. Rate of wire feed and speed of travel in the automated operation
9. Method of joint inspection, extent of tolerable defects, method of defect removal compatible with the pieces to be welded
10. Preheat or postweld heat treatment required, if any
11. Qualified welding procedure
12. Manual, semiautomatic, or automatic operation

10.88 Can any two metals be welded together?

While it is true for most metals, titanium and aluminum cannot be welded to any other but to themselves.

10.89 Can welding be done under any ambient conditions?

Section IX of the ASME Code (American Society of Mechanical Engineers, 1980b) prescribes the welding procedures. It does not permit any welding when the metal is below 0°F (–17.8°C). Between 0° and 32°F (–17.8° and 0°C) ambient, the metal adjacent to the weld should be heated to at least 60°F (15.6°C) before welding. The welding surface should not be wet or covered with ice. It must be protected from falling snow and high winds. It should be free of grease, dirt, and slag.

10.90 Why is the welding generally done in an inert atmosphere?

This is done to exclude oxygen, nitrogen, and moisture from the welding zone so that oxides and nitrides are not formed and no hydrogen is produced by the decomposition of moisture, which would otherwise enter the weld. These adversely affect the weldment ductility and sometimes the strength also and may lead to embrittlement. The shielding of the welding zone can be done by (1) using flux coated electrodes; (2) heaping powdered flux over the welding zone; or (3) flow of inert gases such as argon, helium, carbon dioxide, and so on, over the welding zone.

10.91 What are electrodes, filler wires, and fluxes?

Electrodes are metal or alloy rods with or without a flux coating. An arc is established between the electrode and the base metals. In most cases, the electrode is consumable and melts in the arc. It enters the weld joint along with the base metals. In TIG process, the tungsten electrode is inert and does not melt.

Filler wires are used to fill the gap between the two pieces to be welded where electrode is either not used as in gas welding or does not melt as in TIG.

Fluxes are used to shield the hot metal from attack by the atmosphere and for cleaning the contaminating surface films.

10.92 Why should the electrode coating have a minimum acceptable moisture content?

As this moisture passes through the arc, it decomposes to release hydrogen that gets into the weld. The hydrogen content can become so high that it will cause cracking of the weldment and affect the integrity of the welded joint. It can also result in the hydrogen embrittlement of the joint with the passage of time.

10.93 What are the properties of weldment that one should consider?

These are the following (Peico Electronics and Electrical, 1980):

1. Tensile strength
2. Yield strength
3. Elongation
4. CVN (Charpy V-notch) impact strength at the temperature of operation and at ±20°C (±68°F) of the operating temperature
5. Transition temperature
6. Composition (C, Mn, Si, Cr, Ni, Mo, Cu, etc.)

10.94 What are the various broad categories of electrodes?

These are the following (Philips India Ltd., 1980):

1. Mild steel electrodes
2. Electrodes for cryogenic temperatures
3. Creep-resistant steel electrodes
4. Hardfacing electrodes
5. Stainless steel rutile-coated electrodes
6. Stainless steel basic-coated electrodes
7. Stainless steel electrodes for corrosion and heat resisting applications
8. Stainless steel electrodes for problem steels
9. Nickel electrodes

The categories could vary somewhat with different manufacturers. All of these have AWS numbers assigned to them by the American Welding Society.

10.95 Why do most electrodes have some coatings or coverings?

The coverings provide several advantages (Baumeister, 1978):

1. Protect the welding zone from air and humidity
2. Provide fluxing of the molten metals and remove surface films, etc.
3. Facilitate the establishment and maintenance of the arc
4. Provide a means of introducing alloying ingredients not present in the core wire

10.96 How are the electrodes classified?

The electrodes are classified according to chemical composition, mechanical properties of the weldment, type of covering required, and usability and soundness in different conditions (Baumeister, 1978).

As an example, the mild steel-covered arc-welding electrodes have 12 classifications, each having a distinct number. Their uses are discussed at length in Baumeister (1978). Low-alloy steel electrodes, copper-alloy electrodes, and others are also discussed there.

Besides the above classification, the electrodes can also be classified as consumable or nonconsumable, rods or wire spool, and coated or bare.

10.97 What is the difference between a welder and a welding operator?

According to the ASME Code (American Society of Mechanical Engineers, 1980b), while a welder performs manual or semiautomatic welding, a welding operator operates automatic welding machines.

10.98 Why and how is a welder certified?

Welder certification is required because it assures the manufacturer as well as the buyer that the welder can perform certain types of jobs and has so demonstrated before a duly authorized person. Only a certified welder can fabricate a vessel to be stamped with the Code symbol. The details of the written and performance qualification procedures for welder certification are given in the Code (American Society Mechanical Engineers, 1980b). Test specimens are prepared and tested to see if they have the satisfactory properties for the intended service. A proper certification form is also available in the Code.

10.99 What tests does a welder have to pass?

A welder has to pass fillet weld test and groove weld test. One passing the latter need not take the former since he is qualified to do both, while the reverse is not true.

10.100 What are the welding qualification positions?

These are shown in Fig. 10.22. These should be done in the same position as required in the actual welding (see also Q. 10.107).

10.101 Why are the welders periodically tested and certified? Is a welder's certificate valid at any location within the given time period?

Welders are periodically tested and certified to meet the Code and local regulations. These regulations have been enacted to ensure that the welders continue to be in their best form to do the required job since correct welding is the single most important part of the fabrication requirements. A welder's certificate is valid for a stipulated period on the particular machine(s) on which he has been tested at a given location. If he moves to a different location, he has to be tested and certified again, even though he might be using a machine similar to the one that he was tested upon and certified at the previous location. This provision may vary in some countries.

10.102 What is the welder's identifying mark?

It is a distinct symbol, letter, or identifying number assigned by the manufacturer to each welder or welding operator so that his work can be identified. The record of such identifying marks, along with the tests undergone by each welder or welding operator should be maintained and be made available to the inspector (American Society of Mechanical Engineers, 1980a). The identifying mark has to be stamped

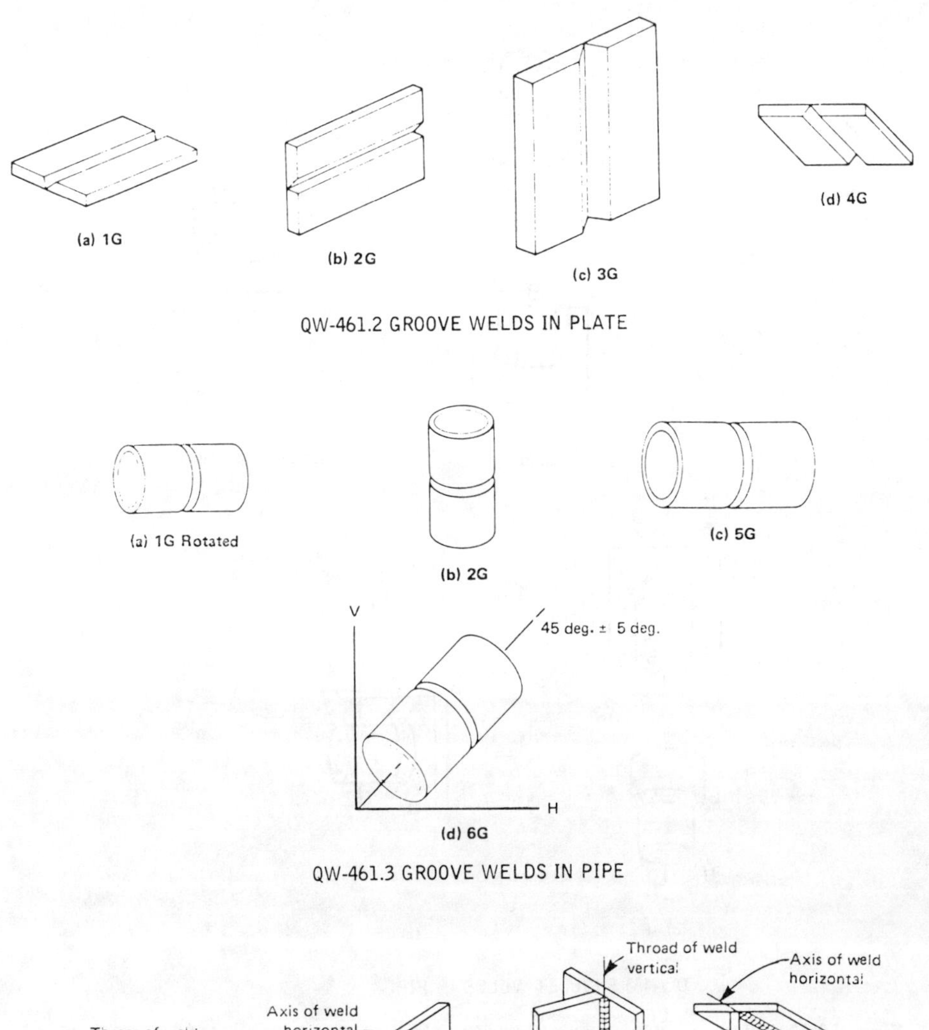

Figure 10.22 Welding qualification positions. *(Reprinted with permission; from American Society of Mechanical Engineers, 1980b.)*

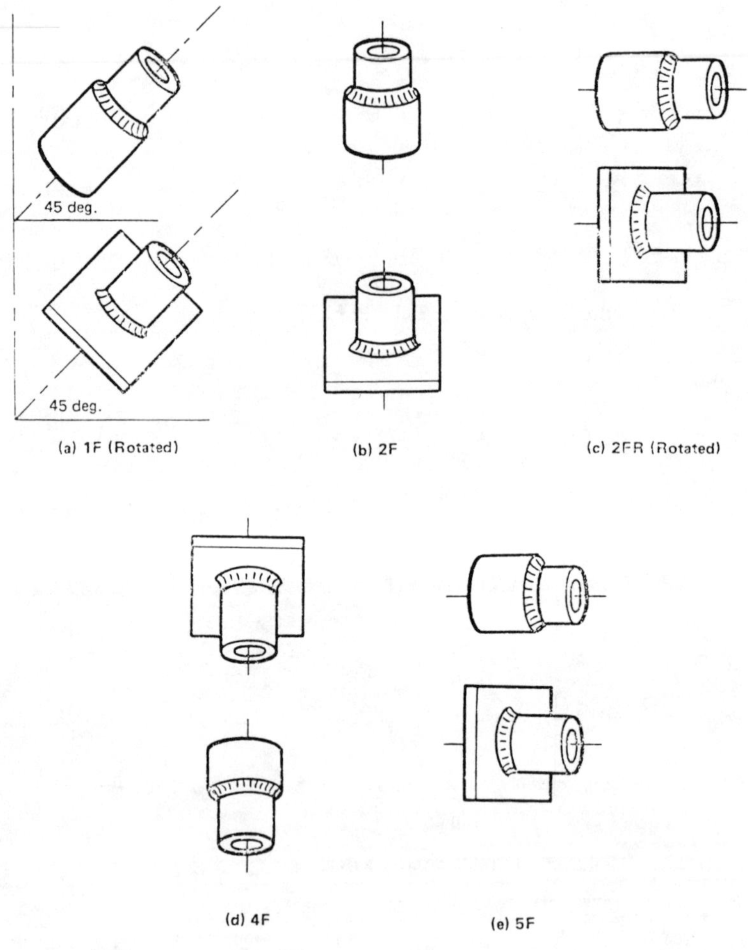

Figure 10.22 Welding qualification positions. *(Reprinted with permission; from American Society of Mechanical Engineers, 1980b.)* (*Continued*)

at regular intervals on the vessel or a record maintained of the welders and welding operators working on each joint.

10.103 When does a welding procedure need to be requalified?

A change in any of the variables that would affect the quality or characteristics of the weld may require requalification of the welding procedure. Amongst such variables are change in procedure; electrode, filler wire, and flux compositions; process (manual, automatic, or semiautomatic); voltage and amperage of electric supply; electrode feed rate; base material; position of base material; shielding gas; preheat or post-weld heat treatment; etc.

10.104 What steps are necessary to make a sound and satisfactory weld?

These are the following:

1. Edges should be properly prepared and parts should be correctly fitted for welding.
2. Shielding gas should be of high purity.
3. Correct type of weld and welding electrodes should be chosen.
4. The working surfaces should be clean and devoid of all paint, rust, scale, slag, grease, or oil.
5. Air drafts at the arc should be avoided.
6. Proper control of the welding parameters, such as voltage, amperage, and feeding of electrode wire, should be maintained.
7. Only adequately trained and properly certified welder and welding procedure should be used.
8. Adequately designed weld joint with regard to its dimensions, angles, and ease of approach should be used.
9. The welded vessel should be stress relieved, if required.
10. The welded joint should be inspected to ensure high quality.

10.105 What are the characteristics of a sound weld?

The examination of the weld should not show any slag inclusion, deep cracks, lack of penetration, heavy porosity, or undercutting at the backing strip.

10.106 How is the base material prepared for welding?

The edges or surfaces of the pieces to be welded may be cut by gas or arc cutting, sawing, machining, or abrasive disc. The cut portion is then cleaned by a metal wire brush or by grinding. All oxide, scale, rust, grease, and oil must be removed from the welding zone.

10.107 What are the various positions of welding?

These are the following (Fig. 10.22) (American Society of Mechanical Engineers, 1980b):

1. Flat (1G, 1F)
2. Horizontal (2G, 2F)
3. Vertical (3G, 3F)
4. Overhead (4G, 4F)
5. Multiple positions (5G, 6G, 5F, 6F)

Depending upon the particular situation, the weld progression in vertical welding may be upward (stainless steel to stainless steel welding) or downward (stainless steel to Incoloy 825 welding).

10.108 Why is welding done in multi-pass?

It improves the mechanical strength of the weldment, and closer inspection is possible after each pass, if so required. It is practiced only in thick sections.

10.109 In multi-pass welding, what is the sequence of passes?

For some cases these are shown in Fig. 10.23. All slag or flux remaining on any pass in welding should be removed by metal wire brush or by grinding before laying the next pass. Similarly, any defects, such as cracks or porosity, also should be removed before laying the next pass.

10.110 When is a welded joint unacceptable?

A welded joint is unacceptable when tests indicate that the defects are beyond the limits permitted by the Code (American Society of Mechanical Engineers, 1980b). These limits vary with the type of defect and the metal thickness. For radiographic examination, the maximum permissible limits are shown in Fig. 10.24 beyond which the welded joint is rejected. Detailed discussion is available in the American Society of Mechanical Engineers (1980b).

10.111 How is a defective weld repaired?

A weld is subjected to nondestructive examination (Chapter 11) to determine whether it is okay. If not, it is a defective weld. It is cut out by an oxyacetylene flame, the scale is chipped out of the cut portion, and new weldment is deposited in place. The new weld is reexamined to determine whether it is okay.

10.112 Why is repair welding different than the fabrication welding for new vessels?

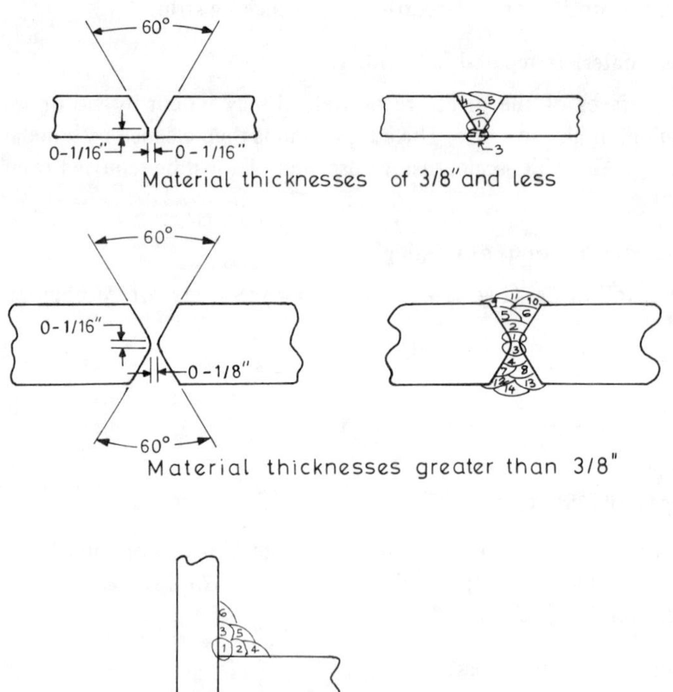

Figure 10.23 Multi-pass welding sequences.

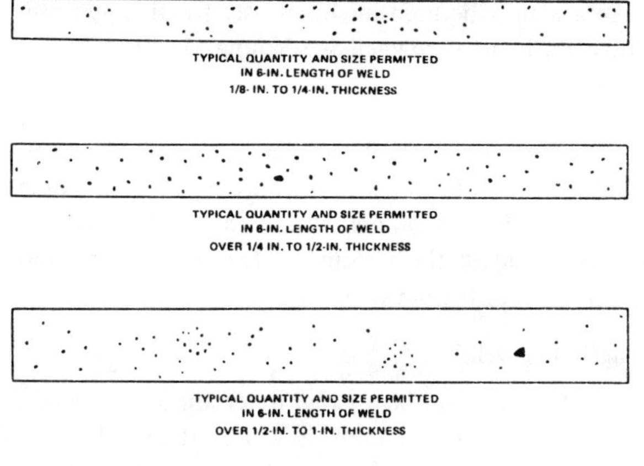

Figure 10.24 Maximum permissible limits on defects revealed by radiography. *(Reprinted with permission, from American Society of Mechanical Engineers, 1980b.)*

In the repair welding, the old weldment has to be removed, partly or totally, and new weldment deposited. The composition of the electrode used may differ from that used in the initial welding of the pieces.

Post-weld heat treatment may not be possible in repair welding, resulting in residual stresses and distortions. In the new vessel, post-weld heat treatment is possible, if required.

The welding position, size of groove, and other items may also be different from the values used in the initial welding of the pieces.

10.113 What is weld shrinkage?

The weldment on cooling contracts or shrinks in all directions distorting the vessel, affecting its dimensions and introducing stresses (Fig. 10.25). The shrinkage

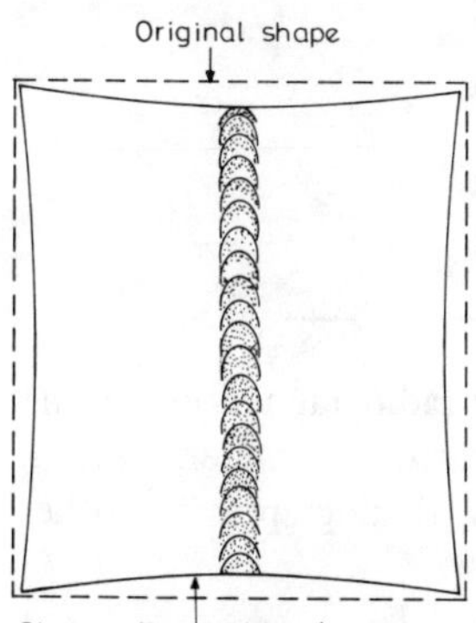

Figure 10.25 Distortions due to weld shrinkage. *(Reprinted with permission; from Houldcroft, 1975.)*

should be kept to a minimum. Preheat may be used in some cases for this purpose. Peening after welding reduces shrinkage stresses (Brownell and Young , 1959).

10.114 What are backing strips?

The backing strips are put at the back of the bevel that is to be single butt-welded (Fig. 10.11). It helps in achieving a full penetration butt welding. In some cases, the backing strip is left in place, while in others it is removed. Removal is a difficult job since the strip also gets joined to the weldment. Different Codes hold different views on the removal or otherwise of the backing strip.

10.115 What determines the strength of a weld?

Composition of the weldment is the main factor that determines its strength. Other factors are the design and dimensions of the joint, how well the welding has been done, and the heat treatment. The composition of the weld is in turn determined by that of the core wire as modified by the composition of the flux coating on it and the composition of the base metals.

10.116 What are the stress values for the weldment?

The allowable stress values for groove and fillet welds as percentages of those for the vessel material are as follows (American Society of Mechanical Engineers, 1980a): groove-weld tension, 74 percent; groove-weld shear, 60 percent; fillet-weld shear, 49 percent.

10.117 What does the efficiency of a welded joint depend upon? Give some typical values.

It depends upon the kind of weld and degree of its radiographic examination. It is defined as the ratio of the allowable stress in the weld to that for the adjacent plate. Some typical values are seen in the tabulation below (American Society of Mechanical Engineers, 1980a; Baumeister, 1978).

	Efficiency (%)
1. Double-welded butt joints	
Fully radiographed (100%)	100
Spot radiographed	85
No radiograph	70
2. Single-welded butt joints (with backing strip left in place)	
Fully radiographed (100%)	90
Spot radiographed	80
No radiograph	65

While the plate thickness is less in the case of fully radiographed joints, the cost of doing the full radiography is usually higher than that of the material saved. Hence, full radiography is used only in critical applications. Radiography and other means of examining a weld are discussed in Chapter 11.

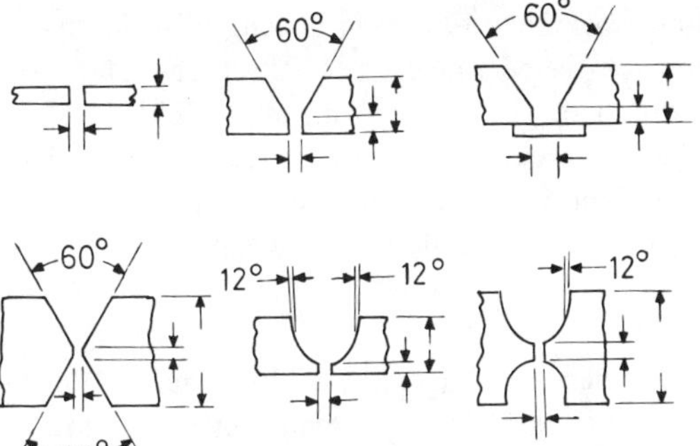

Figure 10.26 Welding groove designs. *(Reprinted with permission; from Baumeister, 1978.)*

10.118 Why is a groove made in welding? What are the typical groove designs?

Grooves are made to provide a shape and size of opening where, under the given conditions, a sound deposition of the filler material can be made with maximum economy. Some typical groove designs are shown in Fig. 10.26.

10.119 How are thick steel plates welded?

A double V-preparation is used (Fig. 10.11, type 1). The welds are visually inspected after each pass, and the entire weld is subjected to nondestructive examination (Chapter 11). Since large weld materials have to be deposited in between the thick plates, it creates thermal stresses and loss of ductility. To reduce the thermal stresses, the ASME Code (American Society of Mechanical Engineers, 1980a) provides guidelines for preheat and post-weld heat treatment of the vessels or parts made of steel. The guidelines take into account the grade of the material, P number, thickness, etc., and provide the temperature, holding time, and cooling rate for the heat treatment (see Q. 9.148–9.150 for post-weld heat treatment and preheating).

10.120 Why should welding between metals of different thermal coefficients of expansion be avoided if possible?

When, after the welding, the metals cool down, they will contract or shrink differently, and this is likely to introduce stresses in the system. However, the design can be such that these stresses are not harmful. One place where the welding of different materials cannot generally be avoided is the tube-to-tubesheet joint.

10.121 How are the tubes welded to the tubesheet?

Tubes and tubesheets are cleaned thoroughly and positioned correctly. Then the welding is done manually, semiautomatically, or automatically. Applications and limitations of each type have been discussed by Emhardt (1969). Some fabricators overlay the tubesheet with weld-deposited cladding to facilitate joining. For example, low carbon-steel classing is used for carbon-steel tubes, and Type 309L or a combination of Type 309L and 308L is used for stainless steel tubes (Winsor, 1978). Depending upon the use of the exchanger, materials of construction and operating conditions, different types of welds are made. These have been discussed at length by Bevevino (1974). After the welds are made, they are tested for leaks (Chapter 11). Then the tubes are expanded throughout the tubesheet thickness (Winsor, 1978). For low chrome alloy materials, the welds are heat treated. A unique heat-treating technique using cartridge heater for inside, split manifold heater for the weld's outer diameter, and fiber optics/infrared temperature control has been discussed by Kay et al. (1979).

10.122 How is a clad vessel welded?

The ASME Code (American Society of Mechanical Engineers, 1980a) describes a clad vessel as one made of integrally clad plates or having applied corrosion resistant lining.

The weldment should also have enough corrosion resistance against the vessel contents as the cladding has, or else it will corrode. Hence, a judicious choice has to be made about the welding electrode, fluxes, and the welding environment. Caution should be exercised in welding because of the different thermal coefficients of expansion of the dissimilar metals used in cladding that might induce unwanted stresses. Similarly, in the post-weld heat treatment to relieve the stresses, one should be careful not to degrade the physical properties of either the base metal or the cladding. Details are available in the Code (American Society of Mechanical Engineers, 1980a).

10.123 What are the relative costs of material, fabrication, and welding in a vessel?

Harris (1971) has given the following rough estimate: (1) material, 37 percent; (2) preparation, assembly, and all mechanical operations, 47 percent; (3) welding, 14 percent; and (4) nondestructive testing, 2 percent.

Item (2) accounts for a large fraction of the cost because accurate assembly, fit-up and preparation of the edges, etc., are very important for the welding operation to be carried out successfully.

● The above should serve as an introduction to the salient features of the three topics mentioned in the title of this chapter. For details, various references mentioned should be consulted.

NOMENCLATURE

A, B, C, D	welded joint categories
CVN	Charpy V-Notch
FOB	free on board
MIG	metal-arc-inert gas
N	normalized
N.C.	numerically controlled
P	P number in material specification
Q	quenched
T	tempered
TIG	tungsten-inert gas

REFERENCES

Agarwal, J. C. Selection of Material for Cryogenic Service, paper presented at Seminar on Pressure Vessels and Heat Exchangers, Bombay, India. April 1977.

American Society of Mechanical Engineers. *ASME Boiler and Pressure Vessel Code, Section VIII Division 1: Pressure Vessels.* ASME: New York. 1980a.

American Society of Mechanical Engineers. *ASME Boiler and Pressure Vessel Code, Section IX: Welding and Brazing Qualifications.* ASME: New York. 1980b.

American Welding Society. *Standard Welding Symbols.* American Welding Society, Inc.: Miami, Florida. 1979.

Baumeister, T. (ed.). *Marks' Standard Handbook for Mechanical Engineers,* 8th ed. McGraw-Hill: New York. 1978.

Bevevino, J. M. Tube to Tubesheet Welding Techniques. *Chem. Eng.* 70(7):71–73. 1974.

Brownell, L. E., and E. N. Young. *Process Equipment Design.* Wiley: New York. 1959.

Chuse, R. *Pressure Vessels,* 5th ed. McGraw-Hill: New York. 1977.

Emhardt, F. W. Heat Exchanger Tube-to-Tubesheet Joints, Paper No. 69-WA/HT-47, presented at the ASME Winter Annual Meeting. 1969.

Foxall, D. M., and P. T. Gilbert. Selecting Tubes for CPI Heat Exchangers–III. *Chem. Eng.* 83(10): 133–136. 1976.

Greene, B. N. Materials of Construction (of Heat Exchangers). *Chem. Eng.* 87(20):149–151. 1980. A suggested reading.

Hagel, W. C., and K. H. Miska. How to Select Alloy Steels for Pressure Vessels, I and II. *Chem. Eng.* 87(15):89–91 and 87(17):105–108. 1980.

Harris, H. Fabrication Aspects. In *Pressure Vessel Engineering Technology.* R. W. Nichols (ed.). Applied Science Publ.: London. 1971.

Heat Exchange Institute. *Standards for Closed Feedwater Heaters,* 3d ed. HEI: Cleveland, Ohio. 1980.

Houldcroft, P. T. *Welding Processes.* Engineering Design Guides No. 06. Oxford University Press: London. 1975.

John, A. Longer Life for Heat Exchangers. *Chem. Eng.* 66(26):77–80. 1959.

Kay, R. C., V. G. McGevna, and H. Schwartzbard. Unique Heat Treating Technique for Tube-to-Tubesheet Welds. *Metal Prog.* 1154:41–43. 1979.

Koopman, K. M. Elements of Joint Design for Welding. In *Pressure Vessel and Piping Handbook-Collected Papers.* ASME: New York, 1960.

Loosemore, G. R., and P. N. Peapell. How Explosive Welding Affects Carbon Steel Plate. *Weld. Met. Fabr.* 47:229–236. 1979.

Mack, W. C. Selecting Steel Tubing for High-Temperature Service. *Chem. Eng.* 83(12):176–179. 1976.

Oak Ridge National Laboratory. *Design Guide for Heat Transfer Equipment in Water-Cooled Nuclear Reactor Systems,* ORNL TM-3578. ORNL: Oak Ridge, Tennessee. 1975.

The Patterson-Kelley Co. *Heat Exchangers, Manual No. 700-A.* The Patterson-Kelley Co. Inc.: East Stroudsberg, Pennsylvania.

Peico Electronics and Electricals. *Welding Electrodes.* Peico Electronics and Electricals Ltd. (Phillips): Bombay, India. 1980.

Philips India Ltd. *A Guide to Classification of Philips Welding Electrodes.* Philips India Ltd.: Bombay, India. 1980.

Schroeder, J. W. *Explosively Free Formed Pass-Partition Plate for a Nuclear Steam Generator.* Foster Wheeler Corp.:Livingston, New Jersey. 1979.

Singh, K. P. Mechanical Design of Pressure Vessels. In *Heat Exchanger Design-II,* J. P. Gupta (ed.). C. S. Enterprises: Delhi, India. 1979.

Tubular Exchanger Manufacturers' Association. *Standards of Tubular Exchanger Manufacturers' Association*, 6th ed. TEMA: Tarrytown, New York. 1978.

Winsor, F. J. Building Heat Exchangers for Power Plants. *Weld. Des. Fabr.* 51:78. 1978.

Yokell, S. Heat-Exchanger Tube-to-Tubesheet Connections. *Chem. Eng.* 89(3):78–94. 1982.

CHAPTER

ELEVEN

QUALITY ASSURANCE, QUALITY CONTROL, INSPECTION, AND TESTING

INTRODUCTION

• Equipment that performs its intended duties for its expected service life with proper maintenance and without failure is said to have been built to "quality." The term quality is a relative one since its requirements would vary on a case by case basis. Pressure vessels and heat exchangers, being pressure-containment vessels, have stringent quality requirements since a failure could lead to loss of property and life besides the loss of production.

In order to ensure that the quality requirements are, in fact, being followed, quality control is necessary. This is accomplished by the two basic mechanisms of quality control, vis. inspection and testing. These cover simple techniques such as visual inspection to very sophisticated testing equipment and also destructive testing of samples.

While quality control has been in existence for a long time in some form or the other, it did not always assure that quality products would be manufactured and it lead to expensive rejections or to acceptance of substandard equipment in the interest of expediency. This happened because of the lack of a proper commitment by the management to the philosophy of quality control and the lack of independence and authority of the manufacturer's inspection department to set up preventive measures. In effect, while the inspectors could reject a bad piece, they generally could not stop its manufacture right in the beginning because of obvious or apparent faults in the design, drawings, etc. This all delayed the delivery of equipment and escalated its cost. Therefore, the need for quality assurance was felt to serve as a "quality control of quality control." It is a management's function with due independence and authority vested in the people involved in it.

In this chapter, we will therefore discuss the four important topics mentioned in the title. As is evident by the above introduction, they are interrelated.

QUALITY ASSURANCE

11.1 What is quality assurance and why is it necessary?

Quality assurance is a management's philosophy that formally defines and controls all company activities that could affect the product quality and customer satisfaction, right from design to the shipping department.

It is necessary because of the following:

1. It provides an assurance to the user by objective evidence that the program of quality surveillance or quality control has been strictly followed and the equipment meets the highest standards and specifications of the user.
2. It results in minimum rejections and repairs during the manufacture of the equipment when checks are made. Thus, it results in meeting the quality specifications at minimum costs and within scheduled time.

11.2 Who is responsible for quality assurance?

The responsibility rests with the top management of the manufacturer who have to:

1. Set the policy and give total independence to the quality assurance group
2. Provide and maintain the necessary infrastructure by way of machines, facilities, and trained personnel
3. Authorize the necessary finances
4. Initiate corrective actions as recommended by the quality assurance group as and when necessary

11.3 What are the advantages of quality assurance?

Some of the advantages of quality assurance are the following:

1. Reduction in cost of rework of defective parts
2. Saving in material
3. Improved productivity
4. Saving in inspection costs
5. Saving in the cost of dealing with the user's complaints

6. Shortening of production cycle and achieving early deliveries
7. Increase user's good will due to greater reliability and enhanced value of the product, leading to increased business
8. Reduction of internal friction within the organization
9. An improvement in the morale of the organization due to conscientiousness of the job done

QUALITY CONTROL

11.4 What is quality control?

Quality control is a detailed procedure adopted by the manufacturer (after discussions with the user, if necessary) to fabricate a given equipment, such as to ensure that the code requirements are met, thereby producing equipment that is fit for use with regard to its performance and reliability, all through its intended life time, with proper and regular maintenance.

11.5 What are the important parts of a quality control program for heat exchangers and pressure vessels?

Each manufacturer has an in-house quality control manual that is generally kept confidential except that the users of equipment and inspectors and certification authorities have access to it. The important parts of such a manual are listed in Table 11.1. There may be some variations from company to company. The quality control chart of a large manufacturer is shown in Fig. 11.1.

Table 11.1 Typical quality control (QC) manual contents

Secton number	Title
1	General
	a. Statement of authority and responsibility
	b. Organization chart
2	Design and document control
3	Procurement and receipt of material
4	Manufacturing process control
	a. Material control
	b. Release of material
	c. Manufacturing control
5	Nonconformities
6	Welding quality assurance
7	Nondestructive examination
8	Gauge and measuring tool control
9	Heat treatment equipment
10	Handling, storage, shipping, and preservation of equipment
11	Records retention
12	Authorized inspector
13	Quality control during field work
14	Periodic audit of QC program
15	Appendices containing revisions and all forms

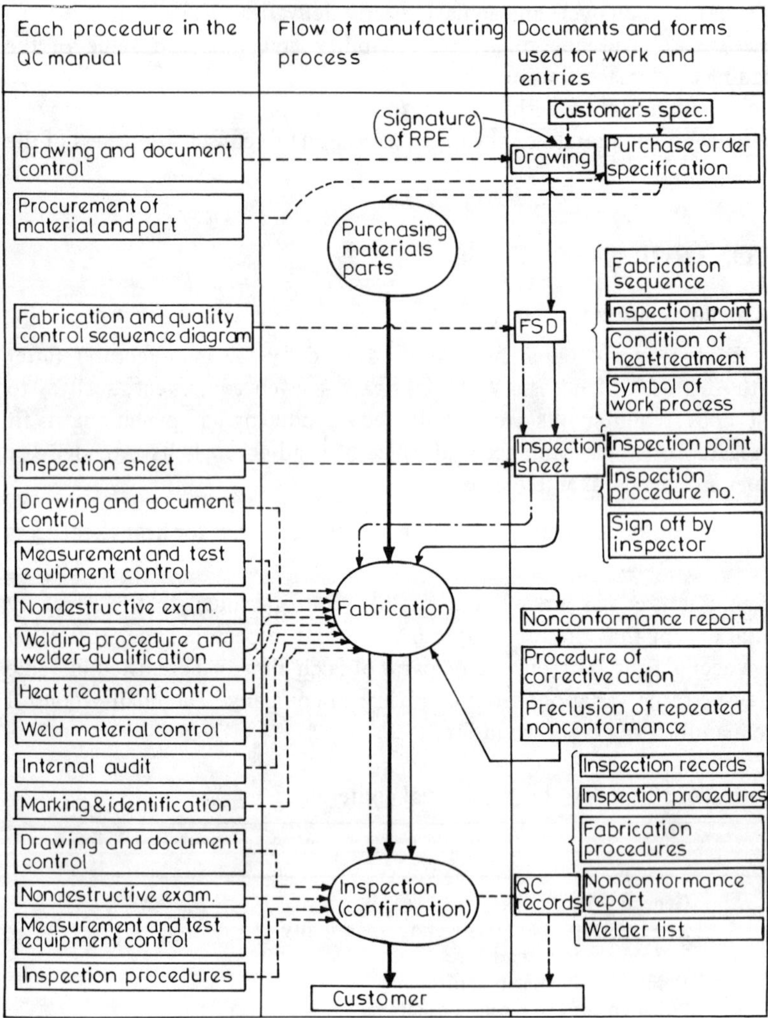

Figure 11.1 Quality control chart for pressure vessels and heat exchangers. *(From Yamada and Minamimoto, 1976.)*

11.6 How does quality assurance differ from quality control?

Quality assurance is a system adopted by manufacturers to ensure that certain minimum and basic requirements are met to attain desired quality and workmanship. This system is a program of various actions that are followed as a built-in system to create an awareness of the necessity of attaining desired quality specifications laid down by the code. On the other hand, quality control is actions carried out as a stagewise requirement tailored to suit a specific equipment to ensure that the quality specifications are maintained.

11.7 How is quality control carried out?

The two basic tools of quality control are inspection and testing, which are extensively used and are discussed at length a little later. Maintenance of detailed report and unambiguous and effective communication between the people involved are also essential for an effective quality control.

11.8 What quality control documents are generally required for a heat exchanger manufactured according to the codes?

These are detailed documents pertaining to the following:

1. Certified mill test reports on the materials of construction detailing tensile tests, chemical analysis, charpy impact tests, etc.
2. Nondestructive examination reports detailing the components tested, data taken, results, etc.
3. Defect and correction reports pertaining to any defect in the material and fabrication procedure
4. Proof testing reports for hydrostatic tests, leak tests, flow tests, thermal acceptance tests, etc. These should include the procedures adopted, data taken, calculations performed, etc.
5. Cleaning reports containing details of cleaning solution, water analysis, temperature, cleaning method used, and decisions reached
6. Mock-up tests, samples of original and welded pieces of tubes, and tubesheet and shell plate with description of the original conditions of materials for unusual units
7. Reports regarding shipment of the unit, blanketing of openings during shipment, details of transport arrangement, etc.
8. Depending upon the local regulations and agreement between the buyer and the manufacturer, more reports may be required and should be specified at the time of signing the contract. These reports may include the stress analysis reports too.

INSPECTION

11.9 Why is inspection necessary?

It is necessary because it ensures that the equipment has been designed and fabricated according to the applicable codes using the correct materials and procedures allowed by the codes.

11.10 Who all have the responsibility for inspection during the fabrication of a pressure vessel or heat exchanger?

The primary responsibility is of the manufacturer who has to carry out the inspections to ensure that the equipment is manufactured according to all the applicable codes.

The buyer can inspect the fabrication at prearranged time, but this does not absolve the manufacturer of his primary responsibility stated above.

Third party inspection may be arranged based upon the buyer's requirements and the local laws. Some codes require third party inspection. The primary responsibility still lies with the manufacturer since his equipment has to perform as per his guarantees given to the buyer.

11.11 Who is an authorized inspector?

An authorized inspector is a person qualified to conduct the inspection and authorized to do so by the state. Authorized inspectors usually belong to various insurance companies or large consulting firms. They have legal authority to inspect and permit the manufacturer to fabricate equipment under the state laws.

11.12 What checks are made by the inspection authority?

These relate to checking the following:

1. Fabricator's quality assurance program
2. Documentation
3. Mechanical design calculations, materials used and their dimensions, corrosion allowance provided, reinforcement of openings, weld details, fabrication drawings, etc.
4. Fabrication procedure
5. Welders' performance tests
6. Various dimensions and welding during actual fabrication
7. Nondestructive examinations and hydrostatic and performance tests
8. Any special requirements of the particular equipment
9. Packaging and shipment to destination

The inspector has the option and freedom to be physically present when most of the procedures of welder certification, fabrication of the equipment, and the nondestructive examinations are being carried out. The above list gives only the broad topics. In each one, a number of items need to be checked. For example, in welding, the inspector has to check the edge preparation, fit-up, selection of the correct type and gauge of the welding electrodes, welding procedure including the fixing of the current/voltage parameters, qualification of the welder, etc. Similarly, the other steps are also quite involved. Therefore, it is essential to have a reliable and reputed inspection authority.

11.13 What vessels are exempt from inspection?

As far as a manufacturer is concerned, no vessel is exempt from inspection. However, certain noncritical vessels, not manufactured according to the codes, i.e., those outside code jurisdictions are exempt from third party inspections. Besides, vessels within the code jurisdiction but not exceeding 5 ft^3 (0.14 m^3) in volume and 250 psi (1.724 MPa) design pressure are also exempt from third party inspection. In all these cases, the inspection is carried out by the manufacturer's authorized persons at predetermined stages to achieve the desired quality and meet the user's requirements.

TESTING

General

11.14 What is meant by testing?

Testing is done to determine the chemical composition, metallurgical structure, mechanical properties, corrosion resistance, etc., of the materials used. These materials should be suitable for the operating conditions they will be subjected to during their service life. It should be possible to subject these materials to the various fabrication processes, such as bending, forming, welding, etc., without unduly impairing their properties (or, if impaired, it should be possible to restore them by some means, e.g., heat treatment). Testing helps determine that the materials being used are suitable for the purpose.

Testing may also be done on the completed equipment to ensure that it will perform as expected.

11.15 What are the different types of testing?

There are two broad types:

1. Destructive testing: This destroys the material or equipment on which the tests are done. These are of four kinds. (a) Chemical analysis test: A piece of material such as a plate or welding electrode is analyzed for its chemical composition. (b) Mechanical properties test: A piece of material is tested to determine its mechanical properties. (c) Metallographic test: It is done to determine its metallurgical structure. (d) Bursting test: The fabricated equipment is pressurized until it bursts. While (a)-(c) are generally supplied by the mill producing the material (or is known from the tables in the codes for a given material), (d) is done by the manufacturer to establish a new manufacturing process or to demonstrate the capability of a new design to withstand high pressures and/or temperature. One case is when a company starts to manufacture multi-wall vessels, it usually confirms the method by testing some units to bursting. High-volume, low-unit-cost vessels are also tested to destruction to demonstrate their viability. Since it is a costly procedure, the destructive testing is generally avoided and recourse is taken to nondestructive testing.
2. Nondestructive testing: This does not destroy the material or equipment. However, if the testing method and/or the equipment are faulty, damage might occur during the test in some cases. There are several types of nondestructive testing, also known as the nondestructive examinations (NDE), which are governed by the ASME Code Section V and Section VIII (American Society of Mechanical Engineers, 1980a,b). These NDE are (a) radiographic examination, (b) ultrasonic examination, (c) liquid penetrant examination, (d) magnetic particle examination, (e) eddy current examination, (f) visual examination, (g) leak testing, (h) hydrostatic testing, and (i) pneumatic testing.

These point out surface and internal faults in plates and tubes, castings, forgings, and welding before, during, and after the fabrication in the workshop and during the maintenance shutdown at the user's site.

11.16 What are the possible defects that NDE try to uncover?

In plates, tubes, castings, and forgings, the NDE uncover the surface and subsurface defects, discontinuities, changes in thickness, scratches, nicks, etc. In welded joints, the NDE uncover slag entrapment, porosity, cracks, lack of penetration, undercutting, splatter, icicles, etc. The various types of NDE supplement each other. They do not necessarily substitute for one another.

11.17 Why and how are the persons performing NDE qualified?

Since NDE is a very important activity of the fabricator and helps in locating the faults as detailed above, it should be done by properly qualified and trained persons. They are qualified according to the procedure outlined in "Recommended Practice for Nondestructive Examining Personnel Qualification and Certification," published by the American Society for Nondestructive Testing (see References). The responsibility for training and qualification of personnel lies with the manufacturer, as is the case with the welder qualification too (Chapter 10).

There are three levels of certification of testing personnel. Level I technicians are the lowest and can perform only under supervision of level II or level III technicians. Level II can train level I and perform the tests. Level III, the highest, can train levels I and II, and can develop techniques and procedures for NDE.

11.18 What is the difference between inspection and examination?

The ASME Code (American Society of Mechanical Engineers, 1980a) defines inspection to be the functions performed by an authorized inspector (Q. 11.11), whereas examination applies to those quality control functions that are performed by the personnel employed by the manufacturer.

11.19 What are the important steps in NDE?

The ASME Code (American Society of Mechanical Engineers, 1980a) lists the steps for each NDE. While some variations occur due to the nature of the particular NDE, the important steps can broadly be classified as the following:

1. Scope of the particular NDE
2. General requirements (e.g., surface preparation, weld preparation, etc.)
3. Equipment requirements
4. Equipment calibration
5. Written procedure of examination
6. Interpretation and evaluation of results
7. Records of NDE
8. Report on NDE

Now we shall discuss each of the NDE in some detail.

Radiographic Examination

11.20 What is radiographic examination?

It is an NDE where radiation is used to examine the welds in metals.

11.21 Under what conditions is it required?

Among the important conditions requiring radiographic examination are the following:

1. Vessels designed to contain lethal or radioactive substances
2. Vessels operating at high pressure
3. Vessels made of thick plates, e.g., $1\frac{1}{2}$ in (38 mm) or above
4. Vessels where special welding method is used, such as electro-slag welding
5. Vessels in which welded joint efficiency is sought to be increased so that their thickness is reduced (Q. 10.117). For details of other situations as required by the Code, see Chuse (1977).

11.22 What radiations are used in the radiographic examinations? What are their properties?

Primarily, X-rays and gamma-rays are used in the radiographic examinations. These are part of the electromagnetic wave spectrum and span the wavelengths from 10^{-7} to 10^{-12} cm and smaller. These have the following properties:

1. Penetration: The radiations can penetrate through all materials.
2. Absorption: The radiations are absorbed differently by materials of different densities or thicknesses.
3. Transmission: All radiation that is not absorbed is transmitted. The transmitted radiations consist of primary radiation and some scatter that cannot be totally avoided
4. Chemical effects: The radiation produces a latent image on a photographic film, which on development gives a record of the inside of the material.
5. Rectilinear propagation: The radiation travels in a straight line between absorbing and scattering events.

11.23 How are the radiations produced?

Bombarding an anode with electrons produces X-ray radiations. The speed of the electron beam increases with the voltage of the cathode tube producing them. The X-rays released by the high-speed electrons striking the anode have greater penetration ability. The intensity of X-rays increases as the number of electrons striking an anode increases.

The gamma-rays are produced from radioactive isotopes, such as cobalt-60 (^{60}Co), cesium-136 (^{136}Cs), thulium-170 (^{170}Tm), iridium-192 (^{192}Ir), and radium-226 (^{226}Ra). Of these, the ^{60}Co radiations can penetrate the deepest, up to approximately $7\frac{1}{2}$ in (190 mm) of steel.

11.24 How does one choose between the two types of radiations?

The choice depends upon the thickness of the material, sharpness of the result desired, portability of the equipment, etc. The gamma-rays can penetrate deeper than the X-rays, and the equipment is portable although the image is not as sharp as produced by the X-rays. Besides, the gamma-ray source always emits radiations and hence has to be shielded even when not in use. On the other hand, the X-ray source emits radiations only when bombarded with high-speed electrons. Hence, the X-ray source needs no shielding when not in use. Furthermore, the energy level of X-rays can be adjusted to suit the material thickness, while no adjustment is possible in gamma-rays emitted by an isotope.

11.25 What is the apparatus required?

The important pieces of apparatus required are the following:

1. Source of radiations for X-rays or gamma-rays as the case may be
2. Control console to control the intensity of the radiations and time of exposure
3. Film to obtain the image, film holder, film processing, and viewing equipment
4. Image quality indicators (I.Q.I.) or penetrameters.

11.26 What are image quality indicators?

Image quality indicators (I.Q.I.) as the name suggests, provide a means for checking the quality of the radiographic image obtained. These help quantify the quality of the image. These are also called "penetrameters" and are of several kinds. These could either be made of thin metal sheets of known thickness with holes of specified diameters (Fig. 11.2), or of wires of specified diameters. The density of the material should be similar to that of the material being radiographed.

11.27 How does radiography work?

The radiation source is placed on one side while the film is placed in a holder on the other side of the weld to be examined (Figs. 11.3 and 11.4). Penetrameter is generally placed on the source side. For tests of welds in pipes of diameter $3\frac{1}{2}$ in (89 mm) or less, the radiations may pass through two walls (Fig. 11.5). After exposure to the radiations transmitted through the weld, the film is developed, and the image so obtained is examined for faults.

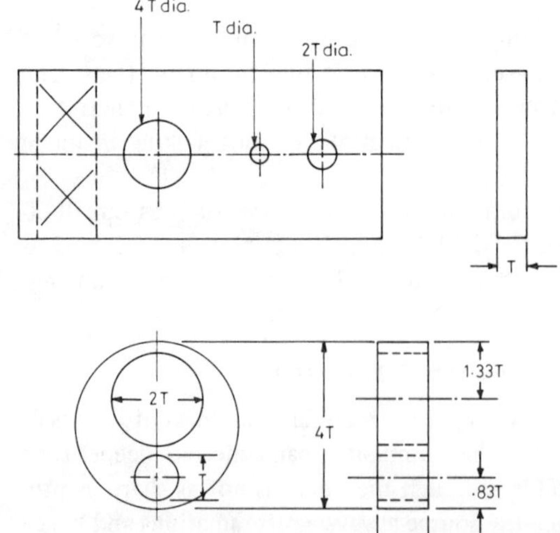

Figure 11.2 Penetrameters. *(From ASME, 1980b.)*

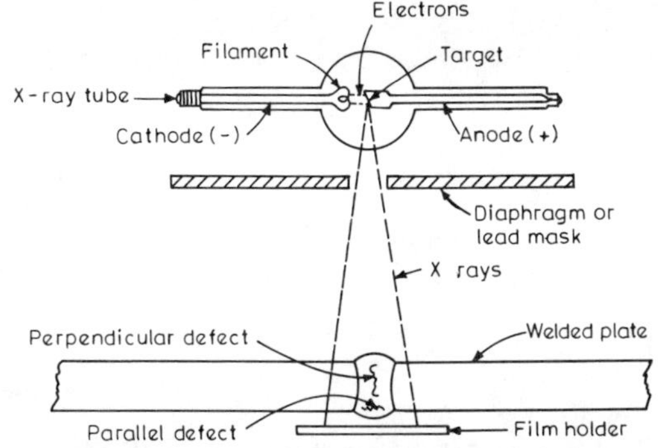

Figure 11.3 X-ray radiography. *(From Chuse, 1977.)*

Filmless radiography (where instead of a film, the image falls on a fluorescent screen) may be used to view a defect immediately since no film processing is required (George, 1981). However, the image is of low quality, and it is not an accepted technique in the inspection of welds where records are required to be maintained.

11.28 What factors affect radiography?

Among the important factors are the following:

1. Surface preparation of the welds
2. Intensity of the radiation source depending upon the material and its thickness. Higher intensity radiations (shorter wave length) penetrate deeper.
3. Direction of radiation (perpendicular or angular)
4. Selection of right density film
5. Exposure time
6. Shielding back scatter by filters
7. Development of film
8. Viewing and interpreting the radiograph
9. Location markers on the weld so that the radiograph can be related to the exact locations

11.29 What is meant by radiographic sensitivity?

It is the capacity of a radiograph to show the discontinuities. It depends upon the size and energy of the radiation source and the quality of the film. In a properly

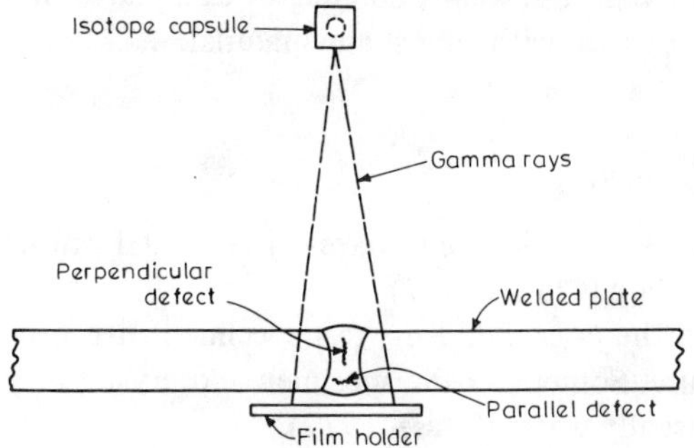

Figure 11.4 Gamma-ray radiography. *(From Chuse, 1977.)*

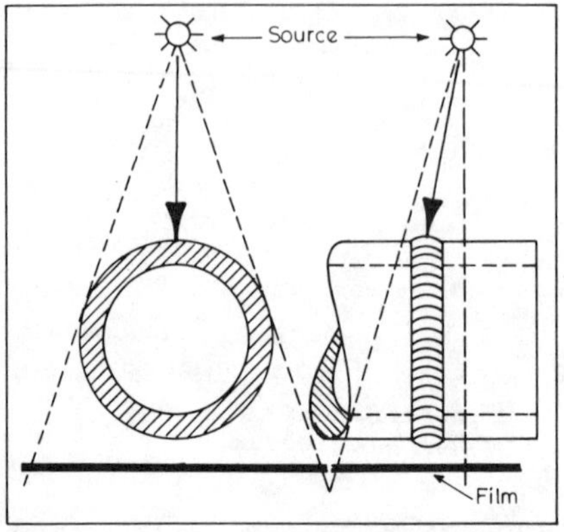

Figure 11.5 Double-wall radiography. *(From ASME, 1980a.)*

aligned testing system, a good radiograph should be able to show defects whose dimensions are as low as 2 percent of the plate thickness.

11.30 How do the various defects show up on a radiograph?

Slag entrapment, porosity, cracks, lack of penetration, and undercutting show up as darker areas compared to the normal density, since more radiations pass through such defective zones. On the other hand, weld reinforcements, welding backing strips, splatter, and icicles show up as lighter areas since there is more metal than in the average plate and hence less radiations pass through such defective zones.

11.31 Are all defects shown on the radiograph unacceptable?

No. Some defects, such as slag inclusions and porosity, are allowed by certain codes within limits,whereas defects such as cracks and lack of penetration are not allowed at all. On the whole, it depends upon the type of defect, its size and extent or density, plate thickness, whether full or spot radiography is done, etc.

Ultrasonic Examination

11.32 What is ultrasonic examination?

It is an NDE conducted using very high-frequency sound waves to detect faults in plates, bars, tubular products, forgings, castings, welds, bonding of clad plates, and to measure the thickness of plates. It can be used with almost any material.

11.33 What is the apparatus required?

The apparatus consists of the following:

1. A piezoelectric transducer crystal to give the ultrasonic waves. The crystal can be made of quartz, ceramic, or lithium sulfate (Chuse, 1977).
2. A couplant to fill the space between the crystal and the test specimen. It can be glycerine, oil, or cellulose gum mixture. Sometimes the specimen and crystal may be completely immersed in a liquid, usually water (Chuse, 1977).

3. A receiving transducer for the ultrasonic waves after they pass through the specimen. It usually has a cathode ray tube (CRT) system associated with it to display the results.

11.34 How does it work?

All substances offer acoustic impedence. The higher the impedence, the less readily do the sound waves pass through it. Metals and liquids have lower impedence than air. When the sound waves traveling through a substance strike a boundary with higher acoustic impedence, such as air, they are reflected back. This interface with air can occur at the back boundary and also at a defect in the specimen if one exists. The reflected wave is displayed on a CRT to determine the location and extent of the defect.

11.35 How is the test carried out?

The piezoelectric transducer crystal is coupled to the test specimen by a couplant, and a variable voltage is applied to it. The crystal expands and contracts, inducing high-frequency mechanical vibrations in the specimen. These vibrations in turn, create ultrasonic waves in the system, which are detected by a receiver. Any defect in the system is also picked up. The crystal and the receiver can be mounted in the following ways:

1. Through transmission: The crystal and the receiver are at the two opposite boundaries of the test specimen (Fig. 11.6). Any defect in the specimen results in a decrease in the amplitude of the signal at the receiver end. This technique requires access to both sides of the specimen.
2. Pulsed echo: The crystal and the receiver are on the same side of the specimen. A pulse sent by the crystal is shown as a pike on the left side (Gumm and Turner,

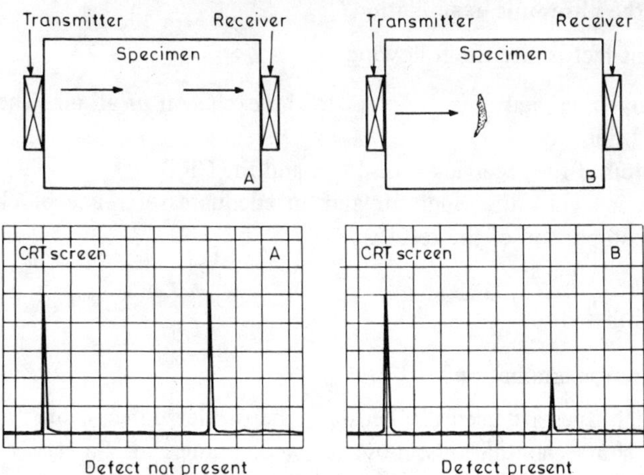

Figure 11.6 Through transmission of ultrasound. *(From Gumm and Turner, 1976. Copyright © 1976 McGraw-Hill, Inc. New York, NY. By permission of the publisher.)*

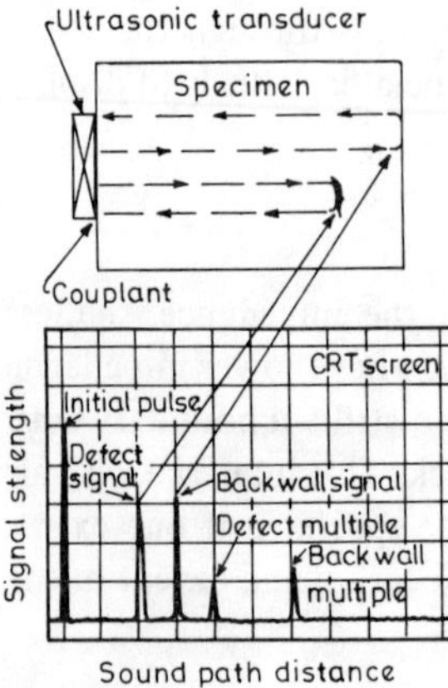

Figure 11.7 Pulsed-echo transmission of ultrasound. *(From Gumm and Turner, 1976. Copyright © 1976 McGraw-Hill, Inc. New York, NY. By permission of the publisher.)*

1976) (Fig. 11.7). The reflected waves from the other side and from the defect are shown on the right side of the CRT. The relative location of the defect can be judged from the initial pulse peak and the one reflected from the rear boundary. The extent of the defect can be judged by the amplitude of the reflection from the defect.

3. Angle beam: Sound waves are transmitted at a predetermined refracted angle, usually 45°, 60°, or 70° from the perpendicular to the surface of the specimen. The transducer crystal is mounted on a wedge of the desired angle (Fig. 11.8). The sound waves are reflected back to the transducer by a defect in the specimen or by a boundary. This technique is mostly used for welds. Another set-up for the ultrasonic weld examination is shown in Fig. 11.9.
4. Pencil beam: These are used to approach areas not otherwise possible to reach by the above-mentioned methods.

11.36 What factors affect the ultrasonic examination?

Among the important factors are the following:

1. Contact between the quartz crystal and the plate to the exclusion of all air. The surface should be ground smooth.
2. Accuracy of the calibration of the receiving transducer and the CRT.
3. Prior training and experience of the operator and an adequate reference block without defects for the comparison of the results.

Liquid Penetrant Examination

11.37 What is liquid penetrant examination?

It is an NDE in which dyes are used to check for the surface defects, such as cracks, seams, laps, cold shuts, laminations, porosity, etc. (American Society of Mechanical Engineers, 1980a).

11.38 What materials are used in this examination?

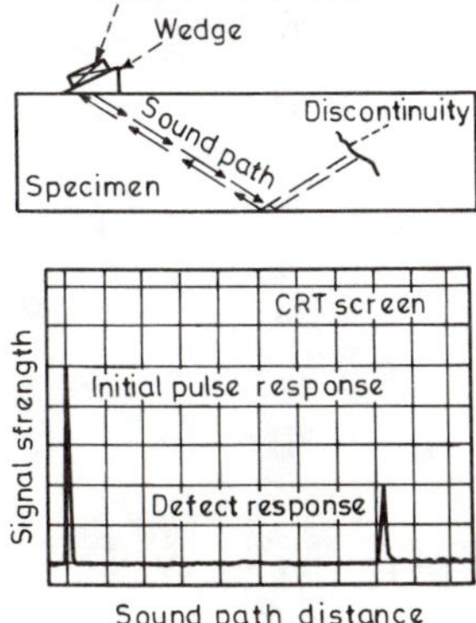

Figure 11.8 Angle-beam transmission of ultrasound. *(From Gumm and Turner, 1976. Copyright © 1976 McGraw-Hill, Inc. New York, NY. By permission of the publisher.)*

Dyes are used to show the defects. These can be color dyes or fluorescent ones. Besides, these dyes can be water-washable, post-emulsifying, and solvent-soluble. The first two are used primarily in the fabrication shop, while the third one is generally used in the field.

Developers are used to draw the dye out of the cracks. The developers can be dry, nonaqueous, or aqueous.

11.39 How does one decide which type of dye to use?

For most sensitive examinations using this method, the post-emulsifying dyes are used. For general work, water-soluble ones are used. For field work, solvent-soluble dyes are used because the system is easily portable.

11.40 How does this method work?

The test specimen is precleaned, degreased, and dried. A suitable penetrant dye is applied on this test specimen. It penetrates into the surface defects (Fig. 11.10*a*). After a certain time, called the dwell time, the excess dye is removed and the surface

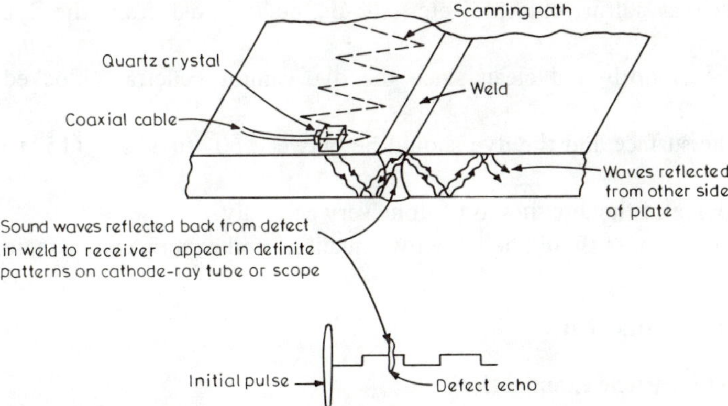

Figure 11.9 Ultrasonic weld examination. *(From Chuse, 1977.)*

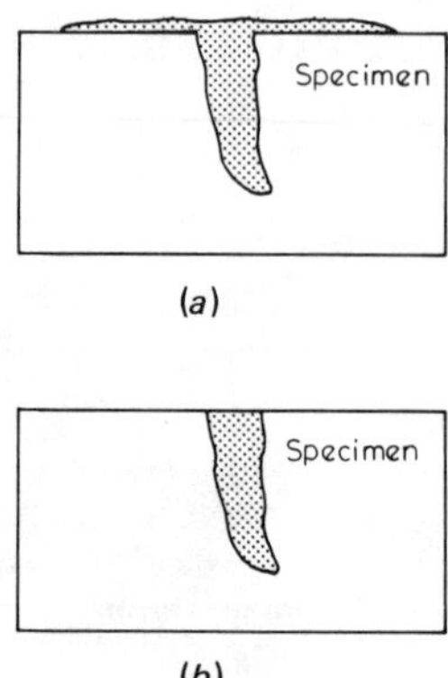

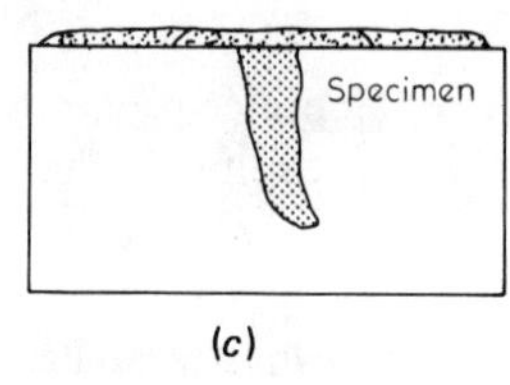

Figure 11.10 Liquid penetrant examination. (*a*) Dye enters crack open to surface. (*b*) Excess dye is removed from surface. (*c*) Developer draws the dye from crack, to present discontinuity indication. *(From Gumm and Turner, 1976. Copyright © 1976 McGraw-Hill, Inc. New York, NY. By permission of the publisher.)*

is allowed to dry (Fig. 11.10*b*). Then the dye is developed by appropriate chemicals to reveal the cracks (Fig. 11.10*c*). Developer is a light-colored powder that is applied on the surface. It draws out a part of the dye that has seeped into the crack. This dye diffuses into the developer powder as it comes out, thus giving indications that are much wider than the surface defects from which the dye is drawn out. These are then photographed, and the results are interpreted. If a fluorescent dye is used, it is photographed under ultraviolet light, since ordinary light does not reveal the fluorescent dye.

11.41 What are the essential requirements for good results?

These are the following:

1. Contaminents, such as sulfur, chloride, etc., should be removed from the dye before applying it.
2. Surface should be smooth and clean since the dye cannot penetrate blocked defects.
3. Temperature of the surface and the dye should be between 60° to 125°F (15° to 52°C).
4. Drying and developing of the dye should be done very carefully.
5. Interpretation of the results should be done by a qualified and experienced person.

Magnetic Particle Examination

11.42 What is magnetic particle examination?

It is an NDE wherein particles, obtained by grinding ferromagnetic materials, are used to detect surface and subsurface defects that are not too deep in the ferromagnetic materials. It is also used to monitor the weld quality as each weld-pass is made.

11.43 What is the apparatus required?

Basically, one needs a magnetizing coil and the magnetic particles or fine powder.

11.44 How does it work?

The part to be tested is magnetized either by passing current directly into it, thus creating a circumferential magnetic field (Fig. 11.11), or by a field induction using an electric coil, which sets up a longitudinal magnetic field (Fig. 11.12). A crack or defect interrupts the continuous magnetic lines of force and creates magnetic poles on either side of it. These poles attract the magnetic particles that are applied on the surface. The magnetic field should be perpendicular to the defect for best results, since a defect parallel to the field gives no indication. Hence, the examination should be done in two directions at 90° to one another by magnetizing the specimen twice, to detect all the defects.

After the test is over, the specimen may be demagnetized using alternating magnetic field whose strength is gradually lowered.

11.45 Are both the surface and subsurface defects detected in the same test?

No. For the better detection of surface defects an ac magnetic field and wet particles are used, while for the better detection of subsurface defects a dc magnetic field and dry particles are used.

11.46 What are the essential requirements for good results using the magnetic particle examination?

These are the following:

1. Surface preparation to clear all oil, grease, rust, scale, dirt, etc.
2. An appropriate current density for the magnetization of the test piece
3. Direction of magnetization
4. Evaluation of results

A number of good photographs of this test are shown in the ASME Code, Section V (American Society of Mechanical Engineers, 1980a).

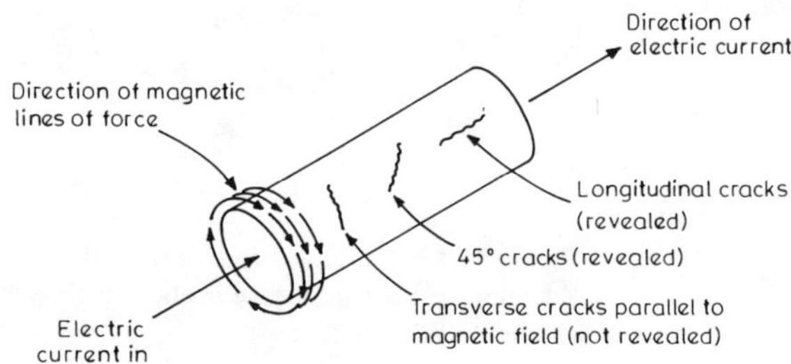

Figure 11.11 Circumferential magnetic field. *(From Chuse, 1977.)*

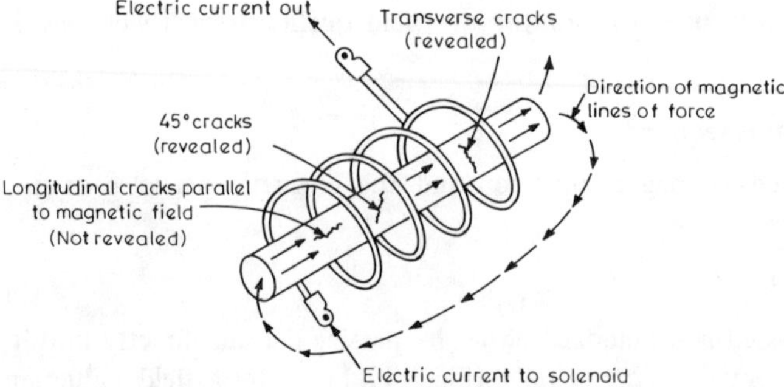

Figure 11.12 Longitudinal magnetic field. *(From Chuse, 1977.)*

Eddy Current Examination

11.47 What is eddy current examination?

It is an NDE where a magnetic field induced by an ac coil or probe is used to detect flaws or variations in thickness of walls of tubes up to 2 in (50.8 mm) diameter and 0.12 in (3 mm) thick.

11.48 What is the apparatus required?

The essential pieces of apparatus are the following:

1. An ac induction coil (Fig. 11.13), either circular doughnut type or surface probe type
2. Power supply
3. Indicator to monitor response

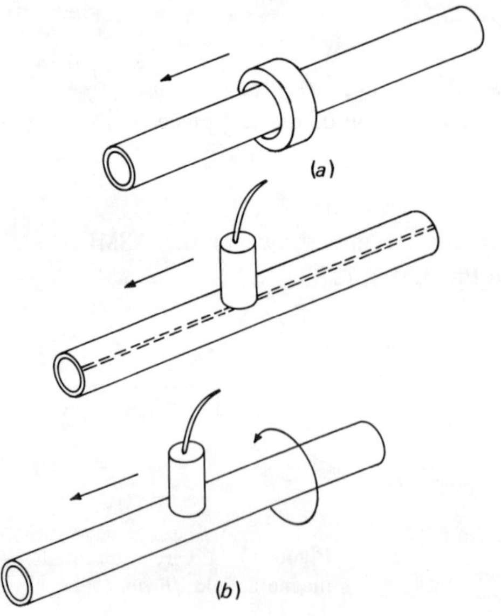

Figure 11.13 Induction coil. (*a*) Circular doughnut type. (*b*) Surface probe type. *(From ASME, 1980a.)*

11.49 How does it work?

The ac coil or probe is moved along the surface of the tube to be tested. This induces eddy currents in the tube that result in a secondary magnetic field opposing the inducing magnetic field of the ac coil. The eddy currents as well as the impedence of the inducing coil change due to flaws or change in composition or thickness of the tube wall. This is a measure of the defect and shows up on a CRT and can also be permanently recorded if required.

11.50 What are the essential requirements for good results?

These are the following:

1. An ac coil or probe with frequency between 1 kHz and 1 MHz (Gumm and Turner, 1976) for deeper penetration of the eddy currents
2. Material properties, such as electrical conductivity and magnetic permeability, should be low. The depth of penetration of the eddy currents decreases as the electrical conductivity and magnetic permeability of the test material rise. A dc magnetic field applied to the specimen compensates for the magnetic permeability.
3. Appropriate speed of movement of the coil
4. Accurate calibration of the detecting equipment
5. Evaluation of results by a qualified person

Visual Examination

11.51 When is visual examination done?

It is done to determine the surface condition and shape of plates, tubes, fabricated parts, etc., to check the alignment of mating surfaces or to look for the evidence of leaks when under leak testing (Q. 11.53–11.56) and in hydrostatic and pneumatic tests (Q. 11.60–11.66) (American Society of Mechanical Engineers, 1980a).

11.52 How is it done?

It is done by viewing the surface or area of interest under appropriate lighting conditions. The examination may be done directly when the surface is approachable within 24 in (610 mm) or remotely by using borescopes, mirrors, telescopes, fiber optics, cameras, etc. Article 28 of the ASME Code (American Society of Mechanical Engineers, 1980a) gives details of classifying visual defects and a host of photographs showing many such defects.

Leak Testing

11.53 What is leak testing?

It is a test to determine the integrity of various joints, such as tube-to-tubesheet, nozzle-to-shell, and flange-to-shell, by finding out if any leaks exist in these joints.

11.54 What are the different methods available for leak testing? What is the apparatus required in each case?

There are at least five different methods for leak testing (American Society of Mechanical Engineers, 1980a). These along with the apparatus and/or material required are listed below, while the test procedures are discussed in the next question.

1. Gas and bubble formation testing: Gas (air, nitrogen, or helium) and bubble-forming solution, pressure indicator
2. Vacuum box testing: Vacuum box, vacuum-creating equipment, vacuum indicator
3. Halogen diode detector testing (sniffer method): Trace gas (refrigerants R-11, 12, 21, 22 or 114; or methylene chloride), detector probe and indicating equipment, pressure indicator
4. Helium mass spectrometer testing (reverse probe-sniffer method): helium gas, portable mass spectrometer, sampling probe
5. Helium mass spectrometer testing (hood method): Helium gas, portable mass spectrometer

11.55 How is the leak testing done?

The procedure varies with the method used (American Society of Mechanical Engineers, 1980a):

1. Gas and bubble-formation testing: The test gas is filled in the part or equipment with all known openings sealed. The pressure is held for a minimum of 15 min. Then the bubble-forming solution is applied by flowing over the surface and visually examined for any bubble formation.
2. Vacuum box testing: A box [say, 6 in (152 mm) by 30 in (762 mm)] with an open bottom is placed on the surface to be tested and is evacuated to seal the open bottom against the surface. A bubble-forming solution is applied to the test surface before the box is placed on it. The leaks are detected by the bubbles forming on the surface. This method is similar to the one above except that instead of creating positive pressure inside the part being tested, a vacuum is created on the outside
3. Halogen diode detector testing (sniffer method): A halogen gas is filled in the part under test. A diode detector is moved over the surface. Any halogen picked up by the probe is ionized by a heated platinum anode and the ions are collected by a cathode. The current indicated by the meter is proportional to the rate of ion formation and hence is indicative of the intensity of the leak produced.
4. Helium mass spectrometer testing (reverse probe-sniffer method): Helium is filled in the test part. A sampling probe connected to the leak detector is moved over the surface. Any helium picked up is detected by the leak detector.
5. Helium mass spectrometer testing (hood method): The part under test is placed in a hood container such as a plastic bag. The space between the hood and the part is filled with helium and its leak rate into the part is determined by the mass spectrometer. Alternately, the helium may be put in the part and its leakage outside into a vacuum encapsulator or bell jar determined.

The helium methods (4) and (5) are very sensitive to minute leakages.

11.56 What are the essential requirements for good results in leak testing?

These are the following:

1. The test surface should be clean and free of oil, grease, paint, or any other items that might mask a leak.
2. All openings should be sealed.
3. All pressure gauges and leak-detecting equipment should be properly calibrated. The calibration should be periodically tested for its accuracy. The equipment should be allowed to warm up before use. Two pressure gauges may be used to ensure that they are functioning correctly.
4. The minimum temperature for leak test should be the same as for the hydrostatic or pneumatic test. The minimum pressure should be 60 psi (414 kPa) or 15 percent of the maximum allowable design pressure, whichever is less.
5. Enough time, called the "soak time" should be allowed as per the ASME Code (American Society of Mechanical Engineers, 1980a) so that the gas used can fill all the parts of the unit under test. Only after that time should the leaks be detected using the proper procedure for each test.
6. The scanning rate for the detector probe in the halogen and helium test methods should be determined by a capillary leak standard so that no leaks are missed by a faster-than-necessary scanning rate.
7. In the above tests, no air draft should be present, or it would blow off the leaking gas and reduce the response of the probe.
8. Proper procedure, as given in the ASME Code (American Society of Mechanical Engineers, 1980a), should be adhered to.

Tube Inspection

11.57 How is a tube inspected?

Inspection of the douments accompanying a shipment of tubes reveals the details regarding the material, heat treatment, dimensions, etc. The outside surface can be visually inspected for scratches, nicks, laps, or flaws in tube drawing or in U-forming. Dye penetrant testing can also be done for more thorough inspection. The inside surface can be inspected by a borescope for up to about 10 ft (3.048 m) length. The borescope shows exactly the condition of the inside surface. The light intensity, however, becomes very low at greater lengths, and therefore the borescope cannot be used at lengths greater than about 10 ft (3.048 m). Very sensitive techniques for tube wall inspection include ultrasonic and eddy current examinations.

Sometimes destructive testing of the tubes is done as per the ASME Code to confirm that it is seamless and not electric resistance welded (ERW) type.

Tube-to-Tubesheet Joints

11.58 What test are performed to ensure that the tubes-to-tubesheet joints are perfect?

The leak detection methods using halogens and helium can be used for this purpose. The probe is placed in each tube end to check for the cracks or splits in the tube wall (American Society of Mechanical Engineers, 1980a). The test gas is put on the shell side. All tube-to-tubesheet welds may be tested by the encapsulator method. It is a funnel with its wide end on the tubesheet and the small end attached to the probe.

The hydrostatic and pneumatic tests (Q. 11.60–11.66) will also reveal a leak at the tube-to-tubesheet joint.

Even before the tubes are joined to the tubesheet, mock-up tests are conducted to confirm the technology of tube-to-tubesheet joining method.

Relative Costs of NDE

11.59 What are the relative costs of various NDE equipment?

Approximate costs in U.S. dollars for testing equipment as given by Gumm and Turner (1976) in 1976 were the following:

1. Radiographic examination: \$3000 for a gamma-ray unit to \$15,000 for a portable X-ray unit
2. Ultrasonic examination: \$3000 to \$5000 for a simple system. The automated systems cost more.
3. Magnetic particle examination: Starting from \$300 for a very simple device, the system can cost a couple of thousand dollars for a more elaborate set.
4. Eddy current examination: A simple portable system costs \$1000.
5. Leak testing equipment: A simple system costs \$100 to \$300.

The above costs, being old, are out of date but are given here to show the relative costs. It is assumed that inflation and technological progress would affect them identically and hence the relative order of their cost would remain approximately the same.

The NDE discussed thus far are conducted to test the materials used and/or the fabrication process. These are governed by the ASME Code Section V (American Society of Mechanical Engineers, 1980a). There are also tests to check the complete heat exchangers and pressure vessels after they have been fabricated. These are hydrostatic and pneumatic tests and are governed by the ASME Code Section VIII (American Society of Mechanical Engineers, 1980b). These will be discussed now.

Hydrostatic Test

11.60 What is a hydrostatic test?

It is a nondestructive test using water to ascertain whether the equipment will withstand the design pressure at the design temperature without developing any leaks.

11.61 How is it done?

The unit is pressurized using water to 1.5 times the maximum design pressure multiplied by the lowest ratio for the vessel construction materials of the stress value at the test temperature to that at the design temperature. This is the minimum test pressure required by the ASME Code (Chuse, 1977). It varies with the code followed. Cast iron vessels are pressurized to two times the design pressure (Chuse, 1977). Units for vacuum service are pressurized to $1\frac{1}{2}$ times the difference between the normal atmospheric pressure and the minimum design internal absolute pressure. For heat exchangers, the shell and the tube sides are pressurized separately so that any leakage on either side can easily be observed. The pressure is held for approximately 30 min as indicated by a calibrated gauge or pressure–time recorder. Any decrease in pressure is indicative of leakages in the system, which are then located and corrected, and the unit is retested.

The hydrostatic test is critical and is generally conducted in the presence of an authorized inspector.

An emerging testing system that is coupled with hydrostatic testing to detect discontinuities in the vessel and prevent overpressurization and damage is the acoustic-emissions testing (Gumm and Turner, 1976). Piezoelectric sensors on the vessel pick up high-frequency sound waves emitted by such discontinuities and display them on a CRT to the examiner before the vessel is fully pressurized. The test can then either be halted or carried out more carefully.

Pneumatic Test

11.62 What is a pneumatic test?

It is a nondestructive test using air to ascertain whether the equipment will withstand the design pressure at the design temperature without developing any leaks. It may be done in lieu of or after the hydrostatic test has been done.

11.63 When is the pneumatic test done?

It is done when the vessel (a) cannot safely be filled with water to conduct a hydrostatic test; (b) cannot be completely drained and dried of water after a hydrostatic test, and when the traces of water are undesirable for the service the vessel is intended for, such as in some nuclear services; (c) is for critical service and the buyer or inspector wants a pneumatic test after a hydrostatic test has been done to locate very minute leaks since the pneumatic test is a very sensitive test; and (d) is handling gaseous products. This checks the permeability or leakage of the gas which is permitted within limits specified in the codes.

It should be kept in mind that the pneumatic test results in a reduced stress level since it is conducted at a pressure lower than that for the hydrostatic test. Hence, if only pneumatic test is used, while it will detect all the minute leaks too, it will not prove the soundness of the equipment to the same level as the hydrostatic test would have.

11.64 How is it done?

The unit is pressurized using air to 1.25 times the maximum design pressure multiplied by the lowest ratio for the vessel construction materials of the stress value for the test temperature to the stress value for the design temperature, according to the ASME Code Section VIII, Division 1 (American Society of Mechanical Engineers, 1980b). The test pressure reduces to 1.15 times in the ASME Code Section VIII, Division 2.

The pressure is gradually increased to one-half the test pressure. Thereafter, it is increased in steps of approximately one-tenth the test pressure until the required test pressure is reached. Then the pressure is reduced to four-fifths the test pressure and held for sufficient time to permit the leak detection and inspection of the vessel (American Society of Mechanical Engineers, 1980b).

11.65 Why is the pneumatic test more dangerous to perform than the hydrostatic test?

Air compresses a lot more than does water when it is pressurized. Hence, should a vessel under pneumatic test burst, air would expand suddenly to many times the volume of the vessel. This sudden release of vast amounts of energy sends pieces of the failed vessel flying and also produces shock waves through the atmosphere, both of which can do much damage to nearby structures and humans. On the other hand, should a vessel under hydrostatic test burst, water will gush out momentarily and no further damage will result. Hence, the pneumatic test is more dangerous and should be done more carefully.

Regulations require that proper bunds be built around an equipment being pneumatically tested to minimize the damage due to possible failure of the equipment under test. Sometimes, the test is carried out under water; i.e., the whole equipment is submerged in water and then air is pumped into it.

11.66 How are the hydrostatic and pneumatic tests different than the leak tests discussed earlier?

The only common point is that all these tests check for the leaks in the system. The differences are many. The hydrostatic and pneumatic tests are done at 1.5 and 1.25 times the design pressure, respectively (corrected for temperature), while the leak tests are done at 0.15 times the design pressure or 60 psi (414 kPa) whichever is less. The first two are done on a complete section or a complete vessel, while the last one is done on small sections such as tube-to-tubesheet joints. The hydrostatic and the pneumatic tests use water and air, respectively, while the leak test uses the gases mentioned in Q. 11.54. The leaks in the former are indicated by a drop in the pressure in the equipment and by visual inspection, while in the latter a leak detector is used at the place where the leak test is being conducted.

Proof Tests

11.67 What are proof tests?

These are the tests performed for the acceptability of a heat exchanger or pressure vessel. These relate to the following:

1. Strength or hydrostatic test
2. Leak tests

3. Hydraulic tests to check the overall fluid flow system
4. Thermal tests to prove the adequacy of the thermal design

The last two tests cost a significant amount. Hence, their requirement should be carefully decided. In some situations, the exchangers may be deliberately oversized and the thermal tests waived.

Choosing between Various NDE

11.68 How does one decide which NDE to use?

Table 11.2 gives in brief the applications and advantages of various NDE techniques. Multiple tests may be used to supplement one another. For example, a weld may be examined both by the radiographic and the ultrasonic techniques. Relative costs of carrying out a particular test may also influence the choice.

Vendor-Supplied Items

11.69 What kind of quality control or tests are carried out over the large variety of materials and welding electrodes purchased from different vendors?

The quality control procedure at each vendor is certified either by the buyer or by a third party. Usually, the buyer does not carry out any tests himself but goes along with the test results and guarantees given by the certified vendors. Only in case of doubt does the buyer do any tests.

11.70 If a vendor-supplied material does not exactly meet the specifications, what are the options open to the fabricator?

The options open to the fabricator are the following:

1. If the deviation is not too much, it may be used with adequate evaluation of performance and the concurrence of the buyer, inspector, and the quality assurance department.
2. Efforts are made in the shop to work on the material to bring it in line with the specifications.
3. The material may be rejected and returned.

Inspection and Tests during Fabrication

11.71 What inspections and tests are carried out at different stages of fabrication of a shell-and-tube exchanger, starting from the fabrication drawings to the final shipment of the unit?

These are given in Table 11.3. This list may vary slightly from one manufacturer to the other.

Table 11.2 Analysis of nondestructive testing techniques

Applications	Advantages	Remarks
	Radiographic inspection (X-ray)	
For examination of internal soundness of welds, castings, forgings, and plate material (see Fig. 11.3)	1. Gives sharper definition and greater contrast for thickness up to 3 in (76.2 mm) 2. Gives a graphic and permanent record indicating the size and nature of defects 3. Offers established standards of interpretation for guidance 4. Radiation source can be turned off when not in use	1. Protective precautions necessary to protect personnel in surrounding area 2. Trained technicians required to take and interpret the film
	Radiographic inspection (gamma-ray)	
For examination of internal soundness of welds, castings, forgings, and plate material (see Fig. 11.4)	1. More suitable for heavy thickness 2. Portable equipment 3. Source holder permits use through small openings 4. Lower initial cost 5. Requires no cooling mechanism 6. Gives a graphic and permanent record 7. No tube replacement	1. Government license is required for possession and use of isotopes 2. Protective precautions necessary 3. Sensitivity is not as sharp as X-ray, and persons not properly trained in interpreting film may underestimate the seriousness of a defect 4. Energy cannot be adjusted, so isotope must be chosen to meet sensitivity requirements and thickness of material
	Magnetic particle inspection	
For locating surface and subsurface defects that are not too deep (Figs. 11.11 and 11.12) Technique employs either electric coils wound around the part or prods to create a magnetic field. A magnetic powder applied to the surface will show up defects as local magnetic fields. The nature of the defects will be revealed by the way the powder is attracted	1. Useful for the inspection of nozzle and manhole welds for which radiography would be difficult at best and most often impossible[a] 2. Reveals small surface defects that cannot be detected radiographically 3. Useful in weld repair to assure removal of defect before rewelding 4. Can be used to detect laminations at plate edges	1. Use only for magnetic material 2. Not suitable for defects parallel to magnetic field 3. Training needed to evaluate visual indications of defects 4. Prod methods of testing may mar the surface of a machined or smooth part 5. Magnetic particle examination methods (see Code)

(*See footnote on p. 488.*)

Table 11.2 Analysis of nondestructive testing techniques (*Continued*)

Applications	Advantages	Remarks
	Penetrant inspection (dye)	
For locating surface defects (Fig. 11.10). A liquid dye penetrant is applied to a dry clean surface and allowed to soak long enough to penetrate only surface defects. After a time interval of up to 1 hr, the excess penetrant is wiped off or cleaned with a cleaner and a thin coating of a developer is applied. The penetrant entrapped in a defect will be drawn to the surface by the developer, and the defect will be indicated by the contrast between the color of the penetrant and that of the developer	1. Especially applicable for nonmagnetic material 2. Useful for the inspection of nozzle and manhole welds when radiography is impossible 3. Easy to apply, accurate, fast, and low cost	1. Detects only defects that are open to the surface 2. Not practical on rough surfaces 3. Liquid penetrant examination methods (see Code) 4. Take care of applying dye penetrant to ensure that excessive amounts of the vapors are not inhaled
	Penetrant inspection (fluorescent)	
For locating defects that run through the surface (Fig. 11.10). Penetrant is applied by spraying, dripping, or brushing. Excess penetrant is washed from surface with a water spray and the surface allowed to dry. Dry powder or water-suspension developer is applied to part to draw penetrant to surface. Penetrant glows under black (ultraviolet) light. For leak test, penetrant is applied to one side, and the other side is examined under black light for indications of glow	1. Easy to perform 2. Defects show clearly under black light 3. Especially applicable for nonmagnetic material 4. Can be used on rough surfaces 5. Detects porosity and defects that would otherwise be hard to find	1. Detects only defects that are open to the surface 2. Black light requires a source of electricity 3. Not very effective as leak test for plates more than $\frac{1}{4}$ in (6.4 mm) thick
	Ultrasonic inspection	
For detecting defects in welds and plate material and for gaging thickness of plate (Figs. 11.6–11.9). High-	1. Portable equipment 2. Access from one side of part being tested is possible	1. Training required for interpretation of visual indications of defects

(*See footnote on p. 488.*)

Table 11.2 Analysis of nondestructive testing techniques (*Continued*)

Applications	Advantages	Remarks
	Ultrasonic inspection (*Continued*)	
frequency sound impulses are transmitted through a search unit, usually a quartz crystal. This search unit is held in close contact with the part being tested, using an intermediary such as oil or glycerine to exclude air. The sound waves pass through the part being tested and are reflected from the opposite side or from a defect The time of travel shows the defect on a CRT or scope. This scope can also measure depth of crack or defect. Although the impulse instrument used to detect defects will also give thickness measurements, the resonance instrument is actually an electronic micrometer designed to measure thickness from one side of material within tolerances of ±0.002 in (0.051 mm)	3. Reveals small root cracks and defects not indicated by radiographic film, especially in thick-walled vessels 4. Thickness measurements are rapidly made 5. Good for detection of laminated plates	2. Rough surfaces must be made smooth if crystal is to make contact with part 3. Photographs must be taken to provide permanent records 4. Ultrasonic test methods when required (see Code) 5. Not very effective on welds with backing rings
	Eddy current testing	
Eddy current testing is based upon the principles involving circulating currents into an electrically conductive article and observing the interaction between the article and the currents (Fig. 11.13)	1. Detects small discontinuities 2. High-speed testing 3. Accurate measurement of conductivity 4. Checks variation in wall thickness 5. Noncontacting	1. Limited to use with conductive materials or conductive base materials 2. Depth of penetration restricts testing to depths of less than $\frac{1}{4}$ in (6.4 mm) in most cases 3. The presence of strong magnetic fields will cause erroneous readings 4. Testing of ferromagnetic metals is sometimes difficult 5. Training required for interpretations of visual defects and indications

Source: Chuse (1977). Used with permission of McGraw-Hill Book Co.
[a]American Society of Mechanical Engineers (1980b).

Table 11.3 Inspection and tests during fabrication of a heat exchanger

Inspection and test of	How done	By whom	Remarks
1. Validity of Code Stamps	From the documents provided by the fabricator	Authorized inspector	QC manager may be present when the authorized inspector is carrying out inspection
2. Welder qualification			
3. Fabrication drawings			
4. Plates, tubes, parts, welding electrodes	From the documents provided by the vendors or, in case of doubt, by one or more of NDE discussed earlier	QC personnel of the fabricator	Similarly, the authorized inspector may be present when the QC personnel are carrying out checks, or he may ask for their records. He may also check records about the suitability of various welding machines and the calibration of the instruments used in testing
5. Welding of shells, heads, nozzles, etc.	Radiographic and/or ultrasonic examination		
6. Tubesheet and baffles' drilling	By measurement		
7. Tube-to-tubesheet joints	Leak test		
8. Final assembled unit			
(a) Dimension check	By measurement	Authorized inspector	
(b) Test for leaks	Hydrostatic and/or pneumatic examination		

- The NDE discussed above are universally used and form standard practices at most fabricators. While these are being further improved, newer techniques are also being evolved for more efficient, fast, and cost-effective detection of the faults. The acoustic emissions testing has already been mentioned in Q. 11.61. Another new technique is called "positron annihilation" (Hughes, 1981). Positrons, emitted by radioactive isotopes of sodium (^{22}Na) or germanium (^{68}Ge), reach a dislocation or void in a material and are annihilated by an electron. In the process, both the positron and the electron disappear by emitting two gamm-rays of 0.511 MeV (8.19×10^{-14} joule). These gamma-rays are detected using germanium detectors, and the information about the position and the type of the defect is obtained from the energy spectrum of the gamma-rays. These defects can forewarn about fatigue and creep damage long before the start and growth of a fatigue crack and long before any other method can detect them.

Yet another new technique is called N-ray method and uses neutrons; it can be used for the nondestructive testing of nonmetallic components where X-rays or gamma-rays cannot be used (Anonymous, 1983).

NOMENCLATURE

CRT cathode ray tube
IQI image quality indicators
kHz kilohertz
MeV million electron volts
MHz megahertz
NDE nondestructive examination
QC quality control

REFERENCES

American Society for Nondestructive Testing. Publication No. SNT-TC-1A. ASNT: Columbus, Ohio.

American Society of Mechanical Engineers. *ASME Boiler and Pressure Vessel Code, Section V: Nondestructive Examination.* ASME: New York. 1980a. A recommended reading for learning the Code requirements for various NDE. It gives details of the equipment required, test procedure, interpretation of results, glossary of terms, etc. In some cases, photos of defects are shown too.

American Society of Mechanical Engineers. *ASME Boiler and Pressure Vessel Code, Section VIII, Division 1: Pressure Vessels.* ASME: New York. 1980b. Gives Code requirements for hydrostatic and pneumatic testing.

Anonymous. Neutron Technique Reveals Hidden Equipment Flaws. *Chem. Eng.* 90(10):35. 1983.

Chuse, R. *Pressure Vessels,* 5th ed. McGraw-Hill: New York. 1977.

George, J. T. Filmless Industrial Radiography for Non-Destructive Testing. *Electronics For You*, 13(2):57–59. February 1981.

Gumm, W. G., and J. E. Turner. The Nondestructive Testing Spectrum. *Chem. Eng.* 83(17):64–78. 1976. A recommended reading for salient features of NDE.

Hughes, A. E. Positron Annihilation: Nondestructive Tool. *Corros. Maint.* 4(1):57–61. 1981.

Yamada, K., and Y. Minamimoto. *Fabrication and Quality Assurance of Heavy Wall Pressure Vessels,* HZTR PVP-6. Hitachi Zosen Technical Report: Osaka, Japan. 1976.

CHAPTER

TWELVE

HIGH–PRESSURE VESSELS

INTRODUCTION

• A heat exchanger is also a pressure vessel. Hence, most of the questions on shell-and-tube heat exchangers in the remainder of this book that do not deal directly with heat transfer and pressure drop apply equally well to pressure vessels designed according to the codes. These vessels are generally restricted to a pressure of 3000 psi (20.7 MPa) if constructed in accordance with the ASME Code Section VIII, Division 1. The limit on pressure could be different with other national codes.

There are, however, many situations in the process and power industries where higher pressures are used. These require special vessels and, to design them, the rules given in the codes have to be supplemented. In this chapter, we shall briefly discuss the salient features of such vessels, covering both the monowall and the multilayer types.

DEFINITION AND USES

12.1 What vessels are called high-pressure vessels?

The ASME Code Section VIII, Division 1 rules are applicable to vessels designed for pressure up to 3000 psig (20.7 MPa). Vessels designed for pressures greater than that are called the high-pressure vessels (see also Q. 12.19). The pressure limits on other national codes could well be different than the one above.

12.2 Where are they used?

Some of the uses of the high-pressure vessels are the following (Rauschenplat, 1981):

1. Urea reactors
2. Ammonia converters
3. Hydrogen, helium, nitrogen storage
4. Polyethylene reactors
5. Hydrocracking reactors
6. Methanol converters
7. High-pressure separators
8. Packages for spent nuclear fuel
9. Absorbers
10. Heat exchangers, condensers, evaporators, etc.

12.3 How does the design of the high-pressure vessels differ from that of the other vessels?

High-pressure vessels have thick walls. The stresses at the inner edge of the cylinder are higher than those at the outer edge (Fig. 12.1), and these should be taken into account in the stress analysis (Harvey, 1974). On the other hand, for a low-pressure vessel, one can safely assume average stress distribution over the wall since it is relatively thin (Fig. 12.2).

The fabrication process for the high-pressure vessels is more complicated and depends on the type of vessel.

TYPES OF HIGH-PRESSURE VESSELS

12.4 What are the different types of high-pressure vessels?

There are basically two types: monowall vessels made out of one thick plate or forged or cast and multilayer vessels made of multiple layers using thin plates. These can be made in three different ways: (1) multilayer vessels made by shrink fitting one shell over the other; (2) multilayer vessels made by weld-shrinkage of one shell over the other; (3) multilayer vessels made by spirally wrapping a continuous sheet over an inner cylinder, then finally covering it with an outer shell to obtain the required thickness.

12.5 What are the advantages of a monowall vessel?

Its construction is less complicated than that of a multilayer vessel. It can be prestressed to reduce maximum tensile stress concentration at the inner edges and thus distribute the stresses more uniformly, resulting in a reduced thickness (Brownell and Young, 1959).

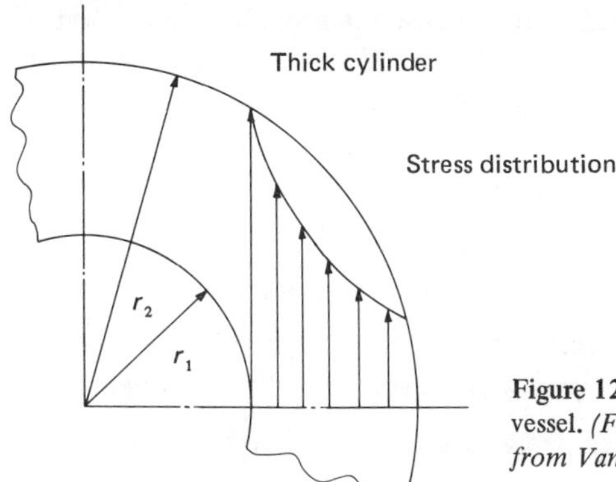

Figure 12.1 Stress distribution in thick-wall cylindrical vessel. *(From Harvey, 1974. Modified with permission from Van Nostrand Reinhold.)*

12.6 What are the advantages of a multilayer vessel?

The following are some of the important advantages (Yasuda et al., 1976):

1. Multilayer construction provides a safe design against any brittle fracture since the fracture would be restricted to one layer and would not propagate through the entire thickness as in the case of a monowall vessel.
2. Better quality control is possible since each layer is very thin compared to the total monowall plate thickness. One hundred percent radiographic examination of all the welds is possible.
3. Materials having different corrosion resistance, strength, high and low creep rates, etc., can be used to obtain the composite vessel with the overall desired characteristics.
4. There is no limit to the thickness of the wall that can be obtained. Conventional equipment using standard procedures is used in the fabrication. The limit on the thickness is only due to the handling and transport limitations.
5. A venting system between each layer allows an immediate detection of any defect in the inside layers, long before the defect has propagated through the full thickness.
6. Since each layer is thin, it is possible to make the layer from a material of uniform and stable quality.

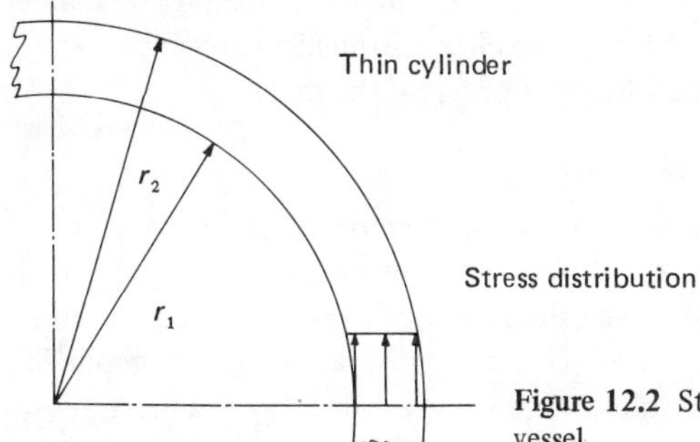

Figure 12.2 Stress distribution in thin-wall cylindrical vessel.

7. Each layer can be made by cold-working and high dimensional accuracy can be obtained.
8. Destructive testing of a monowall and a multilayer shell (of same total thickness) showed that the two had similar pressure versus strain relationship (Rauschenplat, 1981). This shows that the multilayer unit has the same strength, yet is easier to fabricate than a monowall unit.

12.7 What factors influence the choice between a monowall and a multilayer vessel?

There are multiple factors that affect the choice between a monowall and a multilayer vessel. Some of these factors are the following (Brownell and Young, 1959):

1. Operating conditions of pressure and temperature and fluctuations in them
2. Temperature gradients
3. Corrosive nature of the contents and the outer environment
4. Dimensions of the equipment
5. Number and size of openings and closures in the vessel
6. Vertical or horizontal installations
7. Available materials and their physical and mechanical properties and cost
8. Type of construction required: forged, welded, or cast
9. Available fabrication facilities

FABRICATION

12.8 How are the monowall vessels made?

These are made in the usual fashion as a thin-wall vessel is made. Stricter quality control is essential. These vessels can be cold-formed or hot-formed. Figure 12.3 shows one sequence of operations. It can be varied depending upon the requirements.

12.9 What are the principal differences between cold-forming and hot-forming?

Cold-forming means bending the plates at room temperature, while hot-forming means bending the plates at an elevated temperature of approximately 900°C (1650°F). The plate thickness that can be hot-formed in a given set of rollers is usually twice that which can be cold-formed. Thus, thicker vessels can be made by hot-forming. However, residual stresses are left in the shell that require postweld heat treatment at a substantial cost. The heat treatment changes the mechanical properties of the material, and it should be kept in mind when ordering the material.

12.10 How are the monowall vessels prestressed?

It is done by hydrostatically stressing the vessel beyond its yield point so that some plastic deformation occurs (Harvey, 1974). Then the overstressing pressure is released. This results in residual compressive stress in the inner and residual tensile stress in the outer portion of the cylinder. When, during its use, pressure is applied in this prestressed cylinder, the stress distribution is more even; the evenness allows

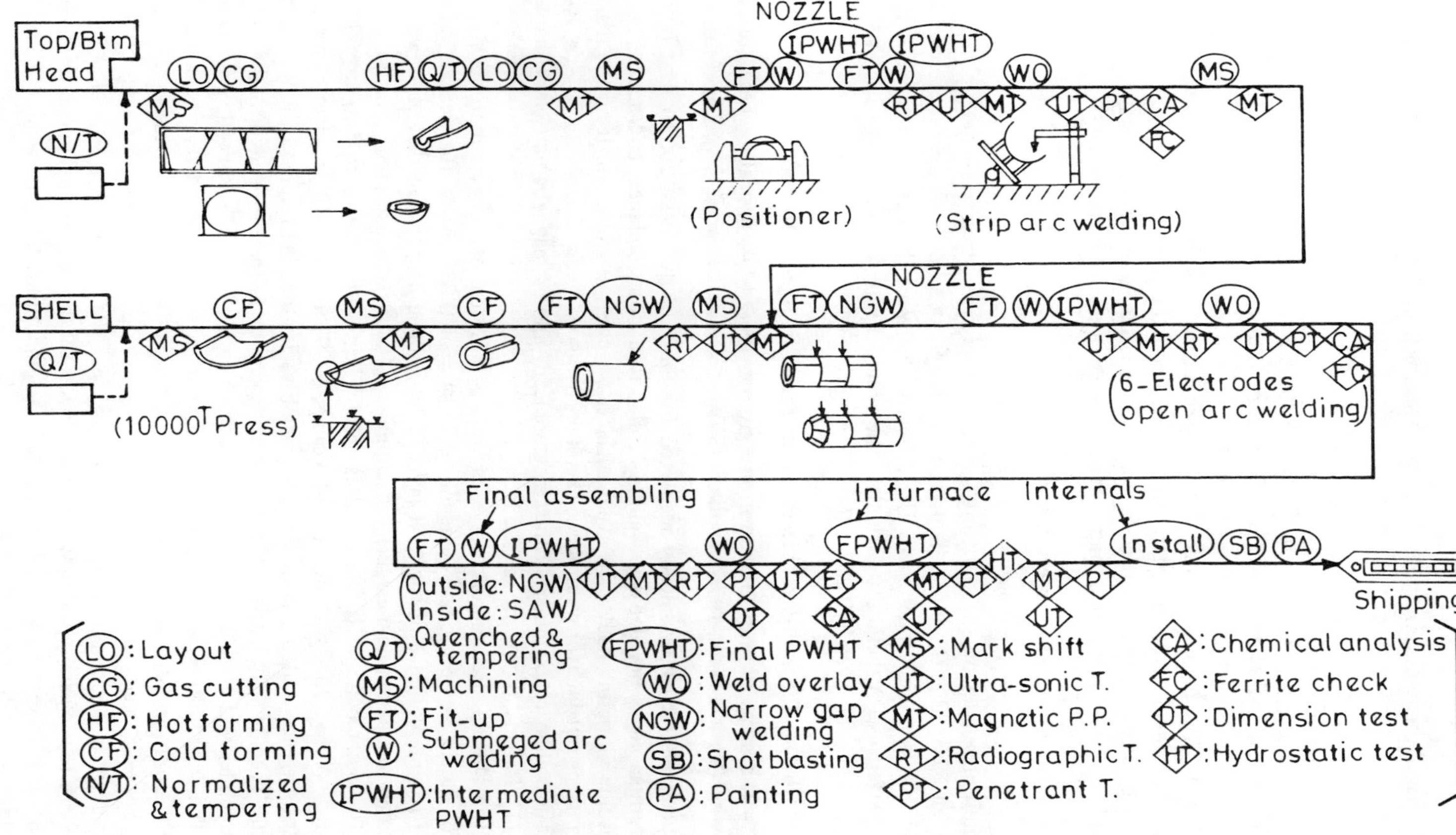

Figure 12.3 A fabrication sequence of heavy monowall pressure vessel. *(From Fujii, 1978.)*

reduced wall thickness. This method of prestressing is known as autofrettage (Brownell and Young, 1959). Another method, thermal prestressing, is not generally used.

12.11 How are the multilayer vessels designed?

These are designed as per the ASME Code Section VIII, Division 1 or Division 2 or other national codes. Division 2 is preferred since it results in a relatively thinner shell (Chapter 8). The thickness of the individual layers varies from 0.5 to 2.5 in (12 to 63 mm). Any material allowed by the code can be used. The outside circumference of the individual layers is determined based upon the plate thickness, interference desired between two consecutive layers, and the anticipated weld shrinkage. There is virtually no limit on the final thickness of the vessel that can be obtained.

12.12 What is meant by "interference?"

In order to make the stress distribution more even, an inner layer is compressed by the outer layer on shrink-fitting so that compressive stresses are set up. This is achieved by making the inner diameter of an outer shell less than the outer diameter of the corresponding inner shell. The difference between the two diameters is called the interference. A method to calculate the required interference along with a solved example has been given by Brownell and Young (1959).

12.13 How are the multilayer vessels of the shrink-fit type fabricated?

The steps in the fabrication are given in Fig. 12.4. Basically, these are the following (Yasuda et al., 1976; Shanker, 1979):

1. Each plate is ultrasonically inspected for uniform thickness, cut to size, longitudinal edges bevelled for welding, rolled to a close tolerance, and double butt-welded longitudinally. The weld surfaces are contour-ground both on the inside and the outside (for welding see Chapter 10).
2. The welds are examined both radiographically and by magnetic particles (Chapter 11). Defects, if any, are repaired.
3. The outside circumference of each layer is measured at three or more locations. It should be very close (generally 0.76 mm or 0.030 in) to the desired value.
4. The circumference of the next layer is determined based upon that of the earlier layer, the plate thickness, plate stretching, and interference desired.
5. The outer cylinder is fabricated as before, heated (usually up to 500° to 600°F or 260° to 320°C) (Brownell and Young, 1959) to expand it to account for the interference and the clearance for assembly. The heating rate and the hold time are specified for each case. The inner cylinder is then slipped into the outer cylinder and cooled in air or in a furnace so that a good shrink-fit is achieved. The longitudinal welds of the consecutive layers are oriented approximately 90° from each other to minimize concentration of the excessive stresses.
6. The weld seam of the outer cylinder is examined after shrink-fitting by 100 percent radiography and by magnetic particles.
7. Steps (4)-(6) are repeated with further layers till the desired thickness is obtained.
8. Both ends of the shell are machined for circumferential butt-welding of additional shells, heads, and/or flanges.

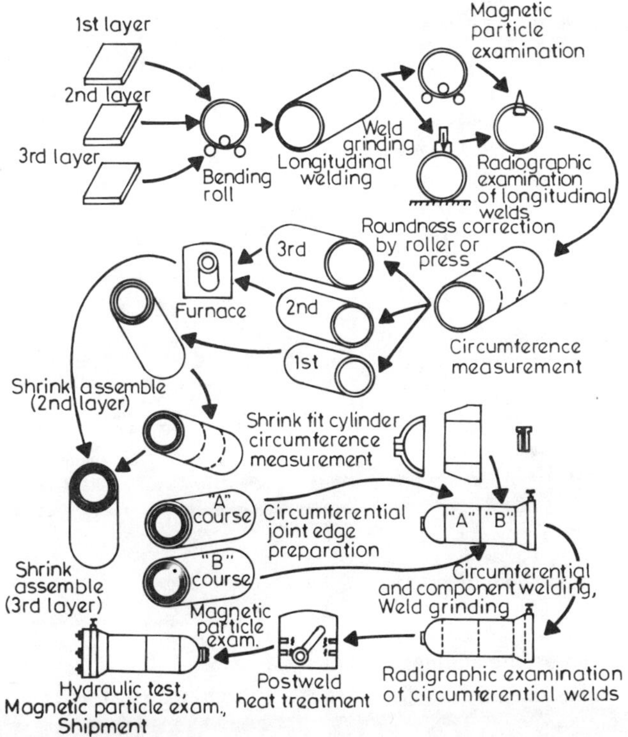

Figure 12.4 Manufacturing sequence of multilayer vessel. *(From Shanker, 1979.)*

9. The circumferential welds are examined both radiographically and by magnetic particles.
10. The completed vessel is post weld heat treated as a whole to relieve the residual stresses.

12.14 How are the multilayer vessels of the weld-shrink type fabricated?

While the shrink-fit type of construction (Q. 12.13) is based on heating the whole shell for ease in assembly and then shrink-fitting, the weld-shrinkage method is based upon heating a small portion of the outer shell circumference to a significantly higher temperature to make a welded joint, and then letting the welded joint cool. Both cooling and the weld shrinkage result in a tight contact between the outer and inner cylinders. Basically it involves the following steps (Brownell and Young, 1959) shown in Fig. 12.5.

1. The innermost shell is prepared as usual.
2. Subsequent shells (usually $\frac{1}{4}$ in or 6.35 mm thick) are wrapped around the inner shell, tightened mechanically, and welded longitudinally. The welds are staggered around the circumference. The welds shrink on cooling and a perfect contact is made with the inner shell.
3. Each weld is examined radiographically and by magnetic particles.

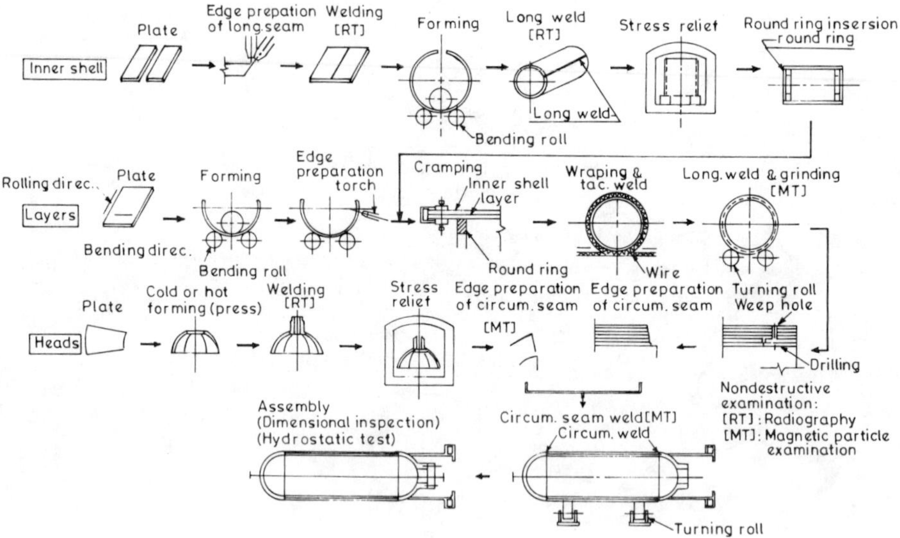

Figure 12.5 Kobe layer construction method. *(Courtesy Kobe Steel, Ltd.)*

4. The welds are contour-ground before adding subsequent layers.
5. After the desired thickness has been achieved, the ends are machined to attach other shells, heads, and/or flanges.
6. The completed unit is post weld heat treated to relieve the stresses.

12.15 How are the multilayer vessels of spirally wrapped plate type made?

Basically, the following steps are involved (Brownell and Young, 1959):

1. In these, as in the earlier cases (Q. 12.13 and 12.14), first an inner shell is made in the usual fashion.
2. Then a plate is spirally wrapped over the inner shell on a lathe machine (Fig. 12.6). The plate is $\frac{1}{4}$ to $\frac{3}{8}$ in (6.35 to 9.50 mm) thick, and its width is approximately ten times the thickness. Figure 12.7 shows a wrapped high-pressure vessel.
3. For prestressing the vessel, the plate is heated to 1100° to 1500°F (600 to 830°C) before wrapping. The wrapped plate is cooled by air and/or water. As it cools, it shrinks and builds up compressive stresses in the vessel.

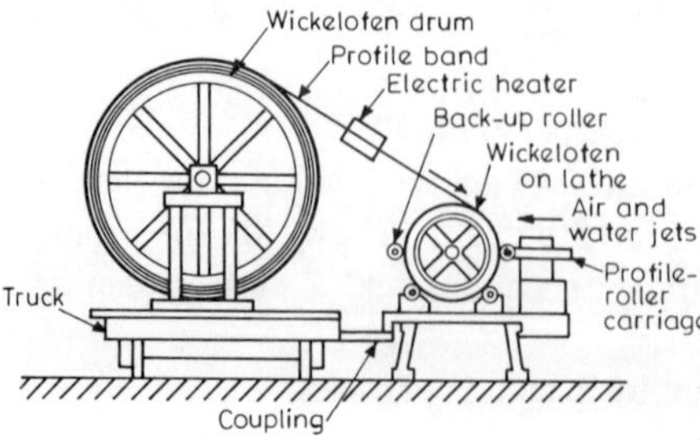

Figure 12.6 Wickelofen-wrapping lathe. *(From Brownell and Young, 1959. With permission from John Wiley and Sons.)*

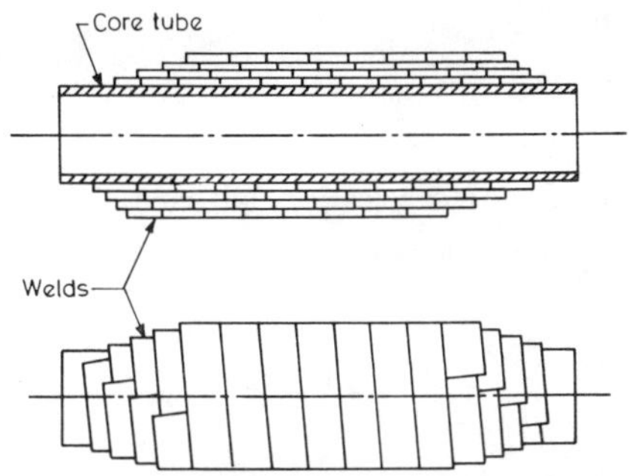

Figure 12.7 Spirally-wrapped high-pressure vessel. *(From Brownell and Young, 1959. With permission from John Wiley and Sons.)*

4. After the desired thickness is achieved, the ends are prepared for welding other shells, heads and/or flanges.

12.16 How are the heads fabricated for high-pressure vessels?

Heads are made from one thick plate or by forging, both for the monowall and the multilayer vessels. Usually, spherical heads are used at high pressures since they have the minimum thickness for a given pressure. These are then joined to the shell by welding or through flanged joints using self-energizing type gaskets that seal better as the pressure increases.

12.17 How are the nozzles fabricated for the high-pressure vessels? How are they installed?

Nozzles for the high-pressure vessels are generally forged and are of self-reinforcing type since reinforcing pads cannot be used in such cases (Chapter 9). These are welded through the full multilayer thickness and radiographically examined.

TESTING

12.18 How are the high-pressure vessels tested?

They are tested like any other vessel. However, greater care is required since any damage during testing can be very costly in terms of money, time and, sometimes, injury to persons nearby. There is still greater risk with monowall units since they can rupture with great force. In a multilayer unit, on the other hand, a weep or vent hole is provided in all but the innermost layer. When a crack develops in the innermost layer, the fluid comes out of the weep hole, and an impending failure is quickly detected and probable disaster prevented.

12.19 What are the ranges of dimensions and operating conditions in which the multilayer construction has been successfully used?

Rauschenplat (1981) reports the range of dimensions as the following:

Wall thickness:	2 to 20 in (51 to 508 mm)
Tangent length:	30 to 130 in (760 to 3300 mm)
Inside diameter:	8 to 100 ft (2.4 to 30.5 m)
Weight:	Up to 1.1×10^6 lbm (500,000 kg)

and, the range of operating conditions as the following:

Pressure: 1500 to 20,000 psig (10.3 to 138 MPa)
Temperature: –50° to 850°F (–45° to 455°C)

As the demand increases and the technology of fabrication advances, larger dimensions and more severe operating conditions can be expected.

ULTRA–HIGH–PRESSURE VESSELS

12.20 How are the ultra-high-pressure vessels fabricated?

Ultra-high-pressure vessels capable of withstanding up to 200,000 psi (1,380 MPa) are used in hydraulic and extrusion presses, powder metallurgy, synthetic gem industries, etc. The stress at the outer wall is considerably less than that at the inner wall. Therefore, the wall material is not used to its full stress value. There are various special techniques used in the fabrication of ultra-high-pressure vessels. Since such pressures are not generally encountered in the chemical process industries, only the names of the techniques are given below:

1. Wedge construction
2. Segment construction
3. Cascade construction

The details are available in Harvey (1974).

● In this chapter, we have discussed the salient features of the high-pressure vessels. The detailed information on this topic is mostly proprietary of a few manufacturers and is scattered in bits and pieces in the open literature. Hopefully, this situation will improve with time.

REFERENCES

Brownell, L. E., and E. N. Young. *Process Equipment Design: Vessel Design.* Wiley: New York. 1959.

Fujii, T. *Fabrication of Heavy Wall Pressure Vessels for Elevated Temperature Service,* Hitachi Zosen Technical Report, HZTR PVP-9. Hitachi Shipbuilding and Engineering Co. Ltd.: Osaka, Japan. 1978.

Harvey, J. F. *Theory and Design of Modern Pressure Vessels,* 2d ed. Van Nostrand Reinhold: New York. 1974.

Kobe Steel Ltd. *Mechanical Petroleum Engineering Seminar: Some Characteristics of Layered Pressure Vessels.* Kobe Steel Ltd.: Tokyo. 1978.

Rauschenplat, H. C. Design and Fabrication of Multiwall Pressure Vessels. In *Pressure Vessel Design,* PVP Vol. 57, G. E. O. Widera (ed.). ASME: New York. 1981.

Shanker, M. R. Stress Analysis of a Multiwall Test Vessel. In *Heat Exchanger Design-II: Mechanical Design, Stress and Seismic Analysis, Corrosion and Cost Analysis,* J. P. Gupta (ed.). C.S. Enterprises: Delhi, India. 1979.

Yasuda, R., K. Wakaizumi, S. Yasui, and K. Tanaka. *Pressure Vessel Design,* Hitachi Zosen Technical Report, HZTR PVP-5. Hitachi Shipbuilding and Engineering Co. Ltd.: Osaka, Japan. 1976.

CHAPTER

THIRTEEN

CORROSION, EROSION, FOULING, AND WATER TREATMENT

INTRODUCTION

• The most frequent causes of failure of a well-designed and fabricated unit operating at the designed conditions are the corrosion and erosion damages to the tubes and fouling deposit over the tubular surface that cuts down the amount of heat transfer. Fouling results in overdesigned units with increased capital and operating costs. Proper water treatment can reduce fouling although not eliminate it totally. Thus the corrosion, erosion, fouling, and water treatment play important roles in the design and operation of an exchanger. This chapter contains questions on these topics.

In this chapter, Nalco Chemical Co. (1979, 1981) and Elayaperumal (1979) have been used very frequently, with permission. Such other references, as used occasionally, have been mentioned at their appropriate places.

CORROSION

13.1 What are the common modes of deterioration of materials?

Materials deteriorate due to the following:

1. Corrosion by electrochemical reactions
2. Wear by mechanical rubbing
3. Erosion by high-velocity fluid, suspended particles in liquid streams or suspended droplets in gaseous streams
4. Fracture by stresses that increase with time

13.2 What is corrosion?

Corrosion is an electrochemical process by which a metal, such as mild steel, returns to its natural state, i.e., iron oxide or rust.

13.3 What is a corrosion cell?

A corrosion cell consists of an anode, a cathode, and an electrolyte. Metal ions dissolve into the electrolyte (water) at the anode. Electrically charged particles (electrons) are left behind. These electrons flow through the metal to other points (cathodes) where electron-consuming reactions occur. The result of this activity is the loss of metal at the anode and often the formation of a deposit. Chemical reactions involved are the following:

1. At the anode

$$\text{Neutral metal} \rightarrow \text{electrically charged metal} + \text{electrons}$$

$$Fe \rightarrow Fe^{2+} + 2e^-$$

2. At the cathode

$$\text{Electrons} + \text{oxygen} + \text{water} \rightarrow \text{charged ion}$$

$$2e^- + \tfrac{1}{2}O_2 + H_2O \rightarrow 2OH^-$$

13.4 What are the various forms of corrosion?

There are eight forms of corrosion that, along with their brief descriptions and examples, are listed in Table 13.1.

13.5 Which are the most common forms of corrosion?

The uniform (or general) corrosion is the most common form followed closely by stress corrosion cracking and pitting corrosion.

13.6 What factors affect corrosion?

These are the following:

1. Metallurgical
   a. Composition of the alloy
   b. Structure of the alloy (single phase, multiple phase, presence of precipitates)
   c. Mechanical state of the alloy (degree of cold work, presence of defects)
2. Environmental
   a. Composition and concentration of the corroding species (H^+, O_2, Cl^-, SO_4^{2-}, HCO_3^-, Ca^{2+})
   b. Temperature

Table 13.1 Eight forms of corrision

Name	Brief description	Examples
1. Uniform corrosion	General attack occurring uniformly over the surface. Metal becomes thin	a. Steel in water and acids b. High-temperature oxidation of metals
2. Galvanic corrosion	Increased corrosion of anodic metal in electrical contact with a nobel metal	Carbon-steel shell in contact with stainless steel tubes and tubesheets
3. Crevice corrosion	Localized corrosion in corners, gasket surfaces, and under deposits	Corrosion of stainless steel tubes and tubesheets
4. Pitting corrosion	Extremely localized corrosion resulting in holes and pits. (Does not require a crevice)	a. Chloride pitting of stainless steel b. Pitting of Al-brass condenser tubes
5. Intergranular corrosion	Localized corrosion along grain boundary regions	Corrosion of stainless steels in the heat affected zones of welds
6. Selective leaching	Preferential removal of one element from an alloy resulting in a porous mass	a. Dezincification in brasses. b. Denickelification in cupronickels
7. Erosion corrosion	Accelerated corrosion due to relative movement between the fluid and the metal	Seawater erosion of condenser tubes. Corrosion of pump impellers. Corrosion of turbine blades
8. Stress corrosion	Synergistic effect of tensile stress and corrosive environment (either static or dynamic)	Cracking of stainless steel in Cl environment. Caustic cracking of boiler tubes. Hydrogen cracking of high strength steels. Corrosion fatigue of Al alloys

Source: Elayaperumal (1979).

c. Flow velocity
d. Formation of a protective film
e. Deposits

3. Operational
   a. Presence of dissimilar metals
   b. Equipment design features (sharp bends, stagnant regions, etc.)
   c. Stresses present (tensile or compressive, static or dynamic, applied or residual)
   d. Thermal and mechanical history

13.7 What are the economic losses due to corrosion?

Corrosion ultimately results in the failure of the equipment. Failure of an important equipment such as heat exchanger causes direct and indirect losses besides proving dangerous to human life in certain cases. The direct losses due to corrosion are the following:

1. Cost of replacement of the equipment or parts
2. Cost of labor for the above
3. Cost of preventive measures

The indirect losses are the following:

1. Plant shutdown
2. Loss of valuable products
3. Loss of efficiency

13.8 Where and how can corrosion occur in a tube-to-tubesheet joint?

The various possibilities are the following:

1. Expansion of the tube into the tubesheet reduces the tube wall thickness and work-hardens the metal. The stresses thus induced can cause stress corrosion.
2. Stress corrosion can also be caused by the stresses developed due to the differential thermal expansion between the tubes and the shell.
3. In order to facilitate the tube removal, the tubes are roller expanded to within $\frac{1}{8}$ in (3.2 mm) of the shell-side face of the tubesheet. Thus, there is a $\frac{1}{8}$ in (3.2 mm) long gap between the tube and the tubesheet (as shown in Fig. 10.7). Corrosion may occur in that gap in some services.
4. The inside edges of the tube holes are chamfered to remove burrs (produced by drilling) to prevent cutting of the tubes. In the lower tubesheet of vertical units, this chamfer collects dirt, etc., and corrosion is likely to occur there.
5. If, in certain services, complete drainage is required when the exchanger is out of service, particular attention should be paid to the upper surface of the tubesheet in the vertical units or else the liquid collected there will corrode the tubesheet.

13.9 What are the main types of corrosion in a steam generator? What causes them?

The types of corrosion encountered in a steam generator are the following:

1. Uniform (or general) corrosion
2. Caustic embrittlement
3. Nonuniform or pitting corrosion

These are caused by the local formation of a concentrated acidic or alkaline solution at the metal surface. (A protective layer of magnetite does not form on mild steel when the water pH is either high or low.)

13.10 What measures are taken to prevent boiler system corrosion?

The most common methods include the following:

1. Removing dissolved oxygen from the feedwater
2. Maintaining slightly alkaline conditions in the boiler water
3. Keeping internal surfaces clean
4. Protecting boilers during out-of-service intervals
5. Counteracting corrosive gases in steam and condensate systems with chemical treatment
6. Controlling solids content in the boiler by blowdown

13.11 What is blowdown in a boiler?

The solids content of water in the boiler is controlled by discharging some boiler water containing concentrated dissolved and suspended solids and replacing this water with fresh low-solids content water. The amount of blowdown depends upon the concentrations of solids that the boiler can tolerate. In smaller boilers, blowdown can be intermittent, whereas in large boilers a continuous blowdown is preferable since it reduces the amount of blowdown and causes less upset in the boiler. The heat rejected in the case of continuous blowdown can be recovered by using it to preheat the feedwater.

13.12 Why is a porous scale deposit conducive to corrosion? How can such corrosion be reduced?

High concentrations of acids and alkalies can collect underneath porous deposits. This accelerates the corrosion of the tubes. It is more pronounced in areas of high heat flux. Adequate boiler water treatment to eliminate free hydroxyl ions (OH^-) will help reduce such corrosion.

13.13 What are the relative resistance to corrosion of the various condenser tube materials?

Condenser tubes are prone to all types of corrosion. The relative resistance to corrosion of different condenser tube materials is given in Table 13.2.

13.14 Why is stainless steel more immune to corrosion than carbon steel?

There are many types of stainless steels, all iron based with 12 to 30 percent chromium, 0 to 22 percent nickel, and minor amounts of carbon, copper, molybdenum, silicon, etc. The popular 18-8 stainless steel has chromium that accumulates at the surface and provides a tightly adherent oxide film that resists further corrosion in an oxidizing medium. Silicon enhances the beneficial effects of

Table 13.2 Relative resistance to corrosion failure of condenser tube materials

Failure mechanism	Admirally brass	Al-brass	90-10 Cu-Ni	70-30 Cu-Ni	Stainless steel	Titanium
General corrosion	2	3	4	4	5	6
Erosion corrosion	2	2	4	5	6	6
Pitting (operating)	4	4	6	5	4	6
Pitting (stagnant)	2	2	5	4	1	6
High-velocity water	3	3	4	5	6	6
Inlet-end corrosion	2	2	3	4	6	6
Steam erosion	2	2	3	4	6	6
Stress corrosion	1	1	6	5	1	6
Chloride attack	3	5	6	5	1	6
Ammonia attack	2	2	4	5	6	6

Note: Numbers indicate relative resistance to the respective failure mechanism on a scale of 1 (lowest) to 6 (highest).

chromium. Addition of nickel does the same also. The choice of different types of stainless steel for various chemicals is given in Davison and Miska (1979a).

13.15 Are copper or aluminum alloys as easily corroded as iron or steel?

In general, copper and aluminum alloys corrode more slowly. However, in some waters, nonferrous metals may be subject to severe pitting or localized attack. In addition, dissolved gases, such as H_2S or NH_3, are generally more destructive to the nonferrous alloys.

13.16 What corrosive effects do low and high effluent gas temperatures have on the exchanger materials?

All fuels have some sulfur content that burns to give SO_2. If the effluent temperature is below about 300°F (150°C), the SO_2 is likely to mix with the water vapors produced during the combustion of the fuel to produce sulfurous or sulfuric acid that will condense and corrode the metal. On the other extreme, steel oxidizes at about 1000°F (540°C) and loses strength.

13.17 What precautions should be exercised when using steam as a heating medium?

Steam condensate is very corrosive due to dissolved O_2 and CO_2. Hence, the steam should preferably be on the tube side. Furthermore, the design of the unit and the piping should be such as to drain the condensate continuously without accumulation, which otherwise will cause more corrosion, and the condensate level will also blank out some heat transfer surface causing reduction in the total heating. However, if the steam velocity in the tubes is going to be very high, such as 60 ft/sec (18 m/sec), it might cause erosion of the tubes. In such a case, the steam should be put on the shell side since it provides more free area and hence less possibility of erosion damage.

In a surface condenser, on the other hand, one is more worried about the corrosion by the cooling water in the tubes since the steam on the shell side does not contain significant amount of dissolved O_2 or CO_2 because these are removed by the boiler water treatment before the steam is produced.

13.18 How can one determine whether a particular material will corrode in a given environment?

Actually, one can assume that all materials will corrode in all environments, except that the extent of corrosion or the rate of corrosion may vary significantly from one environment to the other. A material may experience negligible corrosion in some environments during the entire expected service life, while in some other environments, it may experience severe corrosion in a relatively short time.

The rate of corrosion is dependent upon the type of material and fluid; concentration of dissolved gases, ions, and salts; temperature; pH; velocity; etc. (Q. 13.6). These rates are tabulated in various books and handbooks (Perry and Chilton, 1973) as well as in charts prepared by various materials and equipment manufacturers (Nooter Corp., 1977). If none of these sources provides the requisite data, experiments have to be carried out simulating the operating environment. Some of these experiments take a long time, while the latest developments have brought forth apparatus for quicker measurements (Anonymous, 1981a).

13.19 How do such experiments work?

The material is put in the environment in which it will be used if found satisfactory. It is periodically removed and cleaned of any deposits and corrosion products and weighed. Thus, the loss in weight at different time intervals is determined. From this, the rate of corrosion is calculated. If the rate of corrosion is low enough so that the strength of the material will not be affected during the entire expected working life of the equipment fabricated out of it, the material is recommended for use or else it is discarded.

13.20 What factors should be kept in mind at the design stage for the corrosion control of plants and equipment?

The important factors are the following:

1. When locating the plant and equipment, choose the least corrosive environment
2. Avoid crevices and complex configurations
3. Avoid contact between dissimilar metals
4. Design for easy drainage and cleaning, and easy component replacement
5. Avoid excessive mechanical stresses and stress concentrations
6. Avoid sharp bends in the piping system
7. Avoid hot spots
8. Exclude air
9. Avoid sharp edges in general, and especially when protective coatings have to be applied
10. Join by welding instead of riveting
11. Keep surface-to-volume ratios low
12. Provide for high enough flow rates in the whole equipment so that no excessive settling of suspended material in dead regions occurs
13. Avoid vapor pockets by suitable vent lines
14. Allow for smooth entry of the fluids into the tubes by flaring the tubes or using tube inserts (Fig. 13.1)

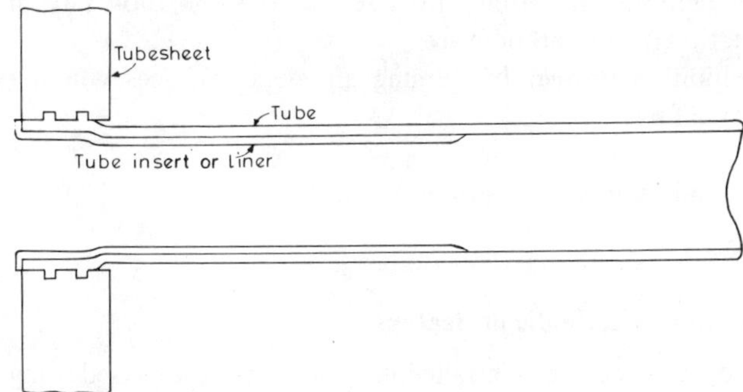

Figure 13.1 Tube inserts for erosion protection.

15. Allow for dirt and sand in cooling water to settle down in settling tanks and use screens to further cut down the dirt and sand entering the system
16. Provide slight slanting of the pipes
17. Equipment supports should not rest in a pool of liquid
18. Porous insulation should be weather-proofed
19. Detailed specifications should be provided for the choice of proper noncorroding materials, fabrication procedure, heat treatment, etc.
20. Detailed inspection is a must at every stage of the fabrication
21. Coatings and linings may be used where feasible

13.21 How is corrosion minimized in an operating unit?

One or more of the following may be used depending upon the unit under consideration:

1. Pretreatment of the metal surface
2. Corrosion inhibitors in the circulating fluid
3. Cathodic protection
4. Water treatment and deaeration
5. Use of corrosion-resistant materials, linings, or coatings

13.22 What is meant by "pretreatment of the metal surface?"

Pretreatment is the application of certain chemicals to the metal surface in order to render it immune to corrosion by establishing an even, tightly adherent, protective film. This film is then maintained by adding corrosion inhibitors to the circulating fluid.

13.23 How do chemical corrosion inhibitors work?

Chemical inhibitors reduce or stop corrosion by interfering with the corrosion mechanism. Inhibitors usually affect one or both parts of the corrosion cell consisting of an anode and a cathode (Q. 13.3).

Anodic corrosion inhibitors interfere with the anodic reactions. Although these inhibitors can be effective, they can also be dangerous since, if insufficient anodic inhibitor is present, the entire corrosion potential occurs at the unprotected anode sites. This causes severe localized or pitting attack. Cathodic corrosion inhibitors interfere with the cathodic reactions. These inhibitors reduce the corrosion rate in direct proportion to the reduction of the cathodic area.

General corrosion inhibitors protect by filming all metal surfaces whether they are anodic or cathodic in nature.

13.24 What are the most commonly used chemical inhibitors?

These are given in Table 13.3.

13.25 Describe corrosion control by cathodic protection.

Sacrifical anodes (zinc, steel, etc.) are installed or a dc current is passed using inert electrodes (platinized titanium, graphite, etc.) to alter the potential of the

Table 13.3 Commonly used chemical inhibitors

Mainly anodic	Mainly cathodic	General
Chromate (most popular)	Bicarbonate	Soluble oils
Orthophosphates	Polyphosphates	Amines
Ferrocyanide	Metal cations	Quaternaries
		Other organics

equipment to be protected. Both these methods reduce corrosion. The initial cost of the first method is lower and is used more often than the second method. For maximum effectiveness, the anodes and the tubes should be cleaned periodically to remove any insulating scale deposit.

13.26 Describe corrosion control by water treatment.

The water used in cooling or in steam generation is treated to remove most of the harmful dissolved substances and gases and to bring the pH to an acceptable level. The various water treatments require one or more of the following: softening, deaeration, addition of amines, phosphates, sulfites, etc., for pH and O_2 control in boiler water; addition of inhibitors, such as chromates, phosphates, nitrites, etc., for corrosion control in cooling water.

A review of cathodic protection of heat exchangers is given in Iyer and Balakrishnan (1981).

13.27 What are corrosion-resistant linings or coatings?

These are linings or coatings on the metal surface to prevent the fluid from touching the surface. There are various types of linings:

1. Explosion clad linings
2. Glass linings
3. PVC linings
4. Rubber linings
5. FRP (fiber reinforced plastic) linings
6. Zinc/aluminum metal spraying

Obviously, the particular lining to be used depends on the equipment and the operating conditions. Details of the usages are given in Anonymous (1979a).

13.28 What properties should the coating or the lining have?

These are the following:

1. Be nonreactive with the fluid streams
2. Prevent oxidation of the metal surface
3. Have good thermal conductivity
4. Bond strongly with the surface and be not easily removed by the chemical and/or mechanical cleaning methods used
5. Have low cost and long life

13.29 Why cannot corrosion be completely prevented in most situations?

Corrosion is essentially an environmental influence. In spite of the best efforts at the design stage, it is not always possible to eliminate completely the corrosion failures in heat exchangers and pressure vessels. This is because the design conditions are at best an approximation of the operating conditions that are not always steady. Hence, a periodic review and upgrading of the corrosion control system is necessary.

13.30 What is corrosion monitoring? How is it done?

Corrosion monitoring means keeping track of the corrosion of various parts of an equipment. This helps in taking preventive measures to forestall a failure and also predicts the service life of the equipment.

There are various ways of corrosion monitoring, such as the following:

1. Visual
2. Corrosion coupons
3. Ultrasonic
4. Radiography
5. Dye penetrant
6. Electrical resistance
7. Polarization resistance
8. Potential measurement
9. Spark test for nonconducting linings
10. Eddy current testing for tubes

Some of these require the shutdown of the unit, while the others do not. Some are slow in response. Details are available in Gole (1979).

- The above should provide an introduction to the important problem of corrosion and its control. In a large plant, there is no substitute for having a qualified, corrosion engineer to supervise the various jobs related to the corrosion control.

EROSION

13.31 What is erosion?

Erosion is the loss of material due to high-velocity impact of liquid streams, suspended drops in gaseous flow, suspended vapor bubbles in liquid flow, or suspended solid particles in gaseous or liquid streams.

13.32 Where is it encountered in a heat exchanger?

It is primarily encountered at the inlet to the tubes and on the outside surface of the tubes under the shell-side nozzles.

13.33 Where else in a chemical plant is erosion encountered?

It is encountered in any equipment or pipeline where two-phase flow occurs and in centrifugal pumps due to low pressure (cavitation-erosion).

13.34 What factors does erosion depend upon?

The important factors are the following:

1. Fluid velocity
2. Whether single or two-phase flow
3. Fluid properties, viz. density, viscosity, and acoustic impedence (Rochester and Brunton, 1973)
4. Material properties and strength

13.35 How is erosion prevented or minimized?

One sure way is to reduce the flow velocity. However, the reduction would adversely affect the exchanger performance. The other ways are the following:

1. Filtering out the suspended solid particles
2. Reducing the amount of liquid droplets in a gaseous stream by passing the stream through a bed of adsorbant or by heating to vaporize the droplets
3. Using plastic tube inserts (Anonymous, 1979b) in the tubes (Fig. 13.1)
4. Using impingement plates below the nozzles (Fig. 1.21)

Since (1) and (2) result in a significant pressure drop and cost, generally (3) and (4) are used.

13.36 What is erosion-corrosion?

Usually, when a metal corrodes, the corrosion film stays on the surface and reduces the rate of further corrosion. However, when that corrosion film is removed by erosion, fresh metal surface gets exposed and corroded. Thus, the cycle repeats continuously. The effect of the two is synergistic and far greater than if either erosion or corrosion were present alone. An example of erosion-corrosion in a converted reformer gas cooler–condenser has been cited by Van der Horst and Sloan (1973) (see also Table 13.1).

13.37 Where is erosion-corrosion encountered and how is it prevented?

It is encountered at all such places that are prone to corrosion and are in contact with a high-velocity fluid. Thus, it can be at any place on the inside of the shell and inside or outside of the tubes. Unlike pure erosion, it is not confined only to the entry zones of the tubes or underneath the nozzles.

It can be prevented by proper treatment to reduce corrosion as discussed in Q. 13.21. If there is no corrosion in the first place, there will not be any erosion-corrosion.

FOULING

13.38 What is fouling?

It is the accumulation of solid material in the equipment in a manner that hampers its operation at the desired level or contributes to its deterioration.

13.39 What is scale?

It usually refers to the deposits from water systems caused by the supersaturation of inorganic salts. However, generally no distinction is made between fouling and scale in the literature, and we shall also use the terms interchangeably.

13.40 What are the effects of fouling?

These are the following:

1. Reduced heat transfer
2. Increased pressure drop
3. Deteriorating product quality
4. Localized pitting and corrosion of the tubes
5. Reduced throughputs
6. Oversized unit at the design stage

All the above lead to economic losses. Therefore, it is important to know the salient features of the fouling of heat exchangers.

13.41 What are the different kinds of fouling?

Fouling has been classified into eight kinds based upon the process by which it is formed (Taborek et al., 1972):

1. Crystallization: Salts dissolved in the fluid stream crystallize out when the fluid is cooled or heated (as in cooling towers and heat exchangers) or when it is evaporated (as in steam generators and reboilers). Salts that crystallize out when the liquid is heated are called the "reverse solubility salts." Their solubility decreases with increasing temperature. Some such salts are $CaCO_3$, $CaSO_4$, etc.
2. Chemical reaction: Some organic products or inorganic salts might get oxidized at high temperature and crystallize out.
3. Polymerization: Some compounds might get polymerized and then deposit on heat transfer surface. This is often found in polymer and textile industries.
4. Coking: It is found extensively in crude oil refineries and hydrocarbon processing units due to high-temperature effects on various hydrocarbons. Since it is a class in itself, it has been separated from the chemical reaction fouling.
5. Sedimentation: Dirt, clay, sand, and corrosion products suspended in the fluid stream settle on the heat transfer surface due to gravitational forces. These particles then act as nucleation sites for other kinds of deposits.
6. Solidification: When the surface temperature is lower than the solidification or freezing temperature of one or more components present in the fluid, such components solidify on the surface leading to solidification fouling. An example is the deposition of wax from crude oil in pipelines.
7. Corrosion: A corrosive fluid corrodes the heat transfer surface. Part of the corrosion product shears off [erosion-corrosion (Q. 13.36)] and gets suspended in the flowing stream only to sediment out downstream as discussed in (5) above. Any corrosion products sticking to the surface reduce the heat transfer due to their low thermal conductivity and also provide nucleation sites for the fouling to deposit by other mechanisms.
8. Microbial fouling: It happens due to the presence of microbial life.

13.42 On what factors does the fouling depend?

These are the following:

1. Physical properties, chemical composition, and pH of the fluid
2. Physical state of the fluid: liquid, vapor, or two-phase
3. Types and amounts of dissolved and suspended impurities
4. Presence of dissolved air and/or carbon dioxide
5. Velocity of the fluid: average velocity as well as velocities of bypass and leakage streams and velocities in baffle windows and in cross-flow over the tube bundle (Q. 2.3)
6. Fluid flow regime: laminar or turbulent
7. Fluid temperature and tube wall temperature
8. Fluctuations in velocity and temperatures, either intentional or accidental
9. Mode of heat transfer: sensible heating or cooling, boiling and evaporation, condensation, or a combination of these
10. Evaporation of liquid leading to supersaturation of the dissolved solids
11. Materials of construction
12. Surface roughness
13. Design of the unit as far as bypass and leakage streams and the stagnant zones are concerned

13.43 What are the relative effects of changing some of the variables on which the fouling depends?

These are the following:

1. With increasing fluid velocity in turbulent flow the deposit decreases. The effect in laminar flow may not be the same.
2. With increasing tube-wall temperature, the deposit increases, specially in the case of cooling water, hydrocarbons, polymerization, and chemical reactions
3. With tube materials, such as copper or its alloys, microbial fouling is reduced
4. Decrease in dissolved oxygen and carbon dioxide reduces corrosion fouling
5. Fluctuations in velocity will momentarily increase the shear stress and hence may dislodge the deposit. Fluctuations in temperature will result in greater expansion (and contraction) of the tubes than that of the deposit layer, thus cracking and dislodging the deposit (Q. 13.62). Thus, the fluctuations in velocity and in temperature are both likely to decrease the fouling deposit. However, their adverse effects on the process and equipment should also be taken into account.
6. Reduction in the amounts of leakage and bypass streams and the elimination of the stagnant zones will also reduce fouling deposit since the cross-flow velocity will be kept high.

13.44 What are the deposits in a cooling water system?

These are the following:

1. Dirt, silt, and sand
2. Corrosion products

3. Natural organics
4. Microbial slime
5. Alumina, aluminum phosphate
6. Iron phosphate
7. Calcium salts
8. Magnesium salts, etc.

13.45 How do the deposits grow?

The whole field of fouling studies is fast changing and so are the concepts about the mechanisms of deposit and growth. However, it is believed that, for the inorganic salts, the three steps involved are the following:

1. Initial nucleation that produces a primary deposit occurs during an induction period where the deposit rate is low.
2. Secondary crystal growth on the primary deposit
3. An asymptotic thickness due to dynamic balance between the rate of deposit and the rate of shear removal (re-entrainment)

Figure 13.2 shows a cross-sectional view of water fouling deposit, and Fig. 13.3 shows a plot of the buildup of the fouling resistance (deposit) with an initial induction period. There are also models proposed that do not have the induction period (Fig. 13.4). A complete discussion of the fourteen different models has been given in a most comprehensive review by Epstein (1978). His later additions to that review are also highly recommended for a complete grasp of the topic (Epstein, 1981a–c).

13.46 What are the differences between the various types of scale and fouling deposits?

The deposits differ from one another in terms of texture, strength, adhesion to the surface, color, thermal conductivity, density, rate of deposition, equilibrium thickness of the deposit, means required for the removal of the deposit, the process of formation of the deposit, or the kind of fouling phenomenon (see Q. 13.41).

13.47 Why is it important to analyze the fouling deposit? How is it done?

Analysis of the fouling deposit tells its nature and points toward both the causes and remedial actions required. Analysis includes the following methods (Tewari and Mehandru, 1977):

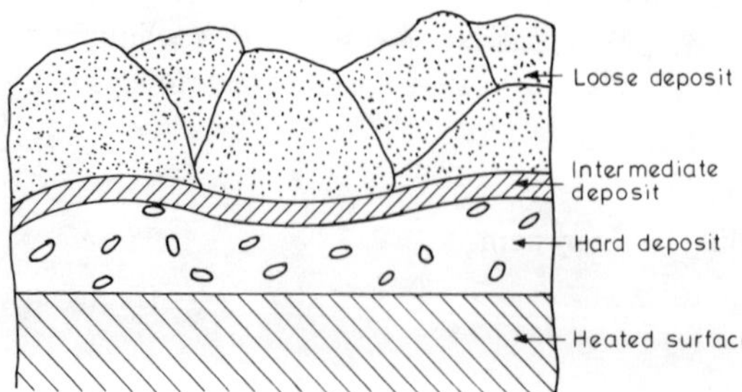

Figure 13.2 Cross section of fouling deposit.

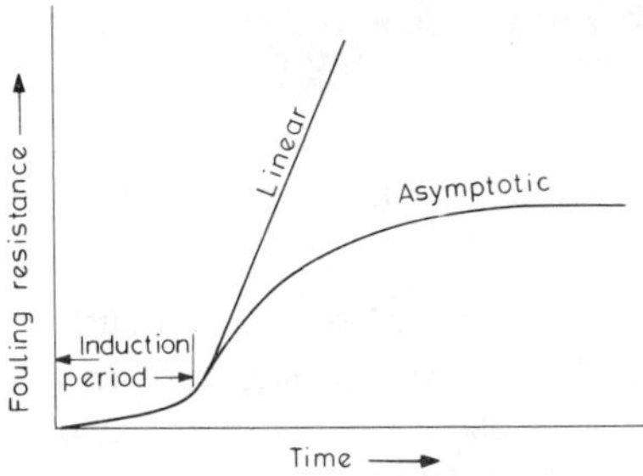

Figure 13.3 Buildup of fouling resistance with an initial induction period.

1. Chemical analysis for various inorganic salts and metal ions and their concentrations
2. X-ray analysis to identify the crystal structure
3. Ignition loss and solvent extraction for organic deposits
4. Differential microbioanalysis to identify the type of microorganisms and their concentrations

13.48 What methods are used to measure the fouling deposits?

The methods used are the following:

1. Measuring the thermal resistance offered by the deposit
2. Direct weighing
3. If direct weighing is not possible, then dissolve the deposit and perform either chemical analysis or gravimetric analysis to determine the amount of the deposit

13.49 How does fouling influence heat transfer?

Fouling increases the resistance to heat transfer thus reducing the capability of a given surface to transfer heat. Usually the fouling layers build up on both sides of the heat exchanger surface. These severely cut down the amount of heat transferred for a given temperature driving potential.

13.50 Can a rough surface lead to higher heat transfer?

Artificially produced rough surfaces have been successfully used to agument heat transfer in turbulent flows of high Prandtl number fluids ($Pr > 5$). Similarly, an uneven fouling deposit on a smooth surface leads to an increase in heat transfer in the initial stages (Bott and Gudmundson, 1978). However, with the passage of

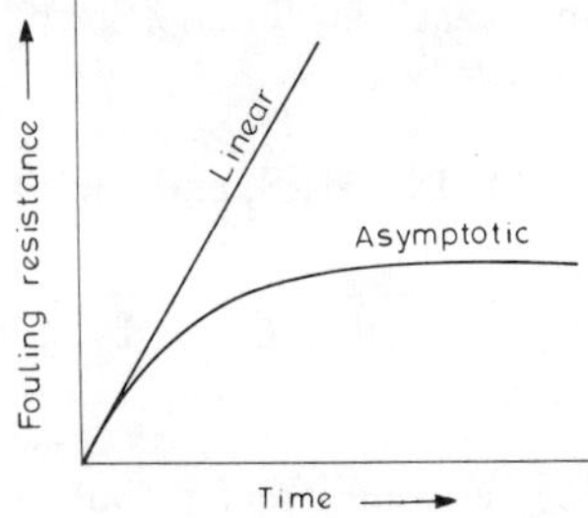

Figure 13.4 Buildup of fouling resistance without induction period.

time, as more deposition occurs, the insulating effect of the deposit overshadows the enhancing effect of its rough surface.

13.51 What are the overall design and clean heat transfer coefficients and how are they related?

The overall clean heat transfer coefficient, U_c, is based on a clean surface without any fouling, while the overall design (or dirty) heat transfer coefficient, U_d, takes into consideration the fouling deposit. The relation between the two is

$$\frac{1}{U_d} = \frac{1}{U_c} + R_{f_o} + R_{f_i} \frac{A_o}{A_i}$$

where U_d and U_c = overall design and clean heat transfer coefficient, respectively
R_{f_o} and R_{f_i} = fouling resistance on the outside and the inside surfaces, respectively
A_o and A_i = area per unit length of the outside and the inside surfaces, respectively

13.52 Is the effect of fouling resistance, R_f, on the overall heat transfer coefficient, U_d, the same regardless of the value of the overall clean heat transfer coefficient, U_c?

No. It varies with the value of U_c as can be shown by

$$\frac{1}{U_d} = \frac{1}{U_c} + R_f \tag{13.1}$$

The percent change in the value of U_c due to R_f is

$$\frac{U_c - U_d}{U_c} \times 100 = \left(1 - \frac{U_d}{U_c}\right) \times 100 \tag{13.2}$$

Substitute for U_d from Eq. (13.1) into Eq. (13.2) and simplify to get the percent change in U_c

$$\frac{100 R_f U_c}{1 + R_f U_c} \tag{13.3}$$

For a constant R_f, the value of Eq. (13.3) increases as U_c increases. Hence, the effect of R_f is high on a system with high U_c. For example, consider two cases with $U_c = 400$ and 60. For a constant $R_f = 0.003$, the change in the overall heat transfer coefficient in the two cases is, respectively, 54.54 and 15.25 percent. Therefore, greater care should be taken to minimize fouling when higher heat transfer coefficient is involved.

13.53 What are the units of the fouling resistance in the British, MKS, and the SI systems?

The British units are hr ft^2 °F/Btu. The MKS units are sec cm^2 °C/cal. The SI units are m^2 K/W.

13.54 What is the difference between specifying a fouling resistance and a fouling layer thickness?

The fouling resistance has the units of the heat transfer resistance [inverse of the units of the heat transfer coefficient (Q. 13.53)] and can be used directly in calculating the overall design heat transfer coefficient (Q. 13.51). It is also called fouling factor. On the other hand, when the fouling layer thickness is specified, one needs to know its thermal conductivity in order to calculate the overall heat transfer coefficient. The thermal conductivity depends upon the temperature, structure, density, and the composition of the deposit and is difficult to evaluate in each case.

13.55 Is the fouling factor the same for the turbulent and the laminar flow of a given liquid?

No, it is significantly higher for the laminar flow than for the turbulent flow since, in the laminar case, the scale has more time available to deposit and the shear force on the deposited scale is less than that in the turbulent case. So the deposited scale cannot easily be dislodged, and its thickness is greater than that in the turbulent case. Kern (1950) suggests that the tabulated fouling factors should be increased 50 to 100 percent for laminar flow conditions.

13.56 Why is air-side fouling factor neglected in an air fin cooler?

It is neglected in the case of clean and dry air since that air does not cause any corrosion or deposit to form. Moist air can lead to erosion-corrosion (Q. 13.36). Dirt-laden air can block the space between fins on the air cooler tubes (Chapter 4) and lead to severe drop in the heat transfer.

13.57 How can the detrimental effects of fouling be minimized?

This can be done in the following ways:

1. By a proper choice of the type of unit, viz. shell-and-tube, plate, or spiral heat exchanger (see also Q. 13.60)
2. By adequate sizing of the flow channels, minimizing stagnant regions, etc., in the chosen unit
3. By reducing the amount of fouling deposit forming on a surface
4. By periodically cleaning the surface to remove deposit that has formed
5. By using antifoulants and dispersants

13.58 How may the scaling and corrosion tendencies of cooling water be reduced?

The following steps have proved helpful (Yadava and Chawla, 1980):

1. Calcium carbonate scale formation may be prevented by reducing water hardness or by adding a mineral acid, such as sulfuric acid.
2. Crystal growth can be inhibited or significantly reduced by adding organic phosphates, polymers, and, in low-severity systems, adding polyphosphates to the cooling system
3. Adding lime to soften the water
4. Microbial growth may be reduced by using biocides (Q. 13.70)
5. Fibrous solids may be reduced by filtration
6. Corrosion may be controlled/reduced by the methods listed in Q. 13.21.

13.59 Can fouling occur even when a corrosion or scale treatment program is being followed?

Yes. The treatment program will definitely reduce the overall extent of fouling and may totally inhibit a particular kind, e.g., microbial fouling, but some fouling deposit will most certainly be formed. This may be only a thin coating or deposit layer or sedimentation of very fine particles. Theoretically, one should probably not allow even these to form. However, the practical costs of such treatment programs and/or pumping power for very high velocities on the shell and tube sides and the related steps to prevent tube vibrations would become enormous. Hence, the treatment program should only aim at reducing the harmful effects until the next scheduled shutdown of the plant for maintenance and repair, so that the desired performance is achieved for the entire expected life with periodic cleanings. There is no virtue in having a very clean heat exchanger at enormous costs when periodic maintenance of the other equipment does require a plant shutdown.

13.60 For a fouling fluid, what points should be kept in mind at the design stage?

The following points generally prove helpful:

1. Place the fouling fluid on the tube side so that cleaning can be done without removing the tube bundle, which is both costly and time consuming. Use straight tubes since U-bends are difficult to clean internally. For heavy fouling fluids, use large diameter tubes.
2. Channels with flat covers should be used since the covers can be easily removed for cleaning the channels and the tubes without disturbing any piping connections.
3. A horizontal installation is easier to clean than a vertical one, which requires tall stairs or scaffolding for cleaning purposes.
4. High tube-side velocity reduces scale deposit. The cost of pumps and pumping power should be compared with the cost of cleaning and shutdown. However, the velocity should not be so high as to cause erosion damage. Fluid velocities of 1 to 2.5 m/sec are usually satisfactory.
5. Spiral plate heat exchangers may be selected if suspended particles are present and the heat transfer surface area required is not very large.
6. Avoid dead or stagnant zones on the shell side. Cross-flow and window-flow velocities should be of the same order of magnitude in order to utilize the whole of the heat transfer surface.
7. Use copper, copper alloys, or stainless steel in cooling services using water to minimize corrosion and microbial fouling.
8. If finned tubes are used, heavy fouling fluid should not be placed on the finned side.
9. If hydraulic cleaning is required on the shell side, large tube pitch and preferably 90° tube layout (Fig. 1.11) should be used.
10. Blockage of the flow area due to the fouling deposit will lead to increased pressure drop. It should be kept in mind at the design stage.
11. Provisions should be made to reverse the flow of water to reduce fouling.

13.61 What are the various ways of cleaning the inside of the tube bundles?

The various ways are the following (Yadava and Chawla, 1980):

1. Circulating hot fresh water, hot wash oil, light distillate, or commercially available compounds or acidic solutions removes salts, sludge, and soft deposits.
2. Back flushing of the tubes *in situ* also helps reduce the deposits.
3. Rodding helps in removing soft deposits.
4. Hydroblasting at 5000 to 10,000 psi (35 to 70 MPa) is used for hard deposits. The force of the water jet loosens and washes away the deposits.
5. Rotating wire brushes or air turbine driven reamers (scraping tool) with annular shaft for pumping flushing water are used for the hard deposits.
6. Sponge balls may be kept in circulation during the operation for continuous cleaning in cooling water systems (Q. 13.68).

13.62 What are the various ways of cleaning the outside of the tubes?

The various ways are the following:

1. Hydroblasting
2. Sandblasting
3. Soaking and flushing with hot gas or oil
4. Chemical cleaning
5. Ultrasonic vibrations
6. Thermal shock by passing steam through the tubes and then spraying cold water on the outside, which causes the deposit to contract suddenly resulting in its cracking and flaking. Then it is flushed out by water.

13.63 How does one decide whether chemical or mechanical cleaning should be used?

The choice between the two methods primarily depends upon the nature and extent of fouling, the construction of exchanger, and ease of access to the fouled surfaces.

13.64 How is chemical cleaning done?

A suitable chemical solvent in appropriate concentration is passed through the exchanger to dissolve all or part of the deposit. It is usually followed by water flushing to remove the deposit and the solvent. No cleaning solution should be left in the unit since it can lead to corrosion. Sometimes, chemical cleaning is done in addition to mechanical cleaning. Usually, shell side or the outer surface of the tube bundle is cleaned chemically. This avoids removal of the tube bundle. In fixed tubesheet exchangers, the tube bundle cannot be removed anyway. The bent portion of the U-tubes cannot easily be cleaned mechanically from inside. Hence, these are also cleaned chemically. As a matter of fact, if chemical cleaning will do the job, it is preferred since the mechanical cleaning requires partial dismantling of the exchanger and hence longer down time. However, in most cases, a little mechanical cleaning is usually required. Depending upon the situation at hand and the costs involved, inhibited acids, such as hydrochloric, sulfuric, nitric, and sulfamic, are widely used for chemical cleaning. Biocides are used to reduce microbial fouling (Q. 13.70).

13.65 How is mechanical cleaning done?

Mechanical cleaning may involve one or more of the following:

1. Flushing with a strong jet of water or high-pressure air
2. Sand blasting
3. Forcing specially designed rubber balls through the tubes by pneumatic or hydraulic pressure (Fig. 13.5)
4. Expanding wire brushes or scrapers with radial actuated blades may be used for soft deposits (Fig. 13.6).
5. Long cutters with hollow shafts are used to cut through the hard deposits. Air or water flows through the hollow shaft. It acts both to flush out the chippings and to cool the cutting tool.
6. After the cutter tool has been used, the tube inside may be polished with an expanding wire brush or else some deposits may be left behind that will act as nuclei for fouling deposit and corrosion after the unit is put back in service.

All the mechanical methods should be used with extreme care in order not to damage the tubes, tube-to-tubesheet joints, or the protective corrosion barrier on the tube surface. For example, hard brushing will break the protective oxide layer on arsenical admirality brass and will expose new metal for more oxidation, resulting in a gradual loss of material and strength. However, stainless steel can be brushed without damaging its corrosion barrier (Davison and Miska, 1979b).

13.66 For ease in the mechanical cleaning of the tube outside surface, compare a square pitch with a triangular one.

The square pitch gives straight cleaning lanes and is preferable to a triangular pitch if the two have the same pitch. However, a wider triangular pitch (say, $1\frac{3}{8}$ in or 34.925 mm for 1 in or 25.4 mm tubes) has several advantages over a smaller square pitch ($1\frac{1}{4}$ in or 31.75 mm for 1 in or 25.4 mm tubes). The ligaments are 50 percent stronger, and cleaning can be done from six directions as compared to four for a

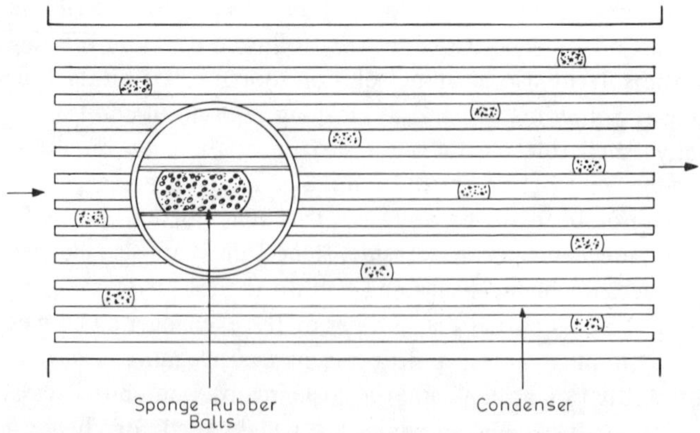

Figure 13.5 Sponge rubber ball cleaning system. *(Courtesy Taprogge U.K. Ltd.)*

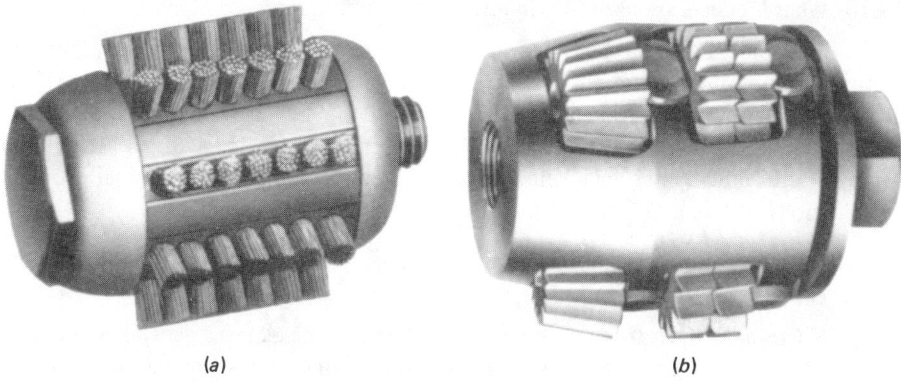

Figure 13.6 (*a*) Wire brush. (*b*) Scraper. *(Courtesy Thomas C. Wilson, Inc., New York.)*

square pitch. The cleaning lanes, number of tubes, baffle spacing, and fluid velocities are nearly the same in both the cases.

13.67 What are the advantages of automatic cleaning?

The advantages are the following:

1. It is continuous, therefore, the heat transfer coefficient is maintained high.
2. There is no build up of fouling between cleanings as in manual cleanings.
3. The unit is not required to be shutdown.

A prominent method of automatically cleaning the tube side is by means of circulating sponge rubber balls that are discussed in Q. 13.68.

13.68 How does the sponge rubber ball cleaning system work?

This is primarily used in cooling water systems. Oversized sponge rubber balls (1-3 mm oversize) are continuously pressed through the tubes by the flow of water (Fig. 13.5). The balls are extracted from the discharge pipes in a strainer section, then drawn off, and pumped through a collector back to the water inlet (Taprogge, 1978). The density of the balls is close to that of water, and hence the balls pass through the tubes at random and in the process clean all the tubes by removing all fouling, debris, etc.

The number of balls is approximately 6-10 percent of the number of tubes. With an average circulation time of 20-30 sec, each tube is cleaned approximately every 5 min. Due to the continuous cleaning, the heat transfer coefficient is kept high. The extra pressure drop due to the circulating balls is more than compensated for by the constantly high heat transfer coefficient because of the low fouling.

13.69 What is back flushing?

It is a provision for periodically reversing the flow through the heat exchangers for removing the dirt and other deposits from the exchanger. This is done by installing a nozzle with a back flush valve in the channel and reversing the flow, which carries the dirt, etc., out of the unit. It has proved very effective on the cooling water side of the exchanger.

13.70 What biocides are used to minimize microbial fouling?

The biocides used are chlorine, sodium hypochlorite, quarternary amines, methyl dibromopropionate, etc. The use of sodium pentochlorophenate as a biocide is decreasing due to its high toxicity.

It has been reported that electrochemical pulsing that changes pH at the tube surface also kills microbial life, thus reducing/eliminating microbial fouling (Anonymous, 1981b). This is only in experimental stage at present.

13.71 Have experiments and/or modeling succeeded in predicting fouling in some cases?

There have been some such studies, although none claims a general application over wide design and operating ranges. These studies relate to the prediction of cooling water fouling (Taborek et al., 1972), evaporator fouling (Epstein, 1979), and solidification fouling (Bott, 1981). This is not an exhaustive listing, and newer results are being published at an ever increasing rate.

13.72 What is the current status of fouling research?

It is recognized that major developments in the heat exchanger design will now come from a better knowledge of fouling. The total effort being exerted is increasing each day, and there are certain groups doing very good work. With their efforts, the problem of fouling has been changed from "the major unresolved one to the major unsolved one" (Epstein, 1981c), the parameters and the behavior trends having been ascertained in a large number of cases.

The importance of fouling studies cannot be overstated. As Fig. 13.7 shows, the percent increase in the surface area to account for the fouling resistance is very high. This leads to higher initial investment as well as higher operating costs. If the unit is considerably overdesigned due to erroneous or excessive fouling factor values, it will actually foul sooner and lead to problems. Thus, an excessive overdesign is not always a safe bet.

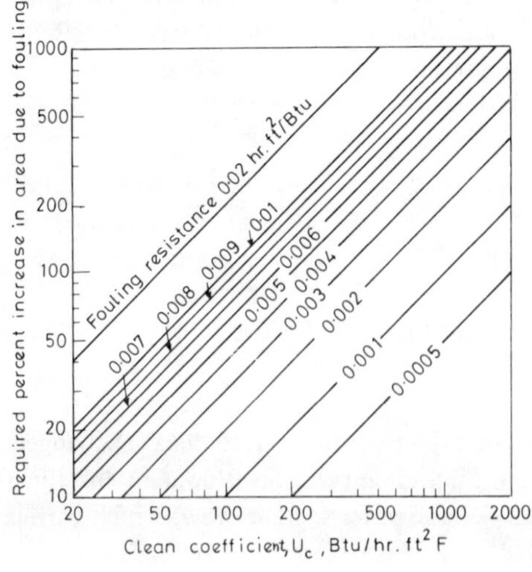

Figure 13.7 Effect of fouling on heat transfer surface requirements.

The reasons why much more effort is not being exerted lie in the variables that affect fouling (Q. 13.42), which give almost infinite possibilities of various combinations, the time extending over months required to collect each data point, the reluctance of the industries to do experimental studies on the on-stream units for fear of disturbing production, and the difficulty in translating laboratory data to industrial use due both to the scaling-up of the size required and the lack of faith by the designers in such data.

The situation is changing, however slowly, due to the interest shown by the cooperative research groups and some groups at the universities around the world. A highly recommended reading for suggested apparatus and techniques for fouling research is Knudson (1981).

WATER TREATMENT

13.73 Why is water used for cooling in chemical and power plants?

Several factors in favor of water as a coolant are the following:

1. It is plentiful, readily available in most locations, and cheap.
2. It is easily handled, and its properties and behavior are well known over a very wide range of operating conditions.
3. It can carry large amounts of heat per unit volume.
4. Its compressibility and coefficient of thermal expansion are quite low.
5. It does not decompose under ordinary conditions.

13.74 What are the common types of cooling water systems?

These are basically two types:

1. Once-through type where the water passes through the unit only once. Large volumes are necessary, and the treatment cost can be very high, although the initial investment is low.
2. Recirculating type where the water passes through the cooler, picks up heat, then is itself cooled in turn in an auxiliary system and is reused. The treatment cost is low but the initial investment is high. The recirculating system can be of open or closed type (Q. 13.82)

13.75 What types of impurities does water contain?

The types of impurities in water from a specific source depend on what it contacts, since water is a universal solvent. The amounts of impurities depend largely on the contact time. In general, water contains dissolved solids, dissolved gases, suspended solids, and bacteria.

13.76 What is the difference between sea water and fresh water?

The main difference is the amount of dissolved minerals. Whereas sea water contains up to 300 to 400 lbs minerals per 1000 U.S. gallons (35 to 50 kg/m^3) of water, the fresh water contains only up to 10 lb per 1000 U.S. gallons (1.2 kg/m^3).

13.77 What are the sources of fresh water, and how do they vary in composition?

Fresh water may be obtained either from surface water (rivers, streams, lakes, and reservoirs) or from groundwater (shallow or deep wells). The former have higher suspended matter and turbidity due to the direct effects of rainfall, soil erosion, and wastes discharged into them, while the latter usually form harder scale due to a longer contact with the ground.

13.78 What is water hardness?

Water contains many dissolved solids and gases as well as suspended solids. Among the dissolved solids, calcium and magnesium compounds contribute to hardness and they make water difficult to wash with. These compounds cause scaling of the heat transfer surfaces. Hence, hard water is not good for use in coolers, boilers, etc., unless properly treated.

13.79 Why are water analysis results expressed as $CaCO_3$?

The molecular weight of $CaCO_3$ is 100. Hence, expressing amounts of compounds in terms of $CaCO_3$ gives a relation between the reacting masses. For example, 1 ppm of Na_2CO_3 (expressed as $CaCO_3$) will react with 1 ppm of $CaSO_4$ (expressed as $CaCO_3$) in water treatment. Actually 106 lbs of Na_2CO_3 (molecular weight 106) react with 136 lbs of $CaSO_4$ (molecular weight 136).

13.80 What are the principal water-side problems of the cooling system?

These are corrosion, scale and fouling deposits, and microbial activity.

13.81 What specific information is needed for determining the proper cooling water treatment?

The information needed is water composition, pH, nature and amounts of suspended particles and dissolved gases, temperature and pressure of operation, once-through or recirculating cooling system, cost of water and of water treatment, local regulations against discharge of used water containing certain treatment chemicals, kinds of metal surfaces the water will contact, etc.

13.82 Does the type of cooling system affect treatment application principles?

Yes. It is basically a matter of economics. In a once-through system, a very large volume of water is used, but it passes through the system only once. Protection can be obtained with a relatively few parts per million of treatment chemicals because the composition does not change significantly while the water passes through the equipment.

In an open recirculating system, more chemicals must be present for protection because water composition changes significantly due to evaporation from a pool or from a cooling tower. Scaling and corrosive constituents get concentrated in water. However, the treatment chemicals also get concentrated by evaporation, thus only moderate dosages will maintain the higher level of treatment needed for these systems.

In a closed recirculating system, the water composition remains fairly constant. There is very little loss of water or treatment chemicals.

13.83 What is stable water?

It is neither scale forming nor corrosive. Most well waters are stable in the well; however, frequently a temperature increase or aeration will make them either scale forming or corrosive. Hence, they require proper treatment for use in industries.

13.84 What are the common water pretreatment steps?

Pretreatment steps vary with the kind of impurities in water and the use to which it is to be put. Some of the common steps are the following:

1. Screening: Removal of large suspended matter and marine life
2. Coagulation: Charge neutralization of colloidal impurities so that they can combine to form large masses that can be filtered or that will settle out.
3. Flocculation: Bringing together of the coagulated particles
4. Precipitation: Adding softening chemicals (lime, soda ash, etc.) to react with the dissolved compounds to produce insoluble substances. This also reduces the hardness, total dissolved solids, alkalinity, and the silica content.
5. Distillation or evaporation: Evaporating and condensing water to obtain pure water without any dissolved salts
6. Reverse osmosis: To remove dissolved solids by passing water under pressure through a semipermeable membrane
7. Ion exchange using both cation and anion exchange resins to remove troublesome ions, such as Ca^{2+}, Mg^{2+}, CO_3^{2-}, SO_4^{2-}, by Na^+, H^+, OH^-, etc. Complete demineralization can be done this way to obtain pure water.
8. Deaeration by heating to remove oxygen and carbon dioxide. Chemicals such as sulfite and hydrazine can be added to react with the remaining dissolved oxygen
9. Adding chemicals, such as caustic and organics (starches), to prevent scale and corrosion in the feed water systems
10. Adding chemicals, such as phosphates and chelates, to react with calcium and magnesium sulfates so that the compounds formed are less sticky to the surface and more soluble in water and hence can be easily removed
11. Adding biocides, such as chlorine, halogenated hydrocarbons, etc., to prevent microbial fouling (Q. 13.70)
12. Adding chromates or other inhibitors for protection against corrosion (Q. 13.24)

13.85 What are the principal classes of troublesome microorganisms?

These are bacteria (filamentous, corrosive, spore-forming and non-spore-forming), fungi (molds, yeast), and algae.

13.86 What is slime?

It is a gelatinous deposit of chemical contaminants or of microbial origin. In the latter case, it has living microscopic organisms and their waste products. These can cause fouling and scale formation and also cause corrosion by sulfate-reducing organisms that generate hydrogen sulfide.

13.87 What factors contribute to the microbial growth?

The degree of microbial contamination, availability of the right nutrients (carbon source), oxygen, moisture, and temperature are all contributory factors.

13.88 How should a microbial control program be formulated?

All the information about the process equipment, operating conditions, water composition, type and quantity of microbial organisms and their adverse effects, sources of contamination, etc., should be known.

Depending upon the system, either oxidizing biocides (chlorine, bromine, etc.) or nonoxidizing biocides (chlorophenates, amines, etc.) are used. Initially, a strong dose is given, then low levels are administered regularly. The treatment may have to be varied so that the organisms do not become immune to the biocides. The biocides should be compatible with the treatment for fouling and corrosion prevention.

13.89 Are there upper limits set on the concentration of various compounds in cooling water?

Generally not, but in specific critical systems, such as the primary cooling circuits in the nuclear power plans, limits are set to reduce their corrosive effects on the materials of construction. Some such compounds and the limits placed on them are the following (Oak Ridge National Laboratory, 1975):

Chloride as chloride ion (Cl^-) < 0.02 ppm
Hydroxyl as hydroxyl ion (OH^-) < 0.001 ppm
Dissolved oxygen (O_2) < 5 ppm

13.90 What is boiler feedwater, and how does it differ from the potable water?

The water added to a boiler is called feedwater. In most cases, condensed steam is returned to the boiler. The water needed to supplement the returned condensate is termed make-up water, which is either raw or treated water. Thus, the composition of feedwater depends upon the quality and quantity of the condensate. With increasing pressure, the maximum limits for total solids, alkalinity, suspended solids, silica, etc., are reduced significantly. The boiler feedwater needs more treatment than the potable water for human use. The boiler cannot tolerate most of the dissolved minerals that are tolerated by the human beings.

13.91 How pure must feedwater be?

It depends on the amount of feedwater used as well as the pressure and temperature of operation. At low pressures, greater hardness can be tolerated than at high pressures. Limits for boiler feedwater composition have been established by the American Boiler Manufacturers Association depending on the operating pressure.

13.92 How are deposits formed in the boilers and what problems do they cause?

Dissolved solids in the raw water deposit on the tubular surface as the water temperature rises. This happens because solubility of some compounds (e.g., $CaSO_4$ and $MgSO_4$) decreases with increasing temperature and because some dissolved compounds break down at high temperature giving insoluble compounds (e.g., $CaHCO_3 \rightarrow CaCO_3$) that then deposit.

Some other compounds that deposit are silica, iron and copper oxides, phosphates, etc.

The deposits act as insulators, causing the tubes to fail due to overheating. Partial obstruction or plugging of the tubes can also occur. Underneath the deposits, corrosion can occur that will damage the tubes.

- The discussion in this chapter points to the fact that the importance of corrosion, erosion, and fouling cannot be overemphasized. The failure due to these problems can occur just about anywhere in a heat exchanger. Some discussion and examples of these along with colored photographs are given in Schwartz (1981). Proper water treatment helps minimize the damage.

NOMENCLATURE

A_i	inside tubular surface area per unit length
A_0	outside tubular surface area per unit length
FRP	fiber reinforced plastic
PVC	polyvinylchloride
R_f	fouling resistance
R_{fi}	inside fouling resistance
R_{f_0}	outside fouling resistance
U_c	overall clean heat transfer coefficient
U_d	overall design (or dirty) heat transfer coefficient

REFERENCES

Anonymous. Special Feature on Linings in the Prevention of Corrosion, *Corros. Mainten.* 2(1):43–67. 1979a. A series of seven articles.

Anonymous. Protects Tube Inlets. Union Carbide, New York. *Chem. Eng. Prog.* 75(7):100. 1979b.

Anonymous. Instantaneous Corrosion Rate Meter. *The Chemical Times*, Bombay, Feb. 9, 1981a, p. 9.

Anonymous. *Chemical and Engineering News*, 59(15):25. 1981b, as reported in *The Chemical Times*, Bombay, August 17, 1981, p. 7.

Bott, T. R. Fouling due to Liquid Solidification. In *Fouling of Heat Transfer Equipment*, E. F. C. Somerscales and J. G. Knudsen (eds.). Hemisphere: Washington, D.C. 1981.

Bott, T. R., and J. S. Gudmundsson. Rippled Silica Deposits in Heat Exchanger Tubes. In *Heat Transfer 1978. Proceedings of the Sixth International Heat Transfer Conference, Toronto, August 1978.* Hemisphere: Washington, D.C., 1979.

Davison, R. M., and K. H. Miska. Stainless Steel Heat Exchangers–Part I. *Chem. Eng.* 86(3):180–184. 1979b.

Davison, R. M., and K. H. Miska. Stainless-Steel Heat Exchangers–Part II. *Chem. Eng.* 86(6):111–114. 1979a.

Elayaperumal, K. Corrosion in Heat-Exchangers–Principles, Forms, Case Histories and Prevention by Design. In *Heat Exchanger Design-II: Mechanical Design, Stress and Seismic Analyses, Corrosion in Heat Exchangers. Cost Analysis.* J. P. Gupta (ed.). C. S. Enterprises: Delhi, India. 1979.

Epstein, N. Fouling in Heat Exchangers. In *Heat Transfer 1978. Proceedings of the Sixth International Heat Transfer Conference, Toronto, August, 1978.* 6: 235–253. Hemisphere: Washington, D.C. 1979.

Epstein, N. Optimum Evaporator Cycles with Scale Formation. *Can. J. Chem. Eng.* 57(5):659. 1979.

Epstein, N. Fouling: Technical Aspects (Afterword to Fouling in Heat Exchangers'). In *Fouling of Heat Transfer Equipment,* E. F. C. Somerscales and J. G. Knudsen (eds.). Hemisphere: Washington, D.C. 1981a.

Epstein, N. Fundamentals of Heat Transfer Equipment Fouling. In *Low Reynolds Number Flow Heat Exchangers.* S. Kakaç, R. K. Shah, and A. E. Bergles (eds.). Hemisphere: Washington, D.C. 1983.

Epstein, N. Fouling of Heat Exchangers. In *Heat Exchangers: Theory and Practice.* J. Taborek. G. Hewitt, and N. H. Afgan (eds.). Hemisphere: Washington, D.C. 1983.

Gole, D. S. Corrosion Monitoring. *Corros. Mainten.* 2(2):117–118, 1979.

Iyer, S. V., and K. Balakrishnan. Cathodic Protection of Heat Exchangers–A Review. *Corros. Mainten.* 4(3):237–242. 1981.

Kern, D. Q. *Process Heat Transfer.* McGraw-Hill: New York. 1950.

Knudsen, J. G. Apparatus and Techniques for Measurement of Fouling of Heat Transfer Surfaces. In *Fouling of Heat Transfer Surfaces,* E. F. C. Somerscales and J. G. Knudsen (eds.). Hemisphere: Washington, D.C. 1981.

NALCO Chemical Co. *An Introduction to Boiler Feedwater Treatment,* Bulletin 30. NALCO Chemical Co.: Oak Brook, Illinois. 1979.

NALCO Chemical Co. *An Introduction to Cooling Water Treatment,* Bulletin 34. NALCO Chemical Co.: Oak Brook, Illinois. 1981.

Nooter Corp. *Corrosion Data Charts,* Bulletin No. 101. Nooter Corp.: St. Louis, Missouri. 1977.

Oak Ridge National Laboratory. *Design Guide for Heat Transfer Equipment in Water-Cooled Nuclear Reactor Systems,* ORNL TM-3578. ORNL: Oak Ridge, Tennessee. 1975.

Perry, R. H., and C. H. Chilton (eds.). *Chemical Engineers Handbook,* 5th ed. McGraw-Hill: New York. 1973.

Rochester, M. C., and J. H. Brunton. Influence of Physical Properties of the Liquid on the Erosion of Solids. In *Erosion, Wear, and Interfaces with Corrosion,* Special Technical Publication 567, ASTM: Philadelphia. 1973.

Schwartz, M. P. Four Types of Heat Exchanger Failures. *Plant Eng.* 35(21):145–150. 1981.

Taborek, J., T. Aoki, R. B. Ritter, J. W. Palen, and J. G. Knudsen. Fouling: The Major Unresolved Problem in Heat Transfer, I and II. *Chem. Eng. Prog.* 68(2):59–67 and 68(7):69–78. 1972.

Taprogge U.K. Ltd. *Condenser Tube Cleaning Systems for Condensers and Heat Exchangers,* Bulletin No. 78/400/19-6. Taprogge U.K. Ltd.: Leamington Spa, United Kingdom. 1978.

Tewari, G. C., and V. K. Mehandru. *Maintenance of Shell and Tube Heat Exchangers,* Memo. Engineers India Ltd.: New Delhi. 1977.

Van der Horst, J. M. A., and C. R. Sloan. Erosion-Corrosion of Finned Heat Exchanger Tubes. In *Erosion, Wear, and Interfaces with Corrosion.* Special Technical Publication 567. ASTM: Philadelphia. 1973.

Yadava, G. S., and O. P. Chawla. Maintenance of Heat Exchangers. *Corros. Mainten.* 3(4):288. 1980.

CHAPTER

FOURTEEN

INSTALLATION, OPERATION, MAINTENANCE, AND REPAIR

INTRODUCTION

- Proper installation and handling and correct start-up, emergency, and shutdown procedures are important for the successful working of a well-designed and fabricated heat exchanger/pressure vessel. Regular cleaning, maintenance, and repairs are necessary to ensure trouble-free operation of the unit for its designed life span. In this chapter we shall discuss these topics.

GENERAL

14.1 On what factors does the successful performance of a heat transfer equipment depend?

It depends upon the proper execution of the following important and critical factors:

1. Selection of proper materials for the service
2. Thermal (or process) design
3. Mechanical design and stress analysis
4. Fabrication, inspection, and testing as per the applicable codes
5. Handling during transportation to site

6. Storage prior to installation
7. Manner of installation including the design of the base and the supports
8. Piping/duct work design and their connection to the nozzles leading the fluids to and from the exchanger
9. Provision for safety valves, vent, and drain lines
10. Methods of start-up and shutdown
11. Methods of steady state operation and monitoring of the exchanger performance
12. Operating conditions being the same as the design conditions with respect to temperatures, pressures, flow rates, and compositions at the inlet
13. An adequate maintenance schedule
14. Materials, tools, and workmanship used in the maintenance, repairs, and testing
15. Provision for corrosion inhibition, if required

14.2 What may cause an exchanger not to perform as designed?

Among the important causes leading to poor performance are the following:

1. Poor quality of materials of construction
2. Inadequate thermal design
3. Poor fabrication, inadequate inspection and testing, and not adhering to the code
4. Mechanical failure due to inadequate analysis of seismic, thermal, and cyclic stresses and nozzle loads
5. Improper installation
6. Incorrect piping connections
7. Operating conditions being different than the design conditions
8. Air or gas binding resulting in less than full utilization of the heat transfer surface
9. Tube failure due to vibration or erosion
10. Incorrect steam trapping
11. Failure to remove preservative materials after storage
12. Excessive clearances between baffles and shell due to corrosion or faulty fabrication, causing flow maldistribution
13. Leakages across pass partition plates
14. Inadequate support design resulting in uneven settling and tilting, which in turn produce unforeseen stresses and flow maldistribution
15. Excessive fouling

INSTALLATION

14.3 What steps should be taken during the installation of a heat exchanger?

Installation consists of two stages (American Standard, 1978):

1. Prior planning: (a) Provide adequate foundations and supports to bear the static weight of the unit, the weight of the fluids in it, and the dynamic loads due to wind, earthquake, etc., making sure the exchanger will not settle causing strains in piping and stresses in the unit. The diameter, length, and pitch of the foundation bolts should be correctly chosen. (b) Provide enough space to remove a tube

bundle in the removable-bundle-type or a tube in a fixed-tubesheet-type exchanger at the stationary end. At the other end, provide sufficient space for removing the shell cover or rolling-in the tubes. (c) Provide space above the exchanger to facilitate its removal from the foundations and cleaning/repairs/replacement. This would need headroom for using a crane for the purpose. (d) Piping should be such as to allow bypassing of the shell side and tube side for inspection, cleaning, and repairs. Cooling water piping should have a provision for backflushing to remove the deposits. (e) Piping should be able to expand or contract freely. It should not transmit fluid pulsations and mechanical vibrations to the exchanger. (f) Provide necessary instrument taps close to the exchanger to measure its temperature and pressure. In critical usage, alarms may be installed that will sound if preset values are exceeded. (g) Do not pipe the drain connections to a common closed manifold. It makes it difficult to determine if an exchanger has been thoroughly drained or not.

2. Actual installation at site: (a) On receipt of an exchanger, inspect it for any damage in transit. (b) Remove all preservative materials if the unit had been stored prior to the installation. Dry with hot and dry air in case of low-temperature service. (c) Set the exchanger square and level so that the piping connections can be made without forcing. (d) Check all the bolts on the channel, cover, nozzles, etc., for tightness since they might relax during transport. (e) Horizontal exchanger should be able to move freely on one of the two supports to accommodate the thermal expansion. This is achieved by using slotted holes at the base of one of the supports that can thus move even with the bolts in place. (f) Inspect all openings for foreign materials, packings, desiccant bags, shipping covers, and supports. (g) Insulate the unit if very hot or very cold compared to the ambient temperature. (h) Pressure test the unit before starting.

14.4 What factors determine if an exchanger should be installed in a horizontal or in a vertical position?

Among the factors involved are the following:

1. Available space for installation and maintenance
2. Piping and layout of the system
3. Ease of operation
4. If phase change is involved (condensers, evaporators, etc.) or whether the process design of the unit is based on horizontal or vertical operation

14.5 How is a flanged joint installed?

Flanged joints are encountered in a heat exchanger at shell-to-channel joints, shell nozzles-to-piping, manholes, channel covers, etc. The proper and leakproof installation of these joints is vital for the functioning of the unit and hence of the plant at large. Various type of flanged joints are shown in Fig. 9.6.

The faces of the two flanges are brought together with a gasket in between them. The gasket seating surface should be thoroughly clean and devoid of any scratches. The bolts or studs are put in place, and the nuts tightened. The optimum degree of tightness occurs when a stress of 30,000 psi (207 MPa) is uniformly reached in each bolt or stud. To achieve the uniform tightness, the bearing surfaces of the nuts should have a good machine finish, and the threads should be properly lubricated if a torque wrench is used (Baumeister, 1978). For proper bolt tightening procedure, see Fig. 10.8. If it is not possible to measure the torque in bolt tightening, another way to ensure that the bolted joints do not leak is to pressure test the unit at 1.5 times the design pressure (see Chapter 11 for pressure testing).

OPERATION

14.6 What steps should be taken during commissioning an exchanger?

In a new exchanger, after it has been installed as discussed in Q. 14.3, the following steps should be taken for commissioning it:

1. Visible dirt and grease should be cleaned
2. Temperature- and pressure-measuring instruments should be installed as required
3. Vent and drain valves should be checked for adequacy
4. Make sure that in exchangers with expansion bellows, the bellows are free to expand
5. Before starting, any instructions accompanying the unit should be read and followed
6. At the start, efforts should be made to introduce fluids such that the differential thermal expansion between the tubes and the shell is minimum. Cold fluid circulation should be introduced gradually. This ensures that the heating rate does not exceed 400°F/hr (0.06°C/sec) and the length-wise temperature gradient 250°F/ 15 ft length (30.4°C/m) as required by the ASME Code for post-weld heat treatment (American Society of Mechanical Engineers, 1978).
7. The vent valves should be kept open so that all the air can escape, no air binding of the exchanger surface occurs, and the exchanger gets filled with the fluids. Then the vent valves should be closed.
8. If hydrocarbons are to be used, the air in the exchanger should be purged by inert gas before letting the hydrocarbons in to avoid the possibility of an explosion.
9. After the unit is brought to a steady operation, all the bolted joints should be re-tightened using a correct bolt tightening procedure.
10. The unit should never be used at pressures, temperatures, and/or flow rates higher than the designed values to avoid undue stresses in the exchanger and vibration damage to the tubes.
11. For shutting down a unit, first the hot fluid flow should be gradually reduced to nil, then the cold fluid flow can be stopped quickly. This is to minimize the differential contraction during cooling. Thermal shocks, overpressure, and/or hydraulic hammer should be avoided at the start-up, shutdown, as well as during the operation.
12. In the case of exchangers using steam, the vent/drain should be kept open after

stopping the flow of steam at shutdown to avoid vacuum formation when the steam condenses on cooling. Vacuum formation can lead to structural failure.
13. All the fluid should be drained at shutdown to avoid freezing and corrosion. The unit should preferably be washed by water and dried using air before storage.

14.7 What procedure should be followed for the start-up of units exposed to cold environmental conditions so that brittle failure is prevented?

The vessel should be gradually heated and a reduced fluid pressure should be maintained until the metal temperature is safely above the transition temperature so that brittle failure will not occur. Then the full operating pressure may be applied. During the fabrication of such vessels, special care should be taken to grind-off any notches and smooth out any sharp edges that might otherwise act as stress concentrators and lead to brittle fracture.

14.8 What controls are usually used on a heat exchanger?

Manual or motor-operated valves are used to control the flow of fluids, and a suitable flow or velocity meter is used to measure the flow rate. While pressure gauges may be installed to measure the pressure, the pressure is controlled by means of a relief valve or rupture disc that does not allow the pressure to rise above a preset value (Q. 1.107-1.110). Some exchangers have pressure drop recorders for the process fluid to monitor fouling of the unit that increases the pressure drop. Generally, no level control is used except in shell-side condenser-subcoolers and kettle reboilers. Temperature is measured by thermocouples and is controlled by adjusting the flow rates, if at all required. These measuring and control devices are of primary importance during the start-up, shutdown, emergency, or upset conditions or when a unit is used for widely different services.

14.9 If steam is used, why is it necessary to remove the condensate effectively from the exchanger?

If the condensate is not removed effectively and completely, it will (i) cause part of the heat transfer surface to be blocked, and it will not be available for heat transfer duties; (ii) corrode the shell and tubes because dissolved oxygen causes oxygen pitting, and dissolved carbon dioxide produces carbonic acid in water; (iii) result in high pressure drop due to condensate slugging; and (iv) lead to water hammer blow and vibrations in the piping.

14.10 What precautions should be observed in high-velocity gaseous flow through the shell side?

The high compressibility of a gas (in contrast to the small compressibility of liquids) affects its local velocity, especially if the pressure drop is high or if the bulk free stream velocity is 20 to 50 percent of the sound velocity under the same conditions of temperature and pressure. In such a case, the local velocity around bends, baffle windows, leakage areas, etc., is likely to break the sound velocity, and result in loud sound, flow instability, and very large pressure drop (Oak Ridge National Laboratory, 1975). Hence, the design should avoid high gaseous velocity in the shell. Furthermore, impingement plate should be provided at the entrance to avoid erosion of the tubes.

14.11 When does an exchanger need to be shutdown?

An exchanger is shutdown when one or more of the following happen:

1. Heat duty is significantly reduced due to fouling and scaling
2. Tubes leak due to corrosion, erosion, vibrations, thermal cycling stresses, seismic load, and so forth
3. Shells leak due to seismic load, foundations settling unevenly, etc.
4. Flanged joints have severe leaks due to gaskets getting crushed or improper connections (Q. 14.5)
5. Tube-to-tubesheet joint and/or expansion joint opens up
6. There is severe interpass leakage at the pass partition groove or along the edges of the longitudinal pass partition plate
7. Fluid flow is suddenly stopped producing hot spots and severe damage to the tubes
8. A change in the process or process conditions that then changes the required heat duty significantly, necessitating the replacement of the exchanger baffles, tube bundle, or the exchanger itself

Except for the last one, the rest of the situations fall under "maintenance and repair," which is discussed next.

MAINTENANCE AND REPAIR

14.12 Why is periodic maintenance of heat exchangers necessary?

Heat exchangers play a significant role in the chemical process, power, and refrigeration industries. Their trouble-free operation is generally necessary for the continuous working and hence for the economic viability of the plant (Lord et al., 1970). If periodic maintenance is not done, a major maintenance may require a long shutdown of the plant with consequential heavy monetary losses. This happens, in most cases, because spare or standby units cannot be stockpiled due to their high costs. The only two exceptions where spare units or capacity is provided are the heavily fouling or polymerization units that require cleaning after every few days, and in nuclear power plants for safety considerations.

14.13 What is done during the periodic maintenance of heat exchangers?

Essentially, the scale and fouling deposits are removed, and corroded or eroded parts are either repaired or replaced. Thus, the maintenance improves the heat transfer coefficient and retards corrosion by removing the corrosion causing deposits and obstructions to smooth flow.

14.14 When should the planning for maintenance begin?

It should begin right at the design stage itself. The design engineer can anticipate problems in the layout, start-up, operation, and shutdown of the exchangers;

and make judicious choices to minimize the maintenance cost and the shutdown time.

14.15 What information should the designer have for the proper planning of heat exchangers' maintenance?

The designer should know the following:

1. Process flowsheet
2. Plant layout and location of other structures and piping in the vicinity of heat exchangers
3. Fouling and corrosion tendencies of the process fluid, minimum velocity required to minimize fouling, and type (chemical, mechanical) and frequency of cleaning required (Chapter 13)
4. Effects of interfluid mixing and leakages into the atmosphere
5. Start-up and shutdown procedures
6. Thermal cycling, upset, and emergency conditions
7. Thermal and pressure stresses in the system
8. Wind and seismic loads
9. Flow-induced vibrations
10. Facilities available for repair and maintenance

14.16 What steps may be taken at the design stage to minimize and ease the maintenance and repair?

The following is a list of some general steps. For specific cases these may be modified or supplemented.

1. Use corrosion-resistant materials for corrosive fluids (Chapters 5 and 10)
2. Put corrosive and highly fouling fluids on the tube side
3. The shell side is easier to clean in a horizontal unit compared to in a vertical unit
4. The velocity of the fouling fluid should be kept high to reduce the fouling deposit and increase the time interval between cleanings
5. For fluids with suspended particles, use screens to remove large particles and evaluate the use of spiral plate heat exchangers (Chapter 6)
6. Avoid dead or stagnant zones in the shell by the proper sizing of baffle cuts and baffle spacing (Chapter 1). Provide nozzles for backflushing, if possible
7. Erosion of the tube ends can be reduced by smooth entrance or using nylon inserts (Fig. 13.1)
8. Erosion and vibration of the tubes should be reduced by the proper sizing and spacing of the impingement baffles, cross-baffles, and the support plates
9. In horizontal cut baffles, notches should be made at the bottom to drain the unit completely
10. Required tools, necessary replacement tubes, and gaskets should be procured in advance of using the unit
11. In case of severe thermal stresses due to differential expansion, use U-bundles or floating-head exchangers instead of the fixed-tubesheet units. If the fixed-tubesheet units must be used, incorporate expansion bellows to account for the differential expansion (Fig. 1.32)

12. Effects of sudden failure of cooling water or heating steam should be evaluated and taken into account
13. For two-phase fluids, vapor belts should be used to reduce the entrance velocity (Fig. 1.22)
14. Standardize the size and materials of the tubes in different exchangers in a given plant to the extent possible to minimize the inventory of the spares and tools and to reduce the time for maintenance. Spare tube bundles may be kept in case of multiple units of the same dimensions
15. Proper arrangement for venting the shell and channel sides, especially in case of multi-passes, should be made
16. Pressure relief valves should be installed if high pressure build-up can occur
17. Restrict cooling water outlet temperature to less than 120°F (49°C) so that microbial fouling is reduced or totally avoided, and inversion of orthophosphate (a water treatment chemical) to metaphosphate does not occur
18. Use seal strips and dummy tubes to reduce bypass flow passages where otherwise excessive fouling will result due to scaling, and heat transfer will be severely reduced (Fig. 1.29)
19. For outside tube cleaning, square layout should preferably be used (Fig. 1.11)
20. For protection against vibration damage, see Chapter 15
21. In horizontal condensers, use vertical cut baffles to avoid damming up the condensate behind the baffles
22. Nozzle locations, head designs, and piping layouts should be such that the heads or covers can easily be removed for inspection/cleaning/removal of the tubes with minimum disturbance to the piping
23. Enough space should be available in front of the exchanger to be able to remove a tube bundle
24. Consider the use of insulation if heat loss to (or heat gain from, in case of cryogenic heat exchangers) the ambient temperature is going to be significant, and also for personnel safety
25. Effects of power failure, especially if it would result in freezing of a stream, should also be considered. Thought should also be given to restarting a frozen system
26. In case of air cooling, the minimum and maximum air temperatures and humidity should be kept in mind
27. Maintenance and repair forms should be devised at the design stage itself. The forms will clarify a number of points
28. If steam is used, the problems associated with condensate control and removal should be kept in mind (Mathur, 1973)
29. Ease of maintenance should be given as much importance as the performance in the design and selection of the equipment (Srinivasan, 1981), otherwise, the performance would decrease with difficult to maintain units
30. Easy availability of equipment and spares for the maintenance is of prime importance. Time taken for imports and possible import restrictions by local government and export restrictions by foreign governments should be kept in mind
31. Discuss maintenance with the maintenance group *before* the final fabrication

drawings are released; it may lead to a more functional design that is easy to maintain

32. Have maintenance personnel work with the manufacturers during the final assembly and fit up and with the construction group during erection of equipment, if possible. This should be settled with the parties before the contract is finalized
33. Interchangeability and rotatability of the tube bundles helps in better performance by a set of units. Since tube bundles do not corrode evenly, rotation would help increase the bundle life without affecting the heat transfer or pressure drop adversely. A spare tube bundle can also be kept if multiple units have similar bundles
34. A device or a marker that ensures that the tube bundle is properly oriented each time it is installed in the shell helps save time during the reassembly after cleaning
35. Ease of access to the units should be provided by the use of permanent ladders, stairs, and platforms
36. Crane facilities should be provided in order to lift heavy units and parts
37. Provision for the isolation of a unit without shutting down the entire plant should be made by using bypass lines and valves
38. In a fixed tubesheet unit, the designer should specify where it may be cut for disassembly and how many times this operation may be performed during the life of unit (Oak Ridge National Laboratory, 1975)
39. If decontamination of the unit is required, adequate provision for it should be made in material selection, piping, and nozzles for the decontaminating solution, etc. (Oak Ridge National Laboratory, 1975)
40. Special tools required for opening, cleaning, maintenance, and repair of the unit should be specified. These may include tube-rolling and welding equipment, eyebolts and wrenches, slings and dollies for the tube bundle, inspection and testing tools, etc. (Oak Ridge National Laboratory, 1975)
41. Lifting lugs should be provided on the shell, tubesheet, and channel covers for ease in lifting them

This is not an exhaustive list. The vast experience of the designers, operators, and the maintenance crew should be combined for the best results.

14.17 What special tools, jigs, and fixtures are required for the maintenance of an exchanger?

These include the following:

1. Flaw detection equipment such as borescopes and ultrasonic ones
2. Various nondestructive testing and examination equipment
3. Tube-to-tubesheet rolling and welding gear
4. Tube plugging, crimping tools
5. Tools to remove the fouling deposit both from the outside and the inside of the tubes
6. Special wrenches to loosen and tighten the bolts
7. Compressed air for soap suds test

14.18 How should the maintenance area be planned?

In the maintenance area, facilities, such as water and high-pressure air supply, electric and pneumatic outlets, properly sized and laid out drains, adequate lighting and ventilation, and space for the location of the required machines to lift the exchangers and the tube bundles and to place the tube bundles should be provided for. Also provide good quality helmets, goggles, safety shoes, gloves, etc., to the maintenance crew and ensure their use.

Maintenance Manual

14.19 What information should the maintenance manual contain?

The maintenance manual is also known as a "service manual." It should contain the following:

1. Information on materials of construction, wear tolerances, corrosion rates, and allowances
2. Assembly and disassembly procedures
3. Perspective drawings and flow sheets using distinctive colors so that correct visual distinction is possible
4. Fabrication drawings and code procedures used in fabrication
5. Thermal design specifications
6. Forms to record the details of all the maintenance carried out
7. Records of previous maintenances, materials and time used, down-time, costs, and detailed observations by the plant personnel
8. Spare parts needed, with details of dimensions and material specifications
9. Maintenance schedule and how it should progress from one piece of equipment to the other
10. Scheduled inspections and inspection procedures to be done during both the operation and the shutdown
11. Leak detection methods
12. Procedures for the repair of the shell, channel, impingement baffles, tube-to-tubesheet joints, tube plugging, or tube replacement
13. Start-up and shutdown procedures

14.20 What kind of cleaning/maintenance records should be maintained for a better planning of the maintenance?

The records should have the following key information:

1. Date of last cleaning/maintenance
2. Nature of deposit
3. Equipment used for cleaning
4. Method adopted for cleaning
5. Number of tubes plugged or replaced and their locations
6. Reasons for plugging the tubes: corrosion or vibration damage, or any other

7. Visual inspection records of how the tubesheets, tubes under the nozzles, tube-to-tubesheet joints looked; and comments on the progress of any corrosion or erosion, and so forth
8. Repairs done to any corroded parts
9. Records of welding repairs that include edge preparation, welding electrodes used, and method of welding

14.21 List the steps in the maintenance of a heat exchanger.

The important steps are the following:

1. Periodic cleaning should be done to remove the fouling/scaling that results in reduced thermal performance and increased pressure drop
2. Tubes should be tested for any leakages by hydraulic/pneumatic or other tests
3. Every time a heat exchanger is reassembled, a new set of gaskets should be used to prevent leakage since old gaskets generally do not seal well the second time
4. Adequate record of maintenance should be kept for future reference

14.22 How can one determine whether the maintenance has been effective or not?

Each industry can establish its own yardstick(s) to measure maintenance effectiveness, which can be assessed in one or more of the following ways (Mitra, 1978):

1. On-stream availability
2. Unit product cost
3. Unitwise maintenance cost (labor, material, spares inventory, tools, wear and replacement, etc.)
4. Value of production and its quality

14.23 What factors contribute to a poor maintenance?

These are the following:

1. Nonplanning of the maintenance work
2. Maintenance jobs not systematically recorded and seldom analyzed
3. Breakdowns not properly documented and make-shift arrangements used to overcome them
4. Refresher training of the maintenance personnel lacking
5. Spare parts and/or proper tools not available when required
6. Maintenance personnel not treated as equals with the operational, administrative, and financial management personnel
7. Lack of coordination at work site between the operating, maintenance, and inspection departments

14.24 What is in-service inspection? Why is it done?

In-service inspection refers to inspecting various parts and components of a unit periodically during its entire service life. Items on the outside can be inspected when the unit is in operation while those on the inside can only be inspected when the unit is shutdown. "In-service" should not be confused with"In-operation."

In-service inspection is done to assess the future serviceability of the exchanger or its components. All data should be preserved for comparison with the data obtained in future inspections.

14.25 What areas of an exchanger require in-service inspection?

Some typical areas are the following:

1. Longitudinal and circumferential welds
2. Nozzle-to-vessel head joints
3. Bolted joints
4. Integrally welded exchanger supports
5. Cladding on the tubesheets and channels
6. Tube-to-tubesheet joints
7. Impingement baffle, if accessible
8. Shell side of the tubes and tubesheets, if accessible
9. Tie-rods and spacers
10. Inner surface of the shell
11. Vent and drain valves
12. Expansion bellows
13. Tubes near the baffles to check any excessive fouling on the outside due to dead or low-velocity zones
14. Welds in critical vessels and pipes must be examined regularly since these might develop cracks due to hydrogen getting into them if moisture was present in the electrode or in the environment, or if the fit-up was not good while welding (Chapter 10). A radiographic examination immediately after the welding may not show any defects since these defects take time to show up.

14.26 What is meant by "integrated maintenance" of a plant?

It means keeping, operating, and maintaining the equipment in an industry so as to ensure high-capacity utilization of the plant, longer life of the equipment, minimum cost of production, high quality of the product, and congenial work environment (Sanyal, 1981).

14.27 How can a computer help in the maintenance?

A computer can help in numerous ways, such as the following:

1. Developing preventive maintenance schedule
2. Keeping track of spare parts
3. Assessing manpower utilization and keeping overtime payments to a minimum
4. Keeping track of pending jobs to be done during shutdown for maintenance
5. Determining maintenance cost for each piece of equipment
6. Evaluating equipment and spare parts from different manufacturers and suppliers
7. Evaluating the performance of maintenance crews
8. Keeping track of all previous maintenance
9. Analyzing frequency of specific problems
10. Analyzing cost of downtime as far as production is concerned

11. Keeping track of revised maintenance procedures
12. Analyzing the performance of the units to anticipate any developing problems and to initiate action to handle them
13. Planning replacement of equipment
14. Collecting and analyzing maintenance information from different plants for the benefit of all, i.e., creating a "maintenance data bank" for similar equipment
15. Preparing the whole schedule of preventive maintenance and maintenance planning (Figs. 14.1 and 14.2)

14.28 In an operating heat exchanger, how can one determine if one or more tubes or tube-to-tubesheet joints are leaking?

One or more of the following events will indicate that there is a leak in the unit:

1. A sample of the low-pressure fluid shows contamination by the high-pressure fluid. This may be observed by eyes alone or by chemical analysis. In case of hazardous liquids, continuous monitoring is required
2. Sudden change in the pressure drop and in the outlet temperatures of the fluids
3. A change in the amounts of outlet fluids compared to the amounts at the inlet

14.29 How are the specific leaky tubes and leaky tube-to-tubesheet joints detected?

After it has been ascertained that there is a leak in the unit, one has to determine which tubes or tube-to-tubesheet joints are leaking. The general procedure to do this is as follows:

1. Empty the unit
2. Remove the channel covers or bonnets
3. Vent the shell side and fill it with water or gas (if water cannot be used)
4. Observe the tube-to-tubesheet joints and tube ends for leaks. To detect gas leaks,

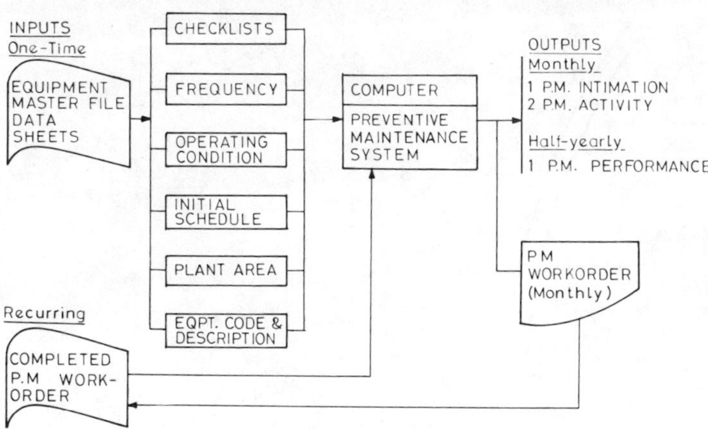

Figure 14.1 Flow diagram for a preventive maintenance system. *(From Patra and Shah, 1981. Reprinted with permission from* Corrosion and Maintenance *4(2):105–108, 1981.)*

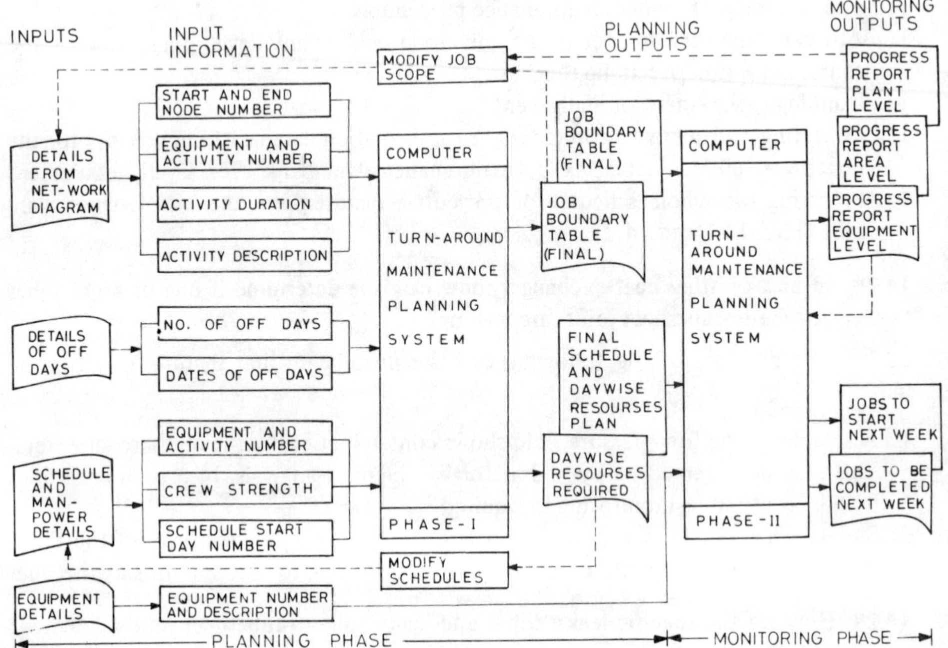

Figure 14.2 Flow diagram for a computerized turn-around maintenance planning system. *(From Patra and Shah, 1981. Reprinted with permission from* Corrosion and Maintenance *4(2):105–108, 1981.)*

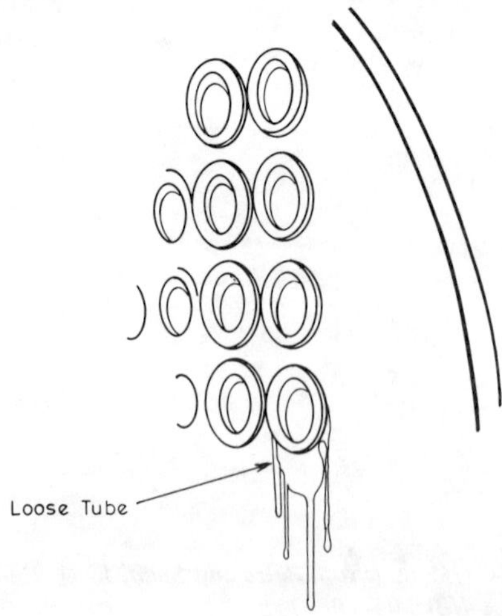

Figure 14.3 Leaky tube-to-tubesheet joint.

soap solution or other agents may have to be used. Figure 14.3 shows a leaky tube-to-tubesheet joint

5. Sometimes both the hydrostatic and pneumatic tests may be required to detect the minutest leaks. Other tests may also be used.
6. In double-tubesheet heat exchangers, it is not easy to locate the leaking tube-to-tubesheet joints for the tubes well inside the bundle, especially if the tubesheets are fixed type. If the tubesheets are the removable type, as in a U-bundle or outside packed-head case, an annular ring can be welded to the two tubesheets with the bundle out of the shell (Fig. 14.4). The annular ring is pressurized and the tube-to-tubesheet joints observed for leaks

14.30 How are the leaky tubes plugged/replaced?

Leaky tubes are plugged by hammering the plugs of brass, steel, Monel, stainless steel, etc., to seal them off. Generally, tapered plugs (Fig. 14.5) are used when the shell-side pressure is less than the tube-side pressure, and cylindrical plugs are used when the tube-side pressure is less. Plugs are introduced at both ends of the leaky tubes (Tewari and Mehandru, 1977). Such plugs are sometimes welded to the tubesheet. Recently, explosive tube plugging has been proposed for quick and effective plugging in high-pressure exchangers (Stanko, 1979). The plugs are hollow, tapered, and have several projecting rings (Fig. 14.6) that provide the necessary seal when the plugs are expanded by means of explosive charges detonated within the plugs. Several plugs can be introduced simultaneously. An explosively expanded plug is shown in cross section inside a tube in Fig. 14.7. As far as the replacement of a leaky tube is concerned, first it has to be removed. Its welded and/or rolled joint with the tubesheet has to be undone. The weld can be cut by rotary seal weld removal cutters. For thin-gauge rolled tubes, a collapsing tool is used which is driven in between the tube O.D. and the tubesheet hole, thus collapsing the tube and loosening it. In order to safeguard any possible damage to the tubesheet hole, the tube wall may be thinned by using a tube wall reducing tool (tube drill) and the tube is then knocked out by a knockout tool. Then the grooves in the tubesheet hole are cleaned by a suitable tool. Thereafter, a new tube is put in the usual fashion. Some of the tools noted above are shown in Fig. 14.8.

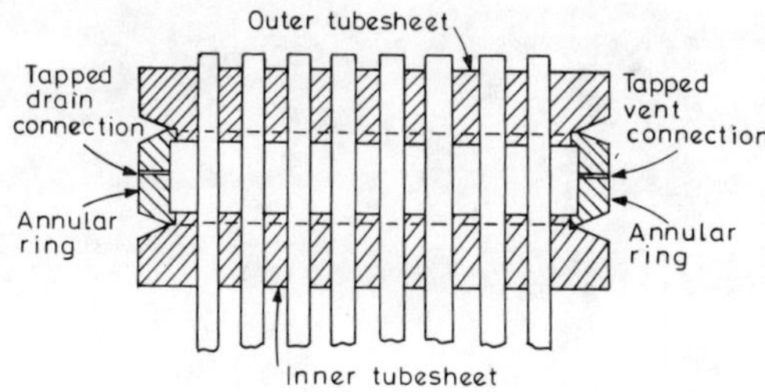

Figure 14.4 Annular ring for testing double tubesheets. *(From Rubin, 1961. Excerpted by special permission from CHEMICAL ENGINEERING, July 24, © 1961, by McGraw-Hill, Inc., New York, NY 10020.)*

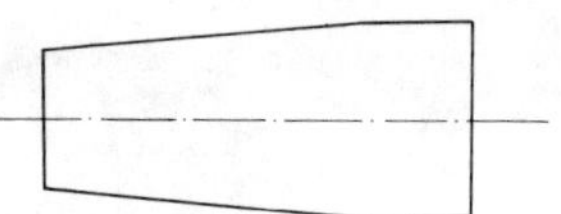

Figure 14.5 Tapered plug.

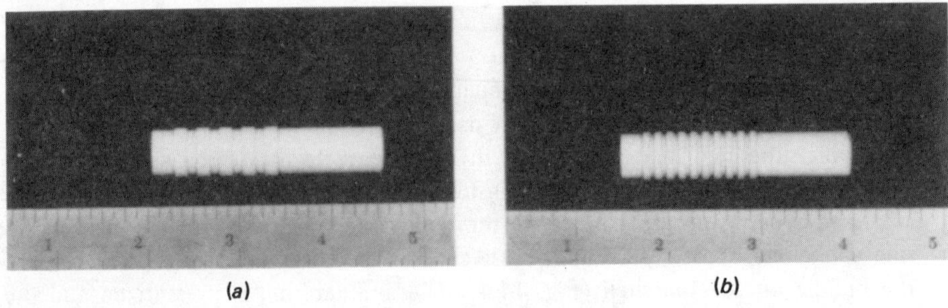

Figure 14.6 Plugs used in explosive tube plugging. *(From Stanko, 1979.)*

14.31 What precaution should be observed when unbolting flanged connections?

When the piping to the nozzle is unbolted, the piping may spring out due to cold stress, misalignment, and/or weight-induced movements. These can hurt the workers and nearby equipment (Perry et al., 1963). Similarly, if manholes are unbolted too quickly, pressure and accumulated liquid will be released. Hence, while unbolting, initially the bolts should only be loosened, but not completely removed, in order to release the pressure and the accumulated liquid, and to see if the piping springs back. Only then should the bolts be carefully removed, taking care to hold the units and prevent them from falling.

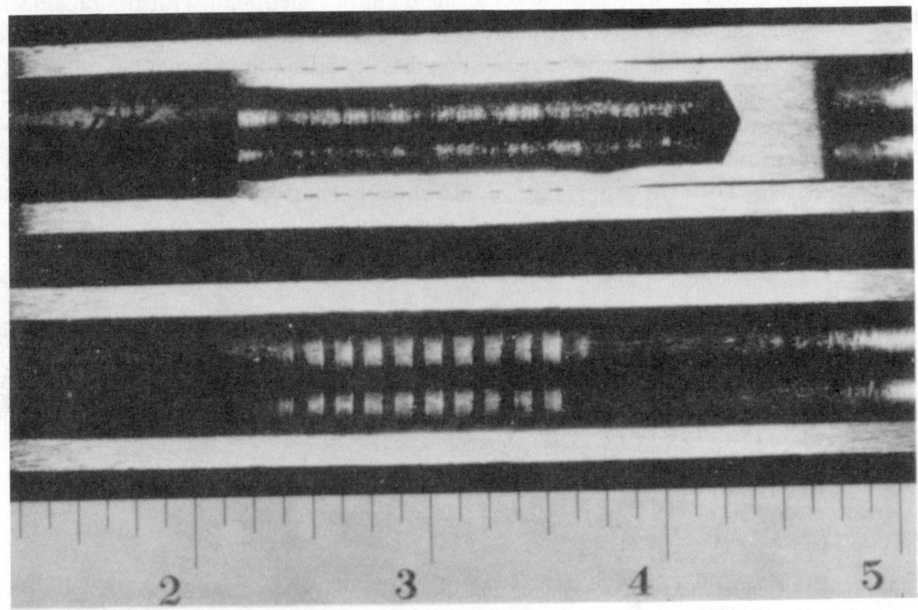

Figure 14.7 A plug in place using explosive plugging (top). Impressions in tube made by projectional rings (bottom, 1.6×). *(From Stanko, 1979.)*

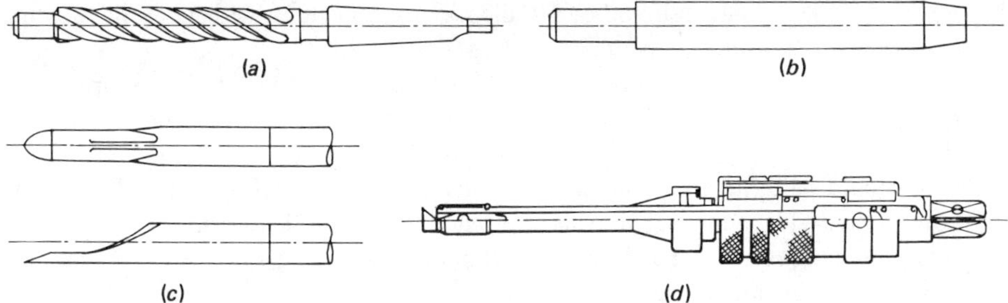

Figure 14.8 Some tools used in maintenance of heat exchangers. (*a*) Tube wall reducing tool (tube drill). (*b*) Knockout tool. (*c*) Collapsing tool, 2 views. (*d*) Cutting tool. *(Courtesy Sugino Machine Ltd.)*

14.32 What precautions should be observed when removing a tube bundle?

The following precautions should be observed while removing or moving a tube bundle:

1. Bundle should be pulled from the shell by means of eyebolts
2. Bundle should be lifted by the tubesheets, baffles, or support plates and never by the tubes since the tubes cannot take the weight of the bundles
3. The bundles should be moved on a cradle with the tubesheet or baffles resting on it.

14.33 How are the inside and the outside of the exchanger tubes cleaned?

The inside of the tubes can be cleaned by (a) flushing with high-velocity water, (b) flushing with chemical solutions or hot wash oil depending on the type of the deposit, (c) long wire brushes, (d) circulating sponge balls that clean at random while the unit is in operation, (e) using cutting tools to cut through hard deposits but taking care not to damage the tubes, and (f) for U-tubes, the "U" portion can be cleaned by using rotating flexible shafts.

For the outside of the tubes in the fixed tubesheet units, the cleaning may be done by (a) flushing with water and (b) flushing with chemical solutions.

For the outside of the tubes in a removable bundle case, the cleaning may be done by (a) flushing with high-velocity water jets and (b) long wire brushes.

Both the inside and outside of the tubes in a removable bundle may be cleaned by hydro-sand blasting in which a mixture of sand and water at a high pressure [1600 psi (11 MPa) or higher] is forced by a hose into or over the tubes in a room specially prepared for the purpose (Butler and Coleman, 1956) (see also Chapter 13).

14.34 What are the necessary steps in the care of a pressure vessel?

These are the following:

1. Protect against excessive pressure by providing relief valve, rupture disc, etc. (Q. 1.107-1.110)

2. Vacuum-breaking valves should be installed if a vessel cannot withstand the vacuum
3. Local hot spots should be avoided
4. Very low temperatures should be avoided since the material becomes brittle. A vessel at low temperature should not be pressurized until it has been heated to about 100°F (38°C)
5. Care should be taken to minimize corrosion on the outside of the vessel. For corrosion on the inside, adequate allowance should be made in the design
6. Bolt tightening should be done according to a schedule procedure so that the gasketed joint is not damaged (Fig. 10.8)
7. The gasksets should not be scratched
8. Any repair work should be stress-relieved depending upon the material of construction, its thickness, and code requirements
9. Periodic in-service inspection is a must for corrosion, erosion, and other kinds of damages, leakages and cracks, etc. (Q. 14.24-14.25). During reassembly, utmost care should be taken

14.35 How is the work of the maintenance group different than that of the operations group?

Unlike operations personnel, maintenance personnel do the following (Sanyal, 1981):

1. Often work in speed/haste and under tension since downtime means lost production
2. Work at odd hours and for long periods at a stretch
3. Work in unfavorable environment and at hazardous locations
4. Often work with improper tools and inadequate spare parts
5. Keep up with changing technology
6. Often tolerate second-class treatment since their efforts in maintaining the profitability of the plant are seldom recognized.

14.36 What areas should the maintenance manager be familiar with for carrying out effective maintenance?

A maintenance manager has to be an expert and should know the latest trends in numerous areas, such as the following (Mitra, 1978):

1. Process flow sheet
2. Mechanical design
3. Material specifications
4. Fabriction and welding techniques
5. Quality control, inspection, and testing
6. Corrosion control and cleaning
7. Maintenance and repair techniques
8. Useful life of equipment
9. Management and supervision of personnel, spare parts, and tools
10. Training facilities for the maintenance personnel
11. Performance analysis (both successes and failures)
12. Objectives of production

Since one person may not be an expert in all the above, it is important that the maintenance manager be assisted by a team of experts.

REFERENCES

American Society of Mechanical Engineers. ASME Pressure Vessel Code Section VIII, Div. 1, as stated in Risks Present in Start-Up of High-Pressure Vessels. *Oil Gas J.*, 74(51):51–54. 1976.

American Standard. *Storage, Installation, Operation and Maintenance of Heat Exchangers,* Bulletin 104-17. Heat Transfer Division, American Standard: Buffalo, New York. 1978.

Baumeister, T. (ed.-in-chief). *Marks' Standard Handbook for Mechanical Engineers,* 8th ed. McGraw-Hill: New York. 1978.

Butler, R. E., and R. P. Coleman. Exchanger Preventive Maintenance Pays. *Petroleum Refiner* 35(1):111–114. 1956.

Lord, R. C., P. E. Minton, and R. P. Slusser. Guide to Trouble-Free Heat Exchangers. *Chem. Eng.* 76(11):153–160. 1970.

Mathur, J. Performance of Steam Heat Exchangers. *Chem. Eng.,* 80(18):101–106. 1973.

Mitra, M. S. Strategy and Action Plan for Maintenance. *Corros. Mainten.* 1(3):191–195. 1978.

Oak Ridge National Laboratory. *Design Guide for Heat Transfer Equipment in Water-Cooled Nuclear Reactor Systems,* ORNL TM-3578. ORNL: Oak Ridge, Tennessee. 1975.

Patra, R. K., and N. C. Shah. Maintenance Information System. *Corros. Mainten.* 4(2):105–109. 1981.

Perry, R. H., C. H. Chilton, and S. D. Kirkpatrick (eds.). *Chemical Engineers Handbook*, 4th. McGraw-Hill: New York. 1963.

Rubin, F. L. Testing for Leaks in Heat Exchangers. *Chem. Eng.* 67(15):160–166. 1961.

Sanyal, A. Maintenance of Conventional Chemical Plants and Process Industries. *Corros. Mainten.* 4(2):87–97. 1981.

Srinivasan, M. R. Thirty Reactor-Years of Maintenance Experience–An Introspection. *Corros. Mainten.* 4(2):79–85. 1981.

Stanko, G. Explosive Tube Plugging. In *Explosive Welding, Forming, Plugging and Compaction-PVP-44.* I. Berman and J. W. Schroeder (eds.). ASME: New York. 1979.

Sugino Machine Ltd. *Sugino Tubework Tools Catalog No. 7505-1.* Sugino Machine Ltd.: Tokyo, Japan. 1975.

Tewari, G. C., and V. K. Mehandru. *Maintenance of Shell-and-Tube Heat Exchangers.* Engineers India Ltd., Heat and Mass Transfer Division Memo: New Delhi, India. 1977.

CHAPTER

FIFTEEN

FLOW-INDUCED VIBRATIONS

INTRODUCTION

All tubes have a fundamental natural frequency (and higher harmonics) of vibration. When the forcing frequency due to the flow of fluid over a tube is close to or equal to the tube's fundamental natural frequency or any of its higher harmonics, resonance sets in and the tube vibrates with an increasing amplitude to a quick failure with adverse consequences, unless sufficient damping exits. Even without resonance, the continuous vibration of a tube can lead to failure due to fatigue.

Flow-induced vibrations have become important since the late 1950s when large heat exchangers were first made. Although there are no mandatory guidelines as yet, the problem is important enough for the TEMA Standards (Tubular Exchanger Manufacturers' Association, 1978) to cover it in the "Recommended Good Practice" section. It has also constituted a subcommittee to come up with mandatory guidelines. A Heat Exchanger Vibration Data Bank has also been created (Chenworth and Taborek, 1980). All these point to the urgency and importance with which the world-wide heat exchanger community views the vibration problem. Therefore, it is appropriate that this chapter be devoted to discuss the important aspects of this problem.

GENERAL

15.1 What is meant by flow-induced vibrations?

Webster's dictionary defines vibrations as a periodic motion of the particles of an elastic body or medium in alternately opposite directions from the position of equilibrium when that equilibrium has been disturbed.

In the case of a heat exchanger tube bundle, the tubes, which are elastic bodies, are disturbed from their equilibrium rest position by the flow of fluid over the tube bundle. The tubes undergo vibratory motion. This motion is called the flow-induced vibrations.

The shell, tie-rods, and the baffles are also prone to vibrations, but these components are much more rigid than the tubes and hence are not easily excited into vibrations.

15.2 Why has the vibration analysis acquired more importance?

Some of the important reasons for this are the following:

1. Emphasis on cost optimization requires designs that give high heat transfer coefficients. This requires high flow rates that may lead to damaging vibrations.
2. Sometimes, instead of the cost, the space may be at a premium as in a nuclear power plant, and hence a smaller unit with higher flow rates may be needed with a higher probability of vibrations.
3. Higher velocity not only gives higher heat transfer coefficient but it also gives lower fouling resistance. Hence, the designs tend toward higher flow rates.
4. Safety considerations in nuclear and toxic chemical plants require protection against possible vibration damage that would otherwise result in the leakage of dangerous fluids.
5. Leakage of expensive heavy water used in heavy-water-moderated nuclear power plants due to tube failure is unacceptable.
6. The downtime due to the vibration damage of the tubes proves very costly due to loss in production.
7. In order to reduce the pressure drop, the indiscriminate use of the baffles is discouraged. This leads to longer tube spans with higher probability of vibrations.

Both the high fluid velocity and the longer tube spans lead to flow-induced vibrations. Hence, it is important to analyze for them at the design stage itself.

15.3 What is the current state of knowledge regarding exchanger designs to prevent vibration damage?

It is now possible in most cases for an experienced person to design heat exchangers that will be free of any flow-induced vibration damage for their intended service life and still not be too conservative economically. It does require proper periodic maintenance and monitoring for the unexpected. This statement should not make one complacent because a lot of work is still required to be done to gain more confidence in the designs and to evaluate the effects of sudden upsets, etc. The encouraging point is that a large number of researchers are working on the problem, and more definitive correlations are on the way.

15.4 Are the current vibration prediction methods completely reliable?

No. In general there is no completely reliable set of data or methods available to predict the damaing vibrations. Hence, one's own experience and judgment must be used to draw conclusions from the prediction methods. The methods themselves vary a lot. Some predict the forcing frequencies (such as vortex shedding, turbulent buffeting, etc.) that excite damaging vibrations when they match with the natural frequencies of the tubes; some others predict the limiting values of parameters (such as critical velocity, midspan deflection, etc.) above which the damaging vibrations occur; yet some others predict the damage that might be done to the tubes (such as baffle damage, collision damage, etc.). The various terms used in this paragraph will be defined at appropriate places in this chapter.

15.5 Are vibrations more severe in nuclear power plants than in the process industries?

No. The vibrations encountered depend upon the design and the operating conditions of the exchanger and not on the type of industry. Hence, one should be as careful in nonnuclear industries to prevent the damaging tube vibrations as in the nuclear ones if, in the former case, the tube failure would lead to a dangerous situation or huge economic losses. It must be mentioned, though, that most of the studies related to the tube vibrations owe their origin to the concern first expressed regarding the damaging tube vibrations in the nuclear power plant heat exchangers.

DEFINITIONS OF INTEREST

15.6 What are (1) simple and continuous beams, (2) natural frequency of vibrations, (3) fundamental and harmonic frequencies, (4) mode, (5) resonance, (6) damping, (7) damping factor, (8) virtual mass, (9) added mass, (10) Strouhal number?

1. Simple and continuous beams: A simple beam has two simple supports (Fig. 15.1*a*). A continuous beam has multiple intermediate supports (Fig. 15.1*b*). A heat exchanger tube is analyzed as a continuous beam.
2. Natural frequency of vibration: All structures such as bridges, tall stacks, cables between poles, beams, and tubes supported at the baffles in a heat exchanger, vibrate when displaced momentarily from their normal position by any force(s) acting on them. The number of cylces of vibrations per second that the structure

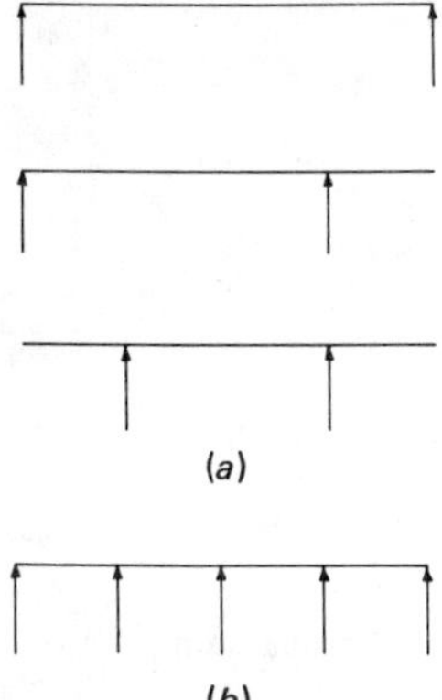

Figure 15.1 (*a*) Simple and (*b*) continuous beams.

undergoes is called the natural frequency of vibrations. A structure may have an infinite number of discrete natural frequencies. The time taken per cycle is called the time period of vibration.

3. Fundamental and harmonic frequencies: Suppose there is a string held tight and tied between two rigid supports (Fig. 15.2*a*). Some of the ways in which it can vibrate are shown in Fig. 15.2*b–e*. Figure 15.2*b* shows it vibrating with a half-wavelength. This is called the fundamental frequency or the first harmonic and has the lowest value. Figure 15.2*c* shows it vibrating with a full wavelength with a node or stationary point in the middle. This is called the second harmonic, and the wavelength in this case is half that in the case of Fig. 15.2*b*. Similarly, Fig. 15.2*d* has two nodes and is called the fourth harmonic, and so on. All of these are natural frequencies.
4. Mode: It is the maximum amplitude shape of a beam or a tube when it is vibrating in one of its natural frequencies (V. Sunderarajan, personal communication). The position related to the fundamental frequency is called the first mode, that related to the second harmonic is called the second mode, and so on (Fig. 15.3). While the natural frequencies are independent of the way the tubes are excited or of the amplitude of vibration, the reverse is not true. The amplitude of vibration for a given mode number depends upon the frequency and strength of the exciting phenomenon and the system damping. For different modes, the related harmonics of the natural frequency also determine the amplitude.
5. Resonance: When the forcing frequency matches with the natural frequency (fundamental or any of its harmonics) of the structure, the amplitude of vibrations of the structure increases. This is called the resonance. It can lead to the failure of the structure, unless the damping is high.
6. Damping: It is the ability of a structure to dissipate energy (Blevins, 1977) (see Q. 15.36–15.40).

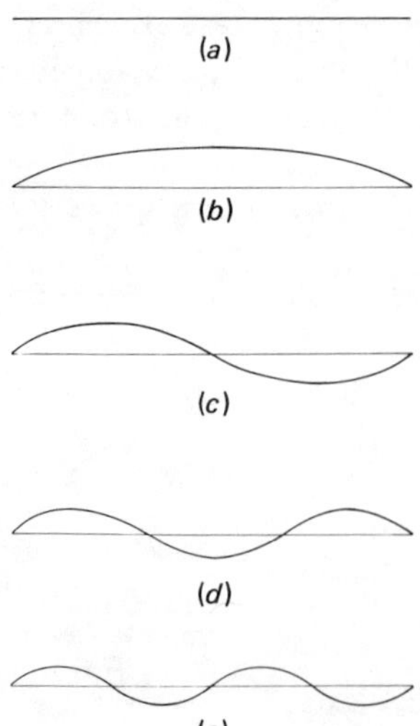

Figure 15.2 Mode shapes of fundamental and harmonic frequencies.

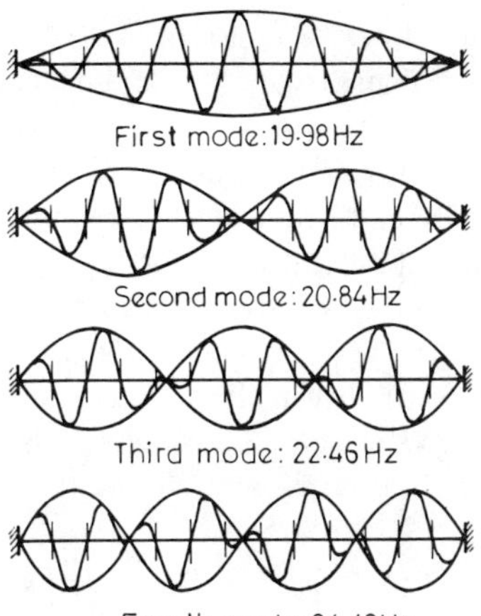

Figure 15.3 Modes of vibrations. *(Reprinted with permission from G. W. Schwarz,* Chem. Eng. *83(14):134, 1976. Copyright © 1976 by McGraw-Hill, New York, NY.)*

7. Damping factor: It is a measure of the energy dissipated per cycle during vibration and is given as

$$\zeta = \frac{\text{energy dissipated per cycle}}{4\pi \times \text{total energy of the structure}} \tag{15.1}$$

where ζ is the damping factor or damping ratio (Blevins, 1977).

For linear, viscously damped structures, a logarithmic decrement of damping, δ, is defined as

$$\delta = 2\pi\zeta = \ln \frac{Y_2}{Y_1} \tag{15.2}$$

where Y_2 and Y_1 are the amplitudes of two consecutive cycles (Fig. 15.4). For multi-span tubes, ζ is generally nonlinear, since it varies with the amplitude but is usually assigned a fixed value for any analysis. Its value for more than five-span heat exchangers, built according to the TEMA Standards, usually lies in the range of 0.02 to 0.05 (Franklin et al., 1978). Damping factor varies with the vibration mode as well as with the other system parameters.

8. Virtual mass: It is the mass of a tube plus the mass of the fluid inside the tube and the mass of the shell-side fluid displaced by the vibrating tube. For gases

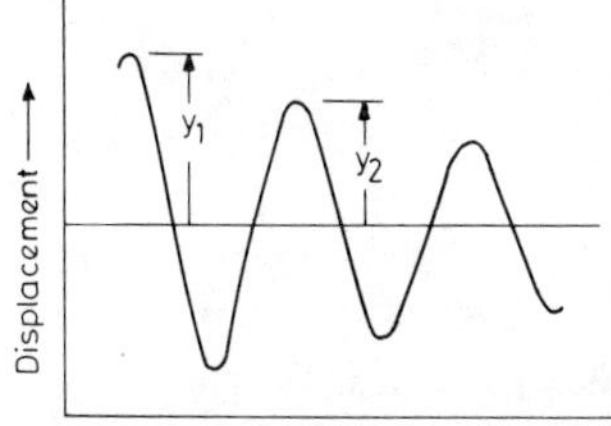

Figure 15.4 Damping of vibrations.

at ordinary pressures, the mass of the fluid is small. However, it is significant for gases at high pressures and for the liquids, since their densities are high and hence have more effect on the inertia of the tubes.

9. Added mass: It is the mass of the fluid entrained by a vibrating tube. The mass displaced by a vibrating tube is greater than that displaced by a stationary tube and depends upon the tube layout, angle, pitch to tube O.D. ratio, and the frequency of vibration. The natural frequency of the tube is lowered due to the added mass.
10. Strouhal number (Sl): It is a proportionality constant between the predominant vortex shedding frequency, f_{vs}, and the free stream velocity of the fluid normal to the tube, V, divided by the tube diameter, D.

$$f_{vs} \propto \frac{V}{D} \tag{15.3}$$

$$f_{vs} = \text{Sl}\,\frac{V}{D} \tag{15.4}$$

$$\text{Sl} = \frac{f_{vs}D}{V} \tag{15.5}$$

For a tube bank, the average fluid velocity at the minimum flow cross section is used in Eqs. (15.3)-(15.5). Plots for the Strouhal numbers for in-line and staggered tube arrays are given in Figs. 15.5 and 15.6. For a single tube or cylinder, the Strouhal number has a constant value close to 0.2 for a wide range of Reynolds numbers form 700 to 5×10^4 (Griffin, 1978; Gupta, 1979). The effect of adjoining tubes in a tube bank is to change the vortex shedding frequency and the Strouhal number, which depends on the tube layout (Figs. 15.5 and 15.6).

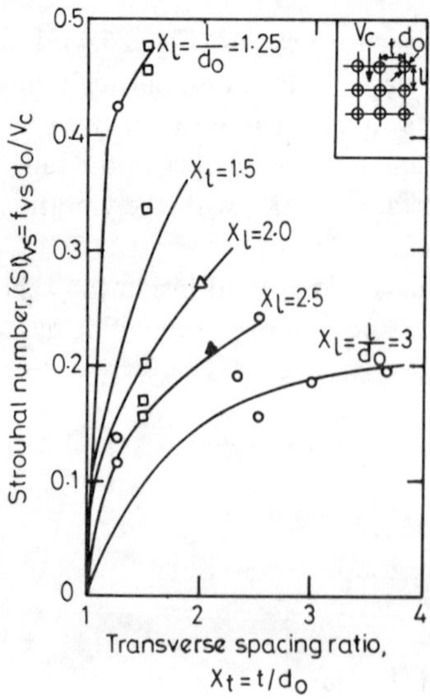

Figure 15.5 Strouhal number for in-line tube layout. *(From Fitz-hugh, 1973.)*

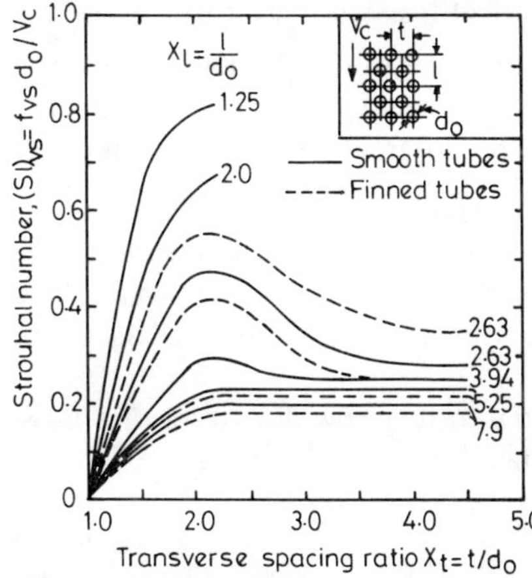

Figure 15.6 Strouhal number for staggered tube layout. *(From Fitz-hugh, 1973.)*

FACTORS CAUSING VIBRATIONS

15.7 What factors cause vibrations in a shell-and-tube heat exchanger?

Vibrations are caused by mechanical vibrations of the support system; hydraulic-coupled noise in the entering stream due to the pump or compressor; tube-side fluid flow; and shell-side fluid flow. We shall primarily be concerned with shell-side fluid flow. The first two factors can be taken into account by the proper design of the piping, support systems, and the auxiliary equipment. The third factor does not contribute significantly to the vibrations in a unit unless parallel flow is encountered over long spans of unsupported tubes.

EFFECTS OF SHELL-SIDE FLOW

15.8 How does the fluid flowing over a tube bundle in a shell result in tube vibrations?

The lift and drag forces exerted by the fluid deform the tube bundle whose orientation to the fluid flow changes dynamically. This may, in turn, result in a change in the force and in the direction in which it acts on the tubes. The tubes thus feel a fluctuating pressure. This results in a dynamic interaction between the tube bundle and the fluid, resulting in the vibrations of the tubes.

Furthermore, the presence of the fluid couples the vibration of one tube with that of the others, and all respond together as a group although each may differ in amplitude and phase. This is called "coupled vibration." If only one tube vibrates while the rest are held stationary, it is called "uncoupled vibration." Formulas accounting for the effect of coupling on the natural frequency are given by Chen (1977a).

15.9 How do the different shell-side velocities affect tube vibrations?

Soon after entering the shell, the fluid subdivides into cross-flow, bypass, and leakage streams (Q. 2.3). All these fluid flows have different velocities and hence each affects the tube vibrations differently. Briefly, the different shell-side velocities are the following:

1. Bundle entrance and escape velocities: These refer to the regions underneath the inlet and exit nozzles. Due to changing area, it is difficult to calculate exactly these velocities as well as their effects on tube vibrations. A number of exchangers are known to have failed in these regions.
2. Cross-flow velocity: This varies from tube row to tube row due to varying cross-flow area. The fluid is constantly being accelerated and decelerated. Hence the fluid force exerted by the cross-flow components keeps varying too.
3. Parallel flow velocities: These relate to the flow through the baffle windows and the bypass and leakage streams. The velocities of the latter streams can be an order of magnitude higher than the cross-flow velocity. These higher velocities can result in vibration damage to the tubes near the outer tube limit (OTL) and near the baffle tips.
4. Turn-around velocities: The fluid turns from cross-flow over the tube bundle to parallel flow in the baffle window. Thus, both the magnitude and direction of the fluid velocity change at each point as does the force exerted by this fluid in the region of the baffle windows. It is difficult to calculate exactly this force and its effects on the tube vibrations.

The situation is still more complicated if the fluid is a two-phase mixture. Difficulties compound further if a change in phase is occurring that results in a change in the amounts of liquid and vapor flowing along the length of the exchanger. Their exact effects on the tube vibrations are extremely difficult to evaluate. A designer should spend some time in identifying the velocity profiles and magnitudes in the shell at various locations. This will help to evaluate better the effects of vibrations.

15.10 What does the response of a tube depend upon?

It depends upon the character of excitation, the natural frequency of the tube, and the damping forces acting on it.

15.11 Is the added mass of a tube the same in a stationary fluid as when the fluid is in cross-flow?

No. Chen (1978) states that the added mass in cross-flow is far less than that in the stationary fluid. This smaller added mass is due to the separation of the flow over the cylinder and the formation of vortices. Hence the values of the natural frequency, f_n, obtained by using the added mass value from stationary fluid will be lower than the actual value, and the resulting design will be conservative.

RESONANCE

15.12 Does the resonance occur only with the fundamental frequency (first mode)?

No. Tubes are flexible, slender cylinders whose fundamental frequency is low. When the fluid flow induces vibrations, the resonance is most likely to occur with their lowest frequency. However, due to the nonuniform flow distribution that may produce higher velocities in some regions (Q. 15.9), higher modes of vibration may occur in those regions instead of the first mode. Hence, in the nonuniform flow situations, one should also calculate the natural frequencies for the higher modes to see if they would get excited by the fluid flow.

15.13 When do the flow-induced vibrations become dangerous?

Whenever there is a flow over a tube bundle, vibrations are induced into each tube. The vibrations become dangerous when they cause resonance; large amounts of energy is put into the tubes from the flowing fluid resulting in large amplitudes of vibrations, leading ultimately to failure. Damage to the tubes is also caused when the tube vibrations result in midspan collision between adjoining tubes and collision between the tubes and the baffle holes. Both kinds of collisions may result in tube rupture. Furthermore, continuous vibrations may result in fatigue failure of the tubes.

15.14 Why cannot the tubes be prevented from vibrating altogether?

Tubes are flexible, slender cylinders. When fluid flows over them, there is a dynamic force exerted on them which results in their vibration. Even rigidly holding a tube between two consecutive baffles will not completely eliminate vibrations, although their amplitude will be greatly reduced. Hence, the attempt in the design should be to keep the vibrations within limits rather than preventing them altogether.

15.15 Should an exchanger be designed so that the forcing frequency is in between two higher harmonics in order to obtain higher shell-side flow velocity?

This plan is obviously based upon the desire to increase the heat transfer coefficient. However, the following points warn against such a design (Fitz-hugh, 1973; Franklin et al., 1978):

1. The natural frequencies of the consecutive harmonics are close enough together not to permit safe operation in between them (Fig. 15.3) (Fitz-hugh, 1973). There is always a bandwidth around a natural frequency in which the resonance can happen. Further, any unforeseen variations in the flow will bring the forcing frequency close to one or the other of the two adjoining harmonic frequencies, thus leading to resonance.
2. Even these values for the natural frequencies as calculated are not definite since there are unknowns and idealizations in the analysis, and the exact effects of the damping and unevenly spaced baffle supports are not known. Thus, it would be difficult to choose a value to operate at between two harmonics.
3. Furthermore, as the flow starts through the unit, the tubes will get increasingly excited and will, at one time, pass through the fundamental frequency on their way to operation between the higher harmonics. It is likely that the resonance, even for a short period at the fundamental frequency, might damage some of the tubes if the damping is not high. Hence, the design should be such that the exciting frequency is always less than the fundamental frequency.

DAMAGING EFFECTS OF VIBRATIONS

15.16 What can be the damaging effects of tubular vibrations?

In any given situation one or more of the following can happen:

1. Collision damage among adjacent tubes: If the amplitude of vibrations is more than half the distance between the adjacent tubes, the tubes will impact against each other, flatten out a bit, and finally rupture.
2. Wear damage to tubes or baffles at the baffle holes: For ease in assembling the tube bundle, the baffle holes are made larger than the tube O.D., thus leaving the tubes to move freely within the baffle holes. If the baffles are of harder material than the tubes and/or if the drilled holes have sharp edges, the tubes will get cut due to the repeated impact, and leakage will result. If the tubes are of harder material than the baffles, the baffle ligament between the adjacent tubes could get cut, thus damaging the baffles and increasing the probability that the tubes would be damaged by vibration.
3. Fatigue failure of tubes or tube-to-tubesheet joints: Due to the continuous vibrations, stresses are set-in which might lead to fatigue failure of (a) tubes where pieces of tubes actually break off, and (b) tube-to-tubesheet joint. The latter can also fail due to the sharp edges of the holes in the tubesheet which will cut the tubes at the joint.
4. Intensive stress corrosion of the tubes: Since the continuous vibrations cause stresses, the tubes become more prone to stress corrosion. This further fatigues the tubes.
5. Excessive acoustic noise: This happens when the acoustic frequency matches a natural or vortex shedding or turbulent buffeting frequency in heat exchangers with gaseous flow (Q. 15.47-15.50). Loud noise of up to 150 db has been reported. This can sometimes lead to vibration of the entire shell.
6. High shell-side pressure drop: Since the vibration of the tubes requires energy input from the flowing stream, the shell-side pressure drop increases as the vibrations increase. Sudden increase in this pressure drop is an indication of the impending damage due to the vibrations.

An important point to keep in mind is that a cure for one vibration problem may enhance the risk of another type of vibration damage. Hence, a careful analysis is required. Furthermore, rupture of one tube can result in a fluid jet impinging on some of the neighboring tubes, and they might start to whirl, vibrate, and also fail; thus compounding the problem. Therefore, consequences of such damages must be analyzed beforehand (Chen, 1977b).

NATURAL FREQUENCY OF TUBES

15.17 How does one calculate the natural frequency of straight tubes on equal, multiple spans?

The equation is given in the RGP-RCB-4.562 Section of the TEMA Standards (Tubular Exchanger Manufacturers' Association, 1978). It takes into account the span length, modulus of elasticity, moment of inertia, effective weight of the tube (consisting of the weight of the empty tube, fluid inside the tube, and the shell-side fluid displaced by the tube), and a mode constant that is calculated. An expanded table of the mode constant is given in MacDuff and Feglar (1957). A nomograph for quick estimates in the case of fixed tubesheets is given by Ganapathy (1980). Other general references for the calculation of the natural frequencies are Gorman (1975) and Blevins (1979a).

15.18 How does one calculate the natural frequency of straight tubes on unequal, multiple spans?

The tubes in an exchanger have unequal spans since the baffle spacings in the inlet and outlet regions are different from the spacings in the central region. There are several ways to calculate their natural frequency:

1. It can be calculated by using the equation in RGP-RCB-4.564 Section, TEMA Standards (TEMA, 1978). This requires solving a determinant, which is time consuming.
2. The same section of the TEMA Standards also recommends that one should calculate the single span frequencies for each span with its own end conditions. The lowest of these will be somewhat lower than the true natural frequency calculated in (1) above.
3. A less conservative estimate than in (2) above will be obtained if one uses the formulation for the equal span case, based upon the longest span in the tube bundle. This longest span may be in the baffle window or in the end regions.

15.19 How is the natural frequency of a U-tube calculated?

A U-tube has two straight legs and a U-bend. The natural frequency of the straight legs is calculated by assuming that one end is rigidly fixed and the other end is simply supported, and then by following the procedure in Q. 15.17 and 15.18. On the other hand, the U-bend has an in-plane and an out-of-plane vibration. The lowest fundamental frequency is for the first out-of-plane mode of vibrations; the next lowest natural frequency could be either the second out-of-plane mode or the first in-plane mode (Blevins, 1977).

The U-bend fundamental frequency can be determined by extending the procedure for the straight tubes (Q. 15.20) or by a method proposed by Gorman (1979).

15.20 How is the natural frequency of a U-bend calculated?

Among the various ways are the following:

1. Replace the unsupported curved span by a straight span 12 percent longer in total length (denHartog, 1956) and use the formula for the straight tube.
2. Replace the unsupported curved span by an equivalent straight length on two simple supports and adjust the natural frequency that is calculated, (f_n), for the in-plane (f_{ni}), and the out-of-plane (f_{no}), vibrations by (Ojalvo and Newman, 1964):

$$f_{ni} = 1.985 f_n$$

$$f_{no} = 0.829 f_n$$

Any intermediate support in the U-bend increases f_n considerably.

Most companies now have computer codes for calculating the natural frequencies of straight and U-tubes as well as for doing the complete vibrational analysis. As more research is done in this direction, these codes will also improve with time.

15.21 How is the natural frequency of a low finned tube calculated?

TEMA Standards (TEMA, 1978) state that the same equation used for a plane tube may be used here, except that the actual weight of the finned tube should be used and that the moment of inertia should be calculated using the actual inside and root diameters. This will lead to a conservative value. A detailed set of very elaborate experimental data has been reported by Welbourne (1978).

15.22 How do the various factors affect the natural frequency, f_n, of a tube?

These are summarized in Table 15.1.

Table 15.1 Effects of various factors on the natural frequency f_n

Factor	Effects on f_n
1. Fluid-filled tube	Lower compared to an empty tube
2. Fluid on shell side	Lower compared to an empty shell
3. Thick wall tube	Higher compared to a thin-walled one with the same O.D.
4. Modulus of elasticity of the tube material	Increases with the increasing modulus of elasticity. It is high for a stainless steel tube compared to the admirality brass[a]
5. Tube-to-baffle hole clearance	Increases only when the clearance is very small since it restrains the tube vibrations
6. Number of tube spans	Higher if the number of spans is less than five since one can assume both the tube-ends clamped (rigid support). For a larger number of spans, the ends are taken to be clamped-supported and the natural frequency is lower. For a tube with a large number (say, 8 to 10) of equally spaced baffles, f_n is approximately equal to that for a simply supported single span tube
7. Size of tube bundle	Decreases as the number of tubes increase (see Q. 15.23)
8. Baffle spacing	Decreases as the baffle spacing increases. Since the spacing in the baffle windows and the end regions is usually longer than the central baffle spacing, the natural frequency of such tubes also decreases
9. Thick baffles	Increases if the tube-to-baffle hole clearance is also tight, or else no effect

(See footnotes on p. 561.)

Table 15.1 Effects of various factors on the natural frequency f_n (*Continued*)

Factor	Effects on f_n
10. Stresses in tubes	Increases with tensile stress, and decreases with compressive stress in the tubes (Fig. 15.7)[b]
11. Rolling of tubes into the baffles	Increases since it isolates the span and increases the stiffness of the tube. However, it is not very practical to do and is limited to about 6 ft (2 m approximately) from the tubesheet end
12. Lacing of tubes	Increases since it stiffens the tubes

[a]Brothman et al. (1978a).
[b]Kissel (1972).

15.23 How does the behavior of a tube bundle in an exchanger differ from that of an isolated tube?

The fundamental frequency of a tube bundle decreases with the increasing number of tubes. In addition, the coupled frequencies become more widely spread as the spacing between the tubes decreases (Chen, 1975).

In a bundle, the tubes move in different directions. Corresponding to each natural frequency there is actually a frequency band with lower and upper boundary frequencies (Chen, 1977b). Thus, for a group of nine tubes, there are eighteen natural frequencies between the two bounds due to the coupled modes including tube/fluid interactions. Figure 15.8 shows the frequencies for nine tubes in a tube bundle. The frequencies are close to one another. The frequency band of the higher harmonics is wider still. The bandwidth depends upon the tube properties, tube spacing, and fluid density; fluid coupling due to the fluid motion broadens the bandwidth.

There is interference or coupling between the adjacent tubes. It has also been observed that the tube vibrations dampen out momentarily and then pick up again.

Not all the tubes in a bundle vibrate in a similar manner because of the different flow conditions and the longer tube spans in the baffle windows. The angle and the location of the fluid impingement on the tube also influence the tube vibrations (Kessel, 1972). Furthermore, the first two rows of tubes are found to behave very differently than the rest of the rows in a tube bundle and hence are not a representative of the total bundle (Zdravkovich and Namork, 1978). They are not representative

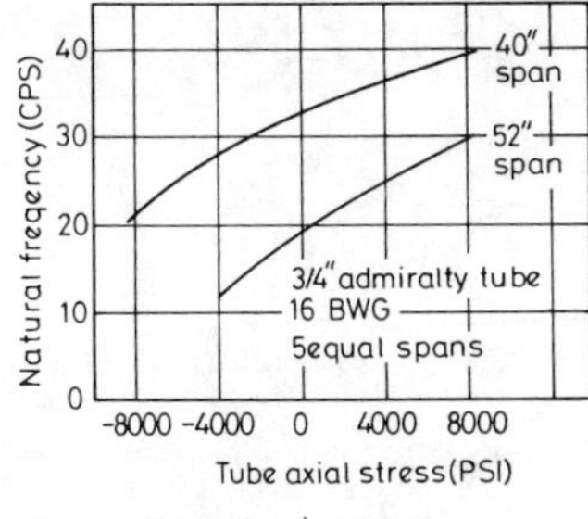

Figure 15.7 Effect of axial stress on natural frequency. *(From Kissel, 1972.)*

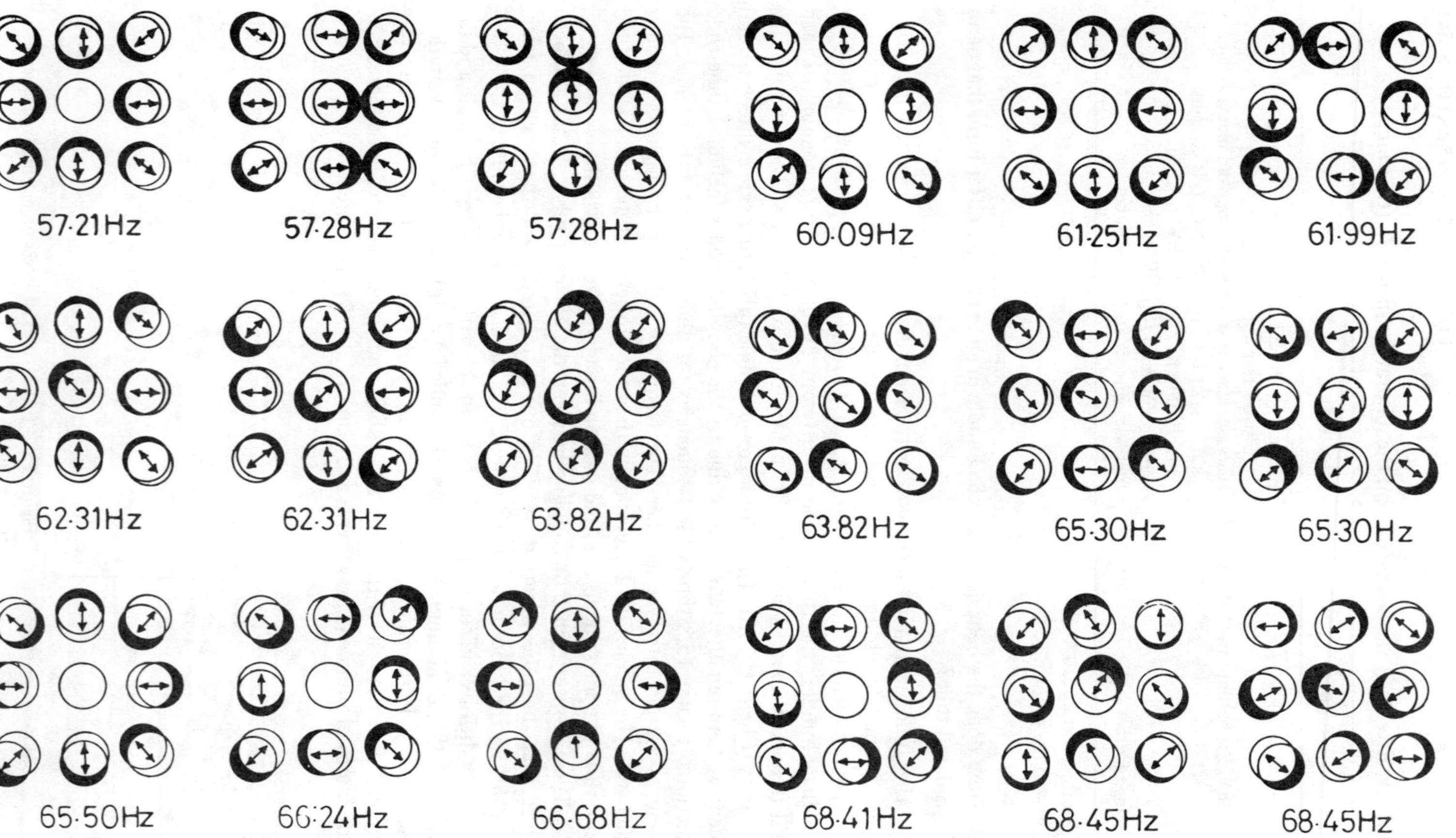

Figure 15.8 Mode shapes for an array of nine tubes. *(From Chen, 1977b.)*

probably because the vortex shedding is important in the first row or two only, while the fluid elastic whirling is important deeper in the bundle (Q. 15.35).

15.24 How do exchanger tubes differ from a true continuous beam on multiple simple supports?

Some of the important differences are the following (Brothman et al., 1978b):

1. The tubesheets invariably get slightly deflected due to pressure and temperature differentials on the two sides. This exerts an end-couple on the tube ends, which is not the case with the continuous beams.
2. The baffles act not only as a simple support for the tubes but also sometimes as an amplitude limiter, whereas the simple support of the continuous beam does not act as amplitude limiter.
3. The tubes have axial loading due to tube-to-tubesheet joining procedures, which is not the case with the continuous beams with rigid end supports.
4. The effective mass of the tube is more than that of the empty tube due to the fluid inside. It is important in the case of gases at high pressure and liquids. Such a situation never arises with a continuous beam.

These factors cause the calculated value of the natural frequency of a tube to spread out in a band, rather than being a definite value as is the case for a continuous beam.

PHENOMENA OF TUBE VIBRATIONS

15.25 What are the various phenomena of tube vibrations caused by shell-side flow?

These are vortex shedding, turbulent buffeting, fluid-elastic whirling or instability, jet swing, parallel flow eddy formation, and acoustic vibrations.

15.26 At the current level of understanding, how would you rate the probability of the occurrence of the various phenomena of tube vibrations in an exchanger?

In the decreasing order of probability are the following (Franklin et al., 1978):

Fluid-elastic whirling or instability
Vortex shedding
Turbulent buffeting
Jet-swing

Acoustic vibrations are highly probable in the case of gases on the shell side, and the parallel flow eddy formation in the case of large sections of the tube bundle under parallel flow.

Vortex Shedding

15.27 How does vortex shedding occur in cross-flow over a tube?

As the fluid in cross-flow reaches the tube, it comes to a stop at the leading edge (Fig. 15.9). Its kinetic energy is converted into pressure energy, and this,

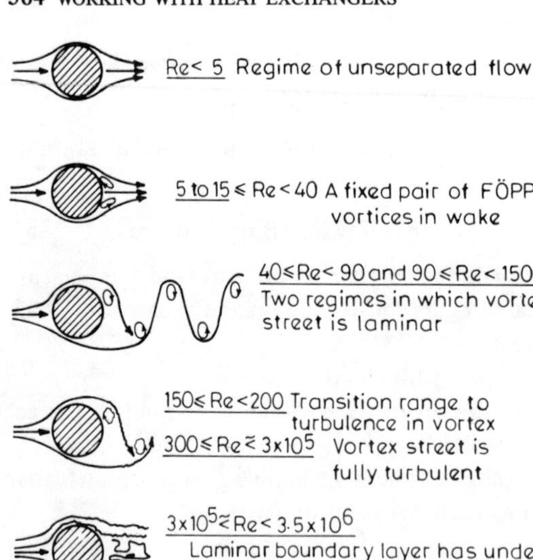

Figure 15.9 Regimes of fluid flow across circular cylinders. *(From Lienhard, 1966.)*

combined with the fluid pressure, results in the highest pressure (stagnation pressure) at the leading edge. From there, the fluid flows around the tube in a boundary layer flow. Near the widest section, the boundary layer separates into two free-shear layers that form the wake at the rear of the tube. The innermost portion of the shear layers travels at a lower velocity than the outermost portion, resulting in the shear layers curling into discrete vortices. This is called vortex shedding. It occurs alternately from each side of the tube. The nature of these vortices, also called Karman vortices or Karman vortex streets, varies with the Reynolds number (Lienhard, 1966).

15.28 How does the vortex shedding across a single tube differ from that across a tube bundle?

In a tube bundle, the upstream tubes affect the downstream ones. The Strouhal number is not the same for all the tubes. The tube layout and pitch also influence in case of a tube bundle. The U-bends behave differently than the straight tubes since the velocity distribution is different in the two cases (Singh, 1976). There are doubts expressed regarding the presence of the vortex shedding deep inside a tube bundle (Chen, 1977c).

15.29 What important effects does the wake of a tube have on a downstream tube?

The important effects are the following (Blevins, 1977):

1. The flow velocity in the wake is less than the free-stream velocity. Hence, the dynamic head and the fluid forces on the downstream tube are reduced.
2. Since the flow and the pressure gradient are nonuniform in the wake, they can produce nonuniform, large, side forces on the downstream tube.

15.30 How does the vortex shedding induce vibrations?

Vortices, shed alternately from each side of the tube, exert forces that also change directions. These oscillating forces result in the vibration of an isolated tube. However, in closely spaced tube arrays with tube pitch-to-diameter ratio of 1.5 or less, as is generally the case in heat exchangers, the vortices degenerate into a broad band of turbulence, and large amplitude vibration is not produced. Hence, in a heat exchanger, vortex shedding is generally not the cause of damaging vibrations.

15.31 Is vortex shedding more probable with gaseous or with liquid flow on the shell side?

It is more probable with the liquids since these are denser than the gases (Franklin et al., 1978). However, in gaseous flow, the vortex shedding frequently interacts with the acoustic field, and the two might combine to produce severe vibrations.

15.32 What criterion should be adopted to avoid resonance with vortex shedding frequency?

The resonance occurs when both the forcing and the natural frequencies are equal or nearly so. However, as discussed in Q. 15.23, there is actually a frequency band around each natural frequency where resonance can occur, and due to the uncertainties in the calculation of the natural and forcing frequencies, it is suggested that the vortex shedding frequency, f_{vs}, be less than 50 percent of the fundamental frequency, f_n, or (Franklin et al., 1978)

$$f_{vs} < 0.5 f_n$$

Fluid-Elastic Whirling

15.33 What is fluid-elastic whirling?

When the fluid in cross-flow over an array of tubes displaces some of them, the force field exerted by the fluid over the tube bundle also changes. The tubes experience lift and drag forces, and this can result in the vibration of the tubes (Fig. 15.10), if the cross-flow velocity is above a critical velocity, i.e., the energy fed to the tubes exceeds that which can be dissipated by available damping. This type of vibratory motion of the tubes is called the fluid-elastic whirling (Connors, 1970) or fluid-elastic instability. Later work by Chen (1981) has demonstrated that, apart from the fluid-elastic instability (called fluid-elastic-stiffness-controlled instability by Chen) based upon the fluid-elastic forces, there is also a fluid-damping-controlled instability based upon the fluid damping forces, and is proportional to the velocities of the tubes. Of the two, though, the former is more dominant.

Another possible motion is called jet switching in which the jets of fluid emerging from in between the tubes alternately couple and uncouple (Fig. 15.10). However, such a situation is not generally encountered in heat exchangers.

15.34 What is critical velocity? What does it depend upon?

Critical velocity, V_c, is the cross-flow velocity at which a tube row or a tube array will become unstable due to fluid-elastic whirling and will vibrate to failure.

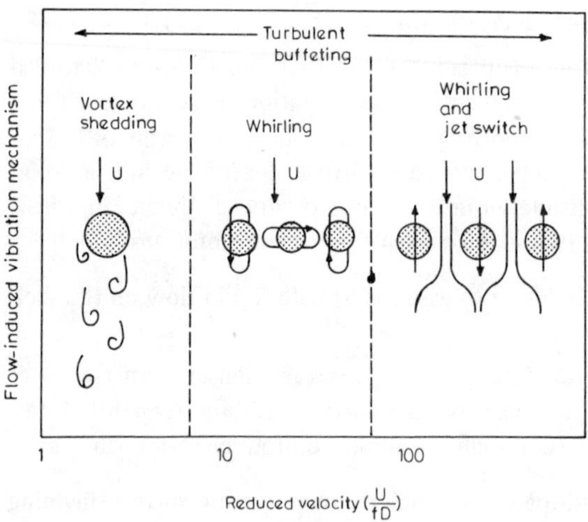

Figure 15.10 Mechanisms of flow-induced vibration.

The amplitude of the orbital motion increases alarmingly above V_c. The critical velocity (Blevins, 1977; Chen, 1977b; Soper, 1980):

1. Depends upon the natural frequencies of the tubes, the system damping, and the mass ratio of fluid to tube
2. Generally increases as the tube pitch increases
3. May be lower for a tube next to the wall or the baffle tips
4. Is the highest for a staggered square tube layout (45°) and the lowest for an in-line square tube layout (90°) with the values for the 30° and 60° layouts lying in between; viz. the decreasing order of V_c with the tube layout angle can be approximated as

$$45^\circ > 30^\circ > 60^\circ > 90^\circ$$

   However, a 45° layout results in acoustic vibrations sooner than the other layouts and hence is not recommended for gaseous flow on the shell side
5. Depends upon the position of the tube in a bundle. Tube arrays have lower V_c than the tube rows for the same pitch since the motion of the lower tube row affects that of an upper one
6. Depends upon the level of upstream turbulence that may be created by an impingement plate or a partly closed valve. In some cases, the critical velocity has increased, while in some it has decreased, and in yet others no significant effect has been observed (Soper, 1981; Mayinger and Gross, 1981). It remains an uncertain factor at present but should be kept in mind.

The calculation of V_c is discussed in several references, such as Blevins (1977), Brothman et al. (1978b), and Gupta (1979).

15.35 What is the difference between tube vibrations caused by vortex shedding and those caused by fluid-elastic whirling?

Vortex shedding is an unsteady phenomenon at the rear of the tube that results in tube vibrations. On the other hand, fluid-elastic whirling is not based on any unsteady phenomenon, but is produced by the interaction of the flow fields between the adjacent tubes (Blevins, 1977). Vortex shedding is fluid mechanical in nature and does not depend upon any motion of the tubes, unlike the fluid-elastic whirling.

Another important point is that if the tube pitch is small, the wake does not have space to break, and vortex shedding is prevented. That is why, while vortex shedding is observed in isolated tubes and in a single tube row, it is not found to any significant degree in a tube bundle with standard pitch-to-diameter ratios. In such bundles, the fluid-elastic whirling dominates.

Damping

15.36 What does damping do?

It dissipates the energy and thus reduces the amplitude of vibrations, which eventually die down if the cause of the vibrations is removed. Without damping, even after the cause of the vibrations is removed, a structure may continue to vibrate and eventually fail due to fatigue. In practice, however, such a zero damping situation would never exist. Figure 15.11 shows the response of a system for small damping, large damping, and critical damping conditions after the applied force has been removed.

15.37 What are the different types of damping?

There are three types (Blevins, 1977):

1. Fluid damping: It is caused by viscous and pressure drag as the structure moves relative to the fluid; e.g., oil in a shock absorber provides most of the damping in a car. A combination of factors may result in a negative damping and the structure may fail (Blevins, 1977). In a heat exchanger the fluid damping consists of the energy losses due to (a) form and frictional drag between the vibrating tubes and the shell-side liquid, and (b) fluid mixing due to the changing slit width between the adjacent tubes. It is predominant when the shell-side fluid is a liquid or a gas under high pressure (Brothman et al., 1978b).
2. Material or internal damping: It is due to the internal energy dissipation within the

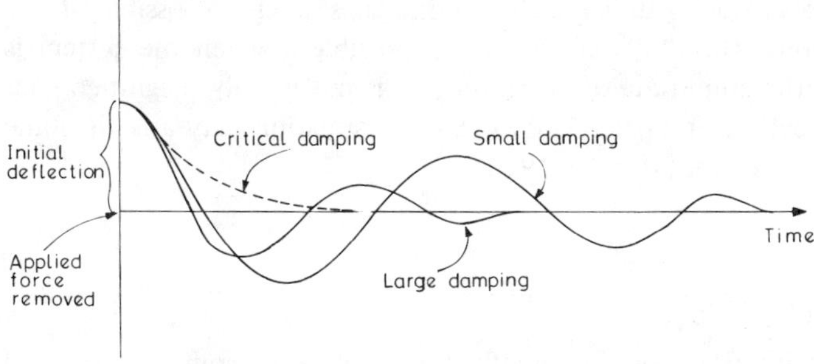

Figure 15.11 System response as a function of damping.

material; e.g., rubber is used for shock mounts. It is negligible in the case of exchanger tubes.

3. Structural or mechanical damping: It is generated by friction, impacting, or the rubbing of two surfaces at a joint; e.g., tall buildings are designed to sway in wind to dissipate the energy imparted by the wind. In a heat exchanger, the structural damping is primarily the energy dissipated due to friction and/or rattling at the bearing points such as the tube-to-tubesheet joints and the baffle holes. It is predominant when the shell-side fluid is a low density gas or vapor (Brothman et al., 1978b).

Total damping in a system is the sum of all the three types. However, in most situations, only one of these plays an important role; e.g., in heat exchangers, the important one is the fluid dynamic damping; while in the tall buildings, it is the structural damping.

It should be pointed out that damping is still one of the most unpredictable elements in vibration analysis. Hence, one should be careful in using the published values and must keep in mind the assumptions and the limitations of the analysis.

15.38 Is the fluid damping the same in the stationary and the flowing fluid?

No. The damping in a flowing fluid is different than in a stationary fluid (Kissel, 1982). In some cases, it may be even less than that in a stationary fluid (Chen and Jendrzejczyk, 1981).

15.39 On what factors do the structural and fluid damping depend in a heat exchanger?

Structural damping depends upon the number of spans, weight of the tube, tube layout, hardness of the tubes and baffles, ratio of the lift to drag forces acting on a span, and lubricating properties, if any, of the shell-side fluid (Brothman et al., 1978a).

Fluid damping depends upon the amplitude and frequency of vibrations of the tubes, fluid velocity, and fluid properties such as the density and the viscosity (Brothman et at., 1978a).

15.40 When is it safe to operate in a region where the forcing frequency is the same as the natural frequency (i.e., resonance is predicted)?

One should avoid operating under such conditions as much as possible. If it is not possible to avoid, then the only safe operation possible is when the system is heavily damped so that the amplitude of vibrations, X is not unduly magnified over the mid-span static deflection, X_s (Fig. 15.12). Such a situation may arise in some condensers and feed water heaters (Moretti, 1974).

Turbulent Buffeting

15.41 What is turbulent buffeting?

A fluid in turbulent flow has random fluctuating velocity components in all directions, with a broad range of frequencies. These turbulent velocity components

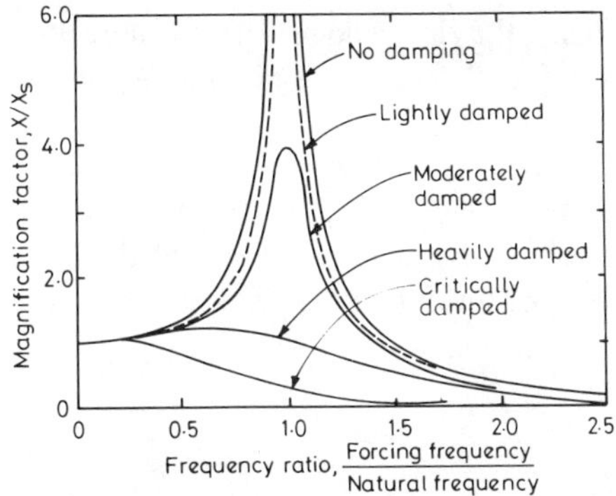

Figure 15.12 Magnification factor. *(Courtesy Oak Ridge National Laboratory, 1975.)*

impart energy to the tubes over which the fluid is flowing in parallel or in cross-flow. This results in random tube vibrations and is called turbulent buffeting. This can cause fretting of the tubes and fatigue them to failure over a period of time. The tubes extract maximum energy from that component of the turbulence whose frequency is close to the fundamental frequency of the tubes. This can set the tubes in resonance, leading to their failure.

Blevins (1977) points out that the staggered tube arrays (45°) will have higher amplitude turbulence-induced vibrations than the comparable in-line (90°) ones. This is in contrast to the fluid elastic whirling (Q. 15.34).

15.42 How does the turbulent buffeting differ from the vortex shedding and fluid-elastic whirling?

The basic differences are the following (Blevins, 1978):

1. The turbulent buffeting is a random process, while the others are basically deterministic.
2. The turbulent buffeting is operative over nearly the entire range of fluid parameters, whereas the others are limited to the parameter ranges that produce resonance and instability.
3. The fluid forces that induce turbulent buffeting are substantially independent of the small tube motions, while those that generate vortex shedding and fluid-elastic whirling are strongly coupled to the fluid motion.

Turbulent flow through a tube bank has velocity fluctuations that result in fluctuating pressures on the surface of the tubes and induce vibrations. These cause fretting wear at the tube supports and fatigue the tubes, leading to their failure.

The amplitude of tube vibrations is significantly higher in the rear of the tube array than at the front due to the higher turbulence at the rear of the array.

Blevins (1978) gives a model that reasonably predicts the amplitude of tube vibrations due to turbulent buffeting.

15.43 What steps should be taken to reduce the tube vibrations due to turbulence?

One or more of the following steps may be taken (Blevins, 1977):

1. Reduce the intensity of turbulence by reducing the flow velocity. This can be done by reducing the flow rate, and by using larger shell, larger tube pitch, longer baffle spacing, and/or using multi-segmental baffles.
2. Increase the natural frequency of the tubes by stiffening them
3. Increase the ratio of the structural mass to the displaced fluid mass $m/\rho D^2$ without decreasing the natural frequency. This can be done by increasing the tube thickness
4. Increase the damping by incorporating energy-absorbing materials, such as rubber, wood, and plastic, in the design

Fretting Wear

15.44 What is fretting wear?

It is the deterioration of the tube due to the collision between tube-to-tube and tube-to-baffle hole. Vibration-induced fretting wear is second only to corrosion as a cause of heat exchanger failures. The fretting wear depends upon the following (Blevins, 1979b):

1. Amplitude of tube vibration
2. Mode shape of tube vibration
3. Frequency of tube vibration
4. Tube dimensions (length, diameter, wall thickness, straight or U-tubes)
5. End supports (tube-to-tubesheet joints)
6. Tube material (Young's modulus, hardness, surface finish, or coating)
7. Dimensions of the baffles and support plates (thickness, hole diameter, spacing)
8. Baffle material (Young's modulus, hardness)
9. Mean reaction (preload on tube) between the tube and the support plates
10. Temperature of the tubes
11. Properties of the surrounding fluid

There is interaction between some of these parameters, e.g., if the amplitude of vibrations is zero, the wear rate is zero also regardless of any other item (Blevins, 1979c).

Fretting wear rate increases as the gap between the tube and the baffle hole increases. Therefore, minimizing the gap will reduce the fretting wear. One cannot weld the tubes to the baffles to reduce the gap to zero since high stresses will result and there will also be vibration damage due to fatigue. Furthermore, no tube replacement would be possible. Results of some experiments on fretting wear by Blevins (1979b) are summarized below:

1. As the gap increases with time due to fretting wear, the wear rate increases too (Fig. 15.13)
2. Fretting wear increases with increasing amplitude and frequency of tube vibrations due probably to an increase in the kinetic energy of the tube (Fig. 15.14).
3. Preload on the tube near the baffle decreases sharply the wear since the amplitude is decreased (Fig. 15.15).
4. Tube thickness and surface roughness have negligible effects.
5. Decreasing the tube support plate thickness increases the wear rate.

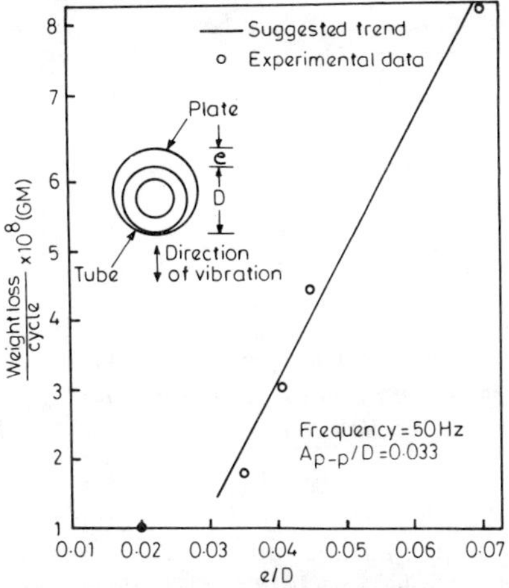

Figure 15.13 Rate of fretting wear as a function of gap between the tube and baffle hole. *(From Blevins, 1976b.)*

6. In controlled experiments, injecting powdered graphite decreased the wear rate by a factor of 30. It is not certain whether the graphite acts as a lubricant or plugs the gap between the tube and the baffle hole.
7. The higher the wear rate, the louder the chattering sound produced at the tube-baffle hole interaction.
8. If the tubes and the baffles are of the same material, both of them wear at approximately the same rate since the wear rate is a function of the material and its hardness.

15.45 What are the criteria to prevent tube-to-tube collision damage and tube-to-baffle hole wear damage?

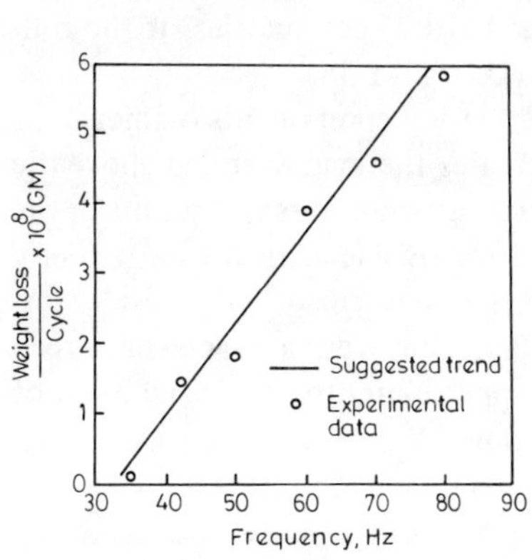

Figure 15.14 Rate of fretting wear as a function of tube vibration frequency. *(From Blevins, 1979b.)*

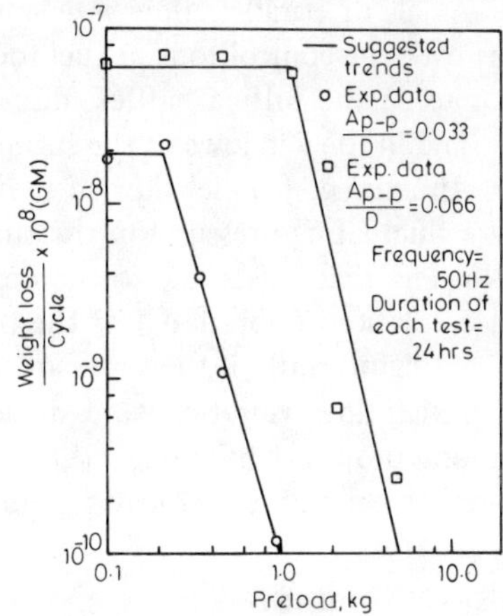

Figure 15.15 Rate of fretting wear as a function of preload on the tube. *(From Blevins, 1979b.)*

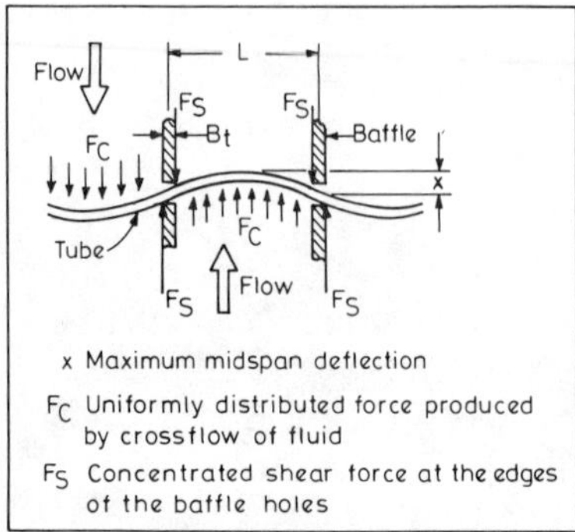

Figure 15.16 Model for Thorngren damage number analysis. *(From Thorngren, 1970. Courtesy Gulf Publishing Company, Houston.)*

The criteria were first enunciated by Thorngren (1970) who used the model in Fig. 15.16 for his analysis. He gave a collision damage number and a baffle damage number that depend on system parameters; each of which should be small for low probability of damage.

Another simple criterion to avoid the tube-to-tube collision is (Brothman et al., 1978b)

$$\frac{y_0}{D} \leqslant 0.05 \tag{15.6}$$

where y_0 = cumulative first mode amplitude response to all forms of excitation
D = tube O.D., and
y_0/D = dimensionless amplitude response

Axial Flow Vibrations

15.46 How does axial flow cause vibrations?

Axial, or parallel, flow is encountered in nuclear power reactor components such as over the control rods or fuel rods, in shell-and-tube exchangers using an L-type shell or using the orifice baffles, disc-and-doughnut baffles, nest baffles or the rod-baffles, and in the windows of the segmental baffles (Q. 1.43–1.48).

An axial or parallel flow in the shell has fluctuating components (eddies) that generate fluctuating pressure on the tubes. This increases the load over and above the static loading that exists due to the design and operating conditions and results in the lateral oscillations (vibrations) of the tubes (Chen, 1977b). The critical axial velocity is lowered significantly if the tubes are under compressive static load.

Axial flow vibrations are difficult to predict. However, a model of forced oscillations proposed by Fung (1980) predicts an upper bound on the magnitude of vibration response of components in turbulent axial flow.

Acoustic Vibrations

15.47 What are acoustic vibrations?

These are standing waves in the gas column perpendicular to the predominant shell-side fluid flow and axial flow directions (Fig. 15.17) (Barrington, 1978). The excitation frequency, f_a, is dependent upon the flow path length, standing wave mode number, fluid molecular weight, compressibility, temperature, and the ratio of the heat capacity at constant pressure to that at constant volume (Brothman et al., 1978a).

Acoustic coupling with the natural frequency, f_n, may be avoided by using detuning baffles that increase f_a by decreasing the standing wave path length (Brothman et al., 1978a). For the condensers, one needs to examine only the outlet zone. For other units, the inlet, central, and the outlet zones should be examined.

15.48 How are the acoustic vibrations excited?

These are excited when the shell-side fluid is a low density gas. The gas column normal to both the tube axis and the flow direction resonates when the vortex shedding, turbulent buffeting, or natural frequency is close to the acoustic frequency of the gas column. Large oscillations of the fluid occur. The worst case occurs when the acoustic frequency, vortex shedding frequency (or the turbulent buffeting frequency), and also the cylinder natural frequency are the same (Chen, 1977c). This reinforces the vibration tendency of the tubes. Furthermore, the fluctuating forces result in higher pressure drop and cause severe vibrations to the connected piping and the foundations. The nature of the gaseous flow changes drastically. While in the no-resonance conditions, the velocity distribution is reasonably uniform; during the resonance, large velocity gradients are found across the tube bank (Fitzpatrick et al., 1978).

The acoustic natural frequency in liquids is very high and hence there are no acoustic vibrations in liquid flow. Alternately, one can say that the compressibility of the liquids is very low, hence the standing waves are not generated in them.

Typical waveforms of the acoustic standing waves in the gases are shown in Fig. 15.17 (Barrington, 1978). Usually the first mode is encountered in practice ($N = 1$).

For finned tubes, the root diameter should be used to calculate the acoustic frequency. For calculation procedures and examples, see Gupta (1979).

15.49 How can the acoustic vibrations be cured?

Barrington recommends several methods including a few supported by field examples (Barrington, 1978). These methods are the following:

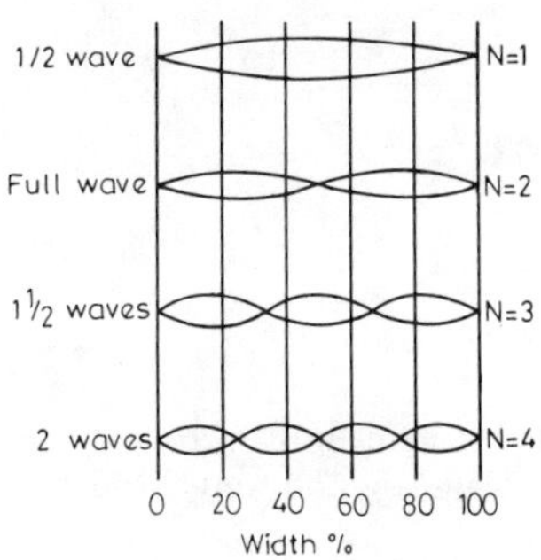

Figure 15.17 Typical wave forms.

1. Reduce flow rate, V, to reduce the vortex shedding frequency, f_{vs}, below the acoustic resonance frequency, f_a. In order to avoid locking-in, f_{vs} should be $<0.8f_a$ or $>1.2f_a$. This, however, affects the thermal performance of the unit.
2. Remove some of the tubes from the bundle at critical areas (nodal points) so that V and hence f_{vs} are reduced. Field experience has proved its utility. This also affects the thermal performance of the unit.
3. Use irregular lateral spacing of the tubes to scramble the vortex shedding frequency. This is difficult to do from the mechanical design and fabrication viewpoints.
4. Place longitudinal detuning baffles normal to both the segmental baffles and the baffle cuts to reduce the dimensions of the gas columns. This increases f_a above the vortex shedding frequency, and hence the acoustic resonance does not set in. Field experience has proved it to be a very useful method (Soper, 1981). Figures 15.18 and 15.19 show the existence of gas columns where acoustic vibrations can occur. Figures 15.20 and 15.21 show how the detuning baffles should be placed to avoid the acoustic vibrations.

15.50 Is all noise emanating from an exchanger due to the acoustic vibrations?

No. The other causes that could produce noise in an exchanger are (Barrington, 1978): collision between tubes, collision between tubes and baffles in the baffle holes, and broad spectrum of flow turbulence. These can be present when the shell-side fluid is a gas or a liquid, whereas the acoustic vibrations are present only when the shell-side fluid is a gas.

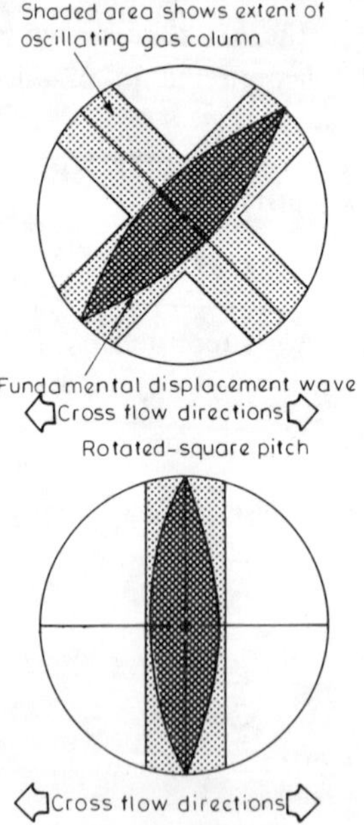

Figure 15.18 Position of diametral standing wave. *(From Barrington, 1978. Courtesy Gulf Publishing Company, Houston.)*

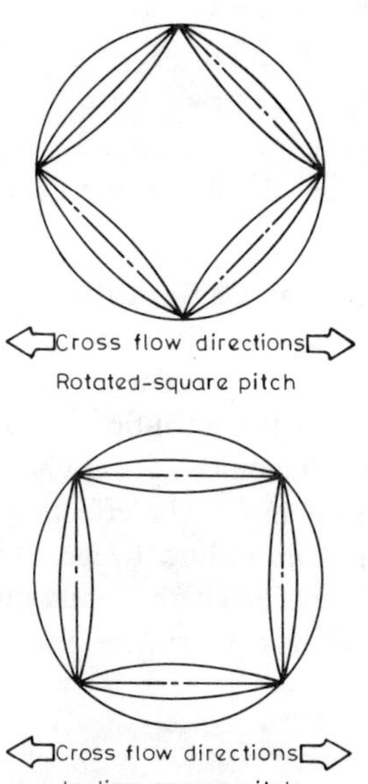

Figure 15.19 Position of inscribed square standing wave. *(From Barrington, 1978. Courtesy Gulf Publishing Company, Houston.)*

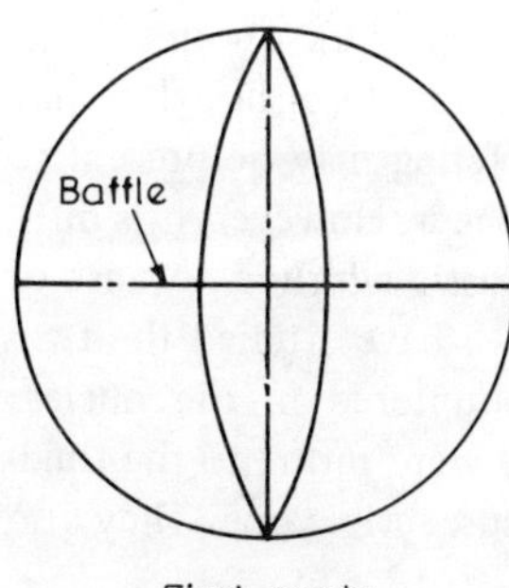

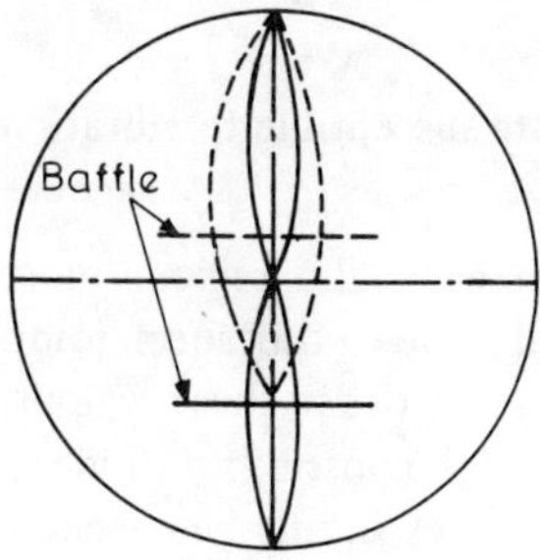

Figure 15.20 Baffle placements for rotated square pitch. *(From Barrington, 1978. Courtesy Gulf Publishing Company, Houston.)*

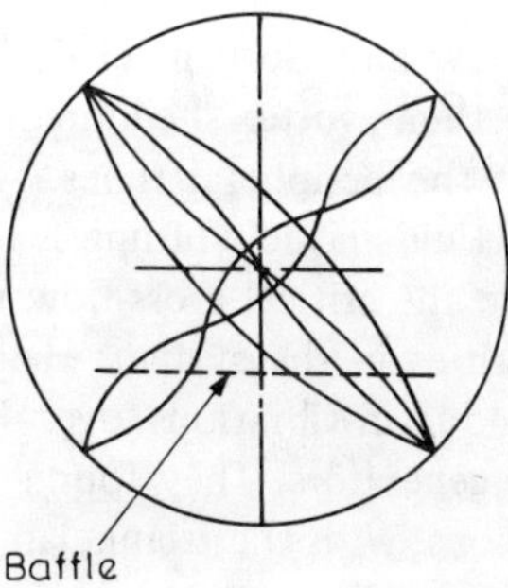

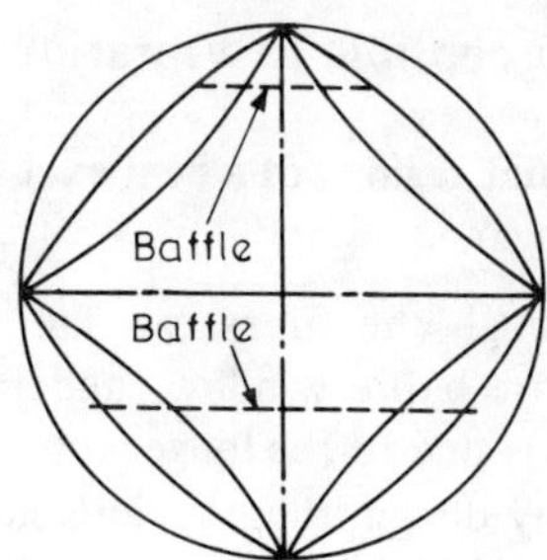

Figure 15.21 Baffle placements for in-line square pitch. *(From Barrington, 1978. Courtesy Gulf Publishing Company, Houston.)*

Combined Phenomena

15.51 Do the various phenomena of tube vibrations act independently or in a combined fashion?

While the question is not fully resolved, it is certain that many of the phenomena act in a combined fashion. For example, the fluid flow over the tubes in the baffle window and close to the baffle tip changes direction all along the tube, subjecting those tubes to vibrations due both to the cross-flow and the axial-flow components. However, what is not certain is whether the different phenomena act in-phase thus amplifying the effects or out-of-phase thus canceling or damping the effects of one another. Brothman et al. (1974) have given an overall procedure to predict whether or not the damaging vibrations will occur. The effects of all the vibration phenomena are summed up. However, this leads to an overly conservative value.

VIBRATIONS IN TWO–PHASE SYSTEMS

15.52 What kinds of vibration phenomena are observed in a two-phase flow over a tube bundle?

Pettigrew and Gorman (1978) have reported substantial work on this topic. According to them, vortex shedding is not observed inside the bundle, although it may be present in the peripheral tubes. Although turbulent buffeting may be present to some extent, fluid-elastic whirling is the one most likely to occur. However, it is difficult to define the critical cross-flow velocity for the fluid-elastic whirling because of the nonlinearities in the system. Heilker and Vincent (1981) have studied the two-phase flow-induced vibrations over U-tube bundles as encountered in the nuclear power steam generators. They found that all the tube arrays were prone to the fluid-elastic vibrations, with the triangular layout being the most susceptible one. They also reported observing the vortex shedding for square pitch arrays.

Regions Susceptible to Vibrations

15.53 In what regions of a heat exchanger are the tubes more susceptible to vibration damage?

Although the tubes can fail anywhere in the exchanger, the regions more susceptible are baffle window, underneath the nozzles, and tube-to-tubesheet joint.

This is due to the longer tube spans and higher fluid velocities in these regions. Hence, in any design, these regions should be checked first for the possibility of vibration damage. If found satisfactory, then the central baffled zones should be checked. In a U-bundle, the U-bend region should also be checked.

15.54 How does a vibrating tube in a heat exchanger develop leaks?

It can happen due to three reasons (Schwarz, 1976):

1. Fatigue cracks at the tube-to-tubesheet joint
2. Notch formation and subsequent crack development due to the continuous hitting of the tube at the baffle hole
3. Flattening and rupturing of the tubes at midspan due to the collision between the neighboring tubes

15.55 If resonance and lock-in phenomena are prevented, can a tube still fail due to vibrations?

Yes. Fatigue at the tube-to-tubesheet joint due to continuous vibrations and wear damage due to tube-to-baffle hole collisions or tube-to-tube collisions can result in the tube failure.

15.56 How long after the start-up of an exchanger does it take for the leaks to develop?

There is no way to predict the time. Leaks occur in new units and in old units that have been operating for months and years in which some tubes may have already been plugged due to the leaks. In some new units, leaks may occur within hours or days of start-up.

If resonance sets in or instability occurs, the tubes may leak within a few hours. If, however, there is no resonance or instability but fatigue and wear, the tubes may take a few months or years to leak. In the latter case, corrosion of the tubes may also contribute to the tube failure.

VIBRATION ANALYSIS

15.57 Discuss the steps in vibration analysis.

A flow-induced vibration analysis of a shell-and-tube heat exchanger should follow the qualitative procedure outlined below (Kissel, 1972):

1. Calculate the lowest natural frequency in the heat exchanger in the end zones and in the central zones. If the end zones have a long unsupported length, use the relationship developed by Kissel (1972) for unequal span lengths. For less than five spans, it should be recognized that this relationship will yield conservative results.
2. Modify the natural frequency to account for any unusual support condition present and to account for possible axial load induced during fabrication and/or in operation (Q. 15.22).
3. Estimate the critical velocity in the end zones and in the central zones utilizing the fluid-elastic whirling theory with a suitable value for the logarithmic decrement.
4. Limit the average cross-flow velocity to a safe value below the predicted critical velocity. The factor of safety will depend on the ratio of the highest calculated local velocity to the average cross-flow velocity. High local velocity will exist at bundle entrance and exit, baffle window, pass lanes, and in the vicinity of any flow obstruction such as the tie-rods.

The important features of vibration analysis and the effects of various parameters remain essentially the same with high-velocity flows such as in steam condensers

(Sebald and Nobles, 1962) and in two-phase flow. However, the damping in the two-phase flow is much higher than in the single-phase flow (Pettigrew and Gorman, 1978).

15.58 What factors make the correct predictions of damaging vibrations difficult?

The important factors are the following:

1. Complex flow patterns in the tube bundle, inlet and the outlet regions, and baffle window. Essentially, there is cross-flow over the tube bundle, parallel flow in the baffle window, and continuous change in the flow direction in between the two regions. Thus, there is a changing angle of attack of the fluid on the tube bundle. There are bypass and leakage streams (Q. 2.3) whose flow rates are difficult to determine. All this complicates the determination of the cross-flow velocity. Furthermore, there is a timewise variation in the flow due to cavitation, unsteadiness, process control, and deliberate changes in the operating conditions and throughputs. All these expose the unit to conditions not generally evaluated at the design stage and reduce the reliability of the vibration analysis (Hartlen and Anderson, 1980).
2. Calculations of the damping factor and its effect are very uncertain.
3. Effects of the tube bundle parameters, viz. transverse and linear pitch, layout angle, pass partition lanes, shell-to-bundle clearance, number of tube rows, etc., cannot be correctly evaluated. Effects of some of the parameters have been studied by Gorman (1977).
4. Rates of tube wear and fatigue due to impacting with other tubes or baffle holes are hard to determine.

15.59 When is the probability of the occurrence of detrimental vibrations high?

When one or more of the criteria in Table 15.2 are satisfied, the probability of the detrimental flow-induced vibrations is high. These checks should be made in the inlet, outlet, and central regions of a heat exchanger, as well as in the baffle windows.

VIBRATION PREVENTION

15.60 What steps should be taken when vibrations are predicted at the design stage?

One or more of the following steps may be taken (Gupta, 1979). These either reduce the forcing frequency or increase the natural frequency, thus separating the two farther.

1. Reduce the tube span length: This will increase the natural frequency as well as the cross-flow velocity. However, the former is dependent upon the inverse of the square of the tube span, while the latter depends upon the inverse of the tube span. Thus, the increase in the natural frequency is significantly more than that in the cross-flow velocity or in the forcing frequency.
2. Reduce the shell-side fluid velocity: This can be achieved by reducing the flow rate or changing the tube pitch and/or the layout angle. This reduces the forcing frequency.

Table 15.2 Probability of vibrations

1. Vortex shedding frequency/lowest natural frequency of tube	$f_{vs}/f_n > 0.5$
2. Cross-flow velocity/critical velocity	$V/V_c \geqslant 1.0$
3. Dominant turbulent buffeting frequency/lowest natural frequency	$f_{tb}/f_n > 0.5$
4. Baffle damage number	$N_{BD} > 1.0$
5. Collision damage number	$N_{CD} > 1.0$
6. Amplitude of midspan tube vibrations due to cross flow and/or parallel flow	$Z > 0.5(P_t - d)$
7. Vortex shedding frequency/fundamental acoustic frequency	$f_{vs}/f_a > 0.8$ and < 1.2
8. Dominant turbulent buffeting frequency/ fundamental acoustic frequency	$f_{tb}/f_a > 0.8$ and < 1.2
9. Fundamental acoustic frequency/lowest natural frequency of tube	$f_a/f_n > 0.8$ and < 1.2
10. Span length/TEMA maximum permissible length	$L/L_{max} > 0.8$
11. Momentum term[a]	$\rho V^2 > 4000$ lb/ft sec² (5950 kg/m sec²) [Impingement protection required for $\rho V^2 >$ 1500 lb/ft sec² (2230 kg/m sec²) for single phase, noncorrosive and nonabrasive fluids, $\rho V^2 > 500$ lb/ ft sec² (745 kg/m sec²) for all others]

[a]Tubular Exchanger Manufacturers' Association (1978).

3. Change the baffle type: The no-tube-in-window design (Q. 1.46) supports all the baffles. Hence, a change to this type of baffles will reduce the longest span and thus increase the natural frequency. Furthermore, support plates (Q. 1.73) can be used with this type of baffles (without affecting the thermal performance) to further decrease the tube span and increase the natural frequency. A baffle cut of approximately 15 percent (diametral) ensures that the maximum number of tubes will be accommodated in a given shell using the no-tube-in-window design. Support plates cannot be used with segmental baffles except in the entry and the exit regions where partial support plates can be installed underneath the nozzles.
4. Reduce the shell inlet velocity: If the region under the inlet nozzle is suspect, use a larger nozzle diameter, impingement plate, and/or provide a jacket (distribution belt) around the shell for a larger entrance to the shell (Fig. 1.22). This reduces the forcing frequency.
5. Increase the baffle thickness.
6. Decrease to a bare minimum the tube-to-baffle hole clearance.
7. The baffle material should not be harder than the tube material.
8. Use thicker tubes and/or place the tubes in tension.
9. Use detuning baffles if acoustic vibrations are predicted.
10. Block any bypass flow passages and pass partition lanes since high velocity in these regions (due to low resistance to fluid flow) may damage the tubes locally even if the overall design looks perfect (Schwarz, 1976).

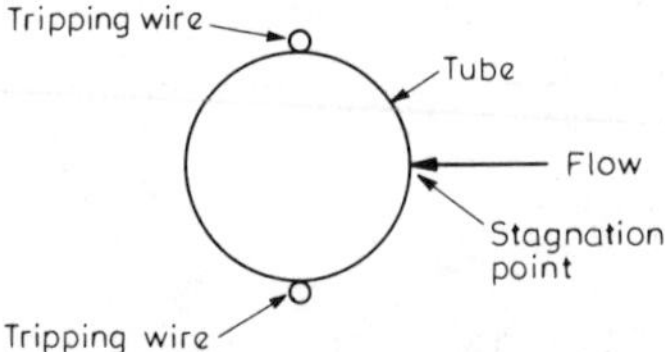

Figure 15.22 Tripping wires.

11. Install tripping wires at 90° from the stagnation point (Fig. 15.22) to change the character of the wake flow and the pressure distribution so that the fluid-elastic vibrations are reduced (Heinecke, 1978). This may not be practical in a large unit, although one could probably replace the wires by longitudinal fins for the purpose.

In Items (1)–(3) above, the unit should be rerated for its thermal performance and pressure drop. Items (4)–(9) do not affect significantly the thermal performance. Items (5)–(8) increase the natural frequency, while item (9) increases the acoustic frequency. Effect of items (10) and (11) on the thermal performance is expected to be positive.

15.61 What steps should be taken if the vibrations are observed in an existing exchanger during operation?

There are not too many options available. Among the steps that one can take are the following (Gupta, 1979):

1. Reduce the shell-side velocity: This reduces the cross-flow velocity and hence also the forcing frequency. The thermal performance is affected too.
2. Lace the tube bundle: Outer tubes, which are most susceptible to vibrations, should be tied all around by a strong wire or lace. Rubber, plastic, or wood wedges can also be inserted in between the tubes. The net result is to increase the natural frequency.
3. Roll the tubes in baffle holes: This stiffens the tubes and increases the natural frequency. However, it is of limited utility and is the least favored. It can only be used for short distances from the tubesheet end due to the tooling problems; tube replacement becomes difficult.
4. Create bypass lanes: Some tubes may be removed to create bypass lanes so that the cross-flow velocity is reduced. Tests conducted at Argonne National Laboratory, Chicago, have shown that the removal of tubes just in the window sections to form a bypass lane is fairly effective. The lane need not extend the full diameter of the shell. This minimizes the loss in the heat transfer (J. H. Kissel, personal communication).
5. Replace the tube bundle: If nothing else works, a new tube bundle may be designed for use in the same shell. In some cases, the whole unit, including the shell, may have to be replaced.

INFORMATION COLLECTION AND FUTURE RESEARCH

15.62 How can the information related to the heat exchanger vibrations be obtained?

There are four sources (Pettigrew and Gorman, 1978):

1. Fundamental studies on tube bundles
2. Model testing
3. Field measurements
4. Operating experience

15.63 What are the desirable future research trends in the vibration analysis?

The following is a brief summary of such desirable research (Chen, 1977c):

1. Characterizing the flow field, effects of the motion of the tubes on it, effects of the Reynolds number, tube layout and pitch, effects of the upstream turbulence, and nonuniform incident flow
2. Characterizing and evaluating the fluid damping and the fluid-elastic forces in the tube arrays
3. Formulating mathematical models for predicting the vibrations based on the tube arrays as used in the real exchangers (multi-span tubes, various pitch-to-diameter ratios, etc.) and not on the single tube and/or with air flow alone
4. Formulating recommendations to avoid vibrations when predicted by analysis. The effects of any changes should be analyzed with regard to the heat transfer and pressure drop and the thermal stresses.

● Thus, in this chapter, we have discussed the various causes of vibrations, and steps to take to prevent damaging vibrations from occurring in a shell-and-tube heat exchanger.

NOMENCLATURE

$A_{p\text{-}p}$ amplitude, peak to peak
B_t baffle thickness
CPS cycles per second
d_0 tube outer diameter
D tube outer diameter
e baffle hole-to-tube diametral clearance
f_a acoustic excitation frequency
f_n natural frequency of vibration of a tube
f_{ni} in-plane natural frequency of vibration of U-bend
f_{no} out-of-plane natural frequency of vibration of U-bend
f_{tb} turbulent buffeting frequency
f_{vs} vortex-shedding frequency
F_C uniformly distributed force produced by cross-flow of fluid
F_S concentrated shear force at the edges of the baffle hole
Hz Hertz
l linear tube pitch
L baffle spacing or tube span
L_{max} TEMA maximum permissible baffle spacing or tube span
m structural mass

N mode number
N_{BD} baffle damage number
N_{CD} collision damage number
O.D. outer diameter of tube
P_t tube pitch
Re Reynolds number
Sl Strouhal number
Sl_{vs} Strouhal number for vortex shedding
t transverse tube pitch
V cross-flow velocity
V_c critical velocity
X maximum midspan dynamic deflection of tube
X_l linear pitch to tube O.D. ratio
X_s midspan static deflection of tube
X_t transverse pitch to tube O.D. ratio
y_0 cumulative first mode amplitude response to all forms of excitation
Y_1 amplitude of vibration in one cycle
Y_2 amplitude of vibration in next cycle
Z amplitude of midspan tube vibrations

Greek Letters

ζ damping factor [Eq. (15.1)]
δ logarithmic decrement of damping [Eq. (15.2)]
ρ fluid density

REFERENCES

Barrington, K. A. Cure Exchanger Acoustic Vibrations. *Hydrocarbon Process.* 57(7):193–198. 1978.

Blevins, R. D. *Flow Induced Vibrations.* Van Nostrand Reinhold: New York. 1977.

Blevins, R. D. Buffeting of Heat Exchanger Tube Arrays in a Cross Flow, Paper 3:4. International Conference on Vibration in Nuclear Plant, Keswick, United Kingdom. 1978.

Blevins, R. D. *Formulas for Natural Frequency and Mode Shape.* Van Nostrand Reinhold: New York. 1979a.

Blevins, R. D. Fretting Wear of Heat Exchanger Tubes, Part I: Experiments. *Trans. ASME, J. Eng. Power* 101(4):625–629. 1979b.

Blevins, R. D. Fretting Wear of Heat Exchanger Tubes, Part II: Models. *Trans. ASME, J. Eng. Power* 101(4):630–633. 1979c.

Brothman, A., A. Devore, G. B. Hollar, H. Horowitz, and H. T. Lee. A Tube Vibration Analysis Method. *AIChE Symp. Ser. No. 138* 70:190–204. 1974.

Brothman, A., W. R. Rowley, A. Devore, G. B. Hollar, and A. Horowitz. Heat Exchanger Vibration Problems. In *Practical Aspects of Heat Transfer.* AIChE: New York. 1978a.

Brothman, A., S. Brannan, A. Devore, G. B. Hollar, and A. Horowitz. A Vibrations Prediction Method. In *Practical Aspects of Heat Transfer.* AIChE: New York. 1978b. A recommended reading.

Chen, S. S. Vibrations of a Row of Circular Cylinders in a Liquid. *Trans. ASME, J. Eng. Industry* 97(4):1212–1218. 1975.

Chen, S. S. Flow Induced Vibrations of Circular Cylindrical Structures, Part I: Stationary Fluids and Parallel Flow. *Shock Vibration Dig.* 9(10):25–38. 1977a.

Chen, S. S. Dynamics of Heat Exchanger Tube Banks. *Trans. ASME, J. Fluids Eng.* 99:462–469. 1977b.

Chen, S. S. Flow Induced Vibrations of Circular Cylindrical Structures, Part II: Cross Flow Considerations. *Shock Vibration Dig.* 9(11):21–27. 1977c.

Chen, S. S. Instability Mechanisms and Stability Criteria of a Group of Circular Cylinders Subjected to Cross-Flow I: Theory; II: Numerical Results and Discussions, Paper Nos. 81-DET-21 and 81-DET-22. ASME Design Engineering Technical Conference, Hartford, Conn. September, 1981.

Chen, S. S., and J. A. Jendrzejczyk. Flow Velocity Dependence of Damping in Tube Arrays Subject to Liquid Cross Flow. *Trans. ASME, J. Press. Vessel Tech.* 103:130–135. 1981.

Chen, Y. N. Criteria for the Cross-Flow Induced Tube Vibrations in Tube Bank Heat Exchangers, Paper 2:6. International Conference on Vibration in Nuclear Plant. Keswick, United Kingdom. 1978.

Chenworth, J. M., and J. Taborek. Flow Induced Tube Vibration Data Banks for Shell and Tube Heat Exchangers. *Heat Trans. Eng.* 2(2):28–42. 1980.

Connors, H. J. Fluid-Elastic Vibration of Tube Arrays Excited by Cross-Flow. In *Flow-Induced Vibration in Heat Exchangers.* ASME: New York. 1970.

denHartog, J. P. *Mechanical Vibrations,* 4th ed. McGraw-Hill: New York. 1956. Noted here from Franklin et al. 1978.

Fitz-hugh, J. S. Flow Induced Vibration in Heat Exchangers. *Oxford University Report RS57.* January, 1973 (AERE-P7238). Noted here from Blevins (1977).

Fitzpatrick, J. A., I. S. Donaldson, and W. Mcknight. Some Observations of the Pressure Distribution in a Tube Bank for Conditions of Self-Generated Acoustic Resonance, Paper No. 3:6. International Conference on Vibration in Nuclear Plant, Keswick, United Kingdom. 1978.

Franklin, R. E., B. M. H. Soper, and R. H. Whittle, Avoiding Vibration Induced Tube Failure in Shell and Tube Heat Exchangers, Paper No. 3:1. International Conference on Vibration in Nuclear Plant, Keswick, United Kingdom. 1978.

Fung, Y. T. A. Mechanism for Parallel Flow-Induced Vibrations. In *Flow-Induced Vibration of Power Plant Components,* PVP-41. M. K. Au-Yang (ed.). ASME: New York. 1980.

Ganapathy, V. Finding the Natural Frequency of Vibration of Exchanger Tubes. In *Process Heat Exchange,* V. Cavaseno (ed.). McGraw-Hill: New York. 1980.

Gorman, D. J. *Free Vibration Analysis of Beams and Shafts.* Wiley: New York. 1975.

Gorman, D. J. Experimental Study of Peripheral Problems Related to Liquid Flow Vibration in Heat Exchangers and Steam Generator, Paper No. F4/g. Proceedings of the Fourth International Conference on Structural Mechanics in Reactor Technology, San Francisco. 1977.

Gorman, D. J. A Comprehensive Analytical Approach to the Free and Forced Vibration Analysis of Large Steam Generator U-Tubes, Paper No. B5/1. Proceedings of the Fifth International Conference on Structural Mechanics in Reactor Technology, Berlin. 1979.

Griffin, O. M. A Universal Strouhal Number for the Locking-On of Vortex Shedding to the Vibrations of Bluff Cylinders. *J. Fluid Mech.* 85:591–606. 1978.

Gupta, J. P. *Heat Exchanger Design–A Pratical Look,* 2d ed. C. S. Enterprises: Delhi, India. 1979.

Hartlen, R. T., and H. L. Anderson. Experience with Heat Exchanger Tube Vibration in a Large Utility: A Viewpoint Regarding Field Incidence, Design Analysis Methods and Pre-Operational Testing. ASME Winter Annual Meeting, Chicago. 1980.

Heilker, W. J., and R. W. Vincent. Vibration in Nuclear Heat Exchangers due to Liquid and Two-Phase Flow. *Trans. ASME, J. Eng. Power* 103:358–366. 1981.

Heinecke, E. Fluid-Elastic Vibrations in Heat Exchangers with Tubes in Cross-Flow, Paper No. 2:4. International Conference on Vibration in Nuclear Plant, Keswick, United Kingdom. 1978.

Kissel, J. H. Flow Induced Vibrations in Heat Exchangers: A Pratical Look. AIChE-ASME Thirteenth National Heat Transfer Conference, Denver, Colorado. August 1972.

Kissel, J. H. Flow Induced Vibration in a Heat Exchanger with Seal Strips. *Trans. ASME, J. Heat Trans.* 104(2):372–378. 1982.

Lienhard, J. H. Synopsis of Lift, Drag and Vortex Frequency Data for Rigid Circular Cylinders. *Washington State University, College of Engineering, Research Division Bulletin 300.* 1966. Noted here from Blevins (1977).

MacDuff, J. N., and R. P. Feglar. Vibration Design Charts. *Trans. ASME* 79:1459–1475. 1957.

Mayinger, F., and H. G. Gross. Vibration in Heat Exchangers. In *Heat Exchangers: Thermal-Hydraulic Fundamentals and Design,* S. Kakac, A. E. Bergles, and F. Mayinger (eds.). Hemisphere: Washington, D.C. 1981.

Moretti, P. M. Critical Review of the Literature and Research on Flow Induced Vibrations in Heat Exchangers. *AIChE Symp. Ser. No. 138* 70:185. 1974.

Oak Ridge National Laboratory. *Design Guide for Heat Transfer Equipment in Water-cooled Nuclear Reactor Systems,* ORNL TM-3578. ORNL: Oak Ridge, Tenn. 1975.

Ojalvo, I. V., and M. Newman. Natural Frequencies of Clamped Ring Segments. *Mach. Des.* 36: 219–222. 1964.

Pettigrew, M. J., and D. J. Gorman. Vibration of Heat Exchanger Components in Liquids and Two-Phase Cross Flow, Paper 2:3. International Conference on Vibration in Nuclear Plant, Keswick, United Kingdom. 1978.

Schwarz, Jr., G. W. Preventing Vibrations in Shell and Tube Heat Exchangers. *Chem. Eng.* 83(14): 134–140. 1976. A recommended reading.

Sebald, J. F., and W. D. Nobles. Control of Tube Vibration in Steam Surface Condensers. *Proc. Am. Power Conf.* 14:630–643. 1962.

Singh, K. P. Predicting Flow Induced Vibration in U-Bend Regions of Heat Exchangers: An Engineering Solution. *J. Franklin Inst.* 302(2):195–205. 1976.

Soper, B. M. H. The Effect of Tube Layout on the Fluid–Elastic Instability of Tube Bundles in Cross Flow, presented at the ASME Winter Annual Meeting, Chicago. 1980.

Soper, B. M. H. The Effect of Grid Generated Turbulence on the Fluid–Elastic Instability of Tube Bundles in Cross Flow. In *Heat Exchangers: Theory and Practice.* J. Taborek, G. Hewitt, and N. H. Afgan (eds.). Hemisphere: Washington, D.C. 1983.

Thorngren, J. T. Predict Exchanger Tube Damage. *Hydrocarbon Process.* 49(4):129–131. 1970.

Tubular Exchanger Manufacturers' Association. *Standards of Tubular Exchanger Manufacturers' Association,* 6th ed. TEMA: Tarrytown, New York. 1978 (abbreviated as TEMA Standards).

Welbourne, M. C. Flow Induced Vibrations of A.G.R. Heat Exchanger Tubes, Paper No. 3:2. International Conference on Vibration in Nuclear Plant, Keswick, United Kingdom. 1978.

Zdravkovich, M. M., and J. E. Namork. Flow Structure within Both Stationary and Vibrating Tube Banks with Triangular Pitch, Paper No. 2:3. International Conference on Vibration in Nuclear Plant, Keswick, United Kingdom. 1978.

AUTHOR INDEX

References are to question number, table number (indicated by T), and/or figure number (indicated by F). Page numbers are given in parentheses.

60984 81800

SUBJECT INDEX

References are to question number, table number (indicated by T), and/or figure number (indicated by F). Page numbers are given in parentheses.